New Directions in Statistical Signal Processing

Neural Information Processing Series
Michael I. Jordan and Thomas Dietterich, editors

New Directions in Statistical Signal Processing: From Systems to Brain

edited by
Simon Haykin
José C. Príncipe
Terrence J. Sejnowski
John McWhirter

The MIT Press
Cambridge, Massachusetts
London, England

Printed and bound in the United States of America

Library of Congress Cataloging-in-Publication Data

New directions in statistical signal processing; from systems to brain / edited by Simon Haykin ... [et al.].
        p. cm.   (Neural information processing series)
Includes bibliographical references and index.
ISBN10: 0-262-08348-5 (alk. paper)
ISBN13: 978-0-262-08348-5
1. Neural networks (Neurobiology) 2. Neural networks (Computer science) 3. Signal processing—Statistical methods. 4. Neural computers. I. Haykin, Simon S., 1931 – II. Series.
QP363.3.N52 2006
612.8'2—dc22   2005056210

10 9 8 7 6 5 4 3 2 1

Contents

Series Foreword

The yearly Neural Information Processing Systems (NIPS) workshops bring together scientists with broadly varying backgrounds in statistics, mathematics, computer science, physics, electrical engineering, neuroscience, and cognitive science, unified by a common desire to develop novel computational and statistical strategies for information processing, and to understand the mechanisms for information processing in the brain. As opposed to conferences, these workshops maintain a flexible format that both allows and encourages the presentation and discussion of work in progress, and thus serves as an incubator for the development of important new ideas in this rapidly evolving field.

The series editors, in consultation with workshop organizers and members of the NIPS Foundation Board, select specific workshop topics on the basis of scientific excellence, intellectual breadth, and technical impact. Collections of papers chosen and edited by the organizers of specific workshops are built around pedagogical introductory chapters, while research monographs provide comprehensive descriptions of workshop-related topics, to create a series of books that provides a timely, authoritative account of the latest developments in the exciting field of neural computation.

Michael I. Jordan
Thomas Dietterich

Preface

In the course of some 60 to 65 years, going back to the 1940s, signal processing and neural computation have evolved into two highly pervasive disciplines. In their own individual ways, they have significantly influenced many other disciplines. What is perhaps surprising to see, however, is the fact that the cross-fertilization between signal processing and neural computation is still very much in its infancy. We only need to look at the brain and be amazed by the highly sophisticated kinds of signal processing and elaborate hierarchical levels of neural computation, which are performed side by side and with relative ease.

If there is one important lesson that the brain teaches us, it is summed up here:

There is much that signal processing can learn from neural computation, and vice versa.

It is with this aim in mind that in October 2003 we organized a one-week workshop on "Statistical Signal Processing: New Directions in the Twentieth Century," which was held at the Fairmont Lake Louise Hotel, Lake Louise, Alberta. To fulfill that aim, we invited some leading researchers from around the world in the two disciplines, signal processing and neural computation, in order to encourage interaction and cross-fertilization between them. Needless to say, the workshop was highly successful.

One of the most satisfying outcomes of the Lake Louise Workshop is that it has led to the writing of this new book. The book consists of 14 chapters, divided almost equally between signal processing and neural computation. To emphasize, in some sense, the spirit of the above-mentioned lesson, the book is entitled *New Directions in Statistical Signal Processing: From Systems to Brain*. It is our sincere hope that in some measurable way, the book will prove helpful in realizing the original aim that we set out for the Lake Louise Workshop.

Finally, we wish to thank Dr. Zhe Chen, who had spent tremendous efforts and time in LaTeX editing and proofreading during the preparation and final production of the book.

Simon Haykin
José C. Príncipe
Terrence J. Sejnowski
John McWhirter

1 Modeling the Mind: From Circuits to Systems

Suzanna Becker

Computational models are having an increasing impact on neuroscience, by shedding light on the neuronal mechanisms underlying information processing in the brain. In this chapter, we review the contribution of computational models to our understanding of how the brain represents and processes information at three broad levels: (1) sensory coding and perceptual processing, (2) high-level memory systems, and (3) representations that guide actions. So far, computational models have had the greatest impact at the earliest stages of information processing, by modeling the brain as a communication channel and applying concepts from information theory. Generally, these models assume that the goal of sensory coding is to map the high-dimensional sensory signal into a (usually lower-dimensional) code that is optimal with respect to some measure of information transmission. Four information-theoretic coding principles will be considered here, each of which can be used to derive unsupervised learning rules, and has been applied to model multiple levels of cortical organization. Moving beyond perceptual processing to high-level memory processes, the hippocampal system in the medial temporal lobe (MTL) is a key structure for representing complex configurations or episodes in long-term memory. In the hippocampal region, the brain may use very different optimization principles aimed at the memorization of complex events or spatiotemporal episodes, and subsequent reconstruction of details of these episodic memories. Here, rather than recoding the incoming signals in a way that abstracts away unnecessary details, the goal is to memorize the incoming signal as accurately as possible in a single learning trial. Most efforts at understanding hippocampal function through computational modeling have focused on sub-regions within the hippocampal circuit such as the CA3 or CA1 regions, using "off-the-shelf" learning algorithms such as competitive learning or Hebbian pattern association. More recently, Becker proposed a global optimization principle for learning within this brain region. Based on the goal of accurate input reconstruction, combined with neuroanatomical constraints, this leads to simple, biologically plausible learning rules for all regions within the hippocampal circuit. The model exhibits the key features of an episodic memory system: the capacity to store a large number of distinct, complex episodes, and to recall a complete episode from a minimal cue, and associate items across time, under extremely high plasticity conditions. Finally, moving beyond the static representation of information, we must consider the brain not simply a passive recipient of information, but as a complex, dynamical system, with internal goals and the ability to select actions based on environmental feedback. Ultimately, models based on the broad goals of prediction and control, using reinforcement-driven learning algorithms, may be the best candidates for characterizing the representations that guide motor actions. Several examples of models are described that begin to address the problem of how we learn representations that can guide our actions in a complex environment.

1.1 Introduction

How does the brain process, represent, and act on sensory signals? Through the use of computational models, we are beginning to understand how neural circuits perform these remarkably complex information-processing tasks. Psychological and neurobiological studies have identified at least three distinct long-term memory systems in the brain: (1) the perceptual/semantic memory system in the neocortex learns gradually to represent the salient features of the environment; (2) The episodic memory system in the medial temporal lobe learns rapidly to encode complex events, rich in detail, characterizing a particular episode in a particular place and time; (3) the procedural memory system, encompassing numerous cortical and subcortical structures, learns sensory-motor mappings. In this chapter, we consider several major developments in computational modeling that shed light on how the brain learns to represent information at three broad levels, reflecting these three forms of memory: (1) sensory coding, (2) episodic memory, and (3) representations that guide actions. Rather than providing a comprehensive review of all models in these areas, our goal is to highlight some of the key developments in the field, and to point to the most promising directions for future work.

1.2 Sensory Coding

At the earliest stages of sensory processing in the cortex, quite a lot is known about the neural coding of information, from Hubel and Wiesel's classic findings of orientation-selective neurons in primary visual cortex (Hubel and Wiesel, 1968) to more recent studies of spatiotemporal receptive fields in visual cortex (DeAngelis et al., 1993) and spectrotemporal receptive fields in auditory cortex (Calhoun and Schreiner, 1998; Kowalski et al., 1996). Given the abundance of electrophysiological data to constrain the development of computational models, it is not surprising that most models of learning and memory have focused on the early stages of sensory coding. One approach to modeling sensory coding is to hand-design filters, such as the Gabor or difference-of-Gaussians filter, so as to match experimentally observed receptive fields. However, this approach has limited applicability beyond the very earliest stages of sensory processing for which receptive fields have been reasonably well mapped out. A more promising approach is to try to understand the developmental processes that generated the observed data. Note that these could include both learning and evolutionary factors, but here our focus is restricted to potential learning mechanisms. The goal is then to discover the general underlying principles that cause sensory systems to self-organize their receptive fields. Once these principles have been uncovered, they can be used to derive models of learning. One can then simulate the developmental process by exposing the model to typical sensory input and comparing the results to experimental observations. More important, one can simulate neuronal functions that might not have been conceived by experimentalists, and thereby generate novel experimental predictions.

Several classes of computational models have been influential in guiding current thinking about self-organization in sensory systems. These models share the general feature of modeling the brain as a communication channel and applying concepts from information theory. The underlying assumption of these models is that the goal of sensory coding is to map the high-dimensional sensory signal into another (usually lower-dimensional) code that is somehow optimal with respect to information content. Four information-theoretic coding principles will be considered here: (1) Linsker's Infomax principle, (2) Barlow's redundancy reduction principle, 3) Becker and Hinton's Imax principle, and (4) Risannen's minimum description length (MDL) principle. Each of these principles has been used to derive models of learning and has inspired further research into related models at multiple stages of information processing.

1.2.1   Linsker's Infomax Principle

How should neurons respond to the sensory signal, given that it is noisy, high-dimensional and highly redundant? Is there a more convenient form in which to encode signals so that we can make more sense of the relevant information and take appropriate actions? In the human visual system, for example, there are hundreds of millions of photoreceptors converging onto about two million optic nerve fibers. By what principle does the brain decide what information to discard and what to preserve?

Infomax principle   Linsker proposed a model of self-organization in sensory systems based on the *Infomax principle*: Each neuron adjusts its connection strengths or weights so as to maximize the amount of Shannon information in the neural code that is conveyed about the sensory input (Linsker, 1988). In other words, the Infomax principle dictates that neurons should maximize the amount of mutual information between their input $\mathbf{x}$ and output y:

$$I_{\mathbf{x};\mathbf{y}} = \langle \ln\left[p(\mathbf{x}|y)/p(\mathbf{x})\right]\rangle$$

Environmental input

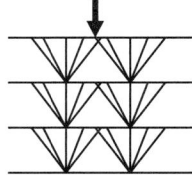

Figure 1.1 Linsker's multilayer architecture for learning center-surround and oriented receptive fields. Higher layers learned progressively more 'Mexican-hat-like' receptive fields. The inputs consisted of uncorrelated noise, and in each layer, center-surround receptive fields evolved with progressively greater contrast between center and surround.

Assuming that the input consists of a multidimensional Gaussian signal with additive, independent Gaussian noise with variance $V(n)$, for a single neuron whose output y is a linear function of its inputs and connection weights $\mathbf{w}$, the mutual information is the log of the signal-to-noise ratio:

$$I_{\mathbf{x};y} = \frac{1}{2} \ln \frac{V(y)}{V(n)}$$

Linsker showed that a simple, Hebb-like weight update rule approximately maximizes this information measure. The center-surround receptive field (with either an on-center and off-surround or off-center and on-surround spatial pattern of connection strengths) is characteristic of neurons in the earliest stages of the visual pathways including the retina and lateral geniculate nucleus (LGN) of the thalamus. Surprisingly, Linsker's simulations using purely uncorrelated random inputs, and a multi-layer circuit as shown in fig 1.1, showed that neurons in successive layers developed progressively more "Mexican-hat" shaped receptive fields (Linsker, 1986a,b,c), reminiscent of the center-surround receptive fields seen in the visual system. In further developments of the model, using a two-dimensional sheet of neurons with local-neighbor lateral connections, Linsker (1989) showed that the model self-organized topographic maps with oriented receptive fields, such that nearby units on the map developed similarly oriented receptive fields. This organization is a good first approximation to that of the primary visual cortex.

independent components analysis

The Infomax principle has been highly influential in the study of neural coding, going well beyond Linsker's pioneering work in the linear case. One of the major developments in this field is Bell and Sejnowksi's Infomax-based independent components analysis (ICA) algorithm, which applies to nonlinear mappings with equal numbers of inputs and outputs (Bell and Sejnowski, 1995). Bell and Sejnowski showed that when the mapping from inputs to outputs is continuous, nonlinear, and invertible, maximizing the mutual information between inputs and outputs is equivalent to simply maximizing the entropy of the output signal. The algorithm therefore performs a form of ICA.

Infomax-based ICA has also been used to model receptive fields in visual cortex. When applied to natural images, in contrast to principal component analysis (PCA), Infomax-based ICA develops oriented receptive fields at a variety of spatial scales that are sparse, spatially localized, and reminiscent of oriented receptive fields in primary visual cortex (Bell and Sejnowski, 1997). Another variant of nonlinear Infomax developed by Okajima and colleagues (Okajima, 2004) has also been applied to modeling higher levels of visual processing, including combined binocular disparity and spatial frequency analysis.

1.2.2 Barlow's Redundancy Reduction Principle

The principle of preserving information may be a good description of the very earliest stages of sensory coding, but it is unlikely that this one principle will capture all levels of processing in the brain. Clearly, one can trivially preserve all the information in the input simply by copying the input to the next level up. Thus, the idea only makes sense in the context of additional processing constraints. Implicit in Linsker's work was the constraint of dimension reduction. However, in the neocortex, there is no evidence of a progressive reduction in the number of neurons at successively higher levels of processing.

redundancy re-
duction principle

Barlow proposed a slightly different principle of self-organization based on the idea of producing a *minimally redundant* code. The information about an underlying signal of interest (such as the visual form or the sound of a predator) may be distributed across many input channels. This makes it difficult to associate particular stimulus values with distinct responses. Moreover, there is a high degree of redundancy across different channels. Thus, a neural code having minimal redundancy should make it easier to associate different stimulus values with different responses.

The formal, information-theoretic definition of redundancy is the information content of the stimulus, less the capacity of the channel used to convey the information. Unfortunately, quantities dependent upon calculation of entropy are difficult to compute. Thus, several different formulations of Barlow's principle have been proposed, under varying assumptions and approximations. One simple way for a learning algorithm to lower redundancy is reduce correlations among the outputs (Barlow and Földiák, 1989). This can remove second-order but not higher-order dependencies.

Atick and Redlich proposed minimizing the following measure of redundancy (Atick and Redlich, 1990):

$$R = 1 - \frac{I_{y;s}}{C_{out}(y)}$$

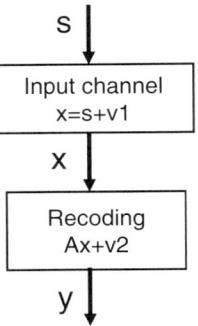

Figure 1.2 Atick and Redlich's learning principle was to minimize redundancy in the output, y, while preserving information about the input, x.

channel capacity

subject to the constraint of zero information loss (fixed $I_{y;s}$). $C_{out}(y)$, the output *channel capacity*, is defined to be the maximum of $I_{y;s}$. The channel capacity is at a maximum when the covariance matrix of the output elements is diagonal, hence Atick and Redlich used

$$C_{out}(y) = \frac{1}{2} \prod_i \left[\frac{R_{yy}}{N_v 2^2} \right]_{ii}$$

Thus, under this formulation, minimizing redundancy amounts to minimizing the the channel capacity. This model is depicted in fig 1.2. This model was used to simulate retinal receptive fields. Under conditions of high noise (low redundancy), the receptive fields that emerged were Gaussian-shaped spatial smoothing filters, while at low noise levels (high redundancy) on-center off-surround receptive fields resembling second spatial derivative filters emerged. In fact, cells in the mammalian retina and lateral geniculate nucleus of the thalamus dynamically adjust their filtering characteristics as light levels fluctuate between these two extremes under conditions of low versus high contrast (Shapley and Victor, 1979; Virsu et al., 1977). Moreover, this strategy of adaptive rescaling of neural responses has been shown to be optimal with respect to information transmission (Brenner et al., 2000).

  Similar learning principles have been applied by Atick and colleagues to model higher stages of visual processing. Dong and Atick modeled redundancy reduction across time, in a model of visual neurons in the lateral geniculate nucleus of the thalamus (Dong and Atick, 1995). In their model, neurons with both lagged and nonlagged spatiotemporal smoothing filters emerged. These receptive fields would be useful for conveying information about stimulus onsets and offsets. Li and Atick (1994) modeled redundancy reduction across binocular visual inputs. Their model generated binocular, oriented receptive fields at a variety of spatial scales, similar to those seen in primary visual cortex. Bell and Sejnowski's Infomax-based ICA algorithm (Bell and Sejnowski, 1995) is also closely related to Barlow's minimal redundancy principal, since the ICA model is restricted to invertible mappings; in

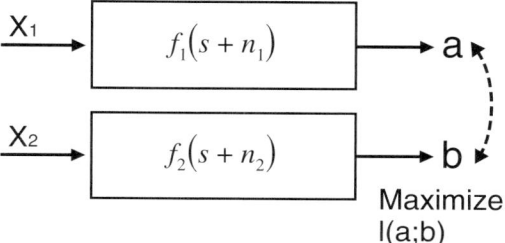

Figure 1.3 Becker and Hinton's Imax learning principle maximizes the mutual information between features a and b extracted from different input channels.

this case, the maximization of mutual information amounts to reducing statistical dependencies among the outputs.

1.2.3 Becker and Hinton's Imax Principle

The goal of retaining as much information as possible may be a good description of early sensory coding. However, the brain seems to do much more than simply preserve information and recode it into a more convenient form. Our perceptual systems are exquisitely tuned to certain regularities in the world, and consequently to irregularities which violate our expectations. The things which capture our attention and thus motivate us to learn and act are those which violate our expectations about the coherence of the world—the sudden onset of a sound, the appearance of a looming object or a predator.

In order to be sensitive to *changes* in our environment, we require internal representations which first capture the regularities in our environment. Even relatively low-order regularities, such as the spatial and temporal coherence of sensory signals, convey important cues for extracting very high level properties about objects. For example, the coherence of the visual signal across time and space allows us to segregate the parts of a moving object from its surrounding background, while the coherence of auditory events across frequency and time permits the segregation of the auditory input into its multiple distinct sources.

Becker and Hinton (1992) proposed the Imax principle for unsupervised learning, which dictates that signals of interest should have high mutual information across different sensory channels. In the simplest case, illustrated in fig 1.3, there are two input sources, x_1 and x_2, conveying information about a common underlying Gaussian signal of interest, s, and each channel is corrupted by independent, additive Gaussian noise:

$$x_1 = s + n_1,$$
$$x_2 = s + n_2.$$

However, the input may be high dimensional and may require a nonlinear transformation in order to extract the signal. Thus the goal of the learning is to transform the two input signals into outputs, y_1 and y_2, having maximal mutual

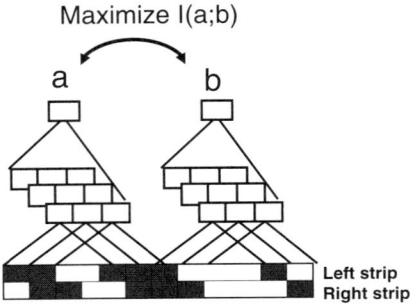

Figure 1.4　Imax architecture used to learn stereo features from binary images.

information. Because the signal is Gaussian and the noise terms are assumed to be identically distributed, the information in common to the two outputs can be maximized by maximizing the following log signal-to-noise ratio (SNR):

$$I_{y_1;y_2} \approx \log\left(\frac{V(y_1 + y_2)}{V(y_1 - y_2)}\right)$$

This could be accomplished by multiple stages of processing in a nonlinear neural circuit like the one shown in fig 1.4. Becker and Hinton (1992) showed that this model could extract binocular disparity from random dot stereograms, using the architecture shown in fig 1.4. Note that this function requires multiple stages of processing through a network of nonlinear neurons with sigmoidal activation functions.

The Imax algorithm has been used to learn temporally coherent features (Becker, 1996; Stone, 1996), and extended to learn multidimensional features (Zemel and Hinton, 1991). A very similar algorithm for binary units was developed by Kay and colleagues (Kay, 1992; Phillips et al., 1998). The minimizing disagreement algorithm (de Sa, 1994) is a probabilistic learning procedure based on principles of Bayesian classification, but is nonetheless very similar to Imax in its objective to extract classes that are coherent across multiple sensory input channels.

1.2.4　Risannen's Minimum Description Length Principle

The overall goal of every unsupervised learning algorithm is to discover the important underlying structure in the data. Learning algorithms based on Shannon information have the drawback of requiring knowledge of the probability distribution of the data, and/or of the extracted features, and hence tend to be either very computationally expensive or to make highly simplifying assumptions about the distributions (e.g. binary or Gaussian variables). An alternative approach is to develop a model of the data that is somehow optimal with respect to coding efficiency. The minimum description length (MDL) principle, first introduced by

minimum description-
tion length

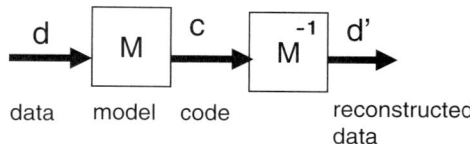

Figure 1.5 Minimum description length (MDL) principle.

Rissanen (1978), favors models that provide accurate encoding of the data using as simple a model as possible. The rationale behind the MDL principle is that the criterion of discovering statistical regularities in data can be quantified by the length of the code generated to describe the data.

A large number of learning algorithms have been developed based on the MDL principle, but only a few of these have attempted to provide plausible accounts of neural processing. One such example was developed by Zemel and Hinton (1995), who cast the autoencoder problem within an MDL framework. They proposed that the goal of learning should be to encode the total cost of communicating the input data, which depends on three terms, the length of the code, c, the cost of communicating the model, M (which depends on the coding cost of communicating how to reconstruct the data), M^{-1}, and the reconstruction error:

$$Cost = Length(c) + Length(M^{-1}) + Length(|d - d'|)$$

as illustrated in fig 1.5. They instantiated these ideas using an autoencoder architecture, with hidden units whose activations were Gaussian functions of the inputs. Under a Gaussian model of the input activations, it was assumed that the hidden unit activations, as a population, encode a point in a lower-dimensional implicit representational space. For example, a population of place cells in the hippocampus might receive very high dimensional multisensory input, and map this input onto a population of neural activations which codes implicitly the animal's spatial location—a point in a two-dimensional Cartesian space. The population response could be decoded by averaging together the implicit coordinates of the hidden units, weighted by their activations. Zemel and Hinton's cost function incorporated a reconstruction term and a coding cost term that measured the fit of the hidden unit activations to a Gaussian model of implicit coordinates. The weights of the hidden units and the coordinates in implicit space were jointly optimized with respect to this MDL cost.

Algorithms which perform clustering, when cast within a statistical framework, can also be viewed as a form of MDL learning. Nowlan derived such an algorithm, called maximum likelihood competitive learning (MLCL), for training neural networks using the expectation maximization (EM) algorithm (Jacobs et al., 1991; Nowlan, 1990). In this framework, the network is viewed as a probabilistic, generative model of the data. The learning serves to adjust the weights so as to maximize the log likelihood of the model having generated the data:

$$L = \log P(data \mid model).$$

If the training patterns, $I^{(\alpha)}$, are independent,

$$L = \log \prod_{\alpha=1}^{n} P(I^{(\alpha)} \mid model)$$
$$= \sum_{\alpha=1}^{n} \log P(I^{(\alpha)} \mid model).$$

The MLCL algorithm applies this objective function to the case where the units have Gaussian activations and form a mixture model of the data:

$$L = \sum_{\alpha=1}^{n} \log \left[\sum_{i=1}^{m} P(I^{(\alpha)} \mid submodel_i) \, P(submodel_i) \right]$$
$$= \sum_{\alpha=1}^{n} \log \left[\sum_{i=1}^{m} y_i^{(\alpha)} \, \pi_i \right],$$

where the π_i's are positive mixing coefficients that sum to one, and the y_i's are the unit activations:

$$y_i^{(\alpha)} = \mathcal{N}(\vec{I}^{(\alpha)}, \vec{\mathbf{w}}_i, \Sigma_i),$$

where $\mathcal{N}(\ )$ is the Gaussian density function, with mean $\vec{\mathbf{w}}_i$ and covariance matrix Σ_i.

The MLCL model makes the assumption that every pattern is independent of every other pattern. However, this assumption of independence is not valid under natural viewing conditions. If one view of an object is encountered, a similar view of the same object is likely to be encountered next. Hence, one powerful cue for real vision systems is the temporal continuity of objects. Novel objects typically are encountered from a variety of angles, as the position and orientation of the observer, or objects, or both, vary smoothly over time. Given the importance of temporal context as a cue for feature grouping and invariant object recognition, it is very likely that the brain makes use of this property of the world in perceptual learning. Becker (1999) proposed an extension to MLCL that incorporates context into the learning. Relaxing the assumption that the patterns are independent, allowing for temporal dependencies among the input patterns, the log likelihood function becomes:

$$L = \log P(data \mid model)$$
$$= \sum_{\alpha} \log P(I^{(\alpha)} \mid I^{(1)}, \dots, I^{(\alpha-1)}, model).$$

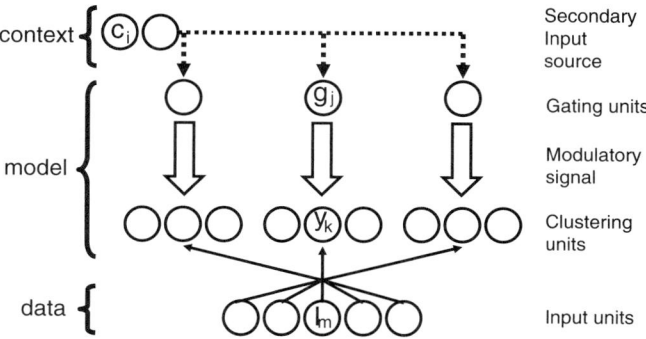

Figure 1.6 Contextually modulated competitive learning.

To incorporate a contextual information source into the learning equation, a contextual input stream was introduced into the likelihood function:

$$L = \log P(data \mid model, context)$$
$$= \sum_{\alpha} \log P(I^{(\alpha)} \mid I^{(1)}, \dots, I^{(\alpha-1)}, model, context),$$

as depicted in fig 1.6. This model was trained on a series of continuously rotating images of faces, and learned a representation that categorized people's faces according to identity, independent of viewpoint, by taking advantage of the temporal continuity in the image sequences.

Many models of population encoding apply to relatively simple, one-layer feed-forward architectures. However, the structure of neocortex is much more complex. There are multiple cortical regions, and extensive feedback connections both within and between regions. Taking these features of neocortex into account, Hinton has developed a series of models based on the Boltzmann machine (Ackley et al., 1985), and the more recent Helmholtz machine (Dayan et al., 1995) and Product of Experts (PoE) model(Hinton, 2000; Hinton and Brown, 2000). The common idea underlying these models is to try to find a population code that forms a causal model of the underlying data. The Boltzmann machine was unacceptably slow at sampling the "unclamped" probability distribution of the unit states. The Helmholtz machine and PoE model overcome this limitation by using more restricted architectures and/or approximate methods for sampling the probability distributions over units' states (see fig 1.7A). In both cases, the bottom-up weights embody a "recognition model"; that is, they are used to produce the most probable set of hidden states given the data. At the same time, the top-down weights constitute a "generative model"; that is, they produce a set of hidden states most likely to have generated the data. The "wake-sleep algorithm" maximizes the log likelihood

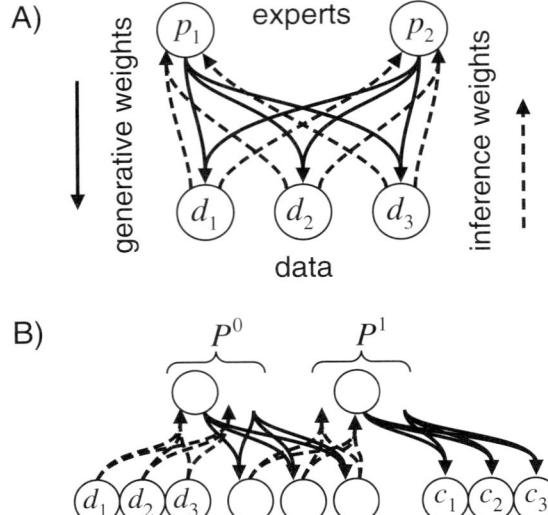

Figure 1.7 Hinton's Product of Experts model, showing (A) the basic architecture, and (B) Brief Gibbs sampling, which involves several alternating iterations of clamping the input units to sample from the hidden unit states, and then clamping the hidden units to sample from the input unit states. This procedure samples the "unclamped" distribution of states in a local region around each data vector and tries to minimize the difference between the clamped and unclamped distributions.

of the data under this model and results in a simple equation for updating either set of weights:

$$\Delta w_{kj} = \varepsilon s_k^\alpha (s_j^\alpha - p_j^\alpha),$$

where p_j^α is the target state for unit j on pattern α, and s_j^α is the corresponding network state, a stochastic sample based on the logistic function of the unit's net input. Target states for the generative weight updates are derived from top-down expectations based on samples using the recognition model, whereas for the recognition weights, the targets are derived by making bottom-up predictions based on samples from the generative model. The Products of Experts model advances on this learning procedure by providing a very efficient procedure called "brief Gibbs sampling" for estimating the most probable states to have generated the data, as illustrated in fig 1.7B).

1.3 Models of Episodic Memory

Moving beyond sensory coding to high-level memory systems in the medial temporal lobe (MTL), the brain may use very different optimization principles aimed at the memorization of complex events or spatiotemporal episodes, and at subsequent reconstruction of details of these episodic memories. Here, rather than recoding the incoming signals in a way that abstracts away unnecessary details, the goal is to memorize the incoming signal as accurately as possible in a single learning trial. The hippocampus is a key structure in the MTL that appears to be crucial for episodic memory. It receives input from most cortical regions, and is at the point of convergence between the ventral and dorsal visual pathways, as illustrated in fig 1.8 (adapted from (Mishkin et al., 1997)). Some of the unique anatomical and physiological characteristics of the hippocampus include the following: (1) the very large expansion of dimensionality from the entorhinal cortex (EC) to the dentate gyrus (DG) (the principal cells in the dentate gyrus outnumber those of the EC by about a factor of 5 in the rat (Amaral et al., 1990)); (2) the large and potent mossy fiber synapses projecting from CA3 to CA1, which are the largest synapses in the brain and have been referred to as "detonator synapses" (McNaughton and Morris, 1987); and (3) the extensive set of recurrent collateral connections within the CA3 region. In addition, the hippocampus exhibits unique physiological properties including (1) extremely sparse activations (low levels of activity), particularly in the dentate gyrus where firing rates of granule cells are about 0.5 Hz (Barnes et al., 1990; Jung and McNaughton, 1993), and (2) the constant replacement of neurons (neurogenesis) in the dentate gyrus: about about 1% of the neurons in the dentate gyrus are replaced each day in young adult rats (Martin Wojtowicz, University of Toronto, unpublished data).

In 1971 Marr put forward a highly influential theory of hippocampal coding (Marr, 1971). Central to Marr's theory were the notions of a rapid, temporary memory store mediated by sparse activations and Hebbian learning, an associative retrieval system mediated by recurrent connections, and a gradual consolidation process by which new memories would be transferred into a long-term neocortical store. In the decades since the publication of Marr's computational theory, many researchers have built on these ideas and simulated memory formation and retrieval in Marr-like models of the hippocampus. For the most part, modelers have focused on either the CA3 or CA1 fields, using variants of Hebbian learning, for example, competitive learning in the dentate gyrus and CA3 (Hasselmo et al., 1996; McClelland et al., 1995; Rolls, 1989), Hebbian autoassociative learning (Kali and Dayan, 2000; Marr, 1971; McNaughton and Morris, 1987; O'Reilly and Rudy, 2001; Rolls, 1989; Treves and Rolls, 1992), temporal associative learning (Gerstner and Abbott, 1997; Levy, 1996; Stringer et al., 2002; Wallenstein and Hasselmo, 1997) in the CA3 recurrent collaterals, and Hebbian heteroassociative learning between EC-driven CA1 activity and CA3 input (Hasselmo and Schnell, 1994) or between EC-driven and CA3-driven CA1 activity at successive points in time (Levy et al., 1990). The key ideas behind these models are summarized in fig 1.9.

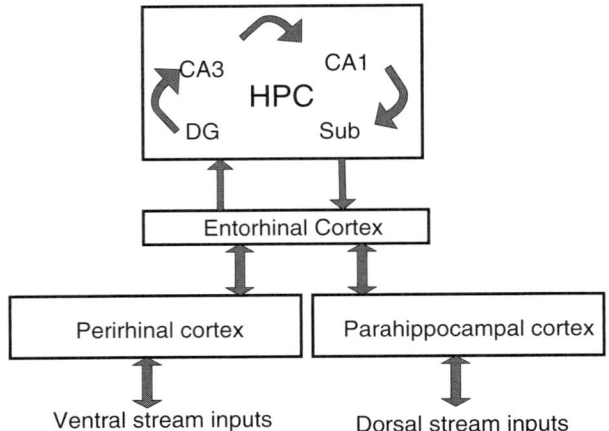

Figure 1.8 Some of the main anatomical connections of the hippocampus. The hippocampus is a major convergence zone. It receives input via the entorhinal cortex from most regions of the brain including the ventral and dorsal visual pathways. It also sends reciprocal projections back to most regions of the brain. Within the hippocampus, the major regions are the dentate gyrus (DG), CA3, and CA1. The CA1 region projects back to the entorhinal cortex, thus completing the loop. Note that the subiculum, not shown here, is another major output target of the hippocampus.

In modeling the MTL's hippocampal memory system, Becker (2005) has shown that a global optimization principle based on the goal of accurate input reconstruction, combined with neuroanatomical constraints, leads to simple, biologically plausible learning rules for all regions within the hippocampal circuit. The model exhibits the key features of an episodic memory system: high storage capacity, accurate cued recall, and association of items across time, under extremely high plasticity conditions.

The key assumptions in Becker's model are as follows:

- During encoding, dentate granule cells are active whereas during retrieval they are relatively silent.

- During encoding, activation of CA3 pyramidals is dominated by the very strong mossy fiber inputs from dentate granule cells.

- During retrieval, activation of CA3 pyramidals is driven by direct perforant path inputs from the entorhinal cortex combined with time-delayed input from CA3 via recurrent collaterals.

- During encoding, activation of CA1 pyramidals is dominated by direct perforant path inputs from the entorhinal cortex.

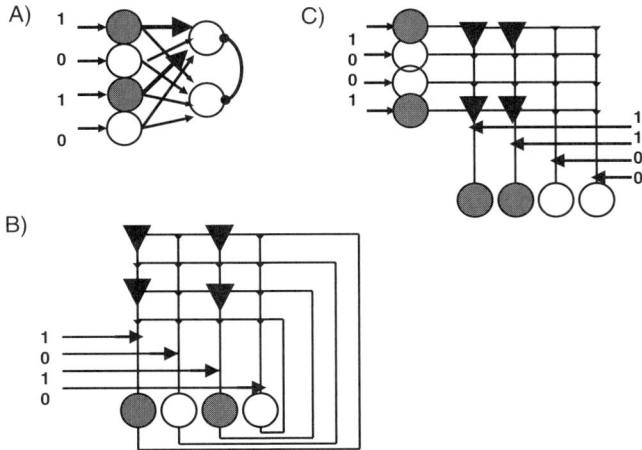

Figure 1.9 Various models have been proposed for specific regions of the hippocampus, for example, (A) models based on variants of competitive learning have been proposed for the dentate gyrus; (B) many models of the CA3 region have been based upon the recurrent autoassociator, and (C) several models of CA1 have been based on the heteroassociative network, where the input from the entorhinal cortex to CA1 acts as a teaching signal, to be associated with the (nondriving) input from the CA3 region.

▪ During retrieval, CA1 activations are driven by a combination of perforant path inputs from the entorhinal cortex and Shaffer collateral inputs from CA3.

Becker proposed that each hippocampal layer should form a neural representation that could be transformed in a simple manner—i.e. linearly—to reconstruct the original activation pattern in the entorhinal cortex. With the addition of biologically plausible processing constraints regarding connectivity, sparse activations, and two modes of neuronal dynamics during encoding versus retrieval, this results in very simple Hebbian learning rules.

It is important to note, however, that the model itself is highly nonlinear, due to the sparse coding in each region and the multiple stages of processing in the circuit as a whole; the notion of linearity only comes in at the point of *reconstructing* the EC activation pattern from any one region's activities. The objective function made use of the idea of an implicit set of reconstruction weights from each hippocampal region, by assuming that the perforant path connection weights could be used in reverse to reconstruct the EC input pattern. Taking the CA3 layer as an example, the CA3 neurons receive perforant path input from the entorhinal cortex, $EC^{(in)}$, associated with a matrix of weights $W^{(EC,CA3)}$. The CA3 region also receives input connections from the dentate gyrus, DG, with associated weights $W^{(DG,CA3)}$ as well as recurrent collateral input from within the CA3 region with connection weights $W^{(CA3,CA3)}$. Using the transpose of the perforant path weights, $(W^{(EC,CA3)})^T$, to calculate the CA3 region's reconstruction of the entorhinal input vector

$$EC^{(reconstructed)} = W^{(EC,CA3)^T} CA3, \tag{1.1}$$

the goal of the learning is to make this reconstruction as accurate as possible. To quantify this goal, the objective function Becker proposed to be maximized here is the cosine angle between the original and reconstructed activations:

$$Perf^{(CA3)} = cos(EC^{(in)}, W^{(EC,CA3)^T} CA3)$$
$$= \frac{(EC^{(in)})^T (W^{(EC,CA3)^T} CA3)}{||EC^{(in)}|| \; ||W^{(EC,CA3)^T} CA3||}. \tag{1.2}$$

By rearranging the numerator, and appropriately constraining the activation levels and the weights so that the denominator becomes a constant, it is equivalent to maximize the following simpler expression:

$$Perf^{(CA3)} = (W^{(EC,CA3)} EC^{(in)})^T CA3, \tag{1.3}$$

which makes use of the locally available information arriving at the CA3 neurons' incoming synapses: the incoming weights and activations. This says that the incoming weighted input from the perforant path should be as similar as possible to the activation in the CA3 layer. Note that the CA3 activation, in turn, is a function of both perforant path and DG input as well as CA3 recurrent input. The objective functions for the dentate and CA1 regions have exactly the same form as equation 1.3, using the DG and CA1 activations and perforant path connection weights respectively. Thus, the computational goal for the learning in each region is to maximize the overlap between the perforant path input and that region's reconstruction of the input. This objective function can be maximized with respect to the connection weights on each set of input connections for a given layer, to derive a set of learning equations.

By combining the learning principle with the above constraints, Hebbian learning rules are derived for the direct (monosynaptic) pathways from the entorhinal cortex to each hippocampal region, a temporal Hebbian associative learning rule is derived for the CA3 recurrent collateral connections, and a form of heteroassociative learning is derived the Shaffer collaterals (the projection from CA3 to CA1).

Of fundamental importance for computational theories of hippocampal coding is the striking finding of neurogenesis in the adult hippocampus. Although there is now a large literature on neurogenesis in the dentate gyrus, and it has been shown to be important for at least one form of hippocampal-dependent learning, surprisingly few attempts have been made to reconcile this phenomenon with theories of hippocampal memory formation. Becker (2005) suggested that the function of new neurons in the dentate gyrus is in the generation of novel codes. Gradual changes in the internal code of the dentate layer were predicted to facilitate the formation of distinct representations for highly similar memory episodes.

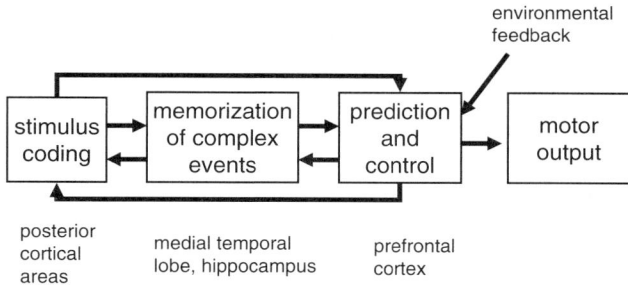

Figure 1.10   Architecture of a learning system that incorporates perceptual learning, episodic memory, and motor control.

Why doesn't the constant turnover of neurons in the dentate gyrus, and hence the constant rewiring of the hippocampal memory circuit, interfere with the retrieval of old memories? The answer to this question comes naturally from the above assumptions about neuronal dynamics during encoding versus retrieval. New neurons are added only to the dentate gyrus, and the dentate gyrus drives activation in the hippocampal circuit only during encoding, not during retrieval. Thus, the new neurons contribute to the formation of distinctive codes for novel events, but not to the associative retrieval of older memories.

1.4   Representations That Guide Action Selection

Moving beyond the question of how information is represented, we must consider the brain not simply a passive storage device, but as a part of a dynamical computational system that acts and reacts to changes within its environment, as illustrated in fig 1.10. Ultimately, models based on the broad goals of prediction and control may be our best hope for characterizing complex dynamical systems which form representations in the service of guiding motor actions.

Reinforcement learning algorithms can be applied to control problems, and have been linked closely to specific neural mechanisms. These algorithms are built upon the concept of a value function, $V(s_t)$, which defines the value of being in the current state s_t at time t to be equal to the expected sum of future rewards:

$$V(t) = r_t + \gamma r_{t+1} + \gamma^2 r_{t_2} + \ldots + \gamma^n r_{t_n} + \ldots$$

The parameter γ, chosen to be in the range $0 \leq \gamma \leq 1$, is a temporal discount factor which permits one to heuristically weight future rewards more or less heavily according to the task demands. Within this framework, the goal for the agent is to choose actions that will maximize the value function. In order for the agent to solve the control problem—how to select optimal actions, it must first solve the prediction problem—how to estimate the value function. The temporal difference (TD) learning algorithm (Sutton, 1988; Sutton and Barto, 1981) provides a rule for incrementally updating an estimate $\hat{V}_t$ of the true value function at time t by an amount called the TD-error: TD-error $= r_{t+1} + \gamma\hat{V}_{t+1} - \hat{V}_t$, which makes use of r_t, the amount of reward received at time t, and the value estimates at the current and the next time step. It has been proposed that the TD-learning algorithm may be used by neurobiological systems, based on evidence that firing of midbrain dopamine neurons correlates well with TD-error (Montague et al., 1996).

TD-learning

Q-learning

The Q-learning algorithm (Watkins, 1989) extends the idea of TD learning to the problem of learning an optimal control policy for action selection. The goal for the agent is to maximize the total future expected reward. The agent learns incrementally by trial and error, evaluating the consequences of taking each action in each situation. Rather than using a value function, Q-learning employs an action-value function, $Q(s_t, a_t)$, which represents the value in taking an action a_t when the state of the environment is s_t. The learning algorithm for incrementally updating estimates of Q-values is directly analogous to TD learning, except that the TD-error is replaced by a temporal difference between Q-values at successive points in time.

Becker and Lim (2003) proposed a model of controlled memory retrieval based upon Q-learning. People have a remarkable ability to encode and retrieve information in a flexible manner. Understanding the neuronal mechanisms underlying strategic memory use remains a true challenge. Neural network models of memory have typically dealt with only the most basic operations involved in storage and recall. Evidence from patients with frontal lobe damage indicates a crucial role for the prefrontal cortex in the control of memory. Becker and Lim's model was developed to shed light on the neural mechanisms underlying strategic memory use in individuals with intact and lesioned frontal lobes. The model was trained to simulate human performance on free-recall tasks involving lists of words drawn from a small set of categories. Normally when people are asked repeatedly to study and recall the same list of words, their recall patterns demonstrate progressively more categorical clustering over trials. This strategy thus appears to be learned, and correlates with overall recall scores. On the other hand, when patients with frontal lobe damage perform such tests, while they do benefit somewhat from the categorical structure of word lists, they tend to recall fewer categories in total, and tend to show lower semantic clustering scores. Becker and Lim (2003) postulated a role for the prefrontal cortex (PFC) in self-organizing novel mnemonic codes that could subsequently be used as retrieval cues to improve retrieval from long-term memory. Their model is outlined in fig 1.11.

The "actions" or responses in this model are actually the activations generated by model neurons in the PFC module. Thus, the activation of each response unit is proportional to the network's current estimate of the Q-value associated with

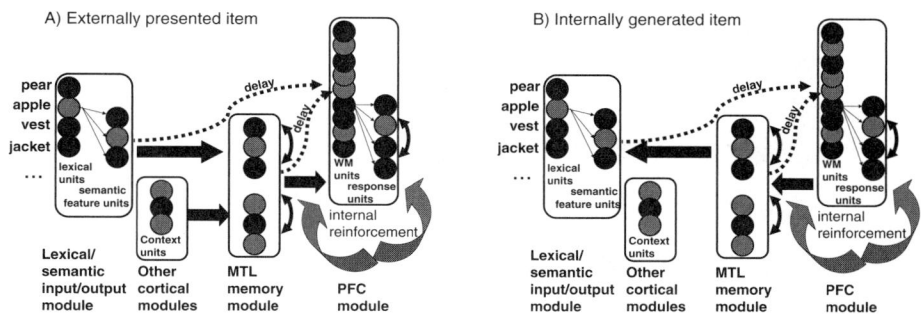

Figure 1.11   Becker and Lim's architecture for modeling the frontal control of memory retrieval. The model operated in two different modes: (A) During perception of an external stimulus (during a study phase) there was bottom-up flow of activation. (B) During free recall, when a response was generated internally, there was a top-down flow of activation in the model. After an item was retrieved, but before a response was generated, the item was used to probe the MTL memory system, and its recency was evaluated. If the recency (based on a match of the item to the memory weight matrix) was too high, the item was considered to be a repetition error, and if too low, it was considered to be an extralist intrusion error. Errors detected by the model were not generated as responses, but were used to generate internal reinforcement signals for learning the PFC module weights. Occasionally, a repetition or intrusion error might go undetected by the model, resulting in a recall error.

that response, and response probabilities are calculated directly from these Q-values. Learning the memory retrieval strategy involved adapting the weights for the response units so as to maximize their associated Q-values. Reinforcement obtained on a given trial was self-generated by an internal evaluation module, so that the PFC module received a reward whenever a nonrepeated study list item was retrieved, and a punishment signal (negative reinforcement) when a nonlist or repeated item was retrieved. The model thereby learned to develop retrieval strategies dynamically in the course of both study and free recall of words. The model was able to capture the performance of human subjects with both intact and lesioned frontal lobes on a variety of types of word lists, in terms of both recall accuracy and patterns of errors.

The model just described addresses a rather high level of complex action selection, namely, the selection of memory retrieval strategies. Most work on modeling action selection has dealt with more concrete and observable actions such as the choice of lever-presses in a response box or choice of body-turn directions in a maze. The advantage of this level of modeling is that it can make contact with a large body of experimental literature on animal behavior, pharmacology, and physiology. Many such models have employed TD-learning or Q-learning, under the assumption that animals form internal representations of value functions, which

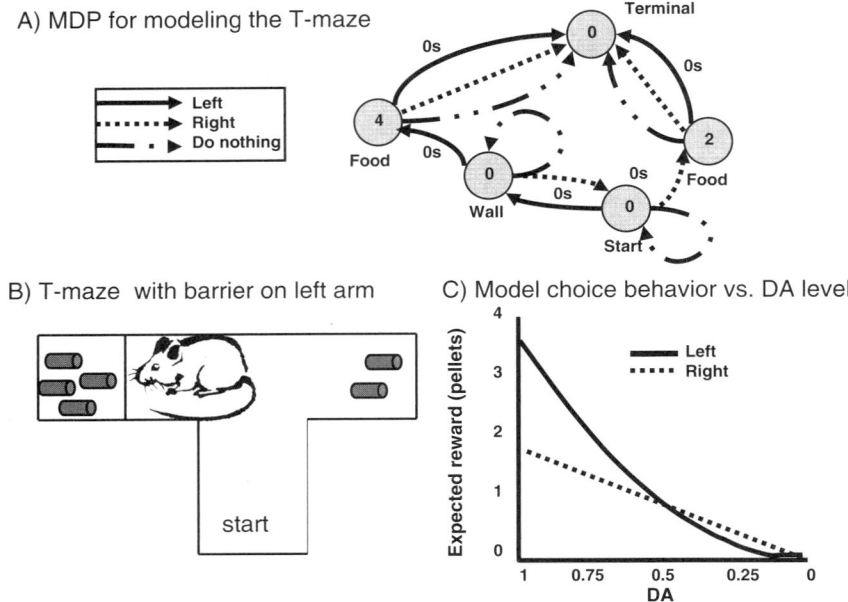

A) MDP for modeling the T-maze

B) T-maze with barrier on left arm

C) Model choice behavior vs. DA level

Figure 1.12 Simulation of Cousins et al.'s T-maze cost-benefit task. The MDP representation of the task is shown in panel A, the T-maze with a barrier and larger reward in the left arm of the maze is shown in panel B, and the performance of the model as a function of dopamine depletion is shown in panel C.

guide action selection. As mentioned above, phasic firing of dopamine neurons has been postulated to convey the TD-error signal critical for this type of learning. However, in addition to its importance in modulating learning, dopamine plays an important role in modulating action choice. It has been hypothesized that tonic levels of dopamine have more to do with motivational value, whereas the phasic firing of dopamine neurons conveys a learning-related signal (Smith et al., 2005).

Rather than assuming that actions are solely guided by value functions, Smith et al. (2005) hypothesized that animals form detailed internal models of the world. Value functions condense the reward value of a series of actions into that of a single state, and are therefore insensitive to the motivational state of the animal (e.g., whether it is hungry or not). Internal models, on the other hand, allow a mental simulation of alternative action choices, which may result in qualitatively different rewards. For example, an animal might perform one set of actions leading to water only if it is thirsty, and another set of actions leading to food only if it is hungry. The internal model can be described by a Markov decision process (MDP) over a set of internal states, with associated transition function and reward function, as in fig 1.12A). The transition function and (immediate) reward value of each state are learned through trial and error. Once the model is fully trained, action selection involves simulating a look-ahead process in the internal model for one or more steps in order to evaluate the consequences of an action. Finally, at the end of the simulation sequence, the animal's internal model reveals whether the outcome is favorable (leads to reward) or not. An illustrative example is shown in fig 1.12B. The choice faced by the animal is either to take the right arm of the T-maze to receive a small reward, or to take the left arm and then jump over a barrier to receive a larger reward. The role of tonic dopamine in this model is to modulate the efficacy of the connections in the internal model. Thus, when dopamine is depleted, the model's ability to simulate the look-ahead process to assess expected future reward will be biased toward rewards available immediately rather than more distal rewards. This implements an online version of temporal discounting.

Cousins et al. (1996) found that normal rats trained in the T-maze task in fig 1.12B) are willing to jump the barrier to receive a larger food reward nearly 100% of the time. Interestingly, however, when rats were administered a substance that destroys dopaminergic (DA) projections to the nucleus accumbens (DA lesion), they chose the smaller reward. In another version of the task, rats were trained on the same maze except that there was no food in the right arm, and then when given DA lesions, they nearly always chose the left arm and jumped the barrier to receive a reward. Thus, the DA lesion was not merely disrupting motor behavior, it was interacting with the motivational value of the behavioral choices. Note that the TD-error account of dopamine only provides for a role in learning, and would have nothing to say about effects of dopamine on behavior subsequent to learning. Smith et al. (2005) argued, based on these and other data, that dopamine serves to modulate the motivational choice of the animals, with high levels of dopamine favoring the selection of action sequences with more distal but larger rewards. In simulations of the model, depletion of dopamine therefore biases the choice in favor of the right arm in this task, as shown in fig 1.12C).

1.5   New Directions: Integrating Multiple Memory Systems

In this chapter, we have reviewed several approaches to modeling the mind, from low-level sensory coding, to high-level memory systems, to action selection. Somehow, the brain accomplishes all of these functions, and it is highly unlikely that they are carried out in isolation from one another. For example, we now know that striatal dopaminergic pathways, presumed to carry a reinforcement learning signal, affect sensory coding even in early sensory areas such as primary auditory cortex (Bao et al., 2001). Future work must address the integration of these various levels of modeling.

2   Empirical Statistics and Stochastic Models for Visual Signals

David Mumford

The formulation of the vision problem as a problem in Bayesian inference (Forsyth and Ponce, 2002; Mumford, 1996, 2002) is, by now, well known and widely accepted in the computer vision community. In fact, the insight that the problem of reconstructing 3D information from a 2D image is ill posed and needs inference can be traced back to the Arab scientist Ibn Al-Haytham (known to Europe as Alhazan) around the year 1000 (Haytham, c. 1000). Inheriting a complete hodgepodge of conflicting theories from the Greeks,[1] Al-Haytham for the first time demonstrated that light rays originated only in external physical sources, and moved in straight lines, reflecting and refracting, until they hit the eye; and that the resulting signal needed to be and was actively decoded in the brain using a largely unconscious and very rapid inference process based on past visual experiences. In the modern era, the inferences underlying visual perception have been studied by many people, notably H. Helmholtz, E. Brunswik (Brunswik, 1956), and J. J. Gibson.

In mathematical terms, the Bayesian formulation is as follows: let I be the observed image, a 2D array of pixels (black-and-white or colored or possibly a stereoscopic pair of such images). Here we are assuming a static image.[2] Let w stand for variables that describe the external scene generating the image. Such variables should include depth and surface orientation information (Marr's 2.5 D sketch), location and boundaries of the principal objects in view, their surface albedos, location of light sources, and labeling of object categories and possibly object identities. Then two stochastic models, learned from past experience, are required: a *prior model* $p(w)$ specifying what scenes are likely in the world we live in and an *imaging model* $p(I|w)$ specifying what images should look like, given the scene. Then by Bayes's rule:

Bayes's rule

$$p(w|I) = \frac{p(I|w)p(w)}{p(I)} \propto p(I|w)p(w).$$

Bayesian inference consists in fixing the observed value of I and inferring that w equals that value which maximizes $p(w|I)$ or equivalently maximizes $p(I|w)p(w)$. This is a fine general framework, but to implement or even test it requires (1) a theory of stochastic models of a very comprehensive sort which can express all the complex but variable patterns which the variables w and I obey, (2) a method of learning from experience the many parameters which such theories always contain, and (3) a method of computing the maximum of $p(w|I)$.

This chapter will be concerned only with problem 1. Many critiques of vision algorithms have failed to allow for the fact that these are three separate problems: if 2 or 3, the methods, are badly implemented, the resulting problems do not imply that the theory itself (1) is bad. For example, very slow algorithms of type 3 may reasonably be used to test ideas of type 1. Progress in understanding vision does not require all these problems to be solved at once. Therefore, it seems to me legitimate to isolate problems of type 1.

In the rest of this chapter, I will review some of the progress in constructing these models. Specifically, I will consider, in section 2.1, models of the empirical probability distribution $p(I)$ inferred from large databases of natural images. Then, in section 2.2, I will consider the problem of the first step in so-called intermediate vision: inferring the regions which should be grouped together as single objects or structures, a problem which includes segmentation and gestalt grouping, the basic grammar of image analysis. Finally in section 2.3, I look at the problem of priors on 2D shapes and the related problem of what it means for two shapes to be "similar". Obviously, all of these are huge topics and I cannot hope to give a comprehensive view of work on any of them. Instead, I shall give my own views of some of the important issues and open problems and outline the work that I know well. As this inevitably emphasizes the work of my associates, I must beg indulgence from those whose work I have omitted.

2.1   Statistics of the Image Alone

The most direct approach to studying images is to ask whether we can find good models for images without any hidden variables. This means first creating a large database of images I that we believe are reasonably random samples of all possible images of the world we live in. Then we can study this database with all the tools of statistics, computing the responses of various linear and nonlinear filters and looking at the individual and joint histograms of their values. "Nonlinear" should be taken in the broadest sense, including order statistics or topological analyses. We then seek to isolate the most important properties these statistics have and to create the simplest stochastic models $p(I)$ that duplicate or approximate these statistics. The models can be further tested by sampling from them and seeing if the resulting artificial images have the same "look and feel" as natural images; or if not, what are the simplest properties of natural images that we have failed to capture. Another recent survey of such models is referred to (Lee et al., 2003b).

2.1.1   High Kurtosis As The Universal Clue To Discrete Structure

kurtosis

The first really striking thing about filter responses is that they always have large kurtosis. It is strange that electrical engineers designing TV sets in the 1950s do not seem to have pointed this out and this fact first appeared in the work of David Field (Field, 1987). By kurtosis, we mean the normalized fourth moment. If x is a random real number, its kurtosis is

$$\kappa(x) = E((x - \bar{x})^4)/E((x - \bar{x})^2)^2.$$

Every normal variable has kurtosis 3; a variable which has no tails (e.g., uniformly distributed on an interval) or is bimodal and small at its mean tends to have kurtosis less than 3; a variable with heavy tails or large peak at its mean tends to have kurtosis larger than 3. The empirical result which is observed for images is that for any linear filter F with zero mean, the values $x = (F * I)(i, j)$ of the filtered image follow a distribution with kurtosis larger than 3. The simplest case of this is the difference of adjacent pixel values, the discrete derivative of the image I. But it has been found (Huang, 2000) to hold even for *random* mean zero filters supported in an 8×8 window.

This high kurtosis is shown in fig 2.1, from the thesis of J. Huang (Huang, 2000). This data was extracted from a large database of high-resolution, fully calibrated images of cities and country taken in Holland by van Hateren (1998). It is important, when studying tails of distributions, to plot the *logarithm* of the probability or frequency, as in this figure, not the raw probability. If you plot probabilities, all tails look alike. But if you plot their logarithms, then a normal distribution becomes a downward facing parabola (since $\log(e^{-x^2}) = -x^2$), so heavy tails appear clearly as curves which do not point down so fast.

stationary
Markov process

It is a well-known fact from probability theory that if X_t is a stationary Markov stochastic process, then the kurtosis of $X_t - X_s$ being greater than 3 means that the process X_t has discrete jumps. In the case of vision, we have samples from an image $I(s, t)$ depending on two variables rather than one and the zero-mean filter is a generalization of the difference $X_t - X_s$. Other signals generated by the world, such as sound or prices, are functions of one variable, time. A nice elementary statement of the link between kurtosis and jumps is given by the following result, taken from Mumford and Desolneux:

Theorem 2.1

Let x be any real random variable which we normalize to have mean 0 and standard deviation 1. Then there is a constant $c > 0$ depending only on x such that if, for some n, x is the sum

$$x = y_1 + y_2 + \cdots + y_n$$

where the y_i are independent and identically distributed, then

$$\text{Prob}\left(\max_i |y_i| \geq \sqrt{(\kappa(x) - 3)/2}\right) \geq c.$$

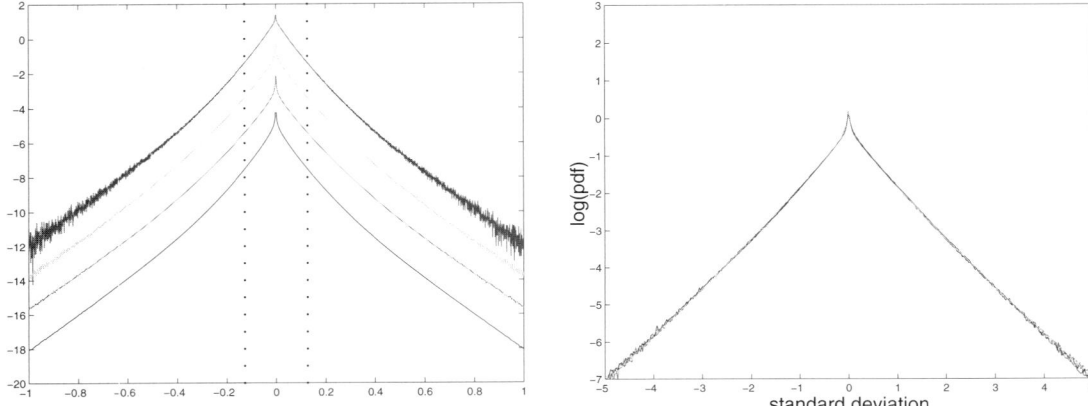

Figure 2.1 Histograms of filter values from the thesis of J. Huang, using the van Hateren database. On the left, the filter is the difference of (a) horizontally adjacent pixels, and (b) of adjacent 2×2, (c) 4×4 and (d) 8×8 blocks; on the right, several *random* mean-zero filters with 8×8 pixel support have been used. The kurtosis of all these filter responses is between 7 and 15. Note that the vertical axis is log of the frequency, not frequency. The histograms on the left are displaced vertically for legibility and the dotted lines indicate one standard deviation.

A striking application of this is to the stock market. Let x be the log price change of the opening and closing price of some stock. If we assume price changes are Markov, as many have, and use the experimental fact that price changes have kurtosis greater than 3, then it implies that stock prices cannot be modeled as a continuous function of time. In fact, in my own fit of some stock-market data, I found the kurtosis of log price changes to be infinite: the tails of the histogram of log price changes appeared to be polynomial, like $1/x^{\alpha}$ with α between 4 and 5.

An important question is, How big are the tails of the histograms of image filter statistics? Two models have been proposed for these distributions. The first is the most commonly used model, the generalized Laplacian distribution:

generalized
Laplacian
distribution

$$p_{\text{laplace}}(x) = \frac{e^{-|x/a|^b}}{Z}, \quad Z = \int e^{-|y/a|^b} dy.$$

Here a is a scale parameter and b controls how large the tails are (larger tails for smaller b). Experimentally, these work well and values of b between 0.5 and 1 are commonly found. However, no rationale for their occurrence seems to have been found. The second is the Bessel distribution (Grenander and Srivastava, 2001; Wainwright and Simoncelli, 2000):

Bessel
distribution

$$p_{\text{bessel}}(x) = \widehat{q(\xi)}, \quad q(\xi) = 1/(1 + (a\xi^2))^{b/2}.$$

Again, a is a scale parameter, b controls the kurtosis (as before, larger kurtosis for smaller b), and the hat means Fourier transform. $p_{\text{bessel}}(x)$ can be evaluated explicitly using Bessel functions. The tails, however, are all asymptotically like those of double exponentials $e^{-|x/a|}$, regardless of b. The key point is that these distributions arise as the distributions of *products* $r \cdot x$ of Gaussian random variables x and an independent positive "scaling" random variable r. For some values of b, the variable r is distributed like $\|\vec{x}\|$ for a Gaussian $\vec{x} \in \mathbb{R}^n$, but in general its square has a gamma (or chi-squared) distribution. The great appeal of such a product is that images are also formed as products, especially as products of local illumination, albedo, and reflectance factors. This may well be the deep reason for the validity of the Bessel models.

Convincing tests of which model is better have not been made. The difficulty is that they differ most in their tails, where data is necessarily very noisy. The best approach might be to use the Kolmogorov-Smirnov statistic and compare the best-fitting models for this statistic of each type.

The world seems to be composed of discrete jumps in time and discrete objects in space. This profound fact about the physical nature of our world is clearly mirrored in the simple statistic *kurtosis*.

2.1.2   Scaling Properties of Images and Their Implications

scale invariance

After high kurtosis, the next most striking statistical property of images is their approximate scale invariance. The simplest way to define scale invariance precisely is this: imagine we had a database of 64×64 images of the world and that this could be modeled by a probability distribution $p_{64}(I)$ in the Euclidean space $\mathbb{R}^{4096}$ of all such images. Then we can form marginal 32×32 images in two different ways: we either extract the central 32×32 set of pixels from the big image I or we cover the whole 64×64 image by 1,024 2×2 blocks of pixels and average each such block to get a 32×32 image (i.e., we "blow down" I in the crudest way). The assertion that images are samples from a scale-invariant distribution is that the two resulting marginal distributions on 32×32 images are the same. This should happen for images of any size and we should also assume that the distribution is stationary, i.e., translating an image gives an equally probable image. The property is illustrated in fig 2.2.

It is quite remarkable that, to my knowledge, no test of this hypothesis on reasonably large databases has contradicted it. Many histograms of filter responses on successively blown down images have been made; order statistics have been looked at; and some topological properties derived from level curves have been studied (Geman and Koloydenko, 1999; Gousseau, 2000; Huang, 2000; Huang and Mumford, 1999). All have shown approximate scale invariance. There seem to be two simple facts about the world which combine to make this scale invariance approximately true. The first is that images of the world are taken from random distances: you may photograph your spouse's face from one inch away or from 100 meters away or anything in between. On your retina, except for perspective distortions, his or her image is scaled up or down as you move closer or farther away. The second is that objects tend to have surfaces on which smaller objects

Figure 2.2 Scale invariance defined as a fixed point under block renormalization. The top is random $2N \times 2N$ image which produces the two $N \times N$ images on the bottom, one by extracting a subimage, the other by 2×2 block averaging. These two should have the same marginal distributions. (Figure from A. Lee.)

cluster: your body has limbs which have digits which have hairs on them, your office has furniture which has books and papers which have writing (a limiting case of very flat objects on its surface) on them, etc. Thus a blowup of a photograph not only shows roughly the same number of salient objects, but they occur with roughly the same contrast.[3]

The simplest consequence of scale invariance is the law for the decay of power at high frequencies in the Fourier transform of images (or better, the discrete cosine transform to minimize edge effects). It says that the expected power as a function of frequency should drop off like

power law

$$\mathrm{E}_I\left(|\hat{I}(\xi,\eta)|^2\right) \approx C/(\xi^2 + \eta^2) = C/f^2,$$

where $f = \sqrt{\xi^2 + \eta^2}$ is the spatial frequency. This power law was discovered in the 1950s. In the image domain, it is equivalent to saying that the autocorrelation of the image is approximated by a constant minus log of the distance:

$$\mathrm{E}_I\left(\sum_{x,y}(I(x,y) - \bar{I}).(I(x+a, y+b) - \bar{I})\right) \approx C - \log(\sqrt{a^2 + b^2}).$$

Note that the models have both infrared[4] and ultraviolet divergences: the total power diverges for both $f \to 0$ and ∞, and the autocorrelation goes to $\pm\infty$ as $a, b \to 0$, and ∞. Many experiments have been made testing this law over moderate ranges of frequencies and I believe the conclusion to draw is this: for small databases of images, especially databases of special sorts of scenes such as forest scenes or city scenes, different powers are found to fit best. These range from $1/f^3$ to $1/f$ but with both a high concentration near $1/f^2$ and a surprisingly large variance[5] (Frenkel et al.; Huang, 2000). But for *large* databases, the rule seems to hold.

Another striking consequence of the approximate scale-invariance is that images, if they have infinitely high resolution, are not functions at all but must be considered generalized functions (distributions in the sense of Schwartz). This means that as their resolution increases, natural images do not have definite limiting numerical values $I(x, y)$ at almost all points x, y in the image plane. I think of this as the "mites on your eyelashes" theorem. Biologists tell us that such mites exist and if you had Superman's X-ray vision, you not only could see them but by the laws of reflectance, they would have high contrast, just like macroscopic objects. This mathematical implication is proven by Gidas and Mumford (Gidas and Mumford, 2001).

This conclusion is quite controversial: others have proposed other function spaces as the natural home for random images. An early model for images (Mumford and Shah, 1989) proposed that observed images were naturally a sum:

$$I(x, y) = u(x, y) + v(x, y),$$

where u was a piecewise smooth "cartoon", representing the important content of the image, and v was some L^2 noise. This led to the idea that the natural function space for images, after the removal of noise, was the space of functions of bounded variation, i.e., $\int \|\nabla I\| dx dy < \infty$. However, this approach lumped texture in with noise and results in functions u from which all texture and fine detail has been removed. More recent models, therefore, have proposed that

$$I(x, y) = u(x, y) + v(x, y) + w(x, y),$$

where u is the cartoon, v is the true texture, and w is the noise. The idea was put forward by DeVore and Lucier (1994) that the true image $u+v$ belongs to a suitable Besov space, spaces of functions $f(x, y)$ for which bounds are put on the L^p norm of $f(x + h, y + k) - f(x, y)$ for (h, k) small. More recently, Carasso has simplified their approach (Carasso, 2004) and hypothesizes that images I, after removal of "noise" should satisfy

$$\int |I(x + h, y + k) - I(x, y)| dx dy < C(h^2 + k^2)^{\alpha/2},$$

for some α as $(h, k) \to 0$.

However, a decade ago, Rosenfeld argued with me that most of what people discard as "noise" is nothing but objects too small to be fully resolved by the resolution of the camera and thus blurred beyond recognition or even aliased. I think

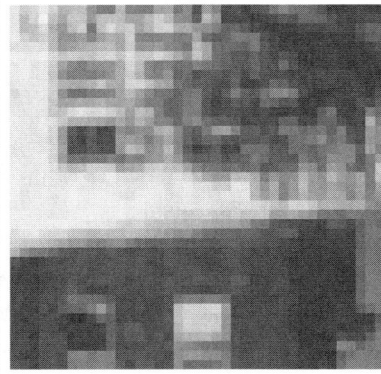

Figure 2.3 This photo is intentionally upside-down, so you can look at it more abstractly. The left photo has a resolution of about 500×500 pixels and the right photo is the yellow 40×40 window shown on the left. Note (a) how the distinct shapes in the road made by the large wet/dry spots gradually merge into dirt texture and (b) the way on the right the bush is pure noise. If the bush had moved relative to the pixels, the pattern would be totally different. There is no clear dividing line between distinct objects, texture, and noise. Even worse, some road patches which ought to be texture are *larger* than salient objects like the dog.

of this as *clutter*. The real world is made up of objects plus their parts and surface markings *of all sizes* and any camera resolves only so many of these. There is an ideal image of infinite resolution but any camera must use sensors with a positive point spread function. The theorem above says that this ideal image, because it carries all this detail, cannot even be a function. For example, it has more and more high-frequency content as the sensors are refined and its total energy diverges in the limit,[6] hence it cannot be in L^2.

In fig 2.3, we illustrate that there is no clear dividing line between objects, texture, and noise: depending on the scale at which you view and digitize the ideal image, the same "thing" may appear as an object, as part of a texture, or as just a tiny bit of noise. This continuum has been analyzed beautifully recently by Wu et al. (2006, in revision).

Is there is a simple stochastic model for images which incorporates both high kurtosis and scale invariance? There is a unique scale-invariant Gaussian model, namely colored white noise whose expected power spectrum conforms to the $1/f^2$ law. But this has kurtosis equal to 3. The simplest model with both properties seems to be that proposed and studied by Gidas and me (Gidas and Mumford, 2001), which we call the *random wavelet* model. In this model, a random image is a countable sum:

random wavelet model

$$I(x, y) = \sum_{\alpha} \psi_{\alpha}(e^{r_{\alpha}} x - x_{\alpha}, e^{r_{\alpha}} y - y_{\alpha}).$$

Here $(r_\alpha, x_\alpha, y_\alpha)$ is a uniform Poisson process in 3-space and ψ_α are samples from the auxiliary Levy process, a distribution on the space of scale- and position-normalized elementary image constituents, which one may call mother wavelets or textons. These expansions converge almost surely in all the Hilbert-Sobolev spaces $H^{-\epsilon}$. Each component ψ_α represents an elementary constituent of the image. Typical choices for the ψ's would be Gabor patches, edgelets or curvelets, or more complex shapes such as ribbons or simple shapes with corners. We will discuss these in section 2.1.4 and we will return to the random wavelet model in section 2.2.3.

2.1.3   Occlusion and the "Dead Leaves" Model

There is, however, a third basic aspect of image statistics which we have so far not considered: occlusion. Images are two-dimensional projections of the three-dimensional world and objects get in front of each other. This means that it is a mathematical simplification to imagine images as *sums* of elementary constituents. In reality, objects are ordered by distance from the lens and they should be combined by the nonlinear operation in which nearer surface patches overwrite distant ones. Statistically, this manifests itself in a strongly non-Markovian property of images: suppose an object with a certain color and texture is occluded by a nearer object. Then, on the far side of the nearer object, the more distant object may reappear, hence its color and texture have a larger probability of occurring than in a Markov model.

This process of image construction was studied by the French school of Matheron and Serra based at the École des Mines (Serra, 1983 and 1988). Their "dead leaves model" is similar to the above random wavelet expansion except that occlusion is used. We imagine that the constituents of the image are tuples $(r_\alpha, x_\alpha, y_\alpha, d_\alpha, D_\alpha, \psi_\alpha)$ where r_α, x_α and y_α are as before, but now d_α is the distance from the lens to the αth image patch and ψ_α is a function only on the set of $(x, y) \in D_\alpha$. We make no a priori condition on the density of the Poisson process from which $(r_\alpha, x_\alpha, y_\alpha, d_\alpha)$ is sampled. The image is then given by

$$I(x, y) = \psi_{\alpha(x,y)}(e^{r_{\alpha(x,y)}}x - x_{\alpha(x,y)}, e^{r_{\alpha(x,y)}}y - y_{\alpha(x,y)}), \qquad \text{where}$$
$$\alpha(x, y) = \operatorname{argmin}\{d_\alpha \,|\, (x, y) \in D_\alpha\}$$

This model has been analyzed by A. Lee, J. Huang and myself (Lee et al., 2001) but has more serious infrared and ultraviolet catastrophes than the additive one. One problem is that nearby small objects cause the world to be enveloped in a sort of fog occluding everything in the distance. Another is the probability that one big nearby object occludes everything. In any case, with some cutoffs, Lee's models are approximately scale-invariant and seem to reproduce *all* the standard elementary image statistics better than any other that I know of, e.g., two-point co-occurrence statistics as well as joint wavelet statistics. Examples of both types of models are shown in fig. 2.4.

I believe a deeper analysis of this category of models entails modeling directly,

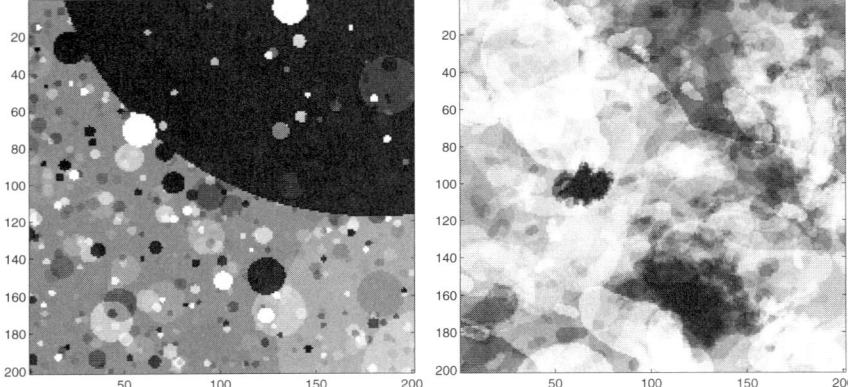

Figure 2.4 Synthetic images illustrating the generic image models from the text. On the left, a sample dead leaves model using disks as primitives; on the right, a random wavelet model whose primitive are short ribbons.

not the objects in 2D projection, but their statistics in 3D. What is evident then is that objects are not scattered in 3-space following a Poisson process, but rather are agglutinative: smaller objects collect on or near the surface of bigger objects (e.g., houses and trees on the earth, limbs and clothes on people, buttons and collars on clothes, etc.). The simplest mathematical model for this would be a random branching process in which an object had "children", which were the smaller objects clustering on its surface. We will discuss a 2D version of this in section 2.2.3.

2.1.4 The Phonemes Of Images

The final component of this direct attack on image statistics is the investigation of its elementary constituents, the ψ above. In analogy with speech, one may call these constituents *phonemes* (or phones). The original proposals for such building blocks were given by Julesz and Marr. Julesz was interested in what made two textures distinguishable or indistinguishable. He proposed that one should break textures locally into *textons* (Julesz, 1981; Resnikoff, 1989) and, supported by his psychophysical studies, he proposed that the basic textons were elongated blobs and their endpoints ("terminators"). Marr (1982), motivated by the experiments of Hubel and Wiesel on the responses of cat visual cortex neurons, proposed that one should extract from an image its "primal sketch", consisting of edges, bars, and blobs. Linking these proposals with raw image statistics, Olshausen and Field (1996) showed that simple learning rules seeking a *sparse* coding of the image, when exposed to small patches from natural images, did indeed develop responses sensitive to edges, bars, and blobs. Another school of researchers have taken the elegant mathematical theory of wavelets and sought to find those wavelets which enabled best image compression. This has been pursued especially by Mallat (1999), Simoncelli (1999), and Donoho and their collaborators (Candes and Donoho, 2005).

Having large natural image databases and powerful computers, we can ask now

texton

for a direct extraction of these or other image constituents from a statistical analysis of the images themselves. Instead of taking psychophysical, neurophysiological, or mathematical results as a basis, what happens if we let images speak for themselves? Three groups have done this: Geman-Koloydenko (Geman and Koloydenko, 1999), Huang-Lee-Pedersen-Mumford (Lee et al., 2003a), and Malik-Shi (Malik et al., 1999). Some of the results of Huang and of Malik et al. are shown in fig. 2.5.

The approach of Geman and Koloydenko was based on analyzing all 3×3 image patches using *order statistics*. The same image patches were studied by Lee and myself using their real number values. A very similar study by Pedersen and Lee (2002) replaced the nine pixel values by nine Gaussian derivative filter responses. In all three cases, a large proportion of such image patches were found to be either low contrast or high contrast cut across by a single edge. This, of course, is not a surprise, but it quantifies the significance of edges in image structure. For example, in the study by Lee, Pedersen and myself, we took the image patches with the top 20% quantile for contrast, then subtracted their mean and divided by their standard deviation, obtaining data points on a seven-dimensional sphere. In this sphere, there is a surface representing the responses to image patches produced by imaging straight edges with various orientations and offsets. Close analysis shows that the data is highly concentrated near this surface, with asymptotic infinite density along the surface itself.

Malik and Shi take small patches and analyze these by a filter bank of 36 wavelet filters. They then apply k-means clustering to find high-density points in this point cloud. Again the centers of these clusters resemble the traditional textons and primitives. In addition, they can adapt the set of textons they derive to individual images, obtaining a powerful tool for representing a single image.

A definitive analysis of images deriving directly the correct vocabulary of basic image constituents has not been made but the outlines of the answer are now clear.

2.2   Grouping of Image Structures

In the analysis of signals of any kind, the most basic "hidden variables" are the labels for parts of the signal that should be grouped together, either because they are homogeneous parts in some sense or because the components of this part occur together with high frequency. This grouping process in speech leads to words and in language leads to the elements of grammar—phrases, clauses, and sentences. On the most basic statistical level, it seeks to group parts of the signal whose probability of occurring together is significantly greater than it would be if they were independent: see section 2.2.3 for this formalism. The factors causing grouping were the central object of study for the Gestalt school of psychology. This school flourished in Germany and later in Italy in the first half of the twentieth century and included M. Wertheimer, K. Koffka, W. Metzger, E. Brunswik, G. Kanizsa, and many others. Their catalog of features which promoted grouping included

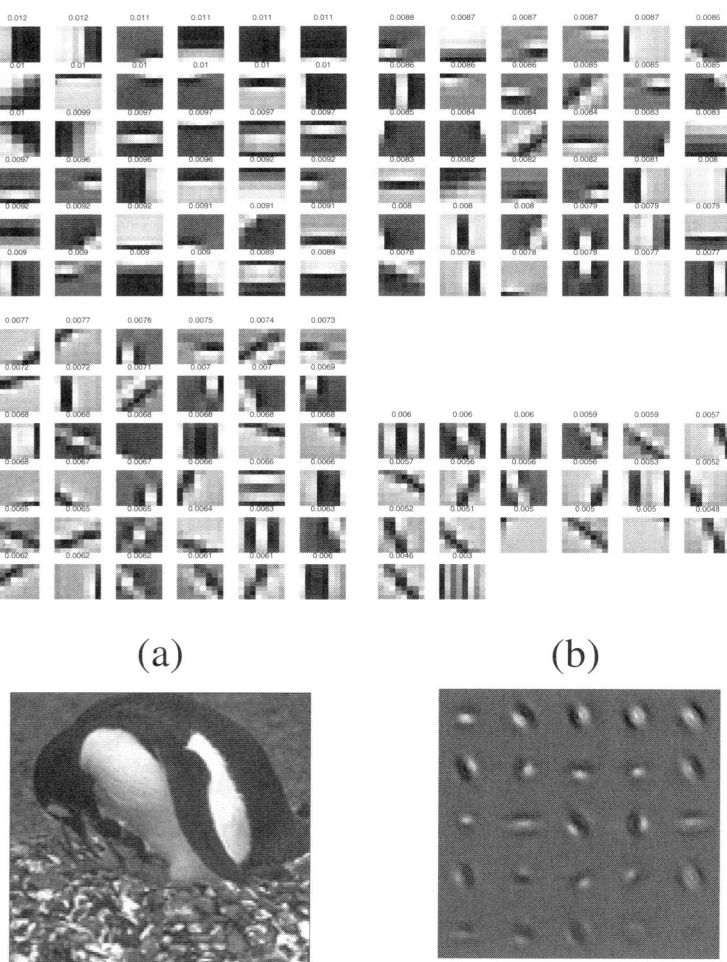

(a) (b)

Figure 2.5 Textons derived by k-means clustering applied to 8×8 image patches. On the top, Huang's results for image patches from van Hateren's database; on the bottom, Malik et al.'s results using single images and filter banks. Note the occasional terminators in Huang's results, as Julesz predicted.

- color and proximity,

- alignment, parallelism, and symmetry,

- closedness and convexity.

Kanizsa was well aware of the analogy with linguistic grammar, titling his last book *Grammatica del Vedere* (Kanizsa, 1980). But they had no quantitative measures for the strength of these grouping principles, as they well knew. This is similar to the situation for traditional theories of human language grammar—a good story to explain what words are to be grouped together in phrases but no numbers. The challenge we now face is to create theories of *stochastic grammars* which can express why one grouping is chosen in preference to another. It is a striking fact that, faced either with a sentence or a scene of the world, human observers choose the same groupings with great consistency. This is in contrast with computers which, given only the grouping rules, find thousands of strange parses of both sentences and images.

2.2.1   The Most Basic Grouping: Segmentation and Texture

grouping

The simplest grouping rules are those of similar color (or brightness) and proximity. These two rules have been used to attack the segmentation problem. The most naive but direct approach to image segmentation is based on the assumption that images break up into regions on which their intensity values are relatively constant and across whose boundaries those values change discontinuously. A mathematical version of this approach, which gives an explicit measure for comparing different proposed segmentations, is the energy functional proposed by Shah and myself (Mumford and Shah, 1989). It is based on a model $I = u + v$ where u is a simplified cartoon of the image and v is "noise":

$$E(I, u, \Gamma) = C_1 \int_D (I - u)^2 + C_2 \int_{D-\Gamma} \|\nabla u\|^2 + C_3 \cdot \text{length}(\Gamma), \qquad \text{where}$$

$$D = \text{ domain of } I,$$

$$\Gamma = \text{ boundaries of regions which are grouped together, and}$$

$$C_i = \text{ parameters to be learned.}$$

In this model, pixels in $D - \Gamma$ have been grouped together by stringing together pairs of nearby similarly colored pixels. Different segmentations correspond to choosing different u and Γ and the one with lower energy is preferred. Using the Gibbs statistical mechanics approach, this energy can be thought of as a probability: heuristically, we set $p(I, u, \Gamma) = e^{-E(I,u,\Gamma)/T}/Z$, where T and Z are constants. Taking this point of view, the first term in E is equivalent to assuming $v = I - u$ is a sample from white noise. Moreover, if Γ is fixed, then the second term in E makes u a sample from the scale-invariant Gaussian distribution on functions, suitably adapted to the smaller domain $D - \Gamma$. It is hard to interpret the third term even heuristically, although Brownian motion $((x(t), y(t))$ is heuristically a sample from the prior $e^{-\int (x'(t)^2 + y'(t)^2) dt}$, which, if we adopt arc length parameterization, becomes $e^{-\text{length}(\Gamma)}$. If we stay in the discrete pixel setting, the Gibbs model corresponding to E makes good mathematical sense; it is a variant of the Ising model of statistical mechanics (Blake and Zisserman, 1987; Geman and Geman, 1984).

The most obvious weakness in this model is its failure to group similarly textured regions together. Textural segmentation is an example of the hierarchical application of gestalt rules: first the individual textons are grouped by having similar colors, orientations, lengths, and aspect ratios. Then these groupings of textons are further grouped into extended textured regions with homogeneous or slowly varying "texture". Ad hoc adaptations of the above energy approach to textural grouping (Geman and Graffigne, 1986; Hofmann et al., 1998; Lee et al., 1992) have been based on choosing some filter bank the similarity of whose responses are taken as a surrogate for the first low-level texton grouping. One of the problems of this approach is that textures are often not characterized so much by an *average* of all filter responses as by the *very large response* of one particular filter, especially by the outliers occurring when this filter precisely matches a texton (Zhu et al., 1997). A careful and very illuminating statistical analysis of the importance of color, textural, and edge features on grouping, based on human segmented images, was given by Malik's group (Foulkes et al., 2003).

2.2.2 Extended Lines and Occlusion

The most striking demonstrations of gestalt laws of grouping come from occlusion phenomena, when edges disappear behind an object and reappear. A typical example is shown in fig 2.6. The most famous example is the so-called Kanizsa triangle, where, to further complicate matters, the foreground triangle has the same color as the background with only black circles of intermediate depth being visible. The grouping laws lead one to infer the presence of the occluding triangle and the completion of the three partially occluded black circles. An amusing variant, the Kanizsa pear, is shown in the same figure.

These effects are not merely psychophysical curiosities. Virtually every image of the natural world has major edges which are occluded one or more times by foreground objects. Correctly grouping these edges goes a long way to finding the correct parse of an image.

A good deal of modeling has gone into the grouping of disconnected edges into extended edges and the evaluation of competing groupings by energy values or probabilities. Pioneering work was done by Parent and Zucker (1989) and Shashua and Ullman (1988). Nitzberg, Shiota, and I proposed a model for this (Nitzberg et al., 1992) which was a small extension of the Mumford-Shah model. The new energy involves explicitly the overlapping regions R_α in the image given by the 3D objects in the scene, both the visible and the occluded parts of these objects. Therefore, finding its minimum involves inferring the occluded parts of the visible objects as well as the boundaries of their visible parts. (These are literally "hidden

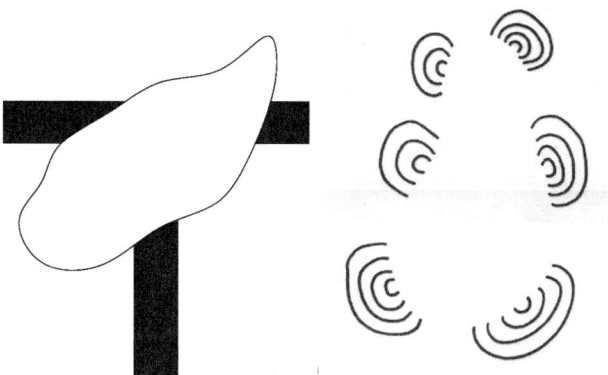

Figure 2.6 Two examples of gestalt grouping laws: on the left, the black bars are continued under the white blob to form the letter T, on the right, the semicircles are continued underneath a foreground "pear" which must completed by contours with zero contrast.

variables".) Moreover, we need the depth order of the objects—which are nearer, which farther away. The cartoon u of the image is now assumed piecewise constant, with value u_α on the region R_α. Then,

$$E(I, \{u_\alpha\}, \{R_\alpha\}) = \sum_\alpha C_1 \int_{R'_\alpha} (I - u_\alpha)^2 + \int_{\partial R_\alpha} \left(C_2 \kappa^2_{\partial R_\alpha} + C_3\right) ds,$$

$$R'_\alpha = \left(R_\alpha - \bigcup_{\text{nearer } R_\beta} R_\alpha \cap R_\beta\right) = \text{ visible part of } R_\alpha,$$

$$\kappa_{\partial R_\alpha} = \text{ curvature of } \partial R_\alpha.$$

This energy allows one to quantify the application of gestalt rules for inferring occluded objects and predicts correctly, for example, the objects present in the Kanizsa triangle. The minima of this E will infer specific types of hidden contours, namely contours which come from the purely geometric variational problem of minimizing a sum of squared curvature and arc length along an unknown curve. This variational problem was first formulated by Euler, who called the resulting curves *elastica*.

To make a stochastic model out of this, we need a stochastic model for the edges occurring in natural images. There are two parts to this: one is modeling the local nature of edges in images and the other is modeling the way they group into extended curves.

Several very simple ideas for modeling curves locally, based on Brownian motion, were proposed in Mumford (1992). Brownian paths themselves are too jagged to be suitable, but one can assume the curves are C^1 and that their orientation $\theta(s)$, as a function of arc length, is Brownian. Geometrically, this is like saying their curvature is white noise. Another alternative is to take 2D projections of 3D curves whose direction of motion, given by a map from arc length to points on the unit sphere, is Brownian. Such curves have more corners and cusps, where the 3D path heads toward or away from the camera. Yet another option is to generate parameterized curves whose velocity $(x'(t), y'(t))$ is given by two Ornstein-Uhlenbeck processes (Brownian functions with a restoring force pulling them to 0). These paths have nearly straight segments when the velocity happens to get large.

A key probability distribution in any such theory is $p(x, y, \theta)$, the probability density that if an image contour passes through $(0, 0)$ with horizontal tangent, then this contour will also pass through (x, y) with orientation θ. This function has been estimated from image databases in (Geisler et al., 2001), but I do not know of any comparison of their results with mathematical models.

Subsequently, Zhu (1999) and Ren and Malik (2002) directly analyzed edges and their curvature in hand-segmented images. Zhu found a high-kurtosis empirical distribution much like filter responses: a peak at 0 showing the prevalence of straight edges and large tails indicating the prevalence of corners. He built a stochastic model for polygonal approximations to these curves using an exponential model of the form

$$p(\Gamma) \propto e^{- \int_\Gamma \psi_1(\kappa(s)) + \psi_2(\kappa'(s)) ds},$$

where κ is the curvature of Γ and the ψ_i are unknown functions chosen so that the model yields the same distribution of κ, κ' as that found in the data. Finding continuum limits of his models under weak convergence is an unsolved problem. Ren and Malik's models go beyond the previous strictly local ones. They are k^{th}-order Markov models in which the orientation θ_{k+1} of a curve at a sample point P_{k+1} is a sample from a joint probability distribution of the orientations θ_k^α of both the curve and smoothed versions of itself at other scales α, all at the previous point P_k.

A completely different issue is finding probabilities that two edges should be joined, e.g., if Γ_1, Γ_2 are two curves ending at points P_1, P_2, how likely is it that in the real world there is a curve Γ_h joining P_1 and P_2 and creating a single curve $\Gamma_1 \cup \Gamma_h \cup \Gamma_2$? This link might be hidden in the image because of either occlusion, noise or low contrast (anyone with experience with real images will not be surprised at how often this happens). Jacobs, Williams, Geiger, and others have developed algorithms of this sort based on elastica and related ideas (Geiger et al., 1998; Williams and Jacobs, 1997). Elder and Goldberg (2002) and Geisler et al. (2001) have carried out psychophysical experiments to determine the effects of proximity, orientation difference, and edge contrast on human judgments of edge completions.

One of the subtle points here (as Ren and Malik make explicit) is that this probability does not depend only on the endpoints P_i and the tangent lines to the Γ_i at these points. So, for instance, if Γ_1 is straight for a certain distance before its endpoint P_1, then the longer this straight segment is, the more likely it is that any continuation it has will also be straight. An elegant analysis of the situation purely for straight edges has been given by Desolneux et al. (2003). It is based on what they call *maximally meaningful alignments*, which come from computing

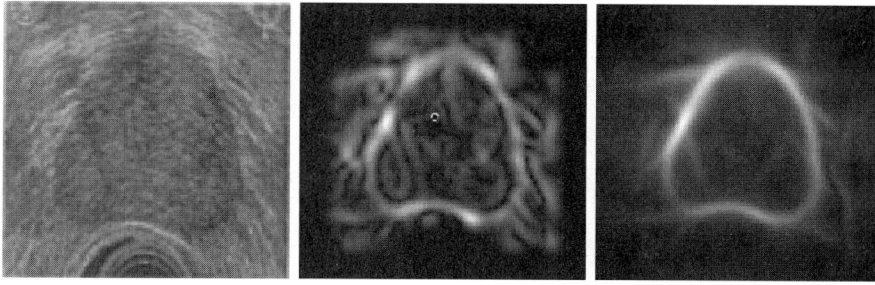

Figure 2.7 An experiment finding the prostate in a MRI scan (from August (2002)). On the left, the raw scan; in the middle, edge filter responses; on the right, the computed posterior of August's *curve indicator random field*, (which actually lives in (x, y, θ) space, hence the boundary of the prostate is actually separated from the background noise).

the probabilities of accidental alignments and no other prior assumptions. The most compelling analysis of the problem, to my mind, is that in the thesis of Jonas August (August, 2001). He starts with a prior on a countable set of true curves, assumed to be part of the image. Then he assumes a noisy version of this is observed and seeks the maximally probable reconstruction of the whole set of true curves. An example of his algorithms is shown in fig 2.7. Another algorithm for global completion of all image contours has been given recently by Malik's group (Ren et al., 2005).

2.2.3 Mathematical Formalisms for Visual Grammars

The "higher level" Gestalt rules for grouping based on parallelism, symmetry, closedness, and convexity are even harder to make precise. In this section, I want to describe a general approach to these questions.

So far, we have described grammars loosely as recursive groupings of parts of a signal, where the signal can be a string of phonemes or an image of pixels. The mathematical structure which these groupings define is a *tree*: each subset of the domain of the image which is grouped together defines a node in this tree and, whenever one such group contains another, we join the nodes by an edge. In the case of sentences in human languages, this tree is called the *parse tree*. In the case of images, it is similar to the *image pyramid* made up of the pixels of the image plus successively "blowndown" images 2^n times smaller. However, unlike the image pyramid, its nodes only stand for natural groupings, so its structure is adaptively determined by the image itself.

To go deeper into the formalism of grammar, the next step is to label these groupings. In language, typical labels are "noun phrase", "prepositional clause," etc. In images, labels might be "edgelet," "extended edge," "ribbon," "T-junction," or

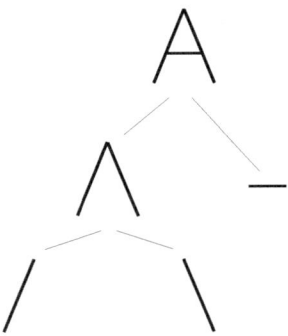

Figure 2.8 The parse tree for the letter A which labels the top node; the lower nodes might be labeled "edge" and "corner." Note that in grouping the two sides, the edge has an attribute giving its length and approximate equality of the lengths of the sides must hold; and in the final grouping, the bar of the A must meet the two sides in approximately equal angles. These are probabilistic constraints involving specific attributes of the constituents, which must be included in B_ℓ.

even "the letter A." Then the grouping laws are usually formulated as *productions*:

$$\text{noun phrase} \longrightarrow \text{determiner} + \text{noun}$$
$$\text{extended edge} \longrightarrow \text{edgelet} + \text{extended edge}$$

where the group is on the left and its constituents are shown on the right. The second rule creates a long edge by adding a small piece, an edgelet, to one end. But now the issue of *agreement* surfaces: one can say "a book" and "some books" but not "a books" or "some book." The determiner and the noun must agree in number. Likewise, to group an edge with a new edgelet requires that the edgelet connect properly to the edge: where one ends, the other must begin. So we need to endow our labeled groupings with a list of attributes that must agree for the grouping to be possible. So long as we can do this, we have created a context-free grammar. Context-freeness means that the possibility of the larger grouping depends only on the labels and attributes of the constituents and nothing else. An example of the parse of the letter A is shown in fig 2.8.

We make the above into a probability model in a top-down generative fashion by assigning probabilities to each production. For any given label and attributes, the sum (or integral) of the probabilities of all possible productions it can yield should be 1. This is called a PCFG (probabilistic context-free grammar) by linguists. It is the same as what probabilists call a random branching tree (except that grammars are usually assumed to almost surely yield *finite* parse trees).

probabilistic context-free grammar

A more general formalism for defining random trees with random data attached to their nodes has been given by Artur Fridman (Fridman, 2003). He calls his models *mixed Markov models* because some of the nodes carry *address variables* whose value is the index of another node. Thus in each sample from the model, this node adds a new edge to the graph. His models include PCFGs as a special case.

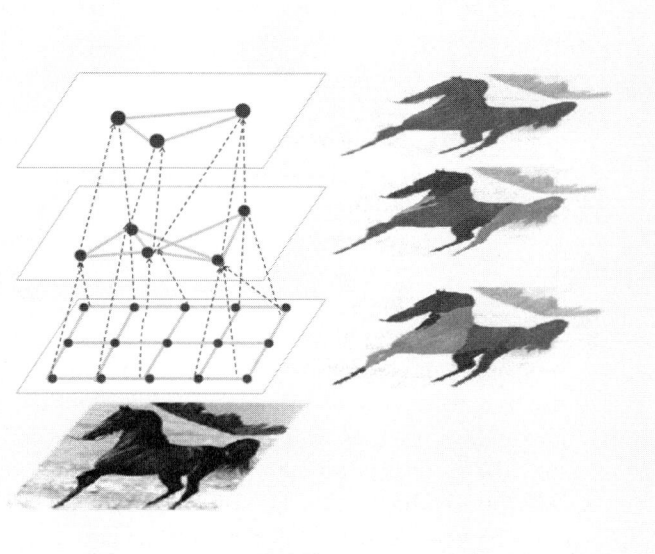

Figure 2.9   A simplification of the parse tree inferred by the segmentation algorithm of Galun et al. (2003). The image is at the bottom and part of its tree is shown above it. On the right are shown some of the regions in the image, grouped by successive levels of the algorithm.

Random trees can be fit naturally into the random wavelet model (or the dead leaves model) described above. To see this, we consider each 4-tuple $\{x_\alpha, y_\alpha, r_\alpha, \psi_\alpha\}$ in the model not merely as generating one elementary constituent of the image, but as the root of a whole random branching tree. The child nodes it generates should add parts to a now compound object, expanding the original simple image constituent ψ_α. For example the root might be an elongated blob representing the trunk of a person and the tree it generates would add the limbs, clothes, face, hands, etc., to the person. Or the root might be a uniform patch and the tree would add a whole set of textons to it, making it into a textured patch. So long as the rate of growth of the random branching tree is not too high, we still get a scale-invariant model.

Two groups have implemented image analysis programs based on computing such trees. One is the multiscale segmentation algorithm of Galun, Sharon, Basri, and Brandt (Galun et al., 2003), which produces very impressive segmentation results. The method follows Brandt's adaptive tree-growing algorithm called *algebraic multi-grid*. In their code, texture and its component textons play the same role as objects and their component parts: each component is identified at its natural scale and grouped further at a higher level in a similar way (see fig 2.9). Their code is fully scale-invariant except at the lowest pixel level. It would be very interesting to fit their scheme into the Bayesian framework.

The other algorithm is an integrated bottom-up and top-down image parsing program from Zhu's lab (Tu et al., 2003). The output of their code is a tree with semantically labeled objects at the top, followed by parts and texture patches in the middle with the pixels at the bottom. This program is based on a full stochastic model.

A basic problem with this formalism is that it is not sufficiently expressive: the grammars of nature appear to be context sensitive. This is often illustrated by contrasting languages that have sentences of the form *abcddcba*, which can be generated recursively by a small set of productions as in

$$s \to asa \to absba \to abcscba \to abcddcba,$$

versus languages which have sentences of the form *abcdabcd*, with two complex repeating structures, which cannot be generated by simple productions. Obviously, images with two identical faces are analogs of this last sentence. Establishing symmetry requires you to reopen the grouped package and examine everything in it to see if it is repeated! Unless you imagine each label given a huge number of attributes, this cannot be done in a context-free setting.

In general, two-dimensional geometry creates complex interactions between groupings, and the strength of higher-order groupings seems to always depend on multiple aspects of each piece. Take the example of a square. Ingredients of the square are (1) the two groupings of parallel edges, each made up of a pair of parallel sides of equal length and (2) the grouping of edgelets adjacent to each vertex into a "right-angle" group. The point is that the pixels involved in these smaller groupings partially intersect. In PCFGs, each group should expand to disjoint sets of primitives or to one set contained in another. The case of the square is best described with the idea of *graph unification*, in which a grouping rule unifies parts of the graph of parts under each constituent.

S. Geman and his collaborators (Bienenstock et al., 1998; Geman et al., 2002) have proposed a general framework for developing such probabilistic context-sensitive grammars. He proposes that for grouping rule ℓ, in which groups $y_1, y_2, \cdots, y_k$ are to be unified into a larger group x, there is a binding function $B_\ell(y_1, y_2, \cdots, y_k)$ which singles out those attributes of the constituents that affect the probability of making the k-tuple of y's into an x. For example, to put two edgelets together, we need to ask if the endpoint of the first is near the beginning of the second and whether their directions are close. The closer are these points and directions, the more likely it is that the two edgelets should be grouped. The basic hypothesis is that the likelihood ratio $p(x, y_1, \cdots, y_k)/\prod_i p(y_i)$ depends only on $B_\ell(y_1, \cdots, y_k)$. In their theory, Geman and colleagues analyze how to compute this function from data.

This general framework needs to be investigated in many examples to further constrain it. An interesting example is the recent work of Ullman and collaborators (Ullman et al., 2002) on face recognition, built up through the recognition of parts: this would seem to fit into this framework. But, overall, the absence of mathematical theories which incorporate all the gestalt rules at once seems to me the biggest gap in our understanding of images.

2.3   Probability Measures on the Space of Shapes

The most characteristic new pattern found in visual signals, but not in one-dimensional signals, are *shapes*, two-dimensional regions in the domain of the image. In auditory signals, one has *intervals* on which the sound has a particular spectrum, for instance, corresponding to some specific type of source (for phonemes, some specific configuration of the mouth, lips, and tongue). But an interval is nothing but a beginning point and an endpoint. In contrast, a subset of a two-dimensional region is much more interesting and conveys information by itself. Thus people often recognize objects by their shape alone and have a rich vocabulary of different categories of shapes often based on prototypes (heart-shaped, egg-shaped, star-shaped, etc.). In creating stochastic models for images, we must face the issue of constructing probability measures on the space of all possible shapes. An even more basic problem is to construct metrics on the space of shapes, measures for the dissimilarity of two shapes. It is striking how people find it quite natural to be asked if some new object has a shape similar to some old object or category of objects. They act as though they carried a clear-cut psychophysical metric in their heads, although, when tested, their similarity judgments show a huge amount of context sensitivity.

2.3.1   The Space of Shapes and Some Basic Metrics on It

What do we mean by the space of shapes? The idea is simply to define this space as the set of 2-dimensional shapes, where a shape is taken to mean an open subset $S \subset \mathrm{R}^2$ with smooth boundary[7]. We let $\mathcal{S}$ denote this set of shapes. The mathematician's approach is to ask: what structure can we give to $\mathcal{S}$ to endow it with a geometry? In particular, we want to define (1) local coordinates on $\mathcal{S}$, so that it is a manifold, (2) a metric on $\mathcal{S}$, and (3) probability measures on $\mathcal{S}$. Having probability measures will allow us to put shapes into our theory as hidden variables and extend the Bayesian inference machinery to include inferring shape variables from images.

manifold

$\mathcal{S}$ itself is not a vector space: one cannot add and subtract two shapes in a way satisfying the usual laws of vectors. Put another way, there is, no obvious way to put global coordinates on $\mathcal{S}$, that is to create a bijection between points of $\mathcal{S}$ and points in some vector space. One can, e.g. describe shapes by their Fourier coefficients, but the Fourier coefficients coming from shapes will be very special sequences of numbers. What we can do, however, is put a *local linear structure* on the space of shapes. This is illustrated in fig 2.10. Starting from one shape S, we erect normal lines at each point of the boundary Γ of S. Then nearby shapes will have boundaries which intersect each normal line in a unique point. Suppose $\psi(s) \in \mathrm{R}^2$ is arc-length parameterization of Γ. Then the unit normal vector is given by $\vec{n}(s) = \psi'^{\perp}(s)$ and each nearby curve is parameterized uniquely in the form

$$\psi_a(s) = \psi(s) + a(s) \cdot \vec{n}(s), \qquad \text{for some function } a(s).$$

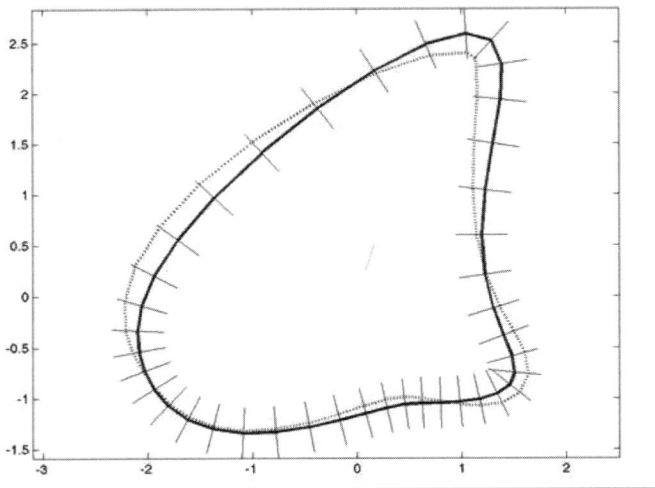

Figure 2.10   The manifold structure on the space of shapes is here illustrated: all curves near the heavy one meet the normal "hairs" in a unique point, hence are described by a *function*, namely, how far this point has been displaced normally.

All smooth functions $a(s)$ which are sufficiently small can be used, so we have created a bijection between an open set of functions a, that is an open set in a vector space, and a neighborhood of $\Gamma \in \mathcal{S}$. These bijections are called *charts* and on overlaps of such charts, one can convert the a's used to describe the curves in one chart into the functions in the other chart: this means we have a manifold. For details, see the paper (Michor and Mumford, 2006). Of course, the function $a(s)$ lies in an infinite-dimensional vector space, so $\mathcal{S}$ is an infinite-dimensional manifold. But that is no deterrent to its having its own intrinsic geometry.

tangent space   Being a manifold means $\mathcal{S}$ has a *tangent space* at each point $S \in \mathcal{S}$. This tangent space consists in the infinitesimal deformations of S, i.e., those coming from infinitesimal $\epsilon a(s)$. Dropping the ϵ, the infinitesimal deformations may be thought of simply as normal vector fields to Γ, that is, the vector fields $a(s) \cdot \vec{n}(s)$. We denote this tangent space as $T_{S,\mathcal{S}}$.

How about metrics? In analysis, there are many metrics on spaces of functions and they vary in two different ways. One choice is whether you make a worst-case analysis or an average analysis of the difference of two functions—or something in between. This means you define the difference of two functions a and b either as the $\sup_x |a(x) - b(x)|$, the integral $\int |a(x) - b(x)| dx$, or as an L^p norm, $(\int |a(x) - b(x)|^p dx)^{1/p}$ (which is in between). The case $p = \infty$ corresponds to the sup, and $p = 1$ to the average. Usually, the three important cases[8] are $p = 1, 2$, or ∞. The other choice is whether to include derivatives of a, b as well as the values of a, b in the formula for the distance and, if so, up to what order k. These distinctions carry

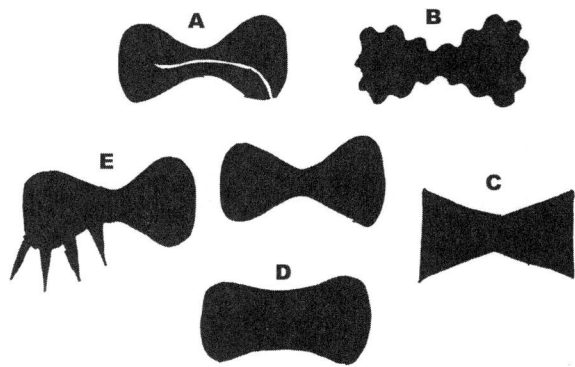

Figure 2.11   Each of the shapes A, B, C, D, and E is similar to the central shape, but *in different ways*. Different metrics on the space of shape bring out these distinctions (adapted from B. Kimia).

over to shapes. The best-known measures are the so-called Hausdorff measure,

$$d_{\infty,0}(S,T) = \max\left(\sup_{x \in S} \inf_{y \in T} \|x - y\|, \sup_{y \in T} \inf_{x \in S} \|x - y\|\right),$$

for which $p = \infty, k = 0$, and the area metric,

$$d_{1,0}(S,T) = \mathrm{Area}(S - S \cap T) \cup \mathrm{Area}(T - S \cap T),$$

for which $p = 1, k = 0$.

   It is important to realize that there is no one *right* metric on $\mathcal{S}$. Depending on the application, different metrics are good. This is illustrated in fig 2.11. The central bow-tie-like shape is similar to all the shapes around it. But different metrics bring out their dissimilarities and similarities in each case. The Hausdorff metric applied to the outsides of the shapes makes A far from the central shape; any metric using the first derivative (i.e., the orientation of the tangent lines to the boundary) makes B far from the central shape; a sup-type metric with the second derivative (i.e., the curvature of the boundary) makes C far from the central shape, as curvature becomes infinite at corners; D is far from the central shape in the area metric; E is far in all metrics, but the challenge is to find a metric in which it is close to the central shape. E has "outliers," the spikes, but is identical to the central shape if they can be ignored. To do this needs what are called *robust* metrics of which the simplest example is $L^{1/2}$ (not a true metric at all).

2.3.2   Riemannian Metrics and Probability Measures via Diffusion

Riemannian
metrics

There are great mathematical advantages to using L^2, so-called Riemannian metrics. More precisely, a Riemannian metric is given by defining a quadratic inner product in the tangent space $T_{S,\mathcal{S}}$. In Riemannian settings, the unit balls are nice and round and extremal problems, such as paths of shortest length, are usually well posed. This means we can expect to have geodesics, optimal deformations of one shape S to a second shape T through a family S_t of intermediate shapes, i.e., we can *morph S to T* in a most efficient way. Having geodesics, we can study the geometry of $\mathcal{S}$, for instance whether its geodesics diverge or converge[9]—which depends on the curvature of $\mathcal{S}$ in the metric. But most important of all, we can define diffusion and use this to get Brownian paths and thus probability measures on $\mathcal{S}$.

A most surprising situation arises here: there are three completely different ways to define Riemannian metrics on $\mathcal{S}$. We need to assign a norm to normal vector fields $a(s)\vec{n}(s)$ along a simple closed plane curve Γ.

local metric

■ In infinitesimal metric, the norm is defined as an integral along Γ. In general, this can be any expression

$$\|a\|^2 = \int_\Gamma F(a(s), a'(s), a''(s), \cdots, \kappa(s), \kappa'(s), \cdots)ds,$$

involving a function F quadratic in a and the derivatives of a whose coefficients can possibly be functions associated to Γ like the curvature and its derivatives. We call these local metrics. We might have $F = a(s)^2$ or $F = (1 + A\kappa^2(s)) \cdot a(s)^2$, where A is a constant; or $F = a(s)^2 + Aa'(s)^2$, etc. These metrics have been studied by Michor and Mumford (Michor and Mumford, 2006, 2005). Globally, the distance between two shapes is then

$$d(S_0, S_1) = \inf_{\text{paths } \{S_t\}} \int_0^1 \|\frac{\partial S_t}{\partial t}\|dt,$$

where $\partial S_t/\partial t$ is the normal vector field given by this path.

diffeomorphism

■ In other situations, a morph of one shape to another needs to be considered as part of a morph of the whole plane. For this, the metric should be a quotient of a metric on the group $\mathcal{G}$ of diffeomorphisms of R^2, with some boundary condition, e.g., equal to the identity outside some large region. But an infinitesimal diffeomorphism is just a vector field $\vec{v}$ on R^2 and the induced infinitesimal deformation of Γ is given by $a(s) = (\vec{v}\cdot\vec{n}(s))$. Let V be the vector space of all vector fields on R^2, zero outside some large region. Then this means that the norm on a is

$$\|a\|^2 = \inf_{\vec{v}\in V, (\vec{v}\cdot\vec{n})=a} \int_{R^2} F(\vec{v}, \vec{v}_x, \vec{v}_y, \cdots)dxdy,$$

where we define an inner product on V using a symmetric positive definite quadratic expression in $\vec{v}$ and its partial derivatives. We might have $F = \|\vec{v}\|^2$ or $F = \|\vec{v}\|^2 + A\|\vec{v}_x\|^2 + A\|\vec{v}_y\|^2$, etc. It is convenient to use integration by parts and write all such F's as $(L\vec{v}, \vec{v})$, where L is a positive definite partial differential operator

Miller's metric

($L = I - A\triangle$ in the second case above). These metrics have been studied by Miller, Younes, and their many collaborators (Miller, 2002; Miller and Younes, 2001) and applied extensively to the subject they call computational anatomy, that is, the analysis of medical scans by deforming them to template anatomies. Globally, the

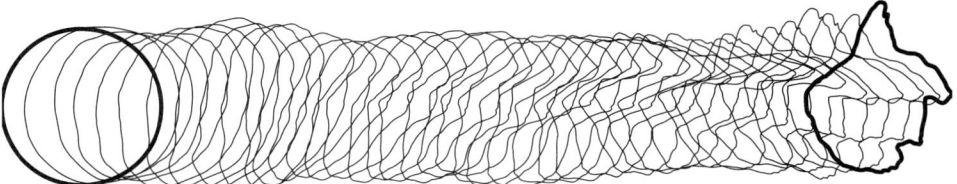

Figure 2.12 A diffusion on the space of shapes in the Riemannian metric of Miller et al. The shapes should be imagined on top of each other, the translation to the right being added in order that each shape can be seen clearly. The diffusion starts at the unit circle.

distance between two shapes is then

$$d_{\text{Miller}}(S,T) = \inf_{\phi} \int_0^1 \left(\int_{\mathbb{R}^2} F(\frac{\partial \phi}{\partial t} \circ \phi^{-1}) dx dy \right)^{1/2} dt, \quad \text{where}$$

$$\phi(t), 0 \leq t \leq 1 \text{ is a path in } \mathcal{G}, \phi(0) = I, \phi(1)(S) = T.$$

Weil-Petersen
metric

■ Finally, there is a remarkable and very special metric on $\bar{\mathcal{S}} = \mathcal{S}$ modulo translations and scalings (i.e., one identifies any two shapes which differ by translation plus a scaling). It is derived from complex analysis and known as the Weil-Petersen (or WP) metric. Its importance is that it makes $\bar{\mathcal{S}}$ into a *homogeneous* metric space, that is, it has everywhere the same geometry. There is a group of global maps of $\mathcal{S}$ to itself which preserve distances in this metric and which can take any shape S to any other shape T. This is not the case with the previous metrics, hence the WP metric emerges as the analog of the standard Euclidean distance in finite dimensions. The definition is more elaborate and we do not give it here, see the paper (Mumford and Sharon, 2004). This metric also has negative or zero curvature in all directions and hence finite sets of shapes as well as probability measures on $\bar{\mathcal{G}}$ should always have a well-defined mean (minimizing the sum of squares of distances) in this metric. Finally, this metric is closely related to the medial axis, which has been frequently used for shape classification.

The next step in each of these theories is to investigate the heat kernel, the solution of the heat equation starting at a delta function. This important question has not been studied yet. But diffusions in these metrics are easy to simulate. In fig 2.12 we show three random walks in $\mathcal{S}$ in one of Miller's metrics. The analog of Gaussian distributions are the probability measures gotten by stopping diffusion at a specific point in time. And analogs of the scale mixtures of Gaussians discussed above are obtained by using a so-called random stopping time, that is, choosing the time to halt the diffusion randomly from another probability distribution. It seems clear that one or more of these diffusion measures are natural general-purpose priors on the space of shapes.

2.3.3   Finite Approximations and Some Elementary Probability Measures

A completely different approach is to infer probability measures directly from data. Instead of seeking general-purpose priors for stochastic models, one seeks special-purpose models for specific object-recognition tasks. This has been done by extracting from the data a finite set of *landmark points*, homologous points which can be found on each sample shape. For example, in 3 dimensions, skulls have long been compared by taking measurements of distances between classical landmark points. In 2 dimensions, assuming these points are on the boundary of the shape, the infinite dimensional space $\mathcal{S}$ is replaced by the finite dimensional space of the polygons $\{P_1, \cdots, P_k\} \in \mathrm{R}^{2k}$ formed by these landmarks. But, if we start from images, we can allow the landmark points to lie in the interior of the shape also. This approach was introduced a long time ago to study faces. More specifically, it was used by Cootes et al. (1993) and by Hallinan et al. (1999) to fit multidimensional Gaussians to the cloud of points in R^{2k} formed from landmark points on each of a large set of faces. Both groups then apply principal component analysis (PCA) and find the main directions for face variation.

However, it seems unlikely to me that Gaussians can give a very good fit. I suspect rather that in geometric situations as well, one will encounter the high kurtosis phenomenon, with geometric features often near zero but, more often than for Gaussian variables, very large too. A first attempt to quantify this point of view was made by Zhu (1999). He took a database of silhouettes of four-legged animals and he computed landmark points, medial axis, and curvature for each silhouette. Then he fit a general exponential model to a set of six scalar variables describing this geometry. The strongest test of whether he has captured some of their essential shape properties is to sample from the model he gets. The results are shown in fig 2.13. It seems to me that these models are getting much closer to the sort of special-purpose prior that is needed in object-recognition programs. Whether his models have continuum limits and of what sort is an open question.

There are really three goals for a theory of shapes adapted to the analysis of images. The first is to understand better the global geometry of $\mathcal{S}$ and which metrics are appropriate in which vision applications. The second is to create the best general-purpose priors on this space, which can apply to arbitrary shapes. The third is to mold special-purpose priors to all types of shapes which are encountered frequently, to express their specific variability. Some progress has been made on all three of these but much is left to be done.

2.4   Summary

Solving the problem of vision requires solving three subproblems: finding the right classes of stochastic models to express accurately the variability of visual patterns in nature, finding ways to learn the details of these models from data, and finding ways to reason rapidly using Bayesian inference on these models. This chapter has

Figure 2.13 Six "animals" that never existed: they are random samples from the prior of S. C. Zhu trained on real animal silhouettes. The interior lines come from his use of medial axis techniques to generate the shapes.

addressed the first. Here a great deal of progress has been made but it must be said that much remains to be done. My own belief is that good theories of groupings are the biggest gap. Although not discussed in this article, let me add that great progress has been made on the second and third problems with a large number of ideas, e.g., the expectation maximization (EM) algorithm, much faster Monte Carlo algorithms, maximum entropy (MaxEnt) methods to fit exponential models, Bayesian belief propagation, particle filtering, and graph-theoretic techniques.

Notes

[1] The chief mistake of the Greeks was their persistent belief that the eye must *emit* some sort of ray in order to do something equivalent to touching the visible surfaces.

[2] This is certainly biologically unrealistic. Life requires rapid analysis of changing scenes. But this article, like much of vision research, simplifies its analysis by ignoring time.

[3] It is the second idea that helps to explain why aerial photographs also show approximate scale invariance.

[4] The infrared divergence is readily solved by considering images *mod constants*. If the pixel values are log of the photon energy, this constant is an irrelevant gain factor.

[5] Some have found an especially large concentration near $1/f^{1.8}$ or $1/f^{1.9}$, especially for forest scenes (Ruderman and Bialek).

[6] Scale invariance implies that its expected power at spatial frequency (ξ, η) is a constant times $1/(\xi^2 + \eta^2)$ and integrating this over (ξ, η) gives ∞.

[7] A set S of points is *open* if S contains a small disk of points around each point $x \in S$. *Smooth* means that it is a curve that is locally a graph of a function with infinitely many derivatives; in many applications, one may want to include shapes with corners. We simplify the discussion here and assume there are no corners.

[8] Charpiat et al., however, have used p-norm as for $p \gg 1$ in order to "tame" L^∞ norms.

[9] This is a key consideration when seeking means to clusters of finite sets of shapes and in seeking principal components of such clusters.

3   The Machine Cocktail Party Problem

Simon Haykin and Zhe Chen

Imagine you are in a cocktail party environment with background music, and you are participating in a conversation with one or more of your friends. Despite the noisy background, you are able to converse with your friends, switching from one to another with relative ease. Is it possible to build an *intelligent machine* that is able to perform like yourself in such a noisy environment? This chapter explores such a possibility.

cocktail party problem

The *cocktail party problem* (CPP), first proposed by Colin Cherry, is a psychoacoustic phenomenon that refers to the remarkable human ability to selectively attend to and recognize one source of auditory input in a noisy environment, where the hearing interference is produced by competing speech sounds or various noise sources, all of which are usually assumed to be independent of each other (Cherry, 1953). Following the early pioneering work (Cherry, 1953, 1957, 1961; Cherry and Taylor, 1954), numerous efforts have been dedicated to the CPP in diverse fields: physiology, neurobiology, psychophysiology, cognitive psychology, biophysics, computer science, and engineering.[1] Over half a century after Cherry's seminal work, however, it is fair to say that a complete understanding of the cocktail party phenomenon is still missing, and the story is far from being complete; the marvelous auditory perception capability of human beings remains enigmatic. To unveil the mystery and thereby imitate human performance by means of a machine, computational neuroscientists, computer scientists, and engineers have attempted to view and simplify this complex perceptual task as a learning problem, for which a tractable computational solution is sought. An important lesson learned from the collective work of all these researchers is that in order to imitate a human's unbeatable audition capability, a deep understanding of the human auditory system is crucial. This does *not* mean that we must duplicate every aspect of the human

machine cocktail party problem

auditory system in solving the *machine cocktail party problem*, hereafter referred to as the machine CPP for short. Rather, the challenge is to expand on what we know about the human auditory system and put it to practical use by exploiting advanced computing and signal-processing technologies (e.g., microphone arrays, parallel computers, and VLSI chips). An efficient and effective solution to the machine CPP will not only be a major accomplishment in its own right, but it will also have a direct impact on ongoing research in artificial intelligence (such as robotics) and human-machine interfaces (such as hearing aids); and these lines of research will, in their own individual ways, further deepen our understanding of the human brain.

There are three fundamental questions pertaining to the CPP:

1. What is the cocktail party problem?
2. How does the brain solve it?
3. Is it possible to build a machine capable of solving it in a satisfactory manner?

The first two questions are human oriented, and mainly involve the disciplines of neuroscience, cognitive psychology, and psychoacoustics; the last question is rooted in machine learning, which involves computer science and engineering disciplines. While these three issues are equally important, this chapter will focus on the third question by addressing a solution to the machine CPP.

To understand the CPP, we may identify three underlying neural processes:[2]

■ *Analysis*: The analysis process mainly involves *segmentation* or *segregation*, which refers to the segmentation of an incoming auditory signal to individual *channels* or *streams*. Among the heuristics used by a listener to do the segmentation, *spatial location* is perhaps the most important. Specifically, sounds coming from the same location are grouped together, while sounds originating from other different directions are segregated.

■ *Recognition*: The recognition process involves analyzing the statistical structure of essential patterns contained in a sound stream. The goal of this process is to uncover the neurobiological mechanisms through which humans are able to identify a segregated sound from multiple streams with relative ease.

■ *Synthesis*: The synthesis process involves the reconstruction of individual sound waveforms from the separated sound streams. While synthesis is an important process carried out in the brain, the synthesis problem is of primary interest to the machine CPP.

From an engineering viewpoint, we may, in a loose sense, regard synthesis as the inverse of the combination of analysis and recognition in that synthesis attempts to uncover relevant attributes of the speech production mechanism. Note also that, insofar as the machine CPP is concerned, an accurate synthesis does not necessarily mean having solved the analysis and recognition problems, although additional information on these two problems might provide more hints for the synthesis process.

Bearing in mind that the goal of solving the machine CPP is to build an intelligent machine that can operate efficiently and effectively in a noisy cocktail party environment, we propose a computational framework for *active audition* that has the potential to serve this purpose. To pave the way for describing this framework, we will discuss the important aspects of *human auditory scene analysis* and *computational auditory scene analysis*. Before proceeding to do so, however, some historical notes on the CPP are in order.

3.1 Some Historical Notes

In the historical notes that follow, we do two things. First, we present highlights of the pioneering experiments performed by Colin Cherry over half a century ago, which are as valid today as they were then; along the way, we also refer to the other related works. Second, we highlight three machine learning approaches: *independent component analysis, oscillatory correlation*, and *cortronic processing*, which have been motivated by the CPP in one form or another.

3.1.1 Cherry's Early Experiments

In the early 1950s, Cherry became interested in the remarkable hearing capability of human beings in a cocktail party environment. He himself raised several questions: What is our selective attention ability? How are we able to select information coming from multiple sources? Some information is still retained even when we pay no attention to it; how much information is retained? To answer these fundamental questions, Cherry (1953) compared the ability of listeners to attend to two different spoken messages under different scenarios. In his classic experimental set-up called *dichotic listening*, the recorded messages were mixed and presented together to the same ear of a subject over headphones, and the listeners were requested to test the intelligibility[3] of the message and repeat each word of the message to be heard, a task that is referred to as *shadowing*. In the cited paper, Cherry reported that when one message is delivered to one ear (the attended channel) and a different message is delivered to the other ear (the unattended channel), listeners can easily attend to one or the other of these two messages, with almost all of the information in the attended message being determined, while very little about the unattended message is recalled. It was also found that the listeners became quite good at the shadowing task after a few minutes, repeating the attended speech quite accurately. However, after a few minutes of shadowing, listeners had no idea of what the unattended voice was about, or even if English was spoken or not. Based on these observations and others, Cherry conjectured that some sort of *spatial filtering* of the concurrently occurring sounds/voices might be helpful in attending to the message. It is noteworthy that Cherry (1953) also suggested some procedures to design a "filter" (machine) to solve the CPP, accounting for the following: (1) the voices come from different directions; (2) lip reading, gesture, and the like; (3) different speaking voices, mean pitches, mean speeds, male vs. female, and so forth; (4) different accents and linguistic factors; and (5) transition probabilities (based on subject matter, voice dynamics, syntax, etc.). In addition, Cherry also speculated that humans have a vast *memory* of transition probabilities that make the task of hearing much easier by allowing prediction of word sequences.

The main findings of the dichotic listening experiments conducted by Cherry and others have revealed that, in general, it is difficult to attend to two sound sources at once; and when we switch attention to an unattended source (e.g., by listening to a spoken name), we may lose information from the attended source. Indeed, our own common experiences teach us that when we attempt to tackle more than one task at a time, we may end up sacrificing performance.

In subsequent joint investigations with colleagues (Cherry, 1961; Cherry and Sayers, 1956, 1959; Sayers and Cherry, 1957), Cherry also studied the *binaural fusion* mechanism and proposed a cross-correlation-based technique for measuring certain parameters of speech intelligibility. Basically, it was hypothesized that the brain performs *correlation* on signals received by the two ears, playing the role of localization and coincidence detection. In the binaural fusion studies, Sayers and Cherry (1957) showed that the human brain does indeed execute short-term correlation analysis for either monaural or binaural listening.

To sum up, Cherry not only coined the term "cocktail party problem," but also was the first experimentalist to investigate the benefits of binaural hearing and point to the potential of lip-reading, etc., for improved hearing, and to emphasize the critical role of correlation in binaural fusion—Cherry was indeed a pioneer of human communication.

3.1.2   Independent Component Analysis

independent component analysis

The development of independent component analysis (ICA) was partially motivated by a desire to solve a cocktail party problem. The essence of ICA can be stated as follows: *Given an instantaneous linear mixture of signals produced by a set of sources, devise an algorithm that exploits a statistical discriminant to differentiate these sources so as to provide for the separation of the source signals in a blind manner* (Bell and Sejnowski, 1995; Comon, 1994; Jutten and Herault, 1991). The key question is how? To address this question, we first recognize that if we are to achieve the blind separation of an instantaneous linear mixture of independent source signals, then there must be a characteristic departure from the simplest possible source model: an *independently and identically distributed* (i.i.d.) *Gaussian model*; violation of which will give rise to a more complex source model. The departure can arise in three different ways, depending on which of the three characteristic assumptions embodied in this simple source model is broken, as summarized here (Cardoso, 2001):

■ *Non-Gaussian i.i.d. model*: In this route to blind source separation, the i.i.d. assumption for the source signals is retained but the Gaussian assumption is abandoned for all the sources, except possibly for one of them. The Infomax algorithm due to Bell and Sejnowski (1995), the natural gradient algorithm due to Amari et al. (1996), Cardoso's JADE algorithm (Cardoso, 1998; Cardoso and Souloumiac, 1993), and the FastICA algorithm due to Hyvärinen and Oja (1997) are all based on the non-Gaussian i.i.d. model. Besides, these algorithms differ from each other in the way in which incoming source information residing in higher-order statistics (HoS) is exploited.

■ *Gaussian non-stationary model*: In this second route to blind source separation, the Gaussian assumption is retained for all the sources, which means that second-order statistics (i.e., mean and variance) are sufficient for characterizing each source signal. Blind source separation is achieved by exploiting the property of nonstationarity, provided that the source signals differ from each other in the ways in which their statistics vary with time. This approach to blind source separation was first described by Parra and Spence (2000) and Pham and Cardoso (2001). Whereas the algorithms focusing on the non-Gaussian i.i.d. model operate in the time domain, the algorithms that belong to the Gaussian nonstationary model operate in the frequency domain, a feature that also makes it possible for the second class of ICA algorithms to work with convolutive mixtures.

■ *Gaussian, stationary, correlated-in-time model*: In this third and final route to blind source separation, the blind separation of Gaussian stationary source signals is achieved on the proviso that their power spectra are *not* proportional to each other. Recognizing that the power spectrum of a wide-sense stationary random process is related to the autocorrelation function via the Wiener-Khintchine theorem, spectral differences among the source signals translate to corresponding *differences in correlated-in-time behavior* of the source signals. It is this latter property that is available for exploitation.

To sum up, Comon's 1994 paper and the 1995 paper by Bell and Sejnowski have been the catalysts for the literature in ICA theory, algorithms, and novel applications. Indeed, the literature is so extensive and diverse that in the course of ten years, ICA has established itself as an indispensable part of the ever-expanding discipline of statistical signal processing, and has had a great impact on neuroscience (Brown et al., 2001). On technical grounds, however, Haykin and Chen (2005) justify the statement that ICA does *not* solve the cocktail party problem; rather, it addresses a blind source separation (BSS) problem.

3.1.3   Temporal Binding and Oscillatory Correlation

temporal binding

Temporal binding theory was most elegantly illustrated by von der Malsburg (1981) in his seminal technical report entitled "Correlation Theory of Brain Function," in which he made two important observations: (1) the binding mechanism is accomplished by virtue of correlation between presynaptic and postsynaptic activities, and (2) strengths of the synapses follow Hebb's postulate of learning. When the synchrony between the presynaptic and postsynaptic activities is strong (weak), the synaptic strength would correspondingly increase (decrease) temporally. Moreover, von der Malsburg suggested a *dynamic link architecture* to solve the temporal binding problem by letting neural signals fluctuate in time and by synchronizing those sets of neurons that are to be bound together into a higher-level symbol/concept. Using the same idea, von der Malsburg and Schneider (1986) proposed a solution to the cocktail party problem. In particular, they developed a *neural cocktail-party processor* that uses synchronization and desynchronization to segment the incoming sensory inputs. Though merely based on simple experiments (where von der Malsburg and Schneider used amplitude modulation and stimulus onset synchrony as the main acoustic cues, in line with Helmholtz's suggestion), the underlying idea is illuminating in that the model is consistent with anatomic and physiologic observations.

The original idea of von der Malsburg was subsequently extended to different sensory domains, whereby phases of neural oscillators were used to encode the binding of sensory components (Brown and Wang, 1997; Wang et al., 1990). Of particular interest is the *two-layer oscillator model* due to Wang and Brown (1999). The aim of this model is to achieve "searchlight attention" by examining the temporal cross-correlation between the activities of pairs (or populations) of neurons. The first layer, *segmentation layer*, acts as a *locally excitatory, globally inhibitory* oscillator; and the second layer, *grouping layer*, essentially performs computational auditory scene analysis (CASA). Preceding the oscillator network, there is an auditory periphery model (cochlear and hair cells) as well as a middle-level auditory representation stage (correlogram). As reported by Wang and Brown (1999), the model is capable of segregating a mixture of voiced speech and different interfering sounds, thereby improving the signal-to-noise ratio (SNR) of the attended speech signal. The correlated neural oscillator is arguably biologically plausible; however, unlike ICA algorithms, the performance of the neural oscillator model appears to deteriorate significantly in the presence of multiple competing sources.

3.1.4 Cortronic Processing

The idea of so-called cortronic network was motivated by the fact that the human brain employs an efficient sparse coding scheme to extract the features of sensory inputs and accesses them through *associative memory* (Hecht-Nielsen, 1998). Specifically, in (Sagi et al., 2001), the CPP is viewed as an aspect of the *human speech recognition problem* in a cocktail party environment, and the solution is regarded as an attended source identification problem. In the experiments reported therein, only one microphone was used to record the auditory scene; however, the listener was assumed to be familiar with the *language of conversation* under study. Moreover, all the subjects were chosen to speak the same language and have the similar voice qualities. The goal of the cortronic network is to *identify* one attended speech of interest, which is the essence of the CPP.

According to the experimental results reported in (Sagi et al., 2001), it appears that the cortronic network is quite robust with respect to variations in speech, speaker, and noise, even under a -8 dB SNR (with a single microphone). Compared to other computational approaches proposed to solve the CPP, the cortronic network distinguishes itself by exploiting prior knowledge pertaining to speech and the spoken language; it also implicitly confirms the validity of Cherry's early speculation in terms of the use of memory (recalling Section 3.1.1)

3.2   Human Auditory Scene Analysis: An Overview

Human auditory scene analysis (ASA) is a general process carried out by the auditory system of a human listener for the purpose of extracting information pertaining to a sound source of interest, which is embedded in a background of noise or interference.

The auditory system is made up of two ears (constituting the organs of hearing) and auditory pathways. In more specific terms, it is a sophisticated information-processing system that enables us to detect not only the frequency composition of an incoming sound wave but also locate the sound sources (Kandel et al., 2000). This is all the more remarkable, given the fact that the energy in the incoming sound waves is exceedingly small and the frequency composition of most sounds is rather complicated.

3.2.1   Where and What

The mechanisms in auditory perception essentially involve two processes: *sound localization* ("where") and *sound recognition* ("what"). It is well known that for localizing sound sources in the azimuthal plane, interaural time difference (ITD) is the main acoustic cue for sound location at low frequencies; and for complex stimuli with low-frequency repetition, interaural level is the main cue for sound localization at high frequencies (Blauert, 1983; Yost, 2000; Yost and Gourevitch, 1987). Spectral differences provided by the head-related transfer function (HRTF) are the main cues used for vertical localization. Loudness (intensity) and early reflections are possible cues for localization as a function of distance. In hearing, the *precedence effect* refers to the phenomenon that occurs during auditory fusion when two sounds of the same order of magnitude are presented dichotically and produce localization of the secondary sound waves toward the outer ear receiving the first sound stimulus (Yost, 2000); the precedence effect stresses the importance of the first wave in determining the sound location.

The what question mainly addresses the processes of sound segregation (streaming) and sound determination (identification). While having a critical role in sound localization, spatial separation is not considered to be a strong acoustic cue for streaming or segregation (Bregman, 1990). According to Bregman's studies, sound segregation consists of a two-stage process: *feature selection* and *feature grouping*. Feature selection invokes processing the auditory stimuli into a collection of favorable (e.g., frequency-sensitive, pitch-related, temporal-spectral-like) features. Feature grouping, on the other hand, is responsible for combining similar elements of incoming sounds according to certain principles into one or more coherent streams, with each stream corresponding to one informative sound source. Sound determination is more specific than segregation in that it not only involves segmentation of the incoming sound into different streams, but also identifies the content of the sound source in question.

3.2.2 Spatial Hearing

From a communication perspective, our two outer ears act as receive antennae for acoustic signals from a speaker or audio source. In the presence of one (or fewer) competing or masking sound source(s), the human ability to detect and understand the source of interest (i.e., target) is degraded. However, the influence of masking source(s) generally decreases when the target and masker(s) are spatially separated, compared to when the target and masker(s) are in the same location; this effect is credited to spatial hearing (filtering).

As pointed out in section 3.1.1, Cherry (1953) suggested that spatial hearing plays a major role in the auditory system's ability to separate sound sources in a multiple-source acoustic environment. Many subsequent experiments have verified Cherry's conjecture. Specifically, directional hearing (Yost and Gourevitch, 1987) is crucial for suppressing the interference and enhancing speech intelligibility (Bronkhorst, 2000; Hawley et al., 1999). Spatial separation of the sound sources is also believed to be more beneficial to localization than segregation (Bregman, 1990). The classic book by Blauert (1983) presents a comprehensive treatment of the psychophysical aspect of human sound localization. Given multiple-sound sources in an enclosed space (such as a conference room), spatial hearing helps the brain to take full advantage of slight differences (e.g., timing, intensity) between the signals that reach the two outer ears. This is done to perform monaural (autocorrelation) and binaural (cross-correlation) processing for specific tasks (such as coincidence detection, precedence detection, localization, and fusion), based on which auditory events are identified and followed by higher-level auditory processing (i.e., attention, streaming, and cognition). Fig. 3.1 provides a functional diagram of the binaural spatial hearing process.

3.2.3 Binaural Processing

One of the key observations derived from Cherry's classic experiment described in section 3.1.1 is that it is easier to separate sources heard binaurally than when they are heard monaurally. Quoting from Cherry and Taylor (1954): "One of the most striking facts about our ears is that we have two of them—and yet we hear one acoustic world; only one voice per speaker." We believe that nature gives us two ears for a reason just like it gives us two eyes. It is the binocular vision (stereovision) and binaural hearing (stereausis) that enable us to perceive the dynamic world and provide the main sensory information sources. Binocular/binaural processing is considered to be crucial in certain perceptual activities (e.g., binocular/binaural fusion, depth perception, localization). Given one sound source, the two ears receive slightly different sound patterns due to a finite delay produced by their physically separated locations. The brain is known to be extremely efficient in extracting and then using different acoustic cues (to be discussed in detail later) to perform specific audition tasks.

An influential binaural phenomenon is the so-called binaural masking (e.g.,

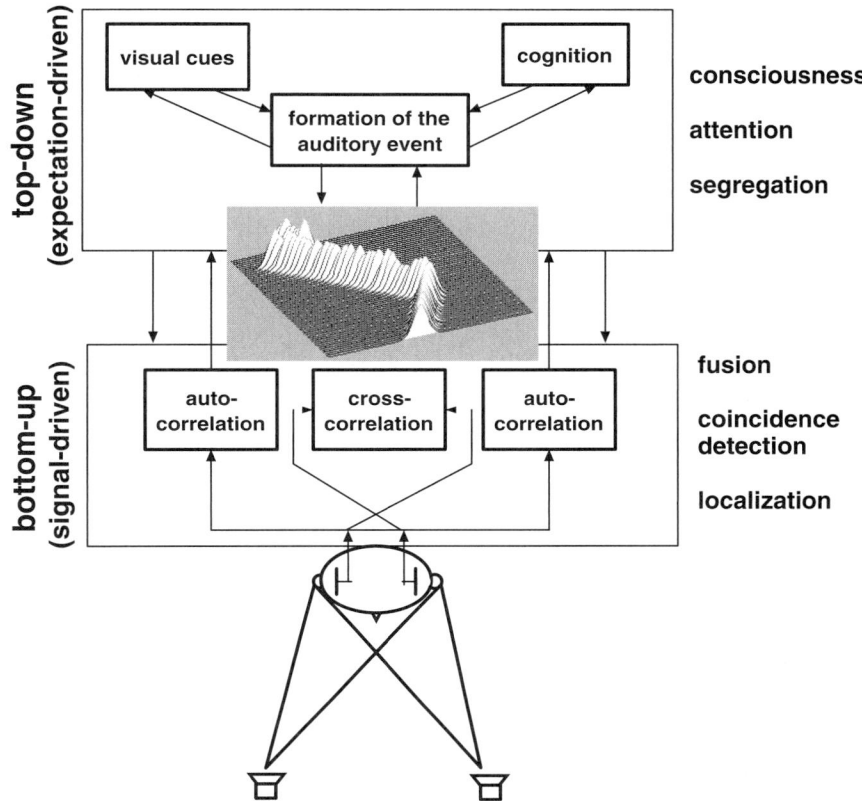

Figure 3.1   A functional diagram of of binaural hearing, which consists of physical, psychophysical, and psychological aspects of auditory perception. Adapted from Blauert (1983) with permission.

Moore, 1997; Yost, 2000). The threshold of detecting a signal masked in noise can sometimes be lower when listening with two ears than it is when listening with only one ear, which is demonstrated by a phenomenon called *binaural masking level difference* (BMLD). It is known (Yost, 2000) that the masked threshold of a signal is the same when the stimuli are presented in a monotic or diotic condition; when the masker and the signal are presented in a dichotic situation, the signal has a lower threshold than in either monotic or diotic conditions. Similarly, many experiments have also verified that binaural hearing increases speech intelligibility when the speech signal and noise are presented dichotically.

Another important binaural phenomenon is *binaural fusion*, which is the essence of directional hearing. As pointed out in section 3.1.1, the fusion mechanism is naturally modeled as performing some kind of correlation (Cherry, 1961; Cherry and Sayers, 1956), for which a binaural fusion model based on the autocorrelogram and cross-correlogram was proposed (as illustrated in fig. 3.1).

3.3 Computational Auditory Scene Analysis

In contrast to the human auditory scene analysis (ASA), computational auditory scene analysis (CASA) relies on the development of a *computational model* of the auditory scene with one of two goals in mind, depending on the application of interest:

- The design of an *intelligent machine*, which, by itself, is able to automatically *extract* and *track* a sound signal of interest in a cocktail party environment.

- The design of an *adaptive hearing system*, which computes the perceptual grouping process missing from the auditory system of a hearing-impaired individual, thereby enabling that individual to attend to a sound signal of interest in a cocktail party environment.

Naturally, CASA is motivated by or builds on the understanding we have of human auditory scene analysis, or even more generally, the understanding of human cognition behavior.

Following the seminal work of Bregman (1990), many researchers (see e.g., Brown and Cooke, 1994; Cooke, 1993; Cooke and Ellis, 2001; Rosenthal and Okuno, 1998) have tried to exploit the CASA in different ways. Representative approaches include the *data-driven* scheme (Cooke, 1993) and the *prediction-driven* scheme (Ellis, 1996). The common feature in these two schemes is to integrate low-level (bottom-up, primitive) acoustic cues for potential grouping. The main differences between them are: Data-driven CASA aims to decompose the auditory scene into time-frequency elements ("strands"), and then run the grouping procedure. On the other hand, prediction-driven CASA views prediction as the primary goal, and it requires only a world model that is consistent with the stimulus; it contains integration of top-down and bottom-up cues and can deal with incomplete or masked data (i.e., speech signal with missing information). However, as emphasized by Bregman (1996), it is important for CASA modelers to take into account psychological data as well as the way humans carry out ASA; namely, modeling the stability of human ASA, making it possible for different cues to cooperate and compete, and accounting for the propagation of constraints across the frequency-by-time field.

3.3.1 Acoustic Cues

acoustic cue

The psychophysical attributes of sound mainly involve three forms of information: *spatial location*, *temporal structure*, and *spectral characterization*. The perception of a sound signal in a cocktail party environment is uniquely determined by this kind of collective information; any difference in any of the three forms of information is believed to be sufficient to discriminate two different sound sources. In sound perception, many acoustic features (cues) are used to perform specific tasks. Table 3.1 summarizes some visual/acoustic features (i.e., the spatial, temporal, or

Table 3.1 The Features and Cues Used in Sound Perception

Feature/Cue	Domain	Task
Visual	Spatial	"Where"
ITD	Spatial	"Where"
IID	Spatial	"Where"
Intensity, loudness	Temporal	"Where" + "What"
Periodicity	Temporal	"What"
Onsets	Temporal	"What"
AM	Temporal	"What"
FM	Temporal-spectral	"What"
Pitch	Spectral	"What"
Timbre, tone	Spectral	"What"
Hamonicity, formant	Spectral	"What"

spectral patterns) used for a single-stream sound perception. A brief description of some important acoustic cues listed in table 3.1 is in order:

- *Interaural time difference* (ITD): A measure of the difference between the time at which the sound waves reach the left ear and the time at which the same sound waves reach the right ear.

- *Interaural intensity difference* (IID): A measure of the difference in intensity of the sound waves reaching the two ears due to head shadow.

- *Amplitude modulation* (AM): A method of sound signal transmission whereby the amplitude of some carrier frequency is modified in accordance with the sound signal.

- *Frequency modulation* (FM): Another method of modulation in which the instantaneous frequency of the carrier is varied with the frequency of the sound signal.

- *Onset*: A sudden increase in the energy of a sound signal; as such, each discrete event in the sound signal has an onset.

- *Pitch*: A property of auditory sensation in which sounds are ordered on a musical scale; in a way, pitch bears a relationship to frequency that is similar to the relationship of loudness to intensity.

- *Timbre*: The attribute of auditory sensation by which a listener is able to discriminate between two sound signals of similar loudness and pitch, but of different tonal quality; timbre depends primarily on the spectrum of a sound signal.

A combination of some or more of these acoustic cues is the key to perform CASA. Psychophysical evidence also suggests that useful cues may be provided by *spectral-temporal correlations* (Feng and Ratnam, 2000).

3.3.2 Feature Binding

feature binding   One other important function involved in CASA is that of *feature binding*, which refs to the problem of representing conjunctions of features. According to von der Malsburg (1999), binding is a general process that applies to all types of knowledge representations, which extend from the most basic perceptual representation to the most complex cognitive representation. Feature binding may be either *static* or *dynamic*. Static feature binding involves a representational unit that stands for a specific conjunction of properties, whereas dynamic feature binding involves conjunctions of properties as the binding of units in the representation of an auditory scene. The most popular dynamic binding mechanism is based on *temporal synchrony*, hence the reference to it as "temporal binding"; this form of binding was discussed in section 3.1.3. König et al. (1996) have suggested that synchronous firing of neurons plays an important role in information processing within the cortex. Rather than being a temporal integrator, the cortical neurons might be serving the purpose of a *coincidence detector*, evidence for which has been addressed by many researchers (König and Engel, 1995; Schultz et al., 2000; Singer, 1993).

Dynamic binding is closely related to the *attention mechanism*, which is used to control the synchronized activities of different assemblies of units and how the finite binding resource is allocated among neuronal assemblies (Singer, 1993, 1995). Experimental evidence has shown that synchronized firing tends to provide the attended stimulus with an enhanced representation.

3.3.3   Dereverberation

For auditory scene analysis, studying the effect of room acoustics on the cocktail party environment is important (Blauert, 1983; MacLean, 1959). A conversation occurring in a closed room often suffers from the multipath effect: *echoes* and *reverberation*, which are almost ubiquitous but rarely consciously noticed. According to the acoustics of the room, a reflection from one surface (e.g., wall, ground) produces reverberation. In the time domain, the reflection manifests itself as smaller, delayed replicas (echoes) that are added to the original sound; in the frequency domain, the reflection introduces a *comb-filter effect* into the frequency response. When the room is large, echoes can sometimes be consciously heard. It is known that the human auditory system is so powerful that it can take advantage of binaural and spatial hearing to efficiently suppress the echo, thereby improving the hearing performance. However, for a machine CPP, the machine design would have to include specific dereverberation (or deconvolution) algorithms to overcome this effect.

Those acoustic cues listed in table 3.1 that are spatially dependent, such as ITD and IID, are naturally affected by reverberation. On the other hand, acoustic cues that are space invariant, such as common onset across frequencies and pitch, are less sensitive to reverberation. On this basis, we may say that an intelligent machine should have the ability to adaptively weight the spatially dependent acoustic cues (prior to their fusion) so as to deal with a reverberant environment in an effective manner.

3.4 Insights from Computational Vision

It is well known to neuroscientists that audition (hearing) and vision (seeing) share substantial common features in the sensory processing principles as well as anatomic and functional organizations in higher-level centers in the cortex. It is therefore highly informative that with the design of an effective and efficient machine CPP as a design goal, we address the issue of deriving insights from the extensive literature on computational vision. We do so in this section by first looking to Marr's classic vision theory.

3.4.1 Marr's Vision Theory and Its Insights for Auditory Scene Analysis

In his landmark book, David Marr presented three levels of analysis of information-processing systems (Marr, 1982):

- *Computation:* What is the goal of the computation, why is it appropriate, and what is the logic of the strategy by which it can be carried out?

- *Representation:* How can this computational theory be implemented? In particular, what is the representation for the input and output, and what is the algorithm for the transformation?

- *Implementation:* How can the representation and the algorithm be realized physically?

In many perspectives, Marr's observations highlight the fundamental questions that need to be addressed in computational neuroscience, not only in the context of vision but also audition. As a matter of fact, Marr's theory has provided many insights into auditory research (Bregman, 1990; Rosenthal and Okuno, 1998).

In a similar vein to visual scene analysis (e.g., Julesz and Hirsh, 1972), auditory scene analysis (Bregman, 1990) attempts to identify the content (what) and the location (where) of the sounds/speech in an auditory environment. In specific terms, auditory scene analysis consists of two stages. In the first stage, the *segmentation process* decomposes a complex acoustic scene into a collection of distinct sensory elements; in the second stage, the *grouping* process combines these elements into a stream according to some principles. Subsequently, the streams are interpreted by a higher-level process for recognition and scene understanding. Motivated by Gestalt psychology, Bregman (1990) has proposed five grouping principles for ASA:

grouping
principle

- *Proximity:* Characterizes the distances between auditory cues (features) with respect to their onsets, pitch, and intensity (loudness).

- *Similarity:* Usually depends on the properties of a sound signal, such as timbre.

- *Continuity:* Features the smoothly-varying spectrum of a sound signal.

- *Closure:* Completes fragmentary features that have a good gestalt; the completion may be viewed as a form of auditory compensation for masking.

- *Common fate*: Groups together activities (e.g., common onsets) that are synchronous.

Moreover, Bregman (1990) has distinguished at least two levels of auditory organization: *primitive streaming* and *schema-based segregation*, with schemas being provided by phonetic, prosodic, syntactic, and semantic forms of information. While being applicable to general sound scene analysis involving speech and music, Bregman's work has focused mainly on primitive stream segregation.

3.4.2   A Tale of Two Sides: Visual and Auditory Perception

Visual perception and auditory perception share many common features in terms of sensory processing principles. According to Shamma (2001), these common features include the following:

- *Lateral inhibition for edge/peak enhancement:* In an auditory task, it aims to extract the profile of the sound spectrum; whereas in the visual system it aims to extract the form of an image.
- *Multiscale analysis:* The auditory system performs the cortical spectrotemporal analysis, whereas the visual system performs the cortical spatiotemporal form analysis.
- *Detecting temporal coincidence:* This process may serve periodicity pitch perception in an auditory scene compared to the perception of bilateral symmetry in a visual task.
- *Detecting spatial coincidence:* The same algorithm captures binaural azimuthal localization in the auditory system (stereausis), while it gives rise to binocular depth perception in the visual system.

Not only sharing these common features and processes, the auditory system also benefits from the visual system. For example, it is well known that there exist interactions between different sensory modalities. Neuroanatomy reveals the existence of corticocortical pathways between auditory and visual cortices. The hierarchical organization of cortices and numerous thalamocortical and corticothalamic feedback loops are speculated to stabilize the perceptual object. Daily life experiences also teach us that a visual scene input (e.g., lip reading) is influential to attention (Jones and Yee, 1996) and beneficial to speech perception. The *McGurk effect* (McGurk and MacDonald, 1976) is an auditory-visual speech illusion experiment, in which the perception of a speech sound is modified by contradictory visual information. The McGurk effect clearly illustrates the important role played by a visual cue in the comprehension of speech.

3.4.3   Active Vision

In the last paragraph of the introduction section, we referred to active audition as having the potential to build an intelligent machine that can operate efficiently and effectively in a noisy cocktail party environment. The proposal to build such a machine has been inspired by two factors: ongoing research on the use of active vision in computational vision, and the sharing of many common sensory principles between visual perception and auditory perception, as discussed earlier. To pave the way for what we have in mind on a framework for active audition, it is in order that we present some highlights on active vision that are of value to a formulation of this framework.

First and foremost, it is important to note that the use of an active sensor is not a necessary requirement for *active sensing*, be that in the context of vision or audition. Rather, a *passive sensor* (which only receives but does *not* transmit information-bearing signals) can perform active sensing, provided that the sensor is capable of changing its own state parameters in accordance with a desired sensing strategy. As such, active sensing may be viewed as an application of intelligent control theory, which includes not only control but also reasoning and decision making (Bajcsy, 1988).[4] In particular, active sensing embodies the use of *feedback* in two contexts:

1. The feedback is performed on complex processed sensory data such as extracted features that may also include relational features.
2. The feedback is dependent on prior knowledge.

active vision   *Active vision* (also referred to as *animated vision*) is a special form of active sensing, which has been proposed by Bajcsy (1988) and Ballard (1988), among others (e.g., Blake and Yuille, 1992). In active vision, it is argued that vision is best understood in the context of visual behaviors. The key point to note here is that the task of vision is *not* to build the model of a surrounding real world as originally postulated in Marr's theory, but rather to use visual information in the service of the real world in real time, and do so efficiently and inexpensively (Clark and Eliasmith, 2003). In effect, the active vision paradigm gives "action" a starring role (Sporns, 2003).

Rao and Ballard (1995) proposed an active vision architecture, which is motivated by biological studies. The architecture is based on the *hierarchical decomposition of visual behavior* involved in scene analysis (i.e., relating internal models to external objects). The architecture employs two components: the "what" component that corresponds to the problem of *object identification*, and the "where" component that corresponds to the problem of *objective localization*. These two visual components or routines are subserved by two separate memories. The central representation of the architecture is a *high-dimensional iconic feature vector*, which is comprised of the responses of different-order derivatives of Gaussian filters; the purpose of the iconic feature vector is to provide an effective photometric description of local intensity variations in the image region about an object of interest.

3.5   Embodied Intelligent Machines: A Framework for Active Audition

Most of the computational auditory scene analysis (CASA) approaches discussed in the literature share a common assumption: the machine merely listens to the environment but does *not* interact with it (i.e., the observer is passive). However, as remarked in the previous section, there are many analogies between the mechanisms that go on in auditory perception and their counterparts in visual perception. In a similar vein to active vision, active audition is established on the premise that the observer (human or machine) interacts with the environment, and the machine (in a way similar to human) should also conduct the perception in an *active* fashion.

embodied cognitive models

According to Varela et al. (1991) and Sporns (2003), *embodied cognitive models* rely on cognitive processes that emerge from interactions between neural, bodily, and environmental factors. A distinctive feature of these models is that they use "the world as their own model." In particular, embodied cognition has been argued to be the key to the understanding of intelligence (Iida et al., 2004; Pfeifer and Scheier, 1999). The central idea of embodied cognitive machines lies in the observation that "intelligence" becomes meaningless if we exclude ourselves from a real-life scenario; in other words, an intelligent machine is a self-reliant and independent agent capable of adapting itself to a dynamic environment so as to achieve a certain satisfactory goal effectively and efficiently, regardless of the initial setup.

Bearing this goal in mind, we may now propose a framework for active audition, which embodies four specific functions: (1) localization and focal attention, (2) segregation, (3) tracking, and (4) learning. In the following, we will address these four functions in turn.

3.5.1   Localization and Focal Attention

sound localization

Sound localization is a fundamental attribute of auditory perception. The task of sound localization can be viewed as a form of binaural depth perception, representing the counterpart to binocular depth perception in vision. A classic model for sound localization was developed by Jeffress (1948) using binaural cues such as ITD. In particular, Jeffress suggested the use of cross-correlation for calculating the ITD in the auditory system and explained how the model represents the ITD that is received at the ears; the sound processing and representation in Jeffress's model are simple yet elegant, and arguably neurobiologically plausible. Since the essential goal of localization is to infer the directions of incoming sound signals, this function may be implemented by using an adaptive array of microphones, whose design is based on *direction of arrival (DOA) estimation* algorithms developed in the signal-processing literature (e.g., Van Veen and Buckley, 1988, 1997).

Sound localization is often the first step to perform the beamforming, the aim of which is to extract the signal of interest produced in a specific direction. For a robot (or machine) that is self-operating in an open environment, sound localization is essential for successive tasks. An essential ingredient in sound localization is *time-delay estimation* when it is performed in a reverberant room environment. To perform this estimation, many signal-processing techniques have been proposed in the literature:

- *Generalized cross-correlation (GCC) method* (Knappand and Carter, 1976): This is a simple yet efficient delay-estimation method, which is implemented in the time domain using maximum-likelihood estimation (however, a frequency-domain implementation is also possible).

- *Cross-power spectrum phase (CSP) method* (Rabinkin et al., 1996): This delay-estimation method is implemented in the frequency domain, which computes the power spectra of two microphone signals and returns the phase difference between the spectra.

- *Adaptive eigenvalue decomposition (EVD)–based methods* (Benesty, 2000; Doclo and Moonen, 2002): It is noted that the GCC and CSP methods usually assume an ideal room model without reverberation; hence they may not perform satisfactorily in a highly reverberant environment. In order to overcome this drawback and enhance robustness, EVD-based methods have been proposed to estimate (implicitly) the acoustic impulse responses using adaptive algorithms that iteratively estimate the eigenvector associated with the smallest eigenvalue. Given the estimated acoustic impulse responses, the time delay can be calculated as the time difference between the main peak of the two impulse responses or as the peak of the correlation function between the two impulse responses.

Upon locating the sound source of interest, the next thing is to focus on the target sound stream and enhance it. Therefore, spatial filtering or beamforming techniques (Van Veen and Buckley, 1997) will be beneficial for this purpose. Usually, with omnidirectional microphone (array) technology, a machine is capable of picking up most if not all of the sound sources in the auditory scene. However, it is hoped that "smart microphones" may be devised so as to adapt their directivity (i.e., autodirective) to the attended speaker considering real-life conversation scenarios. Hence, designing a robust beamformer in a noisy and reverberant environment is crucial for localizing the sound and enhancing the SNR. The *adaptivity* also naturally brings in the issue of *learning*, to be discussed in what follows.

3.5.2 Segregation

In this second functional module for active audition, the target sound stream is segregated and the sources of interference are suppressed, thereby focusing attention on the target sound source. This second function may be implemented by using several acoustic cues (e.g., ITD, IID, onset, and pitch) and then combining them in a *fusion algorithm*.

In order to emulate the human auditory system, a computational strategy for acoustic-cue fusion should dynamically resolve the ambiguities caused by the simple-cue segregation. The simplest solution is the "winner-take-all" competition, which

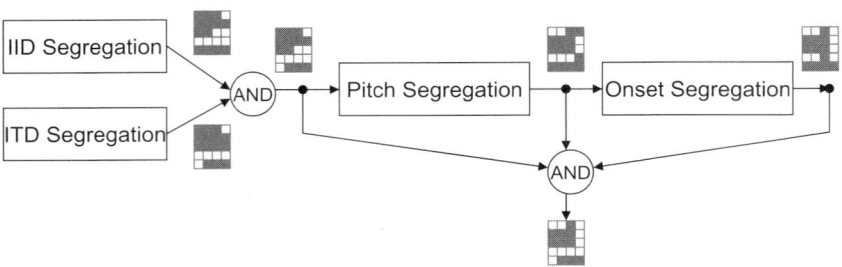

Figure 3.2 A flowchart of multiple acoustic cues fusion process (courtesy of Rong Dong).

essentially chooses the cue that has the highest (quantitatively) confidence (where the confidence values depend on the specific model used to extract the acoustic cue). When several acoustic cues are in conflict, only the dominant cue will be chosen based on some criterion, such as the *weighted-sum mechanism* (Woods et al., 1996) that was used for integrating pitch and spatial cues, or the Bayesian framework (Kashino et al., 1998).

Recently, Dong (2005) proposed a simple yet effective fusion strategy to solve the multiple cue fusion problem (see fig. 3.2). Basically, the fusion process is performed in a cooperative manner: in the first stage of fusion, given IID and ITD cues, the time-frequency units are grouped into two streams (target stream and interference stream), and the grouping results are represented by two binary maps. These two binary maps are then passed through an "AND" operation to obtain a spatial segregation map, which is further utilized to estimate the pitch of the target signal or the pitch of the interference. Likewise, a binary map is produced from the pitch segregation. If the target is detected as an unvoiced signal, onset cue is integrated to group the components into separate streams. Finally, all these binary maps are pooled together by a second "AND" operation to yield the final segregation decision. Empirical experiments on this fusion algorithm reported by Dong (2005) have shown very promising results.[5]

3.5.3 Tracking

state-space model   The theoretical development of sound tracking builds on a *state-space model* of the auditory environment. The model consists of a *process equation* that describes the evolution of the state (denoted by $\mathbf{x}_t$) at time t, and a *measurement equation* that describes the dependence of the observables (denoted by $\mathbf{y}_t$) on the state. More specifically, the *state* is a vector of acoustic cues (features) characterizing the

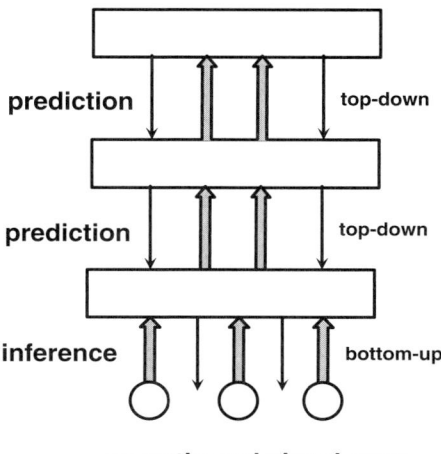

Figure 3.3 An information flowchart integrating "bottom-up" (shaded arrow) and "top-down" flows in a hierarchical functional module of an intelligent machine (in a similar fashion as in the human auditory cortex).

target sound stream and its direction. Stated mathematically, we have the following state-space equations:

$$\mathbf{x}_t = \mathbf{f}(t, \mathbf{x}_{t-1}, \mathbf{d}_t), \tag{3.1}$$

$$\mathbf{y}_t = \mathbf{g}(t, \mathbf{x}_t, \mathbf{u}_t, \mathbf{v}_t), \tag{3.2}$$

where $\mathbf{d}_t$ and $\mathbf{v}_t$ denote the dynamic and measurement noise processes, respectively, and the vector $\mathbf{u}_t$ denotes the action taken by the (passive) observer. The process equation 3.1 embodies the state transition probability $p(\mathbf{x}_t|\mathbf{x}_{t-1})$, whereas the measurement equation 3.2 embodies the likelihood $p(\mathbf{y}_t|\mathbf{x}_t)$. The goal of optimum filtering is then to estimate the posterior probability density $p(\mathbf{x}_t|\mathbf{y}_{0:t})$, given the initial prior $p(\mathbf{x}_0)$ and $\mathbf{y}_{0:t}$ that denotes the measurement history from time 0 to t. This classic problem is often referred to as "state estimation" in the literature. Depending on the specific scenario under study, such a hidden state estimation problem can be tackled by using a Kalman filter (Kalman, 1960), an extended Kalman filter, or a particle filter (e.g., Cappé et al., 2005; Doucet et al., 2001).

In Nix et al. (2003), a particle filter is used as a statistical method for integrating temporal and frequency-specific features of a target speech signal. The elements of the state represent the azimuth and elevation of different sound signals as well as the band-grouped short-time spectrum for each signal; whereas the observable measurements contain binaural short-time spectra of the *superposed* voice signals. The state equation, representing the spectral dynamics of the speech signal, was learned off-line using vector quantization and lookup table in a large codebook, where the codebook index for each pair of successive spectra was stored in a Markov transition matrix (MTM); the MTM provides statistical information about the transition probability $p(\mathbf{x}_t|\mathbf{x}_{t-1})$ between successive short-time speech spectra. The measurement equation, characterized by $p(\mathbf{y}_t|\mathbf{x}_t)$, was approximated as a multidimensional Gaussian mixture probability distribution. By virtue of its very design, it is reported in Nix et al. (2003) that the tracker provides a one-step prediction of the underlying features of the target sound.

In the much more sophisticated neurobiological context, we may envision that the hierarchical auditory cortex (acting as a predictor) implements an online tracking task as a basis for *dynamic feature binding* and *Bayesian estimation*, in a fashion similar to that in the hierarchical visual cortex (Lee and Mumford, 2003; Rao and Ballard, 1999). Naturally, we may also incorporate "top-down" expectation as a feedback loop within the hierarchy to build a more powerful inference/prediction model. This is motivated by the generally accepted fact that the hierarchical architecture is omnipresent in sensory cortices, starting with the primary sensory cortex and proceeding up to the highest areas that encode the most complex, abstract, and stable information.[6] A schematic diagram illustrating such a hierarchy is depicted in fig 3.3, where the bottom-up (data-driven) and top-down (knowledge-driven) information flows are illustrated with arrows. In the figure, the feedforward pathway carries the inference, given the current and past observations; the feedback pathway conducts the prediction (expectation) to lower-level regions. To be specific, let $\mathbf{z}$ denote the top-down signal; then the conditional joint probability of hidden state $\mathbf{x}$ and bottom-up observation $\mathbf{y}$, given $\mathbf{z}$, may be written as

$$p(\mathbf{x}, \mathbf{y}|\mathbf{z}) = p(\mathbf{y}|\mathbf{x}, \mathbf{z})p(\mathbf{x}|\mathbf{z}), \qquad (3.3)$$

Bayes's rule

and the posterior probability of the hidden state can be expressed via Bayes's rule:

$$p(\mathbf{x}|\mathbf{y}, \mathbf{z}) = \frac{p(\mathbf{y}|\mathbf{x}, \mathbf{z})p(\mathbf{x}|\mathbf{z})}{p(\mathbf{y}|\mathbf{z})}, \qquad (3.4)$$

where the denominator is a normalizing constant term that is independent of the state $\mathbf{x}$, the term $p(\mathbf{x}|\mathbf{z})$ in the numerator characterizes a top-down *contextual prior*, and the other term $p(\mathbf{y}|\mathbf{x}, \mathbf{z})$ describes the likelihood of the observation, given all available information. Hence, feedback information from a higher level can provide useful context to interpret or disambiguate the lower-level patterns. The same inference principle can be applied to different levels of the hierarchy in fig. 3.3. To sum up, the top-down *predictive coding* and bottom-up *inference* cooperate for learning the statistical regularities in the sensory environment; the top-down and bottom-up mechanisms also provide a possible basis for *optimal action control* within the framework of active audition in a way similar to active vision (Bajcsy, 1988).

3.5.4   Learning

Audition is a sophisticated, dynamic information-processing task performed in the human brain, which inevitably invokes other tasks almost simultaneously (such

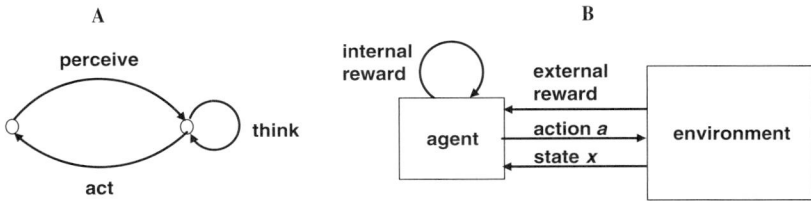

Figure 3.4   (A) The sensorimotor feedback loop consisting of three distinct functions: *perceive*, *think*, and *act*. (B) The interaction between the agent and environment.

as action). Specifically, it is commonly believed that perception and action are mutually coupled, and integrated via a sensorimotor interaction feedback loop, as illustrated in fig 3.4A. Indeed, it is this unique feature that enables the human to survive in a dynamic environment. For the same reason, it is our belief that an intelligent machine that aims at solving the CPP must embody a learning capability, which must be of a kind that empowers the machine to take action whenever changes in the environment call for it.

In the context of embodied intelligence, an autonomous agent is also supposed to conduct a goal-oriented behavior during its interaction with the dynamic environment; hence, the necessity for taking *action* naturally arises. In other words, the agent has to continue to adapt itself (it terms of action or behavior) to maximize its (internal or external) reward, in order to achieve better perception of its environment (illustrated in fig. 3.4B). Such a problem naturally brings in the theory of *reinforcement learning* (Sutton and Barto, 1998). For example, imagine a maneuverable machine is aimed at solving a computational CPP in a noisy room environment. The system then has to learn how to adjust its distance and the angle of the microphone array with respect to attended audio sources (such as speech, music, etc.). To do so, the machine should have a built-in rewarding mechanism when interacting with the dynamic environment, and it has to gradually adapt its behavior to achieve a higher (internal and external) reward.[7]

reinforcement learning

To sum up, an autonomous intelligent machine self-operating in a dynamic environment will always need to conduct optimal action control or decision making. Since this problem bears much resemblance to the Markov decision process (MDP) (Bellman and Dreyfus, 1962), we may resort to the well-established theory of dynamical programming and reinforcement learning.[8]

3.6   Concluding Remarks

In this chapter, we have discussed the machine cocktail party problem and explored possible ways to solve it by means of an intelligent machine. To do so, we have briefly reviewed historical accounts of the cocktail party problem, as well as the important aspects of human and computational auditory scene analysis. More important, we have proposed a computational framework for active audition as an inherent part of an embodied cognitive machine. In particular, we highlighted the essential functions of active audition and discussed its possible implementation. The four functions identified under the active audition paradigm provide the basis for building an embodied cognitive machine that is capable of human-like hearing in an "active" fashion. The central tenet of active audition embodying such a machine is that an observer may be able to understand an auditory environment more effectively and efficiently if the observer interacts with the environment than if it is a passive observer. In addition, in order to build a maneuverable intelligent machine (such as a robot), we also discuss the issue of integrating different sensory (auditory and visual) features such that active vision and active audition can be combined in a single system to achieve the true sense of *active perception*.

Appendix: Reinforcement Learning

Mathematically, a Markov decision process [9] is formulated as follows:

Definition 3.1
A Markov decision process (MDP) is defined as a 6-tuple $(\mathcal{S}, \mathcal{A}, \mathcal{R}, p_0, p_s, p_r)$, where

- $\mathcal{S}$ is a (finite) set of (observable) environmental states (state space), $s \in \mathcal{S}$;

- $\mathcal{A}$ is a (finite) set of actions (action space), $a \in \mathcal{A}$;

- $\mathcal{R}$ is a (finite) set of possible rewards;

- p_0 is an initial probability distribution over $\mathcal{S}$; it is written as $p_0(s_0)$;

- p_s is a transition probability distribution over $\mathcal{S}$ conditioned on a value from $\mathcal{S} \times \mathcal{A}$; it is also written as $p^a_{ss'}$ or $p_s(s_t|s_{t-1}, a_{t-1})$;[10]

- p_r is a probability distribution over $\mathcal{R}$, conditioned on a value from $\mathcal{S}$; it is written as $p_r(r|s)$.

Definition 3.2
A policy is a mapping from states to probabilities of selecting each possible action. A policy, denoted by π, can be deterministic, $\pi : \mathcal{S} \mapsto \mathcal{A}$, or stochastic, $\pi : \mathcal{S} \mapsto P(\mathcal{A})$. An optimal policy π^* is a policy that maximizes (minimizes) the expected total reward (cost) over time (within finite or infinite horizon).

Given the above definitions, the goal of reinforcement learning is to find an optimal policy $\pi^*(s)$ for each s, which maximizes the expected reward received over time. We assume that the policy is *stochastic*, and Q-learning (a special form of reinforcement learning) is aimed at learning a stochastic world in the sense that the p_s and p_r are both nondeterministic. To evaluate reinforcement learning, a common measure of performance is the infinite-horizon discounted reward, which can be represented by the *state-value function*

$$V^\pi(s) = \mathrm{E}_\pi\left\{ \sum_{t=0}^{\infty} \gamma^t r(s_t, \pi(s_t)) \Big| s_0 = s \right\}, \tag{3.5}$$

where $0 \le \gamma < 1$ is a discount factor, and E_π is the expectation operator over the policy π. The value function $V^\pi(s)$ defines the expected discounted reward at state s, as shown by

$$V^\pi(s) = \mathrm{E}_\pi\{R_t | s_t = s\}, \tag{3.6}$$

where

$$\begin{aligned}
R_t &= r_{t+1} + \gamma r_{t+2} + \gamma^2 r_{t+3} + \cdots \\
&= r_{t+1} + \gamma(r_{t+2} + \gamma r_{t+2} + \cdots) \\
&= r_{t+1} + \gamma R_{t+1}.
\end{aligned}$$

Similarly, one may define a *state-action value function*, or the so-called Q-function

$$Q^\pi(s, a) = \mathrm{E}_\pi\{R_t | s_t = s, a_t = a\}, \tag{3.7}$$

for which the goal is not only to achieve a maximal reward but also to find an optimal action (supposing multiple actions are accessible for each state). It can be shown that

$$V^\pi(s) = \mathrm{E}_\pi\{r(s, \pi(s))\} + \gamma \sum_{s'} \pi(s, s') V^\pi(s'),$$

$$Q^\pi(s, a) = \sum_{s' \in \mathcal{S}} p_{ss'}^a [R(s, a, s') + \gamma V^\pi(s')], \quad \text{and}$$

$$V^\pi(s) = \sum_{a \in \mathcal{A}} \pi(s, a) Q^\pi(s, a) = \sum_{a \in \mathcal{A}} \pi(s, a)\left(r_s^a + \gamma \sum_{s'} p_{ss'}^a V^\pi(s')\right),$$

which correspond to different forms of the *Bellman equation* (Bellman and Dreyfus, 1962). Note that if the state or action is continuous-valued, the summation operations are replaced by corresponding integration operations.

The optimal value functions are then further defined as:

$$V^{\pi^*}(s) = \max_\pi V^\pi(s), \quad Q^{\pi^*}(s, a) = \max_\pi Q^\pi(s, a).$$

Therefore, in light of dynamic programming theory (Bellman and Dreyfus, 1962), the optimal policy is deterministic and greedy with respect to the optimal value functions. Specifically, we may state that given the state s and the optimal policy π^*, the optimal action is selected according to the formula

$$a^* = \arg\max_{a \in \mathcal{A}} Q^{\pi^*}(s, a),$$

such that $V^{\pi^*}(s) = \max_{a \in \mathcal{A}} Q^{\pi^*}(s, a)$.

A powerful reinforcement learning tool for tackling the above-formulated problem is the Q-learning algorithm (Sutton and Barto, 1998; Watkins, 1989). The classic Q-learning is an *asynchronous, incremental,* and *approximate* dynamic programming method for stochastic optimal control. Unlike the traditional dynamic programming, Q-learning is model-free in the sense that its operation requires neither the state transition probability nor the environmental dynamics. In addition, Q-learning is computationally efficient and can be operated in an online manner (Sutton and Barto, 1998). For finite state and action sets, if each (s, a) pair is visited infinitely and the step-size sequence used in Q-learning is nonincreasing, then Q-learning is assured to converge to the optimal policy with probability 1 (Tsitsiklis, 1994; Watkins and Dayan, 1992). For problems with continuous state, functional approximation methods can be used for tackling the *generalization* issue; see Bertsekas and Tsitsiklis (1996) and Sutton and Barto (1998) for detailed discussions.

Another model-free reinforcement learning algorithm is the actor-critic model (Sutton and Barto, 1998), which describes a bootstrapping strategy for reinforcement learning. Specifically, the actor-critic model has separate memory structures to represent the policy and the value function: the policy structure is conducted within the *actor* that selects the optimal actions; the value function is estimated by the *critic* that criticizes the actions made by the actor. Learning is always on-policy in that the critic uses a form of temporal difference (TD) error to maximize the reward, whereas the actor will use the estimated value function from the critic to bootstrap itself for a better policy.

A much more challenging but more realistic reinforcement learning problem is the so-called *partially observable Markov decision process* (POMDP). Unlike the MDP that assumes the full knowledge of observable states, POMDP addresses the stochastic decision making and optimal control problems with only partially observable states in the environment. In this case, the elegant Bellman equation does not hold since it requires a completely observable Markovian environment (Kaelbling, 1993). The literature of POMDP is intensive and ever growing, hence it is beyond the scope of the current chapter to expound this problem; we refer the interested reader to the papers (Kaelbling et al., 1998; Lovejoy, 1991; Smallwood and Sondik, 1973) for more details.

Acknowledgments

This chapter grew out of a review article (Haykin and Chen, 2005). In particular, we would like to thank our research colleagues R. Dong and S. Doclo for valuable feedback. The work reported here was supported by the Natural Sciences and Engineering Research Council (NSERC) of Canada.

Notes

[1] For tutorial treatments of the cocktail party problem, see Chen (2003), Haykin and Chen (2005), and Divenyi (2004).

[2] Categorization of these three neural processes, done essentially for research-related studies, is somewhat artificial; the boundary between them is fuzzy in that the brain does not necessarily distinguish between them as defined herein.

[3] In Cherry's original paper, "intelligibility" is referred to as the probability of correctly identifying *meaningful* speech sounds; in contrast, "articulation" is referred to as a measure of *nonsense* speech sounds (Cherry, 1953).

[4] As pointed out by Bajcsy (1988), the proposition that active sensing is an application of intelligent control theory may be traced to the PhD thesis of Tenenbaum (1970).

[5] At McMaster University we are currently exploring the DSP hardware implementation of the fusion scheme that is depicted in fig. 3.2.

[6] The "top-down" influence is particularly useful for (1) synthesizing missing information (e.g., the auditory "fill-in" phenomenon); (2) incorporating contextual priors and inputs from other sensory modalities; and (3) resolving perceptual ambiguities whenever lower-level information leads to confusion.

[7] The reward can be a measure of speech intelligibility, signal-to-interference ratio, or some sort of utility function.

[8] Reinforcement learning is well known in the machine learning community, but, regrettably, not so in the signal processing community. An appendix on reinforcement learning is included at the end of the chapter largely for the benefit of readers who may not be familiar with this learning paradigm.

[9] For the purpose of exposition simplicity, we restrict our discussion to finite discrete state and action spaces, but the treatment also applies to the more general continuous state or action space.

[10] In the case of finite discrete state, p_s constitutes a transition matrix.

4   Sensor Adaptive Signal Processing of Biological Nanotubes (Ion Channels) at Macroscopic and Nano Scales

Vikram Krishnamurthy

Ion channels are biological nanotubes formed by large protein molecules in the cell membrane. All electrical activities in the nervous system, including communications between cells and the influence of hormones and drugs on cell function, are regulated by ion channels. Therefore understanding their mechanisms at a molecular level is a fundamental problem in biology. This chapter shows how dynamic stochastic models and associated statistical signal-processing techniques together with novel learning-based stochastic control methods can be used to understand the structure and dynamics of ion channels at both macroscopic and nanospatial scales. The unifying theme of this chapter is the concept of sensor adaptive signal processing, which deals with sensors dynamically adjusting their behavior so as to optimize their ability to extract signals from noise.

4.1   Introduction

All living cells are surrounded by a cell membrane, composed of two layers of phospholipid molecules, called the lipid bilayer. Ion channels are biological nanotubes formed by protein macromolecules that facilitate the diffusion of ions across the cell membrane. Although we use the term *biological nanotube*, ion channels are typically the size of angstrom units (10^{-10} m), i.e., an order of magnitude smaller in radius and length compared to carbon nanotubes that are used in nanodevices.

In the past few years, there have been enormous strides in our understanding of the structure-function relationships in biological ion channels. These advances have been brought about by the combined efforts of experimental and computational biophysicists, who together are beginning to unravel the working principles of these exquisitely designed biological nanotubes that regulate the flow of charged particles across the cell membrane. The measurement of ionic currents flowing through single ion channels in cell membranes has been made possible by the gigaseal *patch-clamp* technique (Hamill et al., 1981; Neher and Sakmann, 1976). This was a major breakthrough for which the authors Neher and Sakmann won the 1991 Nobel Prize in Medicine (Neher and Sakmann, 1976). More recently, the 2003 Nobel Prize in Chemistry was awarded to MacKinnon for determining the structure of several different types of ion channels (including the bacterial potassium channel; Doyle et al. (1998)) from crystallographic analyses. Because all electrical activities in the nervous system, including communications between cells and the influence of hormones and drugs on cell function, are regulated by membrane ion channels, understanding their mechanisms at a molecular level is a fundamental problem in biology. Moreover, elucidation of how single ion channels work will ultimately help neurobiologists find the causes of, and possibly cures for, a number of neurological and muscular disorders.

We refer the reader to the special issue of IEEE Transactions on NanoBio-Science Krishnamurthy et al. (2005) for an excellent up-to-date account of ion channels written by leading experts in the area. This chapter addresses two fundamental problems in ion channels from a statistical signal processing and stochastic control (optimization) perspective: the *gating problem* and the *ion permeation problem*.

The *gating problem* (Krishnamurthy and Chung, 2003) deals with understanding how ion channels undergo structural changes to regulate the flow of ions into and out of a cell. Typically a gated ion channel has two states: a "closed" state which does not allow ions to flow through, and an "open" state which does allow ions to flow through. In the open state, the ion channel currents are typically of the order of pico-amps (i.e., 10^{-12} amps). The measured ion channel currents (obtained by sampling typically at 10 kHz, i.e, 0.1 millisecond time scale) are obfuscated by large amounts of thermal noise. In sections 4.2 4.3 of this chapter, we address the following issues related to the gating problem: (1) We present a hidden Markov model (HMM) formulation of the observed ion channel current. (2) We present in section 4.2 a discrete stochastic optimization algorithm for controlling a patch-clamp experiment to determine the Nernst potential of the ion channel with minimal effort. This fits in the class of so-called experimental design problems. (3) In section 4.3, we briefly discuss dynamic scheduling algorithms for activating multiple ion channels on a biological chip so as to extract maximal information from them.

The *permeation problem* (Allen et al., 2003; O'Mara et al., 2003) seeks to explain the working of an ion channel at an Å(10^{-10} m) spatial scale by studying the propagation of individual ions through the ion channel at a femto (10^{-15}) second time scale. This setup is said to be at a *mesoscopic scale* since the individual ions (e.g., Na$^+$ions) are of the order of a few Åin radius and are comparable in radius to the ion channel. At this mesoscopic level, point-charge approximations and continuum electrostatics break down. The discrete finite nature of each ion needs to be taken into consideration. Also, failure of the mean field approximation in narrow channels implies that any theory that aspires to relate channel structure to its function must treat ions explicitly. In sections 4.4, 4.5, and 4.6 of this chapter, we discuss the permeation problem for ion channels. We show how Brownian dynamics simulation can be used to model the propagation of individual ions. We also show how stochastic gradient learning based schemes can be used to control the evolution of Brownian dynamics simulation to predict the molecular structure of an ion channel. We refer the reader to our recent research (Krishnamurthy and Chung, a,b) where a detailed exposition of the resulting adaptive Brownian dynamics simulation algorithm is given. Furthermore numerical results presented in Krishnamurthy and Chung (a,b) for antibiotic Gramicidin-Aion channels show that the estimates obtained from the adaptive Brownian dynamics algorithm are consistent with the known molecular structure of Gramicidin-A.

sensor adaptive signal processing

An important underlying theme of this chapter is the ubiquitous nature of *sensor adaptive signal processing*. This transcends standard statistical signal processing, which deals with extracting signals from noisy observations, to examine the deeper problem of how to dynamically adapt the sensor to optimize the performance of the signal-processing algorithm. That is, the sensors dynamically modify their behavior to optimize their performance in extracting the underlying signal from noisy observations. A crucial aspect in sensor adaptive signal processing is feedback—past decisions of adapting the sensor affect future observations. Such sensor adaptive signal processing has recently been used in defense networks (Evans et al., 2001; Krishnamurthy, 2002, 2005) for scheduling sophisticated multimode sensors in unattended ground sensor networks, radar emission control, and adaptive radar beam allocation. In this chapter we show how the powerful paradigm of sensor adaptive signal processing can be successfully applied to biological ion channels both at the macroscopic and nano scales.

4.2   The Gating Problem and Estimating the Nernst Potential of Ion Channels

In this section we first outline the well-known hidden Markov model (HMM) for modeling the ion channel current in the gating problem. We refer the reader to the paper by Krishnamurthy and Chung (2003) for a detailed exposition. Estimating the underlying ion channel current from the noisy HMM observations is a well-studied problem in HMM signal processing (Ephraim and Merhav, 2002; James et al., 1996; Krishnamurthy and Yin, 2002). In this section, consistent with the theme of sensor adaptive signal processing, we address the deeper issue of how to dynamically control the behavior of the ion channels to extract maximal information about their behavior. In particular we propose two novel applications of stochastic control for adapting the behavior of the ion channel. Such ideas are also relevant in other applications such as sensor scheduling in defense networks (Krishnamurthy, 2002, 2005).

4.2.1   Hidden Markov Model Formulation of Ion Channel Current

The patch clamp is a device for isolating the ion channel current from a single ion channel. A typical trace of the ion channel current measurement from a patch-clamp experiment (after suitable anti-aliasing filtering and sampling) shows that the channel current is a piecewise constant discrete time signal that randomly jumps between two values—zero amperes, which denotes the *closed state* of the channel, and $I(\theta)$ amperes (typically a few pico-amperes), which denotes the *open state*. $I(\theta)$ is called the *open-state current level*. Sometimes the current recorded from a single ion channel dwells on one or more intermediate levels, known as conductance substates.

Chung et al. (1990, 1991) first introduced the powerful paradigm of hidden Markov models (HMMs) to characterize patch-clamp recordings of small ion channel currents contaminated by random and deterministic noise. By using sophisticated HMM signal-processing methods, Chung et al. (1990, 1991) demonstrated that the underlying parameters of the HMM could be obtained to a remarkable precision despite the extremely poor signal-to-noise ratio. These HMM parameter estimates yield important information about the dynamics of ion channels. Since the publications of Chung et al. (1990, 1991), several papers have appeared in the neurobiological community that generalize the HMM signal models in Chung et al. (1990, 1991) in various ways to model measurements of ion channels (see the paper of Venkataramanan et al. (2000) and the references therein). With these HMM techniques, it is now possible for neurobiologists to analyze not only large ion channel currents but also small conductance fluctuations occurring in noise.

Markov Model for Ion Channel Current Suppose a patch-clamp experiment is conducted with a voltage θ applied across the ion channel. Then, as described in Chung et al. (1991) and in Venkataramanan et al. (2000), the ion channel current $\{i_n(\theta)\}$ can be modeled as a three-state homogeneous first-order Markov chain. The state space of this Markov chain is $\{0_g, 0_b, I(\theta)\}$, corresponding to the physical states of *gap mode*, *burst-mode-closed*, and *burst-mode-open*. For convenience, we will refer to the burst-mode-closed and burst-mode-open states as the *closed* and *open* states, respectively. In the gap mode and the closed state, the ion channel current is zero. In the open state, the ion channel current has a value of $I(\theta)$.

The 3×3 transition probability matrix $A(\theta)$ of the Markov chain $\{I_n^{(\theta, \lambda)}(\theta)\}$, which governs the probabilistic behavior of the channel current, is given by

$$A(\theta) = \begin{array}{c|c|c|c} & 0_g & 0_b & I(\theta) \\ \hline 0_g & a_{11}(\theta) & a_{12}(\theta) & 0 \\ \hline 0_b & a_{21}(\theta) & a_{22}(\theta) & a_{23}(\theta) \\ \hline I(\theta) & 0 & a_{32}(\theta) & a_{33}(\theta) \end{array} \qquad (4.1)$$

The elements of $A(\theta)$ are the transition probabilities $a_{ij}(\theta) = P(I_{n+1}^{(\theta,\lambda)}(\theta) = j | I_n^{(\theta,\lambda)}(\theta) = i)$ where $i, j \in \{0_g, 0_b, I(\theta)\}$. The zero probabilities in the above

hidden   Markov
model (HMM)

matrix $A(\theta)$ reflect the fact that a ion channel current cannot directly jump from the gap mode to the open state; similarly an ion channel current cannot jump from the open state to the gap mode. Note that in general, the applied voltage θ affects both the transition probabilities and state levels of the ion channel current $\{I_n^{(\theta,\lambda)}(\theta)\}$.

Hidden Markov Model (HMM) Observations Let $\{y_n(\theta)\}$ denote the measured noisy ion channel current at the electrode when conducting a patch-clamp experiment:

$$y_n(\theta) = i_n(\theta) + w_n(\theta), \quad n = 1, 2, \ldots \tag{4.2}$$

Here $\{w_n(\theta)\}$ is thermal noise and is modeled as zero-mean white Gaussian noise with variance $\sigma^2(\theta)$. Thus the observation process $\{y_n(\theta)\}$ is a hidden Markov model (HMM) sequence parameterized by the model

$$\lambda(\theta) = \{A(\theta), I(\theta), \sigma^2(\theta)\} \tag{4.3}$$

where θ denotes the applied voltage. We remark here that the formulation trivially extends to observation models where the noise process $w_n(\theta)$ includes a time-varying deterministic component together with white noise — only the HMM parameter estimation algorithm needs to be modified as in Krishnamurthy et al. (1993).

HMM Parameter Estimation of Current Level $I(\theta)$ Given the HMM mode for the ion channel current above, estimating $I(\theta)$ for a fixed voltage θ involves processing the noisy observation $\{y_n(\theta)\}$ through a HMM maximum likelihood parameter estimator. The most popular way of computing the maximum likelihood estimate (MLE) $I(\theta)$ is via the expectation maximization (EM) algorithm (Baum Welch equations). The EM algorithm is an iterative algorithm for computing the MLE. It is now fairly standard in the signal-processing and neurobiology literature—see Ephraim and Merhav (2002) for a recent exposition, or Chung et al. (1991), which is aimed at neurobiologists.

Let $\hat{I}_\Delta(\theta)$ denote MLE of $I(\theta)$ based on the Δ-point measured channel current sequence $(y_1(\theta), \ldots, y_\Delta(\theta))$. For sufficiently large batch size Δ of observations, due to the asymptotic normality of the MLE for a HMM (Bickel et al., 1998),

$$\sqrt{\Delta}\left(\hat{I}_\Delta(\theta) - I(\theta)\right) \sim N(0, \Sigma(\theta)), \tag{4.4}$$

where $\Sigma^{-1}(\theta)$ is the Fisher information matrix. Thus asymptotically $\hat{I}_\Delta(\theta)$ is an unbiased estimator of $I(\theta)$, i.e., $\mathbf{E}\left\{\hat{I}_\Delta(\theta)\right\} = I(\theta)$ where $\mathbf{E}\{\cdot\}$ denotes the mathematical expectation operator.

4.2.2   Nernst Potential and Discrete Stochastic Optimization for Ion Channels

To record currents from single ion channels, the tip of an electrode, with the diameter of about 1 μm, is pushed against the surface of a cell, and then a tight seal is formed between the rim of the electrode tip and the cell membrane. A patch of the membrane surrounded by the electrode tip usually contains one or more single ion channels. The current flowing from the inside of the cell to the tip of the electrode through a single ion channel is monitored. This is known as cell-attached configuration of patch-clamp techniques for measuring ion channel currents through a single ion channel. Figure 4.1 shows the schematic setup of the cell in electrolyte and the electrode pushed against the surface of the cell.

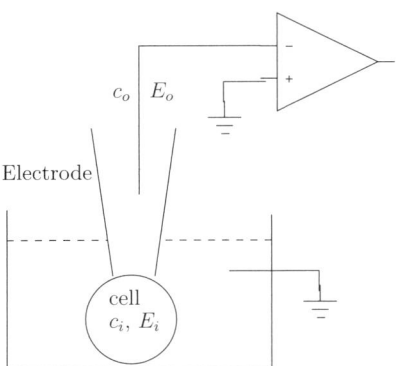

Figure 4.1   Cell-attached patch experimental setup.

In a living cell, there is a potential difference between its interior and the outside environment, known as the membrane potential. Typically, the cell interior is about 60 mV more negative with respect to outside. Also, the ionic concentrations (mainly Na^+, Cl^-, and K^+) inside of a cell are very different from outside of the cell. In the cell-attached configuration, the ionic strength in the electrode is usually made the same as that in the outside of the cell. Let E_i and E_o, respectively, denote the resting membrane potential and the potential applied to the electrode. If E_o is identical to the membrane potential, there will be no potential gradient across the membrane patch confined by the tip of the electrode. Let c_i denote the intracellular ionic concentration and c_o the ionic concentration in the electrode. Here the intracellular concentration c_i inside the cell is unknown as is the resting membrane potential E_i. c_o and E_o are set by the experimenter and are known.

Let $\theta = E_o - E_i$ denote the potential gradient. Both the potential gradient θ and concentration gradient $c_o - c_i$ drive ions across an ion channel, resulting in an ion channel current $\{I_n^{(\theta,\lambda)}(\theta)\}$. This ion channel current is a piecewise constant signal that jumps between the values of zero and $I(\theta)$, where $I(\theta)$ denotes the current when the ion channel is in the open state.

The potential E_o (and hence potential difference θ) is adjusted experimentally until the current $I(\theta)$ goes to zero. This voltage θ^* at which the current $I(\theta^*)$ vanishes is called the *Nernst potential* and satisfies the so-called Nernst equation

Nernst potential

$$\theta^* = -\frac{kT}{e} \ln \frac{c_o}{c_i} = -59 \log_{10} \frac{c_o}{c_i} (\text{mV}), \qquad (4.5)$$

where $e = 1.6 \times 10^{-19}$ C denotes the charge of an electron, k denotes Boltzmann's constant, and T denotes the absolute temperature. The Nernst equation (4.5) gives the potential difference θ required to maintain electrochemical equilibrium when the concentrations are different on the two faces of the membrane.

Estimating the Nernst potential θ^* requires conducting experiments at different values of voltage θ. In patch-clamp experiments, the applied voltage θ is usually chosen from a finite set. Let

$$\theta \in \Theta = \{\theta(1), \dots, \theta(M)\}$$

denote the finite set of possible voltage values that the experimenter can pick. For example, in typical experiments, if one needs to determine the Nernst potential to a resolution of 4 mV, then $M = 80$ and $\theta(i)$ are uniformly spaced in 4 mV steps from $\theta(1) = -160$ mV and $\theta(M) = 160$ mV.

Note that the Nernst potential θ^* (zero crossing point) does not necessarily belong to the discrete set Θ—instead we will find the point in Θ that is closest to θ^* (with resolution $\theta(2) - \theta(1)$). With slight abuse of notation we will denote the element in Θ closest to the Nernst potential as θ^*. Thus determining $\theta^* \in \Theta$ can be formulated as a discrete optimization problem:

$$\theta^* = \arg\min_{\theta \in \Theta} |I(\theta)|^2.$$

discrete stochastic approximation

Discrete Stochastic Approximation Algorithm Learning the Nernst Potential can be formulated as the following discrete stochastic optimization problem

$$\text{Compute } \theta^* = \arg\min_{\theta \in \Theta} \left[\mathbf{E} \left\{ \hat{I}(\theta) \right\} \right]^2, \tag{4.6}$$

where $\hat{I}(\theta)$ is the MLE of the parameter $I(\theta)$ of the HMM. Since for a HMM, no closed-form expression is available for $\Sigma^{-1}(\theta)$ in equation 4.4, the above expectation cannot be evaluated analytically. This motivates the need to develop a simulation-based (stochastic approximation) algorithm. We refer the reader to the paper by Krishnamurthy and Chung (2003) for details.

The idea of discrete stochastic approximation (Andradottir, 1999) is to design a plan of experiments which provides more observations in areas where the Nernst potential is expected and less in other areas. More precisely what is needed is a dynamic resource allocation (control) algorithm that dynamically controls (schedules) the choice of voltage at which the HMM estimator operates in order to efficiently obtain the zero point and deduce how the current increases or decreases as the applied voltage deviates from the Nernst potential. We propose a discrete stochastic approximation algorithm that is both *consistent* and *attracted* to the Nernst potential. That is, the algorithm should spend more time gathering observations $\{y_n(\theta)\}$ at the Nernst potential $\theta = \theta^*$ and less time for other values of $\theta \in \Theta$. Thus in discrete stochastic approximation the aim is to devise an *efficient* (Pflug, 1996, chapter 5.3) adaptive search (sampling plan) which allows finding the minimizer θ^* with as few samples as possible by not making unnecessary observations at nonpromising values of θ. Here we construct algorithms based on the random search procedures of Andradottir (1995, 1999). The basic idea is to generate a homogeneous Markov chain taking values in Θ which spends more time at the global optimum than at any other element of Θ. We will show that these algorithms can be modified for tracking time-varying Nernst potentials. Finally, it is worthwhile mentioning that there are other classes of simulation-based discrete stochastic optimization algorithms, such as nested partition methods (Swisher et al., 2000), which combine partitioning, random sampling, and backtracking to create a Markov chain that converges to the global optimum.

Let $n = 1, 2, \ldots$ denote discrete time. The proposed algorithm is recursive and requires conducting experiments on batches of data. Since experiments will be conducted over batches of data, it is convenient to introduce the following notation. Group the discrete time into batches of length Δ—typically $\Delta = 10,000$ in experiments. We use the index $N = 1, 2, \ldots$ to denote batch number. Thus batch N comprises the Δ discrete time instants $n \in \{N\Delta, N\Delta + 1, \ldots, (N + 1)\Delta - 1\}$. Let $D_N = (D_N(1), \ldots, D_N(M))'$ denote the vector of duration times the algorithm spends at the M possible potential values in Θ.

Finally, for notational convenience define the M dimensional unit vectors, e_m, $m = 1, \ldots, M$ as

$$e_m = \begin{bmatrix} 0 & \cdots & 0 & 1 & 0 & \cdots & 0 \end{bmatrix}', \tag{4.7}$$

with 1 in the mth position and zeros elsewhere.

The discrete stochastic approximation algorithm of Andradottir (1995) is not directly applicable to the cost function 4.6, since it applies to optimization problems of the form $\min_{\theta \in \Theta} \mathbf{E}\{C(\theta)\}$. However, equation 4.6 can easily be converted to this form as follows: Let $\hat{I}_1(\theta)$, $\hat{I}_2(\theta)$ be two statistically independent unbiased HMM estimates of $I(\theta)$. Then defining $\hat{C}(\theta) = \hat{I}_1(\theta)\hat{I}_2(\theta)$, it straightforwardly follows that

$$\mathbf{E}\left\{\hat{C}(\theta)\right\} = \left[\mathbf{E}\left\{\hat{I}(\theta)\right\}\right]^2 = |I(v)|^2. \tag{4.8}$$

The discrete stochastic approximation algorithm we propose is as follows:

Algorithm 4.1
Algorithm for Learning Nernst Potential
Step 0 Initialization: At batch-time $N = 0$, select starting point $X_0 \in \{1, \ldots, M\}$ randomly. Set $D_0 = e_{X_0}$, Set initial solution estimate $\widehat{\theta_0^*} = \theta(X_0)$.
Step 1 Sampling: At batch-time N, sample $\tilde{X}_N \in \{X_N - 1, X_N + 1\}$ with uniform distribution.
Step 2 Evaluation and acceptance: Apply voltage $\tilde{\theta} = \theta(\tilde{X}_N)$ to patch-clamp experiment. Obtain two Δ length batches of HMM observations. Let $\hat{I}_N^{(1)}(\tilde{\theta})$ and $\hat{I}_N^{(2)}(\tilde{\theta})$ denote the HMM-MLE estimates for these two batches, which are computed using the EM algorithm (James et al., 1996; Krishnamurthy and Chung, 2003). Set $\hat{C}_N(\tilde{\theta})) = \hat{I}_N^{(1)}(\tilde{\theta})\hat{I}_N^{(2)}(\tilde{\theta})$.

Then apply voltage $\theta = \theta(X_N)$. Compute the HMM-MLE estimates for these two batches, denoted as $\hat{I}_N^{(1)}(\theta)$ and $\hat{I}_N^{(2)}(\theta)$. Set $\hat{C}_N(\theta)) = \hat{I}_N^{(1)}(\theta)\hat{I}_N^{(2)}(\theta)$.

If $\hat{C}_N(\tilde{\theta}) < \hat{C}_N(\theta)$, set $X_{N+1} = \tilde{X}_N$, else, set $X_{N+1} = X_N$.

Step 3 Update occupation probabilities of X_N: $D_{N+1} = D_N + e_{X_{N+1}}$.

Step 4 Update estimate of Nernst potential: $\widehat{\theta}_N^* = \theta(m^*)$ where

$$m^* = \arg \max_{m \in \{1, \ldots, M\}} D_{N+1}(m).$$

Set $N \to N + 1$. Go to step 1.

The proof of convergence of the algorithm is given in theorem 4.1 below. The main idea behind the above algorithm is that the sequence $\{X_N\}$ (or equivalently $\{\theta(X_N)\}$) generated by steps 1 and 2 is a homogeneous Markov chain with state space $\{1, \ldots, M\}$ (respectively, Θ) that is designed to spend more time at the global maximizer θ^* than any other state. In the above algorithm, $\hat{\theta}_N^*$ denotes the estimate of the Nernst potential at batch N.

Interpretation of Step 3 as Decreasing Step Size Adaptive Filtering Algorithm Define the occupation probability estimate vector as $\hat{\pi}_N = D_N/N$. Then the update in step 3 can be reexpressed as

$$\hat{\pi}_{N+1} = \hat{\pi}_N + \mu_{N+1}\left(e_{X_{N+1}} - \hat{\pi}_N\right), \quad \hat{\pi}_0 = e_{X_0}. \tag{4.9}$$

This is merely an adaptive filtering algorithm for updating $\hat{\pi}_N$ with decreasing step size $\mu_N = 1/N$.

Hence algorithm 4.1 can be viewed as a decreasing step size algorithm which involves a least mean squares (LMS) algorithm (with decreasing step size) in tandem with a random search step and evaluation (steps 1 and 2) for generating X_m. Figure 4.2 shows a schematic diagram of the algorithm with this LMS interpretation for step 3.

In Andradottir (1995), the following stochastic ordering assumption was used for convergence of the algorithm 4.1.

(O) For any $m \in \{1, \ldots, M-1\}$,

$$I^2(\theta(m+1)) > I^2(\theta(m)) \implies P\left(\hat{C}(\theta(m+1)) > \hat{C}(\theta(m))\right) > 0.5,$$

$$I^2(\theta(m+1)) < I^2(\theta(m)) \implies P\left(\hat{C}(\theta(m+1)) > \hat{C}(\theta(m))\right) < 0.5$$

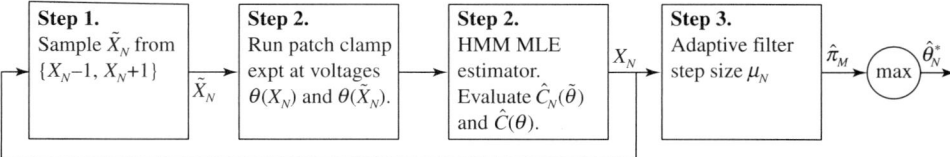

Figure 4.2   Schematic of algorithm 4.1.

Theorem 4.1

Under the condition (O) above, the sequence $\{\theta(X_N)\}$ generated by algorithm 4.1 is a homogeneous, aperiodic, irreducible Markov chain with state space Θ. Furthermore, algorithm 4.1 is attracted to the Nernst potential θ^*, i.e., for sufficiently large N, the sequence $\{\theta(X_N)\}$ spends more time at θ^* than at another state. (Equivalently, if $\theta(m^*) = \theta^*$, then $D_N(m^*) > D_N(j)$ for $j \in \{1, \ldots, M\} - \{m^*\}$.)

The above discrete stochastic approximation algorithm can be viewed as the discrete analog of the well-known LMS algorithm. Recall that in the LMS algorithm, the new estimate is computed from the previous estimate by moving along a desirable search direction (based on gradient information). In complete analogy with the above discrete search algorithm, the new estimate is obtained by moving along a discrete search direction to a desirable new point. We refer the reader to Krishnamurthy and Chung (2003) and our recent papers (Krishnamurthy et al., 2004; Yin et al., 2004) for complete convergence details of the above discrete stochastic approximation algorithm.

4.3   Scheduling Multiple Ion Channels on a Biological Chip

In this section, we consider dynamic scheduling and control of the gating process of ion channels on a biological chip. Patch clamping has rapidly become the "gold standard" (Fertig et al., 2002) for study of the dynamics of ion channel function by neurobiologists. However, patch clamping is a laborious process requiring precision micromanipulation under high-power visual magnification, vibration damping, and an experienced, skillful experimenter. Because of this, high-throughput studies required in proteomics and drug development have to rely on less valuable methods such as fluorescence-based measurement of intracellular ion concentrations (Xu et al., 2001). There is thus significant interest in an automated version of the whole patch-clamp principle, preferably one that has the potential to be used in parallel on a number of cells.

In 2002, Fertig et al. (2002) made a remarkable invention—the first successful demonstration of a *patch clamp on a chip*—a planar quartz-based biological chip that consists of several hundred ion channels (Sigworth and Klemic, 2002). This patch-clamp chip can be used for massively parallel screens for ion channel activity, thereby providing a high-throughput screening tool for drug discovery efforts.

Typically, because of their high cost, most neurobiological laboratories have only one patch-clamp amplifier that can be connected to the patch-clamp chip. As a result, only one ion channel in the patch-clamp chip can be monitored at a given time. It is thus of significant interest to devise an adaptive scheduling strategy that dynamically decides which single ion channel to activate at each time instant in order to maximize the throughput (information) from the patch-clamp experiment. Such a scheduling strategy will enable rapid evaluation and screening of drugs. Note that this problem directly fits into our main theme of sensor adaptive signal

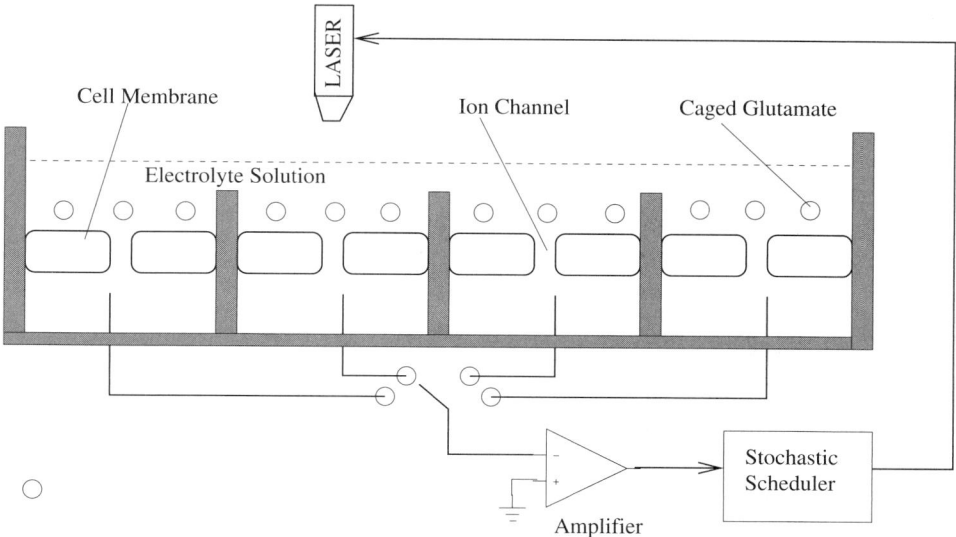

Figure 4.3 One-dimensional section of planar biological chip.

processing.

Here we consider the problem of how to dynamically schedule the activation of individual ion channels using a laser beam to maximize the information obtained from the patch-clamp chip for high-throughput drug evaluation. We refer the reader to Krishnamurthy (2004) for a detailed exposition of the problem together with numerical studies. The ion channel activation scheduling algorithm needs to dynamically plan and react to the presence of uncertain (random) dynamics of the individual ion channels in the chip. Moreover, excessive use of a single ion channel can make it desensitized. The aim is to answer the following question: *How should the ion channel activation scheduler dynamically decide which ion channel on the patch clamp chip to activate at each time instant in order to minimize the overall desensitization of channels while simultaneously extracting maximum information from the channels?*

We refer the reader to Fertig et al. (2002) for details on the synthesis of a patch-clamp chip. The chip consists of a quartz substrate of 200 micrometers thickness that is perforated by wet etching techniques resulting in apertures with diameters of approximately 1 micrometer. The apertures replace the tip of glass pipettes commonly used for patch-clamp recording. Cells are positioned onto the apertures from suspension by application of suction.

A schematic illustration of the ion channel scheduling problem for the patch-clamp chip is given in fig. 4.3. The figure shows a cross section of the chip with 4 ion channels. The planar chip could, for example, consist of 50 rows each containing 4 ion channels. Each of the four wells contains a membrane patch with an ion channel. The external electrolyte solutions contain caged ligands (such as caged glutamate). When a beam of laser is directed at the well, the inert caged ligands become active ligands that cause a channel to go from the closed conformation to an open conformation. Ions then flow across the open channel, and the current generated by the motion of charged particles is monitored with a patch-clamp amplifier. The amplifier is switched to the output of one well to another electronically. Typically, the magnitude of currents across each channel, when it is open, is about 1 pA (10^{-12} A).

The design of the ion channel activation scheduling algorithm needs to take into account the following subsystems.

Heterogeneous ion channels (macro-molecules) on chip: In a patch-clamp chip, the dynamical behavior of individual ion channels that are activated changes with time since they can become desensitized due to excessive use. Deactivated ion channels behave quite differently from other ion channels. Their transition to the open state becomes less frequent when they are de-sensitized due to excessive use.

Patch-clamp amplifier and heterogeneous measurements: The channel current of the activated ion channel is of the order of pico-amps and is measured in large amounts of thermal noise. Chung et al. (1990, 1991), used the powerful paradigm of HMMs to characterize these noisy measurements of single ion channel currents. The added complexity in the patch-clamp chip is that the signal-to-noise ratio is different at different parts of the chip—meaning that certain ion channels have higher SNR than other ion channels.

Ion channel activation scheduler: The ion channel activation scheduler uses the noisy channel current observations of the activated ion channel in the patch-clamp chip to decide which ion channel to activate at the next time instant to maximize a reward function that comprises the information obtained from the experiment. It needs to avoid activating desensitized channels, as they yield less information.

4.3.1 Stochastic Dynamical Models for Ion Channels on Patch-Clamp Chip

In this section we formulate a novel Markov chain model for the ion channels that takes into account both the ion channel current state and the ion channel sensitivity. The patch-clamp chip consists of P ion channels arranged in a two-dimensional grid indexed by $p = 1, \ldots, P$. Let $k = 0, 1, 2, \ldots$, denote discrete time. At each time instant k the scheduler decides which single ion channel to activate by directing a laser beam on the ion channel as described above. Let $u_k \in \{1, \ldots, P\}$ denote the ion channel that is activated by the scheduler at time k. The remaining $P - 1$ ion channel channels on the chip are inactive. It is the job of the dynamic scheduler to dynamically decide which ion channel should be activated at each time instant k in order to maximize the amount of information that can be obtained from the chip. If channel p is active at time k, i.e., $u_k = p$, the following two mechanisms determine the evolution of this active ion channel.

Ion Channel Sensitivity Model The longer the channel is activated, the more probably it becomes desensitized. Let $d_k^{(p)} \in \{normal, de\text{-}sens\}$ denote the sensitivity of ion channel p at any time instant k. If ion channel p is activated at time

k, i.e, $u_k = p$, then $d_k^{(p)}$ can be modeled as a two-state Markov chain with state transition probability matrix

$$
D = \quad
\begin{array}{c|c|c}
 & \text{normal} & \text{de-sens} \\
\hline
\text{normal} & d_{11} & 1 - d_{11} \\
\hline
\text{de-sens} & 0 & 1
\end{array}
\quad, \quad 0 \le d_{11} \le 1. \tag{4.10}
$$

The above transition probabilities reflect the fact that if the channel is overused it becomes desensitized with probability d_{12}. The $0, 1$ in the second row imply that once the channel is desensitized, it remains de-sensitized. Note that the sensitivity of the inactive channels remains fixed, i.e., $d_{k+1}^{(q)} = d_k^{(q)}$, $q \ne p$.

Ion Channel Current Model Suppose channel p is active at time k, i.e., $u_k = p$. Let $i_k^{(p)} \in \{0, I\} = \{\text{closed, open}\}$ denote the channel current. As is well known (Chung et al., 1991), the channel current is a binary valued signal that switches between zero "closed state" and the current level I "open state." The open-state current level I is of importance to neurobiologists since it quantifies the effect of a drug on the ion channel. Moreover, $i_k^{(p)}$ can be modeled as a two-state Markov chain (Chung et al., 1991) conditional on $d_k^{(p)}$ with transition probability matrix $Q = (P(i_{k+1}^{(p)} | i_k^{(p)}, d_{k+1}^{(p)}))$ given by

$$
Q(d_{k+1}^{(p)} = \text{normal}) = \quad
\begin{array}{c|c|c}
 & \text{closed} & \text{open} \\
\hline
\text{closed} & q_{11} & q_{12} \\
\hline
\text{open} & q_{21} & q_{22}
\end{array}
$$

$$
Q(d_{k+1}^{(p)} = \text{de-sens}) = \quad
\begin{array}{c|c|c}
 & \text{closed} & \text{open} \\
\hline
\text{closed} & \bar{q}_{11} & \bar{q}_{12} \\
\hline
\text{open} & \bar{q}_{21} & \bar{q}_{22}
\end{array}
\quad. \tag{4.11}
$$

For each ion channel $p \in \{1, \ldots, P\}$ on the patch clamp chip, define the ion channel state as the vector Markov process $s_k^{(p)} \triangleq (d_k^{(p)}, i_k^{(p)})$ with state space $\{(\text{normal,closed}), (\text{normal,open}), (\text{de-sens,closed}), (\text{de-sens,open})\} = \{1, 2, 3, 4\}$ where for notational convenience we have mapped the four states to $\{1, 2, 3, 4\}$. It is clear that only the state $s_k^{(u_k)}$ of the ion channel that is activated evolves with time. Since

$$
P(s_{k+1}^{(p)} | s_k^{(p)}) = P(d_{k+1}^{(p)} | d_k^{(p)}) P(i_{k+1}^{(p)} | i_k^{(p)}, d_{k+1}^{(p)})
$$

if channel p is active at time k, i.e., $u_k = p$, then $s_k^{(p)}$ has transition probability matrix

$$
A^{(p)} = \begin{bmatrix}
d_{11}q_{11} & d_{11}q_{12} & (1 - d_{22})\bar{q}_{11} & (1 - d_{11})\bar{q}_{12} \\
d_{11}q_{21} & d_{11}q_{22} & d_{12}\bar{q}_{21} & d_{12}\bar{q}_{22} \\
0 & 0 & \bar{q}_{11} & \bar{q}_{12} \\
0 & 0 & \bar{q}_{21} & \bar{q}_{22}
\end{bmatrix}. \tag{4.12}
$$

More generally one can assume that the state $s_k^{(p)}$ of each ion channel p has a finite number of values $\mathcal{N}_p$ (instead of just four states). If $u_k = p$, the state $s_k^{(p)}$ of ion channel p evolves according to an $\mathcal{N}_p$-state homogeneous Markov chain with transition probability matrix

$$A^{(p)} = (a_{ij}^{(p)})_{i,j \in \mathcal{N}_p} = P\left(s_{k+1}^{(p)} = j \mid s_k^{(p)} = i\right) \qquad (4.13)$$

if ion channel p is active at time k. The states of all the other $(P-1)$ ion channels that are not activated are unaffected, i.e., $s_{k+1}^{(q)} = s_k^{(q)}$, $q \neq p$. To complete our probabilistic formulation, assume the initial states of all ion channels on the chip are initialized with prior distributions: $s_0^{(p)} \sim x_0^{(p)}$ where $x_0^{(p)}$, are specified initial distributions for $p = 1, \dots, P$.

The above formulation captures the essence of an activation controlled patch-clamp chip—the channel activation scheduler dynamically decides which single ion channel to activate at each time instant.

4.3.2   Patch-Clamp Amplifier and Hidden Markov Model Measurements

The state of the active ion channel $s_k^{(p)}$ on the chip is not directly observed. Instead, the output of the patch-clamp amplifier is the ion channel current $i_k^{(p)}$ observed in large amounts of thermal noise. This output is quantized to an $\mathcal{M}$ symbol alphabet set $y_k^{(p)} \in \{O_1, O_2, \dots, O_\mathcal{M}\}$. The probabilistic relationship between the observations $y_k^{(p)}$ and the actual ion channel state $s_k^{(p)}$ of the active ion channel p is summarized by the $(\mathcal{N}_p \times \mathcal{M})$ state likelihood matrix:

$$B^{(p)} = (b_{im}^{(p)})_{i \in \mathcal{N}_p, m \in \mathcal{M}}, \qquad (4.14)$$

where $b_{im}^{(p)} \triangleq P(y_{k+1}^{(p)} = O_m | s_{k+1}^{(p)} = i, u_k = p)$ denotes the conditional probability (symbol probability) of the observation symbol $y_{k+1}^{(p)} = O_m$ when the actual state is $s_{k+1}^{(p)} = i$ and the active ion channel is $u_k = p$. Note that the above model allows for the state likelihood probabilities $(b_{im}^{(p)})$ to vary with p, i.e., to vary with the spatial location of the ion channel on the patch-clamp chip, thus allowing for spatially heterogeneous measurement statistics.

Let $Y_k = (y_1^{(u_0)}, \dots, y_k^{(u_{k-1})})$ denote the observed history up to time k. Let $U_k = (u_0, \dots, u_k)$ denote the sequence of past decisions made by the ion channel activation scheduler regarding which ion channels to activate from time 0 to time k.

4.3.3   Ion Channel Activation Scheduler

The above probabilistic model for the ion channel, together with the noisy measurements from the patch-clamp amplifier, constitute a well-known type of dynamic Bayesian network called a hidden Markov model (HMM) (Ephraim and Merhav, 2002). The problem of state inference of a HMM, i.e., estimating the state $s_k^{(p)}$ given

(Y_k, U_k), has been widely studied. (see e.g., Chung et al. (1991); Ephraim and Merhav (2002)). In this chapter we address the deeper and more fundamental issue of how the ion channel activation scheduler should dynamically decide which ion channel to activate at each time instant in order to minimize a suitable cost function that encompasses all the ion channels. Such dynamic decision making based on uncertainty (noisy channel current measurements) transcends standard sensor-level HMM state inference, which is a well-studied problem (Chung et al., 1991).

The activation scheduler decides which ion channel to activate at time k, based on the optimization of a discounted cost function which we now detail: The instantaneous cost incurred at time k due to all the ion channels (both active and inactive) is

$$C_k = -c_0(u_k) + c_1(s_k^{(u_k)}, u_k) + \sum_{p \neq u_k} r(s_k^{(p)}, p), \qquad (4.15)$$

where $-c_0(u_k) + c(s_k^{(u_k)}, u_k)$ denotes the cost incurred by the active ion channel u_k, and $\sum_{p \neq u_k} r(s_k^{(p)}, p)$ denotes the cost of remaining $P - 1$ inactive ion channels. The three components in the above cost function 4.15, can be chosen by the neurobiologist experimenter to optimize the information obtained from the patch-clamp experiment. Here we present one possible choice of costs:

Ion channel quality of service (QoS): $c_0(p)$ denotes the quality of service of the active ion channel p. The minus signs in equation 4.15 reflects the fact that the lower the QoS the higher the cost and vice versa.

State information cost: The final outcome of the patch-clamp experiment is often the estimate of the open-state level I. The accuracy of this estimate increases linearly with the number of observations obtained in the open state (since the covariance error of the estimate decreases linearly with the data length according to the central limit theorem). Maximizing the accuracy I requires maximizing the utilization of the patch-clamp chip, i.e., maximizing the expected number of measurements made from ion channels that are in the open normal state. That is, preference should be given to activating ion channels that are normal (i.e., not desensitized) and that quickly switch to the open state compared to other ion channels.

Desensitization cost of inactive channels: The instantaneous cost $r(s_k^{(p)}, p)$ in equation 4.15 incurred by each of the $P - 1$ inactive ion channels $p \in \{1, 2, \ldots, P\} - \{u_k\}$ should be chosen so as to penalize desensitized channels.

Based on the observed history $Y_k = (y_1^{(u_0)}, \ldots, y_k^{(u_{k-1})})$, and the history of decisions $U_{k-1} = (u_0, \ldots, u_{k-1})$, the scheduler needs to decide which ion channel on the chip to activate at time k. The scheduler decides which ion channel to activate at time k based on the stationary policy $\mu : (Y_k, U_{k-1}) \to u_k$. Here μ is a function that maps the observation history Y_k and past decisions U_{k-1} to the choice of which ion channel u_k to activate at time k. Let $\mathcal{U}$ denote the class of admissible stationary policies, i.e., $\mathcal{U} = \{\mu : u_k = \mu(Y_k, U_{k-1})\}$. The total expected discounted reward over an infinite time horizon is given by

$$J_\mu = \mathbf{E}\left\{\sum_{k=0}^{\infty}\beta^k C_k\right\}, \tag{4.16}$$

where C_k is defined in equation 4.15 and $\mathbf{E}\{\cdot\}$ denotes mathematical expectation. *The aim of the scheduler is to determine the optimal stationary policy $\mu^* \in \mathcal{U}$ which minimizes the cost in equation 4.16.*

infinite   horizon
discounted   cost
POMDP

The above problem of minimizing the infinite horizon discounted cost 4.16 of stochastic dynamical system 4.13 with noisy observations (equation 4.14) is a partially observed Markov decision process (POMDP) problem. Developing numerically efficient ion channel activation scheduling algorithms to minimize this cost is the subject of the rest of this section.

4.3.4 Formulation of Activation Scheduling as a Multiarmed Bandit

The above stochastic control problem (eq. 4.16) is an infinite-horizon partially observed Markov decision process with a multiarmed bandit structure which considerably simplifies the solution. But first, as is standard with partially observed stochastic control problems—we convert the partially observed multiarmed bandit problem to a fully observed multiarmed bandit problem defined in terms of the *information state* (Bertsekas, 1995a).

4.3.5 Information State Formulation

For each ion channel p, the information state at time k—which we will denote by $x_k^{(p)}$ (column vector of dimension $\mathcal{N}_p$)—is defined as the conditional filtered density of the Markov chain state $s_k^{(p)}$ given $Y_k^{(p)}$ and U_{k-1}:

information state

$$x_k^{(p)}(i) \triangleq P\left(s_k^{(p)} = i \mid Y_k, U_{k-1}\right), \qquad i = 1, \ldots, \mathcal{N}_p. \tag{4.17}$$

The information state can be computed recursively by the HMM state filter, which is also known as the forward algorithm or Baum's algorithm (James et al., 1996), according to equation 4.18 below.

In terms of the information state formulation, the ion channel activation scheduling problem described above can be viewed as the following dynamic scheduling problem: Consider P parallel HMM state filters, one for each ion channel on the chip. The pth HMM filter computes the state estimate (filtered density) $x_k^{(p)}$ of the pth ion channel, $p \in \{1, \ldots, P\}$. At each time instant, only one of the P ion channels is active, say ion channel p, resulting in an observation $y_{k+1}^{(p)}$. This is processed by the pth HMM state filter, which updates its Bayesian estimate of the

HMM filter

ion channel's state as

$$x_{k+1}^{(p)} = \frac{B^{(p)}(y_{k+1}^{(p)})A^{(p)\prime}x_k^{(p)}}{\mathbf{1}'B^{(p)}(y_{k+1}^{(p)})A^{(p)\prime}x_k^{(p)}} \qquad \text{if ion channel } p \text{ is active,} \tag{4.18}$$

where if $y_{k+1}^{(p)} = O_m$, then $B^{(p)}(m) = \text{diag}[b_{1m}^{(p)}, \dots, b_{\mathcal{N}_p,m}^{(p)}]$ is the diagonal matrix formed by the mth column of the observation matrix $B^{(p)}$ and $\mathbf{1}$ is an $\mathcal{N}_p$-dimensional column unit vector (we use $'$ to denote transpose).

The state estimates of the other $P - 1$ HMM state filters remain unaffected, i.e., if ion channel q is inactive,

$$x_{k+1}^{(q)} = x_k^{(q)}, \quad q \in \{1, \dots, P\}, \ q \neq p. \tag{4.19}$$

Let $\mathcal{X}^{(p)}$ denote the state space of information states $x^{(p)}$ for ion channels $p \in \{1, 2, \dots, P\}$. That is, for all $i \in \{1, \dots, \mathcal{N}_p\}$

$$\mathcal{X}^{(p)} = \left\{ x^{(p)} \in \mathrm{R}^{\mathcal{N}_p} : \ \mathbf{1}' x^{(p)} = 1, \qquad 0 < x^{(p)}(i) < 1 \right\}. \tag{4.20}$$

Note that $\mathcal{X}^{(p)}$ is an $(\mathcal{N}_p - 1)$-dimensional simplex. Using the smoothing property of conditional expectations, the cost function 4.16 can be rewritten in terms of the information state as

$$J_\mu = \mathbf{E} \left\{ \sum_{k=0}^{\infty} \beta^k \left(c'(u_k) x_k^{(u_k)} + \sum_{p \neq u_k} r'(p) x_k^{(p)} \right) \right\} \tag{4.21}$$

where $c(u_k)$ denotes the $\mathcal{N}_{u_k}$-dimensional reward vector $[c(s_k^{(p)} = 1, u_k), \dots, c(s_k^{(p)} = \mathcal{N}_{u_k}, u_k)]'$, and $r(p)$ is the $\mathcal{N}_{u_k}$-dimensional reward vector $[r(s_k^{(p)} = 1, p), \dots, c(s_k^{(p)} = \mathcal{N}_p, p)]'$. The aim is to compute the optimal policy $\arg\min_{\mu \in \mathcal{U}} J_\mu$. In terms of equations 4.18 and 4.21, the multiarmed bandit problem reads thus: Design an optimal dynamic scheduling policy to choose which ion channel to activate and hence which HMM Bayesian state estimator to use at each time instant.

As it stands the POMDP problem of equations 4.18, 4.19, and 4.21 or equivalently that of equations 4.16, 4.13, and 4.14 has a special structure:

(1) Only one Bayesian HMM state estimator operates according to 4.18 at each time k, or equivalently, only one ion channel is active at a given time k. The remaining $P - 1$ Bayesian estimates $x_k^{(q)}$ remain frozen, or equivalently, the remaining $P - 1$ ion channels remain inactive.

(2) The active ion channel incurs a cost depending on its current state and QoS. Since the state estimates of the inactive ion channels are frozen, the cost incurred by them is a fixed constant depending on the state when they were last active.

on-going multi-armed bandit   The above two properties imply that equations 4.18, 4.19, and 4.21 constitute what Gittins (1989) terms as an *ongoing multiarmed bandit*. It turns out that by a straightforward transformation an ongoing bandit can be formulated as a standard multiarmed bandit.

It is well known that the multiarmed bandit problem has a rich structure which results in the ion channel activation scheduling problem decoupling into P independent optimization problems. Indeed, from the theory of multiarmed bandits it follows that the optimal scheduling policy has an *indexable rule* (Whittle, 1980): for each channel p there is a function $\gamma^{(p)}(x_k^{(p)})$ called the *Gittins index*, which is only a function of the ion channel p and its information state $x_k^{(p)}$, whereby the optimal ion channel activation policy at time k is to activate the ion channel with the largest Gittins index, i.e.,

$$\text{activate ion channel } q \text{ where } q = \max_{p \in \{1,\dots,P\}} \left\{ \gamma^{(p)}(x_k^{(p)}) \right\}. \qquad (4.22)$$

A proof of this index rule for general multiarmed bandit problems is given by Whittle (1980). Computing the Gittins index is a key requirement for devising an optimal activation policy for the patch-clamp chip. We refer the reader to the paper of Krishnamurthy (2004) for details on how the Gittins index is computed and numerical examples of the performance of the algorithm.

Remarks: The indexable structure of the optimal ion channel activation policy (eq. 4.22) is convenient for two reasons:

(1) Scalability: Since the Gittins index is computed for each ion channel independently of every other ion channel (and this computation is off-line), the ion channel activation problem is easily scalable in that we can handle several hundred ion channels on a chip. In contrast, without taking the multiarmed bandit structure into account, the POMDP has $\mathcal{N}_p^P$ underlying states, making it computationally impossible to solve—e.g., for $P = 50$ channels with $\mathcal{N}_p = 2$ states per channel, there are 2^{50} states!

(2) Suitability for heterogeneous ion channels: Notice that our formulation of the ion channel dynamics allows for them to have different transition probabilities and likelihood probabilities. Moreover, since the Gittins index of an ion channel does not depend on other ion channels, we can meaningfully compare different types of ion channels.

4.4   The Permeation Problem: Brownian Stochastic Dynamical Formulation

In the previous section we dealt with ion channels at a macroscopic level—both in the spatial and time scales. The permeation problem considered in this and the following two sections seeks to explain the working of an ion channel at an Å(angstrom unit $= 10^{-10}$m) spatial scale by studying the propagation of individual ions through the ion channel at a femto second (10^{-15} timescale). This setup is said to be at a *mesoscopic scale* since the individual ions (e.g., Na$^+$ions) are of the order of a few Åin radius and are comparable in radius to the ion channel. At this mesoscopic level, point charge approximations and continuum electrostatics break down. The discrete finite nature of each ion needs to be taken into consideration. Also, failure of the mean field approximation in narrow channels implies that any theory that aspires to relate channel structure to its function must treat ions gramicidin-A   explicitly.

For convenience we focus in this section primarily on gramicidin-A channels—which are one of the simplest ion channels. Gramicidin-A is an antibiotic produced by *Bacillus brevis*. It was one of the first antibiotics to be isolated in the 1940s (Finkelstein, 1987, p. 130). In submicromolar concentrations it can increase the conductance of a bacterial cell membrane (which is a planar lipid bilayer membrane) by more than seven orders of magnitude by the formation of cation selective channels. As a result the bacterial cell is flooded and dies. This property of dramatically increasing the conductance of a lipid bilayer membrane has recently been exploited by Cornell et al. (1997) to devise gramicidin-A channel based biosensors with extremely high gains.

The aim of this section and the following two sections is to develop a stochastic dynamical formulation of the permeation problem that ultimately leads to estimating a *potential of mean force* (PMF) profile for an ion channel by optimizing the fit between the simulated current and the experimentally observed current. In the mesoscopic simulation of an ion channel, we propagate each individual ion using Brownian dynamics (Langevin equation), and the force experienced by each ion is a function of the PMF. As a result of the PMF and external applied potential to the ion channel there is a drift of ions from outside to inside the cell via the ion channel resulting in the simulated current.

Determining the PMF profile that optimizes the fit between the mesoscopic simulated current and observed current yields useful information and insight into how an ion channel works at a mesoscopic level. Determining the optimal PMF profile is important for several reasons: First, it yields the effective charge density in the peptides that form the ion channel. This charge density yields insight into the crystal structure of the peptide. Second, for theoretical biophysicists, the PMF profile yields information about the permeation dynamics including information about where the ion is likely to be trapped (called *binding sites*), the mean velocity of propagation of ions through the channel, and the average conductance of the ion channel.

We refer the reader to Krishnamurthy and Chung (a,b) for complete details of the Brownian dynamics algorithm and adaptively controlled Brownian dynamics algorithms for estimating the PMF of ion channels. Also the tutorial paper by Krishnamurthy and Chung (2005) and references therein give a detailed overview of Brownian dynamics simulation for determining the structure of ion channels.

4.4.1   Levels of Abstraction for Modeling Ion Channels at the Nanoscale

The ultimate aim of theoretical biophysicists is to provide a comprehensive physical description of biological ion channels. At the lowest level of abstraction is the ab initio quantum mechanical approach, in which the interactions between the atoms are determined from first-principles electronic structure calculations. Due to the extremely demanding nature of the computations, its applications are limited to very small systems at present. A higher level of modeling abstraction is to use classical molecular dynamics. Here, simulations are carried out using empirically-determined pairwise interaction potentials between the atoms, via ordinary differential equations (Newton's equation of motion). However, it is not computationally feasible to simulate the ion channel long enough to see permeation of ions across a model channel. For that purpose, one has to go up one further step in abstraction to

molecular dynamics vs. Brownian dynamics

stochastic dynamics, of which Brownian dynamics (BD) is the simplest form, where water molecules that form the bulk of the system in ion channels are stochastically averaged and only the ions themselves are explicitly simulated. Thus, instead of considering the dynamics of individual water molecules, one considers their average effect as a random force or Brownian motion on the ions. This treatment of water molecules can be viewed as a functional central limit theorem approximation. In BD, it is further assumed that the protein is rigid. Thus, in BD, the motion of each individual ion is modeled as the evolution of a stochastic differential equation, known as the Langevin equation.

A still higher level of abstraction is the Poisson-Nernst-Planck (PNP) theory, which is based on the continuum hypothesis of electrostatics and the mean-field approximation. Here, ions are treated not as discrete entities but as continuous charge densities that represent the space-time average of the microscopic motion of ions. For narrow ion channels—where continuum electrostatics does not hold—the PNP theory does not adequately explain ion permeation.

Remark: Bio-Nanotube Ion Channel vs. Carbon Nanotube There has recently been much work in the nanotechnology literature on carbon nanotubes and their use in field effect transistors (FETs). BD ion channel models are more complex than that of a carbon nanotube. Biological ion channels have radii of between 2 Å and 6 Å. In these narrow conduits formed by the protein wall, the force impinging on a permeating ion from induced surface charges on the water-protein interface becomes a significant factor. This force becomes insignificant in carbon nanotubes used in FETs with radius of approximately 100 Å, which is large compared to the debye length of electrons or holes in Si. Thus the key difference is that while in carbon nanotubes point charge approximations and continuum electrostatics holds, in ion channels the discrete finite nature of each ion needs to be considered.

ion channel vs.
carbon nanotube

4.4.2   Brownian Dynamics (BD) Simulation Setup

Figure 4.4 illustrates the schematic setup of Brownian dynamics simulation for permeation of ions through an ion channel. The aim is to obtain structural information, i.e., determine channel geometry and charges in the protein that forms the ion channel.

Figure 4.4 shows a schematic illustration of a BD simulation assembly for a particular example of an antibiotic ion channel called a gramicidin-A ion channel. The ion channel is placed at the center of the assembly. The atoms forming the ion channel are represented as a homogeneous medium with a dielectric constant of 2. Then, a large reservoir with a fixed number of positive ions (e.g., K^+ or Na^+ ions) and negative ions (e.g., Cl^- ions) is attached at each end of the ion channel. The electrolyte in the two reservoirs comprises 55 M (moles) of H_2O, and 150 mM concentrations of Na^+ and Cl^- ions.

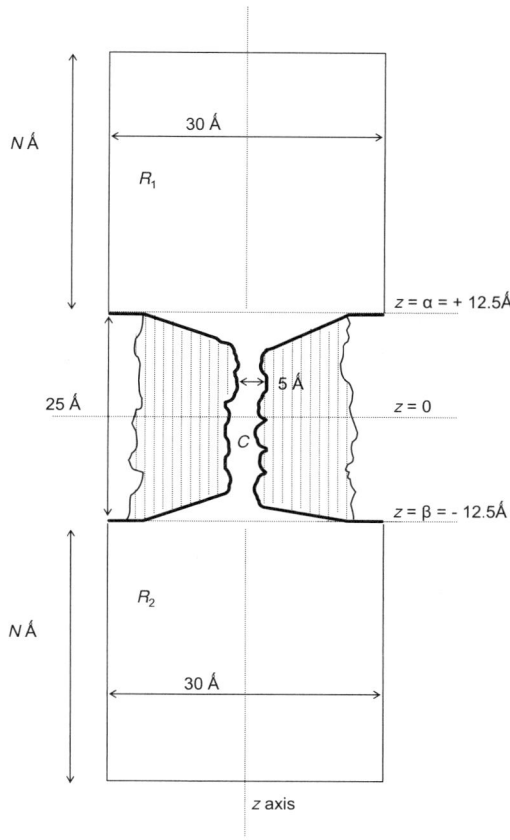

Figure 4.4   Gramicidin-Aion channel model Gramicidin-Acomprising $2N$ ions within two cylindrical reservoirs $\mathcal{R}_1$, $\mathcal{R}_2$, connected by the ion channel $\mathcal{C}$.

4.4.3 Mesoscopic Permeation Model of Ion Channel

Our permeation model for the ion channel comprises 2 cylindrical reservoirs $\mathcal{R}_1$ and $\mathcal{R}_2$ connected by the ion channel $\mathcal{C}$ as depicted in fig. 4.4, in which $2N$ ions are inserted (N denotes a positive integer). In fig. 4.4, as an example we have chosen a gramicidin-Aantibiotic ion channel—although the results below hold for any ion channel. These $2N$ ions comprise (1) N positively charged ions indexed by $i = 1, 2, \dots, N$. Of these, $N/2$ ions indexed by $i = 1, 2, \dots N/2$ are in $\mathcal{R}_1$, and $N/2$ ions indexed by $i = N/2 + 1, \dots, 2N$ are in $\mathcal{R}_2$. Each Na$^+$ion has charge q^+, mass $m^{(i)} = m^+ = 3.8 \times 10^{-26}$ kg and frictional coefficient $m^+\gamma^+$, and radius r^+; and (2) N negatively charge ions indexed by $i = N + 1, N + 2, \dots, 2N$. Of these, $N/2$ ions indexed by $i = N = 1, \dots 3N/2$ are placed in $\mathcal{R}_1$ and the remaining $N/2$ ions indexed by $i = 3N/2 + 1, \dots, 2N$ are placed in $\mathcal{R}_2$. Each negative ion has charge $q^{(i)} = q^-$, mass $m^{(i)} = m^-$, frictional coefficient $m^-\gamma^-$, and radius r^-. $\mathcal{R} = \mathcal{R}_1 \cup \mathcal{R}_2 \cup \mathcal{C}$ denotes the set comprised of the interior of the reservoirs and ion channel.

Let $t \geq 0$ denote continuous time. Each ion i, moves in three-dimensional space over time. Let $\mathbf{x}_t^{(i)} = (x_t^{(i)}, y_t^{(i)}, z_t^{(i)})' \in \mathcal{R}$ and $\mathbf{v}_t^{(i)} \in \mathrm{R}^3$ denote the position and velocity of ion i and time t. The three components $x_t^{(i)}, y_t^{(i)}, z_t^{(i)}$ of $\mathbf{x}_t^{(i)} \in \mathcal{R}$ are, respectively, the x, y, and z position coordinates. An external potential $\Phi_\lambda^{\mathrm{ext}}(\mathbf{x})$ is applied along the z-axis of fig. 4.4, i.e., with $\mathbf{x} = (x, y, z)$, $\Phi_\lambda^{\mathrm{ext}}(\mathbf{x}) = \lambda z, \lambda \in \Lambda$. Here Λ denotes a finite set of applied potentials. Typically $\Lambda = \{-200, -180, \dots, 0, \dots, 180, 200\}$ mV/m. Due to this applied external potential, the Na$^+$ions drift from reservoir $\mathcal{R}_1$ to $\mathcal{R}_2$ via the ion channel $\mathcal{C}$ in fig. 4.4. Let $\mathbf{X}_t = \left(\mathbf{x}_t^{(1)\prime}, \mathbf{x}_t^{(2)\prime}, \mathbf{x}_t^{(3)\prime}, \dots, \mathbf{x}_t^{(2N)\prime}\right)' \in \mathcal{R}^{2N}$ and $\mathbf{V}_t = \left(\mathbf{v}_t^{(1)\prime}, \mathbf{v}_t^{(2)\prime}, \mathbf{v}_t^{(3)\prime}, \dots, \mathbf{v}_t^{(2N)\prime}\right)' \in \mathrm{R}^{6N}$ denote the velocities of all the $2N$ ions. The position and velocity of each individual ion evolves according to the following continuous-time stochastic dynamical system:

$$\mathbf{x}_t^{(i)} = \mathbf{x}_0^{(i)} + \int_0^t \mathbf{v}_s^{(i)} ds, \tag{4.23}$$

$$m^+\mathbf{v}_t^{(i)} = m^+\mathbf{v}_0^{(i)} - \int_0^t m^+\gamma^+(\mathbf{X}_s^{(i)})\mathbf{v}_s^{(i)} ds + \int_0^t F_{\theta,\lambda}^{(i)}(\mathbf{X}_s) ds + b^+\mathbf{w}_t^{(i)},$$
$$i \in\in \{1, 2, \dots, N\}, \tag{4.24}$$

$$m^-\mathbf{v}_t^{(i)} = m^-\mathbf{v}_0^{(i)} - \int_0^t m^-\gamma^-(\mathbf{X}_s^{(i)})\mathbf{v}_s^{(i)} ds + \int_0^t F_{\theta,\lambda}^{(i)}(\mathbf{X}_s) ds + b^-\mathbf{w}_t^{(i)},$$
$$i \in \{N + 1, N + 2, \dots, 2N\}. \tag{4.25}$$

Langevin equation

Equations 4.24 and 4.25 constitute the well-known Langevin equations and describe the evolution of the velocity $\mathbf{v}_t^{(i)}$ of ion i as a stochastic dynamical system. The random process $\{\mathbf{w}_t^{(i)}\}$ denotes a three-dimensional Brownian motion, which is component-wise independent. The constants b^+ and b^- are, respectively, $b^{+2} = 2m^+\gamma^+kT$, $b^{-2} = 2m^-\gamma^-kT$. Finally, the noise processes $\{\mathbf{w}_t^{(i)}\}$ and $\{\mathbf{w}_t^{(j)}\}$, that drive any two different ions, $j \neq i$, are assumed to be statistically independent.

In equations 4.24 and 4.25, $F_{\theta,\lambda}^{(i)}(\mathbf{X}_t) = -q^{(i)}\nabla_{\mathbf{x}_t^{(i)}}\Phi_{\theta,\lambda}^{(i)}(\mathbf{X}_t)$ represents the *systematic force* acting on ion i, where the scalar-valued process $\Phi_{\theta,\lambda}^{(i)}(\mathbf{X}_t)$ is the total electric potential experienced by ion i given the position $\mathbf{X}_t$ of the $2N$ ions. The subscript λ is the applied external potential. The subscript θ is a parameter that characterizes the potential of mean force (PMF) profile, which is an important component of $\Phi_{\theta,\lambda}^{(i)}(\mathbf{X}_t)$.

It is convenient to represent the above system (equations 4.23, 4.24, and 4.25) as a vector stochastic differential equation. Define the following vector-valued variables:

$$\mathbf{V}_t = \begin{bmatrix} \mathbf{V}_t^+ \\ \mathbf{V}_t^- \end{bmatrix}, \quad \mathbf{V}_t^+ = \begin{bmatrix} \mathbf{v}_t^{(1)} \\ \vdots \\ \mathbf{v}_t^{(N)} \end{bmatrix}, \quad \mathbf{V}_t^- = \begin{bmatrix} \mathbf{v}_t^{(N+1)} \\ \vdots \\ \mathbf{v}_t^{(2N)} \end{bmatrix}, \quad \mathbf{w}_t = \begin{bmatrix} \mathbf{0}_{2N\times 1} \\ \mathbf{w}_t^{(1)} \\ \vdots \\ \mathbf{w}_t^{(2N)} \end{bmatrix}, \quad \zeta_t = \begin{bmatrix} \mathbf{X}_t \\ \mathbf{V}_t^+ \\ \mathbf{V}_t^- \end{bmatrix},$$

$$\mathbf{F}_{\theta,\lambda}^+(\mathbf{X}_t) = \begin{bmatrix} F_{\theta,\lambda}^{(1)}(\mathbf{X}_t) \\ \vdots \\ F_{\theta,\lambda}^{(N)}(\mathbf{X}_t) \end{bmatrix}, \quad \mathbf{F}_{\theta,\lambda}^-(\mathbf{X}_t) = \begin{bmatrix} F_{\theta,\lambda}^{(N+1)}(\mathbf{X}_t) \\ \vdots \\ F_{\theta,\lambda}^{(2N)}(\mathbf{X}_t) \end{bmatrix}, \quad \mathbf{F}_{\theta,\lambda}(\mathbf{X}_t) = \begin{bmatrix} \frac{1}{m^+}\mathbf{F}_{\theta,\lambda}^+(\mathbf{X}_t) \\ \frac{1}{m^-}\mathbf{F}_{\theta,\lambda}^-(\mathbf{X}_t) \end{bmatrix}.$$

$$(4.26)$$

Then equations 4.23, 4.24, and 4.25 can be written compactly as

$$d\zeta_t = \mathbf{A}\zeta_t dt + \mathbf{f}_{\theta,\lambda}(\zeta_t)dt + \Sigma^{1/2}d\mathbf{w}_t, \qquad (4.27)$$

where $\Sigma^{1/2} = \text{block diag}(\mathbf{0}_{6N\times 6N}, b^+/m^+\mathbf{I}_{3N\times 3N}, b^-/m^-\mathbf{I}_{3N\times 3N})$,

$$\mathbf{A} = \left[\begin{array}{c|cc} \mathbf{0}_{6N\times 6N} & \multicolumn{2}{c}{\mathbf{I}_{6N\times 6N}} \\ \hline \mathbf{0}_{6N\times 6N} & -\gamma^+\mathbf{I}_{3N\times 3N} & \mathbf{0}_{3N\times 3N} \\ & \mathbf{0}_{N\times N} & -\gamma^-\mathbf{I}_{N\times N} \end{array}\right], \quad \mathbf{f}_{\theta,\lambda}(\zeta_t) = \begin{bmatrix} \mathbf{0}_{6N\times 1} \\ \mathbf{F}_{\theta,\lambda}(\mathbf{X}_t) \end{bmatrix}. \qquad (4.28)$$

We will subsequently refer to equations 4.27 and 4.28 as the Brownian dynamics equations for the ion channel.

Remark: The BD approach is a stochastic averaging theory framework that models the average effect of water molecules:

1. The friction term $m\gamma\mathbf{v}_t^{(i)}dt$ captures the average effect of the ions driven by the applied external electrical field bumping into the water molecules every few femtoseconds. The frictional coefficient is given from Einstein's relation.

2. The Brownian motion term $\mathbf{w}_t^{(i)}$ also captures the effect of the random motion of ions bumping into water molecules and is given from the *fluctuation-dissipation* theorem.

4.4.4   Systematic Force Acting on Ions

As mentioned after equation 4.25, the systematic force experienced by ion i is

$$F^{(i)}_{\theta,\lambda}(\mathbf{X}_t) = -q^{(i)} \nabla_{\mathbf{x}^{(i)}_t} \Phi^{(i)}_{\theta,\lambda}(\mathbf{X}_t),$$

where the scalar valued process $\Phi^{(i)}_{\theta,\lambda}(\mathbf{X}_t)$ denotes the total electric potential experienced by ion i given the position $\mathbf{X}_t$ of all the $2N$ ions. We now give a detailed formulation of these systematic forces.

The potential $\Phi^{(i)}_{\theta,\lambda}(\mathbf{X}_t)$ experienced by each ion i comprises the following five components:

$$\Phi^{(i)}_{\theta,\lambda}(\mathbf{X}_t) = U_\theta(\mathbf{x}^{(i)}_t) + \Phi^{\text{ext}}_\lambda(\mathbf{x}^{(i)}_t) + \Phi^{IW}(\mathbf{x}^{(i)}_t) + \Phi^{C,i}(\mathbf{X}_t) + \Phi^{SR,i}(\mathbf{X}_t). \quad (4.29)$$

Just as $\Phi^{(i)}_{\theta,\lambda}(\mathbf{X}_t)$ is decomposed into five terms, we can similarly decompose the force $F^{(i)}_{\theta,\lambda}(\mathbf{X}_t) = -q\nabla_{\mathbf{x}^{(i)}_t}\Phi^{(i)}_{\theta,\lambda}(\mathbf{X}_t)$ experienced by ion i as the superposition (vector sum) of five force terms, where each force term is due to the corresponding potential in equation 4.29—however, for notational simplicity we describe the scalar-valued potentials rather than the vector-valued forces.

Note that the first three terms in equation 4.29, namely $U_\theta(\mathbf{x}^{(i)}_t)$, $\Phi^{\text{ext}}_\lambda(\mathbf{x}^{(i)}_t)$, $\Phi^{IW}(\mathbf{x}^{(i)}_t)$, depend only on the position $\mathbf{x}^{(i)}_t$ of ion i, whereas the last two terms in equation 4.29, $\Phi^{C,i}(\mathbf{X}_t)$, $\Phi^{SR,i}(\mathbf{X}_t)$, depend on the distance of ion i to all the other ions, i.e., the position $\mathbf{X}_t$ of all the ions. The five components in equation 4.29 are now defined.

PMF

Potential of mean force (PMF), denoted $U_\theta(\mathbf{x}^{(i)}_t)$ in equation 4.29, comprises electric forces acting on ion i when it is in or near the ion channel (nanotube $\mathcal{C}$ in fig. 4.4). The PMF U_θ is a smooth function of the ion position $\mathbf{x}^{(i)}_t$ and depends on the structure of the ion channel. Therefore, estimating $U_\theta(\cdot)$ yields structural information about the ion channel. In section 4.6, we outline an adaptive Brownian dynamics approach to estimate the PMF $U_\theta(\cdot)$. The PMF U_θ originates from two different sources; see Krishnamurthy and Chung (2005) for details. First, there are fixed charges in the channel protein, and the electric field emanating from them renders the pore attractive to cations and repulsive to anions, or vice versa. Some of the amino acids forming the ion channels carry the unit or partial electronic charges. For example, glutamate and aspartate are acidic amino acids, being negatively charged at pH 6.0, whereas lysine, arginine, and histidine are basic amino acids, being positively charged at pH 6.0. Second, when any of the ions in the assembly comes near the protein wall, it induces surface charges of the same polarity at the water-protein interface. This is known as the induced surface charge.

External applied potential: In the vicinity of the cell, there is a strong electric field resulting from the membrane potential, which is generated by diffuse, unpaired, ionic clouds on each side of the membrane. Typically, this resting potential across a cell membrane, whose thickness is about 50 Å, is 70 mV, the cell interior negative with respect to the extracellular space. In simulations, this field is mimicked by applying a uniform electric field across the channel. This is equivalent to placing a pair of large plates far away from the channel and applying a potential difference between the two plates. Because the space between the electrodes is filled with electrolyte solutions, each reservoir is in isopotential. That is, the average potential anywhere in the reservoir is identical to the applied potential at the voltage place on that side. For ion i at position $\mathbf{x}_t^{(i)} = \mathbf{x} = (x, y, z)$, $\Phi_\lambda^{\mathrm{ext}}(\mathbf{x}) = \lambda z$ denotes the potential on ion i due to the applied external field. The electrical field acting on each ion due to the applied potential is therefore $-\nabla_{\mathbf{x}_t^{(i)}} \Phi_\lambda^{\mathrm{ext}}(\mathbf{x}) = (0, 0, \lambda)$ V/m at all $\mathbf{x} \in \mathcal{R}$. It is this applied external field that causes a drift of ions from the reservoir $\mathcal{R}_1$ to $\mathcal{R}_2$ via the ion channel $\mathcal{C}$. As a result of this drift of ions within the electrolyte in the two reservoirs, eventually the measured potential drop across the reservoirs is zero and all the potential drop occurs across the ion channel.

Inter-ion Coulomb potential: In equation 4.29, $\Phi^{C,i}(\mathbf{X}_t)$ denotes the Coulomb interaction between ion i and all the other ions

$$\Phi^{C,i}(\mathbf{X}_t) = \frac{1}{4\pi\epsilon_0} \sum_{j=1, j\neq i}^{2N} \frac{q^{(j)}}{\epsilon_w \|\mathbf{x}_t^{(i)} - \mathbf{x}_t^{(j)}\|}. \tag{4.30}$$

Ion-wall interaction potential: The ion-wall potential Φ^{IW}, also called the $(\sigma/r)^9$, potential ensures that the position $\mathbf{x}_t^{(i)}$ of all ions $i = 1, \ldots, 2N$ lie in $\mathcal{R}^o$. With $\mathbf{x}_t^{(i)} = (x_t^{(i)}, y_t^{(i)}, z_t^{(i)})'$, it is modeled as

$$\Phi^{IW}(\mathbf{x}_t^{(i)}) = \frac{F_0}{9} \frac{(r^{(i)} + r_w)^9}{\left[r_c + r_w - \left(\sqrt{{x_t^{(i)}}^2 + {y_t^{(i)}}^2} \right) \right]^9}, \tag{4.31}$$

where for positive ions $r^{(i)} = r^+$ (radius of Na^+ atom) and for negative ions $r^{(i)} = r^-$ (radius of Cl^- atom); $r_w = 1.4$ Åis the radius of atoms making up the wall; r_c denotes the radius of the ion channel, and $F_0 = 2 \times 10^{-10}$N, which is estimated from the ST2 water model used in molecular dynamics (Stillinger and Rahman, 1974). This ion-wall potential results in short-range forces that are only significant when the ion is close to the wall of the reservoirs $\mathcal{R}_1$ and $\mathcal{R}_2$ or anywhere in the ion channel $\mathcal{C}$ (since the ion channel is comparable in radius to the ions).

Short-range potential: Finally, at short ranges, the Coulomb interaction between two ions is modified by adding a potential $\Phi^{SR,i}(\mathbf{X}_t)$, which replicates the effects of the overlap of electron clouds. Thus,

$$\Phi^{SR,i}(\mathbf{X}_t) = \frac{F_0}{9} \sum_{j=1, j\neq i}^{2N} \frac{(r^{(i)} + r^{(j)})}{\|\mathbf{x}_t^{(i)} - \mathbf{x}_t^{(j)}\|^9}. \tag{4.32}$$

Similar to the ion-wall potential, $\Phi^{SR,i}$ is significant only when ion i gets very close to another ion. It ensures that two opposite-charge ions attracted by inter-ion Coulomb forces 4.30 cannot collide and annihilate each other. Molecular dynamics simulations show that the hydration forces between two ions add further structure to the $1/\|\mathbf{x}_t^{(i)} - \mathbf{x}_t^{(j)}\|^9$ repulsive potential due to the overlap of electron clouds in the form of damped oscillations (Guàrdia et al., 1991a,b). Corry et al. (2001) incorporated the effect of the hydration forces in equation 4.32 in such a way that the maxima of the radial distribution functions for Na^+-Na^+, Na^+-Cl^-, and Cl^--Cl^- would correspond to the values obtained experimentally.

4.5 Ion Permeation: Probabilistic Characterization and Brownian Dynamics (BD) Algorithm

Having given a complete description of the dynamics of individual ions that permeate through the ion channel, in this section we give a probabilistic characterization of the ion channel current. In particular, we show that the mean ion channel current satisfies a boundary-valued partial differential equation. We then show that the Brownian dynamics (BD) simulation algorithm can be viewed as a randomized multiparticle-based algorithm for solving the boundary-valued partial differential equation to estimate the ion channel current.

4.5.1 Probabilistic Characterization of Ion Channel Current in Terms of Mean Passage Time

The aim of this subsection is to give a probabilistic characterization of the ion channel current in terms of the mean first passage time of the diffusion process (see equation 4.27). This characterization also shows that the Brownian dynamical system 4.27 has a well-defined, unique stationary distribution.

A key requirement in any mathematical construction is that the concentration of ions in each reservoir $\mathcal{R}_1$ and $\mathcal{R}_2$ remains approximately constant and equal to the physiological concentration. The following probabilistic construction ensures that the concentration of ions in reservoir $\mathcal{R}_1$ and $\mathcal{R}_2$ remain approximately constant.

Step 1: The $2N$ ions in the system are initialized as described above, and the ion channel $\mathcal{C}$ is closed. The system evolves and attains stationarity. Theorem 4.2 below shows that the probability density function of the $2N$ particles converges geometrically fast to a unique stationary distribution. Theorem 4.3 shows that in the stationary regime, all positive ions in $\mathcal{R}_1$ have the same stationary distribution and so are statistically indistinguishable (similarly for $\mathcal{R}_2$).

Step 2: After stationarity is achieved, the ion channel is opened. The ions evolve according to equation 4.27. As soon as an ion from $\mathcal{R}_1$ crosses the ion channel $\mathcal{C}$ and enters $\mathcal{R}_2$, the experiment is stopped. Similarly if an ion from $\mathcal{R}_2$ cross $\mathcal{C}$ and enters $\mathcal{R}_1$, the experiment is stopped. Theorem 4.3 gives partial differential equations for the mean minimum time an ion in $\mathcal{R}_1$ takes to cross the ion channel and reach $\mathcal{R}_2$ and establishes that this time is finite. From this a theoretical expression for the mean ion channel current is constructed (eq. 4.42).

Note that if the system was allowed to evolve for an infinite time with the channel open, then eventually, due to the external applied potential, more ions would be in $\mathcal{R}_2$ than $\mathcal{R}_1$. This would violate the condition that the concentration of particles in $\mathcal{R}_1$ and $\mathcal{R}_2$ remain constant. In the BD simulation algorithm 4.2 presented later in this chapter, we use the above construction to restart the simulation each time an ion crosses the channel—this leads to a regenerative process that is easy to analyze.

Let

$$\pi_t^{(\theta,\lambda)}(\mathbf{X}, \mathbf{V}) = p^{(\theta,\lambda)}\left(\mathbf{x}_t^{(1)}, \mathbf{x}_t^{(2)}, \dots, \mathbf{x}_t^{(2N)}, \mathbf{v}_t^{(1)}, \mathbf{v}_t^{(2)}, \dots, \mathbf{v}_t^{(2N)}\right)$$

denote the joint probability density function (pdf) of the position and velocity of all the $2N$ ions at time t. We explicitly denote the θ, λ dependence of the pdfs since they depend on the PMF U_θ and applied external potential λ. Then the joint pdf $\pi_t^{(\theta,\lambda)}(\mathbf{X}) = p^{(\theta,\lambda)}\left(\mathbf{x}_t^{(1)}, \mathbf{x}_t^{(2)}, \dots, \mathbf{x}_t^{(2N)}\right)$ of the positions of all $2N$ ions at time t is

$$\pi_t^{(\theta,\lambda)}(\mathbf{X}) = \int_{\mathrm{R}^{6N}} \pi_t^{(\theta,\lambda)}(\mathbf{X}, \mathbf{V}) d\mathbf{V}.$$

The following result, proved in Krishnamurthy and Chung (a), states that for the above stochastic dynamical system, $\pi_t^{(\theta,\lambda)}(\mathbf{X}, \mathbf{V})$ converges exponentially fast to its stationary (invariant) distribution $\pi_\infty^{(\theta,\lambda)}(\mathbf{X}, \mathbf{V})$.

Theorem 4.2
For the Brownian dynamics system (4.27, 4.28), with $\zeta = (\mathbf{X}, \mathbf{V})$, there exists a unique stationary distribution $\pi_\infty^{(\theta,\lambda)}(\zeta)$, and constants $K > 0$ and $0 < \rho < 1$, such that

$$\sup_{\zeta \in \mathcal{R}^{2N} \times \mathrm{R}^{6N}} |\pi_t^{(\theta,\lambda)}(\zeta) - \pi_\infty^{(\theta,\lambda)}(\zeta)| \leq K\mathcal{V}(\zeta)\rho^t. \tag{4.33}$$

Here $\mathcal{V}(\zeta) > 1$ is an arbitrary measurable function on $\mathcal{R}^{2N} \times \mathrm{R}^{6N}$.

The above theorem on the exponential ergodicity of $\zeta_t = (\mathbf{X}_t, \mathbf{V}_t)$ has two consequences that we will subsequently use. First, it implies that as the system evolves, the initial coordinates $\mathbf{x}_0^{(i)}, \mathbf{v}_0^{(i)}$ of all the $2N$ ions are forgotten exponentially fast. This allows us to efficiently conduct BD simulations in section 4.5.2 below. Second, the exponential ergodicity also implies that a strong law of large numbers holds—this will be used below to formulate a stochastic optimization problem in terms of the stationary measure $\pi_\infty^{(\theta,\lambda)}$ for computing the potential mean force.

Notation is as follows: For $\zeta = (\zeta^{(1)}, \dots \zeta^{(4N)})'$, define

$$\nabla_\zeta = \left(\frac{\partial}{\partial \zeta^{(1)}}, \frac{\partial}{\partial \zeta^{(2)}}, \dots, \frac{\partial}{\partial \zeta^{(4N)}}\right)'.$$

For a vector field $\mathbf{f}_{\theta,\lambda}(\zeta) = \left[f^{(1)}(\zeta) \; f^{(2)}(\zeta) \; \cdots \; f^{(4N)}(\zeta) \right]'$ defined on $\mathbf{R}^{4N}$, define the divergence operator

$$\operatorname{div}(\mathbf{f}_{\theta,\lambda}) = \frac{\partial f^{(1)}}{\partial \zeta^{(1)}} + \frac{\partial f^{(2)}}{\partial \zeta^{(2)}} + \cdots + \frac{\partial f^{(4N)}}{\partial \zeta^{(4N)}}.$$

For the stochastic dynamical system 4.27, comprising $2N$ ions, define the backward elliptic operator (infinitesimal generator) $\mathcal{L}$ and its adjoint $\mathcal{L}^*$ for any test function $\phi(\zeta)$ as

$$\mathcal{L}(\phi) = \frac{1}{2}\operatorname{Tr}[\Sigma \nabla_\zeta^2 \phi(\zeta)] + (\mathbf{f}_{\theta,\lambda}(\zeta) + \mathbf{A}\zeta)' \nabla_\zeta \phi(\zeta), \tag{4.34}$$

$$\mathcal{L}^*(\phi) = \frac{1}{2}\operatorname{Tr}\left[\nabla_\zeta^2(\Sigma\phi(\zeta))\right] - \operatorname{div}[(\mathbf{A}\zeta + \mathbf{f}_{\theta,\lambda}(\zeta))\phi(\zeta)].$$

Here, $\mathbf{f}_{\theta,\lambda}$ and Σ are defined in equation 4.28.

It is well known that the probability density function $\pi_t^{(\theta,\lambda)}(\cdot)$ of $\zeta_t = (\mathbf{X}_t', \mathbf{V}_t')'$ satisfies the *Fokker-Planck equation* (Wong and Hajek, 1985):

$$\frac{d\pi_t^{(\theta,\lambda)}}{dt} = \mathcal{L}^* \pi_t^{(\theta,\lambda)}. \tag{4.35}$$

Fokker-Planck equation

Also the stationary probability density function $\pi_\infty^{(\theta,\lambda)}(\cdot)$ satisfies

$$\mathcal{L}^*(\pi_\infty^{(\theta,\lambda)}) = 0, \quad \int_{\mathbf{R}^{6N}} \int_{\mathcal{R}^{2N}} \pi_\infty^{(\theta,\lambda)}(\mathbf{X}, \mathbf{V})\, d\mathbf{X}\, d\mathbf{V} = 1. \tag{4.36}$$

We next show that once stationarity has been achieved, the N positive ions behave statistically identically, i.e., each ion has the same stationary marginal distribution. Define the stationary marginal density $\pi_\infty^{(\theta,\lambda)}(\mathbf{x}^{(i)}, \mathbf{v}^{(i)})$ of ion i as

$$\pi_\infty^{(\theta,\lambda)}(\mathbf{x}^{(i)}, \mathbf{v}^{(i)}) = \int_{\mathbf{R}^{6N-3}} \int_{\mathcal{R}^{2N-1}} \pi_\infty^{(\theta,\lambda)}(\mathbf{X}, \mathbf{V}) \prod_{j=1, j\neq i}^{2N} d\mathbf{x}^{(j)} d\mathbf{v}^{(j)}. \tag{4.37}$$

The following result states that the ions are statistically indistinguishable—see the paper of Krishnamurthy and Chung (a) for proof.

Theorem 4.3
Assuming that the ion channel $\mathcal{C}$ is closed, the stationary marginal densities for the positive ions in $\mathcal{R}_1$ are identical:

$$\pi_\infty^{(\theta,\lambda),\mathcal{R}_1} \triangleq \pi_\infty^{(\theta,\lambda)}(\mathbf{x}^{(1)}, \mathbf{v}^{(1)}) = \pi_\infty^{(\theta,\lambda)}(\mathbf{x}^{(2)}, \mathbf{v}^{(2)}) = \cdots = \pi_\infty^{(\theta,\lambda)}(\mathbf{x}^{(N)}, \mathbf{v}^{(N/2)}).$$

Similarly, the stationary marginal densities for the positive ions in $\mathcal{R}_2$ are identical:

$$\pi_\infty^{(\theta,\lambda),\mathcal{R}_2} \triangleq \pi_\infty^{(\theta,\lambda)}(\mathbf{x}^{(N/2+1)}, \mathbf{v}^{(N/2+1)}) = \pi_\infty^{(\theta,\lambda)}(\mathbf{x}^{(N/2+2)}, \mathbf{v}^{(N/2+2)})$$
$$= \cdots = \pi_\infty^{(\theta,\lambda)}(\mathbf{x}^{(N)}, \mathbf{v}^{(N)}). \tag{4.38}$$

Theorem 4.3 is not surprising: equations 4.23, 4.24, and 4.25 are symmetric in *i*, therefore intuitively one would expect that once steady state as been attained, all the positive ions behave identically—similarly with the negative ions. Due to above result, in our probabilistic formulation below, once the system has attained steady state, any positive ion is representative of all the N positive ions, and similarly for the negative ions.

Assume that the system 4.27 comprising $2N$ ions has attained stationarity with the ion channel $\mathcal{C}$ closed. Then the ion channel is opened so that ions can diffuse into it. Let $\tau_{\mathcal{R}_1,\mathcal{R}_2}^{(\theta,\lambda)}$ denote the mean first-passage time for any of the $N/2$ Na$^+$ions in $\mathcal{R}_1$ to travel to $\mathcal{R}_2$ via the gramicidin-Achannel $\mathcal{C}$, and $\tau_{\mathcal{R}_2,R_1}^{(\theta,\lambda)}$ denote the mean first-passage time for any of the $N/2$ Na$^+$ions in $\mathcal{R}_2$ to travel to R_1:

mean   first-passage time

$$\tau_{\mathcal{R}_1,\mathcal{R}_2}^{(\theta,\lambda)} = \mathbf{E}\{t_\beta\} \text{ where } t_\beta \triangleq \inf\left\{t : \max\left(z_t^{(1)}, z_t^{(2)}, \ldots, z_t^{(N/2)}\right) \geq \beta\right\},$$

$$\tau_{\mathcal{R}_2,R_1}^{(\theta,\lambda)} = \mathbf{E}\{t_\alpha\} \text{ where } t_\alpha \triangleq \inf\left\{t : \min\left(z_t^{(N/2+1)}, z_t^{(N/2+2)}, \ldots, z_t^{(2N)}\right) \leq \alpha\right\}.$$

$$(4.39)$$

Note that for ion channels such as gramicidin-A, only positive Na$^+$ions flow through the channel to cause the channel current—so we do not need to consider the mean first-passage times of the Cl$^-$ions. In order to give a partial differential equation for $\tau_{\mathcal{R}_1,\mathcal{R}_2}^{(\theta,\lambda)}$ and $\tau_{\mathcal{R}_2,R_1}^{(\theta,\lambda)}$, it is convenient to define the closed sets

$$\mathcal{P}_2 = \left\{\zeta : \{z^{(1)} \geq \beta\} \cup \{z^{(2)} \geq \beta\} \cup \cdots \cup \{z^{(N/2)} \geq \beta\}\right\},$$

$$\mathcal{P}_1 = \left\{\zeta : \{z^{(N/2+1)} \leq \alpha\} \cup \{z^{(N/2+2)} \leq \alpha\} \cup \cdots \cup \{z^{(2N)} \leq \alpha\}\right\}. \quad (4.40)$$

Then it is clear that $\zeta_t \in \mathcal{P}_2$ is equivalent to $\max\left(z_t^{(1)}, z_t^{(2)}, \ldots, z_t^{(N/2)}\right) \geq \beta$ since either expression implies that at least one ion has crossed from $\mathcal{R}_1$ to $\mathcal{R}_2$. Similarly $\zeta_t \in \mathcal{P}_1$ is equivalent to $\min\left(z_t^{(N/2+1)}, z_t^{(N/2+2)}, \ldots, z_t^{(2N)}\right) \leq \alpha$. Thus t_β and t_α defined in system 4.39 can be expressed as $t_\beta = \inf\{t : \zeta_t \in \mathcal{P}_2\}$, $t_\alpha = \inf\{t : \zeta_t \in \mathcal{P}_1\}$. Hence system 4.39 is equivalent to

$$\tau_{\mathcal{R}_1,\mathcal{R}_2}^{(\theta,\lambda)} = \mathbf{E}\{\inf\{t : \zeta_t \in \mathcal{P}_2\}\}, \quad \tau_{\mathcal{R}_2,R_1}^{(\theta,\lambda)} = \mathbf{E}\{\inf\{t : \zeta_t \in \mathcal{P}_1\}\}. \quad (4.41)$$

In a gramicidin-Achannel, typically $\tau_{\mathcal{R}_2,R_1}^{(\theta,\lambda)}$ is much larger compared to $\tau_{\mathcal{R}_1,\mathcal{R}_2}^{(\theta,\lambda)}$.

In terms of the mean first-passage times $\tau_{\mathcal{R}_1,\mathcal{R}_2}^{(\theta,\lambda)}, \tau_{\mathcal{R}_2,R_1}^{(\theta,\lambda)}$ defined in equations 4.39 and 4.41, the mean current flowing from $\mathcal{R}_1$ via the gramicidin-Aion channel $\mathcal{C}$ into $\mathcal{R}_2$ is defined as

$$I^{(\theta,\lambda)} = q^+\left(\frac{1}{\tau_{\mathcal{R}_1,\mathcal{R}_2}^{(\theta,\lambda)}} - \frac{1}{\tau_{\mathcal{R}_2,R_1}^{(\theta,\lambda)}}\right). \quad (4.42)$$

The following result adapted from (Gihman and Skorohod, 1972, p. 306) shows that $\tau_{\mathcal{R}_1,\mathcal{R}_2}^{(\theta,\lambda)}, \tau_{\mathcal{R}_2,R_1}^{(\theta,\lambda)}$ satisfy a boundary-valued partial differential equation.

Theorem 4.4

The mean first-passage times $\tau_{\mathcal{R}_1,\mathcal{R}_2}^{(\theta,\lambda)}$ and $\tau_{\mathcal{R}_2,\mathcal{R}_1}^{(\theta,\lambda)}$ in 4.42 are obtained as

$$\tau_{\mathcal{R}_1,\mathcal{R}_2}^{(\theta,\lambda)} = \int_{\Xi_{\mathcal{R}_1}} \tau_{\mathcal{R}_1,\mathcal{R}_2}^{(\theta,\lambda)}(\zeta) \pi_\infty^{(\theta,\lambda)}(\zeta) d\zeta, \tag{4.43}$$

$$\tau_{\mathcal{R}_2,\mathcal{R}_1}^{(\theta,\lambda)} = \int_{\Xi_{\mathcal{R}_2}} \tau_{\mathcal{R}_2,\mathcal{R}_1}^{(\theta,\lambda)}(\zeta) \pi_\infty^{(\theta,\lambda)}(\zeta) d\zeta, \tag{4.44}$$

where

$$\tau_{\mathcal{R}_1,\mathcal{R}_2}^{(\theta,\lambda)}(\zeta) \mathbf{E}\left\{\inf\{t : \zeta_t \in \mathcal{P}_2 | \zeta_0 = \zeta\}\right\}, \tag{4.45}$$

$$\tau_{\mathcal{R}_2,\mathcal{R}_1}^{(\theta,\lambda)}(\zeta) \mathbf{E}\left\{\inf\{t : \zeta_t \in \mathcal{P}_1 | \zeta_0 = \zeta\}\right\}. \tag{4.46}$$

Here $\tau_{\mathcal{R}_1,\mathcal{R}_2}^{(\theta,\lambda)}(\zeta)$ and $\tau_{\mathcal{R}_2,\mathcal{R}_1}^{(\theta,\lambda)}(\zeta)$ satisfy the following boundary-valued partial differ-

boundary-valued
PDE for mean
first passage time

ential equations

$$\mathcal{L}(\tau_{\mathcal{R}_1,\mathcal{R}_2}^{(\theta,\lambda)}(\zeta)) = -1 \quad \zeta \notin \mathcal{P}_2, \quad \tau_{\mathcal{R}_1,\mathcal{R}_2}^{(\theta,\lambda)}(\zeta) = 0 \quad \zeta \in \mathcal{P}_2,$$

$$\mathcal{L}(\tau_{\mathcal{R}_2,\mathcal{R}_1}^{(\theta,\lambda)}(\zeta)) = -1 \quad \zeta \notin \mathcal{P}_1, \quad \tau_{\mathcal{R}_2,\mathcal{R}_1}^{(\theta,\lambda)}(\zeta) = 0 \quad \zeta \in \mathcal{P}_1, \tag{4.47}$$

where $\mathcal{L}$ denotes the backward operator defined in equation 4.34. Furthermore, $\tau_{\mathcal{R}_1,\mathcal{R}_2}^{(\theta,\lambda)}$ and $\tau_{\mathcal{R}_2,\mathcal{R}_1}^{(\theta,\lambda)}$ are finite.

The proof of the equations 4.47 directly follows from corollary 1, p. 306 in Gihman and Skorohod (1972), which shows that the mean first-passage time from any point ζ to a closed set $\mathcal{P}_2$ satisfies equations 4.47. The proof that $\tau_{\mathcal{R}_1,\mathcal{R}_2}^{(\theta,\lambda)}$ and $\tau_{\mathcal{R}_2,\mathcal{R}_1}^{(\theta,\lambda)}$ are finite follows directly from p. 145 of Friedman (1975).

Remark: Equation 4.42 specifies the mean current as the charge per mean time it takes for an ion to cross the ion channel. Instead of equation 4.42, an alternative definition of the mean current is the expected rate of charge across the ion channel, i.e.,

$$\tilde{I}(\theta,\lambda) = q^+\left(\mu_{\mathcal{R}_1,\mathcal{R}_2}^{(\theta,\lambda)} - \mu_{\mathcal{R}_2,\mathcal{R}_1}^{(\theta,\lambda)}\right), \tag{4.48}$$

where with t_α and t_β defined in equation 4.39, the mean rates $\mu_{\mathcal{R}_1,\mathcal{R}_2}^{(\theta,\lambda)}$ and $\mu_{\mathcal{R}_2,\mathcal{R}_1}^{(\theta,\lambda)}$ are defined as

$$\mu_{\mathcal{R}_1,\mathcal{R}_2}^{(\theta,\lambda)} \triangleq \mathbf{E}\left\{\frac{1}{t_\beta}\right\}, \quad \mu_{\mathcal{R}_2,\mathcal{R}_1}^{(\theta,\lambda)} \triangleq \mathbf{E}\left\{\frac{1}{t_\alpha}\right\}. \tag{4.49}$$

It is important to note that the two definitions of current—namely $I^{(\theta,\lambda)}$ in 4.42 and $\tilde{I}(\theta,\lambda)$ in 4.48 are not equivalent, since $\mathbf{E}\{1/t_\beta\} \neq 1/\mathbf{E}\{t_\beta\}$. Similar to the proof of theorem 4.4, partial differential equations can be obtained for $\mu_{\mathcal{R}_1,\mathcal{R}_2}^{(\theta,\lambda)}$ and $\mu_{\mathcal{R}_2,\mathcal{R}_1}^{(\theta,\lambda)}$—however, the resulting boundary conditions are much more complex than equations 4.47.

4.5.2   Brownian Dynamics Simulation for Estimation of Ion Channel Current

It is not possible to obtain explicit closed-form expressions for the mean first passage times $\tau_{\mathcal{R}_2,R_1}^{(\theta,\lambda)}$ and $\tau_{\mathcal{R}_2,R_1}^{(\theta,\lambda)}$ and hence the current $I^{(\theta,\lambda)}$ in equation 4.42. The aim of BD simulation is to obtain estimates of these quantities by directly simulating the stochastic dynamical system 4.27. In this subsection we show that the current estimates $\hat{I}^{(\theta,\lambda)}(L)$ (defined below) obtained from an L-iteration BD simulation are statistically consistent, i.e., $\lim_{L\to\infty} \hat{I}^{(\theta,\lambda)}(L) = I^{(\theta,\lambda)}$ almost surely.

Due to the applied external potential $\Phi_\lambda^{\text{ext}}$ (see equation 4.29), ions drift from reservoir $\mathcal{R}_1$ via the ion channel $\mathcal{C}$ to the reservoir $\mathcal{R}_2$ thus generating an ion channel current. In order to construct an estimate for the current flowing from $\mathcal{R}_1$ to $\mathcal{R}_2$ in the BD simulation, we need to count the number of upcrossings of ions (i.e., the number of times ions cross from $\mathcal{R}_1$ to $\mathcal{R}_2$ across the region $\mathcal{C}$) and downcrossings (i.e., the number of times ions cross from $\mathcal{R}_2$ to $\mathcal{R}_1$ across the region $\mathcal{C}$). Recall from fig. 4.3 that $z = \alpha = -12.5\text{Å}$ denotes the boundary between $\mathcal{R}_1$ and $\mathcal{C}$, and $z = \beta = 12.5\text{Å}$ denotes the boundary between $\mathcal{R}_2$ and $\mathcal{C}$.

Time Discretization of Ion Dynamics   To implement the BD simulation algorithm described below on a digital computer, it is necessary to discretize the continuous-time dynamics (see e.g. 4.27) of the $2N$ ions. The BD simulation algorithm typically uses a sampling interval of $\Delta = 10^{-15}$, i.e., 1 femtosecond for this time discretization, and propagates the $2N$ ions over a total time period of $T = 10^{-4}$ seconds. The time discretization proceeds as follows: Consider a regular partition $0 = t_0 < t_1 < \cdots < t_{k-1} < t_k < \cdots < T$ with discretization interval $\Delta = t_k - t_{k-1} = 10^{-15}$ seconds. There are several possible methods for time discretization of the stochastic differential equation 4.27; see Kloeden and Platen (1992) for a detailed exposition. Here we briefly present a zero-order hold and first-order hold approximation. The first-order hold approximation was derived by van Gunsteren et al. (1981).

It is well known (Wong and Hajek, 1985) that over the time interval $[t_k, t_{k+1})$, the solution of equation 4.27 satisfies

$$\zeta_{t_{k+1}} = e^{\mathbf{A}\Delta}\zeta_{t_k} + \int_{t_k}^{t_{k+1}} e^{\mathbf{A}(t_{k+1}-\tau)}\mathbf{f}_{\theta,\lambda}(\zeta_\tau)d\tau + \int_{t_k}^{t_{k+1}} e^{A(t_{k+1}-\tau)}\Sigma^{1/2}d\mathbf{w}_\tau. \quad (4.50)$$

In the zero-order hold approximation, $\mathbf{f}_{\theta,\lambda}(\zeta_\tau)$ is assumed to be approximately constant over the short interval $[t_k, t_{k+1})$ and is set to the constant $\mathbf{f}_{\theta,\lambda}(\zeta_{t_k})$ in equation 4.50. This yields

$$\zeta_{t_{k+1}} = e^{\mathbf{A}\Delta}\zeta_{t_k} + \int_{t_k}^{t_{k+1}} e^{\mathbf{A}(t_{k+1}-\tau)}\mathbf{f}_{\theta,\lambda}(\zeta_{t_k})d\tau + \int_{t_k}^{t_{k+1}} e^{A(t_{k+1}-\tau)}\Sigma^{1/2}d\mathbf{w}_\tau. \quad (4.51)$$

In the first-order hold, the following approximation is used in equation 4.50:

$$\mathbf{f}_{\theta,\lambda}(\zeta_\tau) \approx f(\zeta_{t_k}) + (\tau - t_k)\frac{\partial \mathbf{f}_{\theta,\lambda}(\zeta_t)}{\partial t}.$$

In van Gunsteren et al. (1981), the derivative above is approximated by

$$\frac{\partial \mathbf{f}_{\theta,\lambda}(\zeta_t)}{\partial t} \approx \frac{\mathbf{f}_{\theta,\lambda}(\zeta_{t_k}) - \mathbf{f}_{\theta,\lambda}(\zeta_{t_{k-1}})}{\Delta}.$$

Thus the first-order hold approximation of van Gunsteren et al. (1981) yields

$$\zeta_{t_{k+1}} = e^{\mathbf{A}\Delta}\zeta_{t_k} + \int_{t_k}^{t_{k+1}} e^{\mathbf{A}(t_{k+1}-\tau)}\mathbf{f}_{\theta,\lambda}(\zeta_{t_k})d\tau$$

$$+ \int_{t_k}^{t_{k+1}} e^{\mathbf{A}(t_{k+1}-\tau)}(\tau - t_k)\frac{\mathbf{f}_{\theta,\lambda}(\zeta_{t_k}) - \mathbf{f}_{\theta,\lambda}(\zeta_{t_{k-1}})}{\Delta}d\tau + \int_{t_k}^{t_{k+1}} e^{A(t_{k+1}-\tau)}\Sigma^{1/2}d\mathbf{w}_{\tau}.$$

$$(4.52)$$

Let $k = 0, 1, \ldots$ denote discrete time where k corresponds to time t_k. Note that the last integral above is merely a discrete-time Gauss-Markov process, which we will denote as $\mathbf{w}_k^{(d)}$. Moreover, since the first block element of $\mathbf{w}$ in (4.26) is $\mathbf{0}$,

$$\mathbf{w}_k^{(d)} = \begin{bmatrix} \mathbf{0}_{6N\times 1} \\ \bar{\mathbf{w}}_k^{(d)} \end{bmatrix}. \tag{4.53}$$

where the $6N$ dimensional vector $\bar{\mathbf{w}}_k^{(d)}$ denotes the nonzero components of $\mathbf{w}_k^{(d)}$.

We now elaborate on the zero-order hold model. Next, due to the simple structure of $\mathbf{A}$ in equation 4.28, the matrix exponentials

$$\Gamma \overset{\triangle}{=} e^{\mathbf{A}\Delta}, \quad \mathbf{B} \overset{\triangle}{=} \int_{t_k}^{t_{k+1}} e^{\mathbf{A}(t_{k+1}-\tau)}d\tau \tag{4.54}$$

in equation 4.50 can be explicitly computed as

$$\Gamma = e^{\mathbf{A}\Delta} = \begin{bmatrix} \mathbf{I}_{6N\times 6N} & \mathbf{L} \\ \mathbf{0}_{6N\times 6N} & e^{\mathbf{D}\Delta} \end{bmatrix}, \quad \mathbf{B} = \int_{t_k}^{t_{k+1}} e^{\mathbf{A}(t_{k+1}-\tau)}d\tau = \begin{bmatrix} \mathbf{I}_{6N\times 6N} & \mathbf{L} \\ \mathbf{0}_{6N\times 6N} & e^{\mathbf{D}\Delta} \end{bmatrix}$$

$$\text{where} \quad \mathbf{D} = \begin{bmatrix} -\gamma^+\mathbf{I}_{3N\times 3N} & \mathbf{0}_{3N\times 3N} \\ \mathbf{0}_{3N\times 3N} & -\gamma^-\mathbf{I}_{3N\times 3N} \end{bmatrix}, \quad \mathbf{L} = \mathbf{D}^{-1}\left(e^{\mathbf{D}\Delta} - \mathbf{I}\right). \tag{4.55}$$

Then the above update for $\zeta_{t_{k+1}}$ in discrete-time notation reads:

$$\zeta_{k+1}^{(d)} = \Gamma\zeta_k^{(d)} + \mathbf{B}\,\mathbf{f}_{\theta,\lambda}(\zeta_k^{(d)}) + \mathbf{w}_k^{(d)}. \tag{4.56}$$

Expanding this out in terms of $\mathbf{X}_k$ and $\mathbf{V}_k$, we have the following discrete time dynamics for the positions and velocities of the $2N$ ions:

$$\mathbf{X}_{k+1} = \mathbf{X}_k + \mathbf{L}\mathbf{V}_k + \mathbf{D}^{-1}(\mathbf{L} - \Delta\mathbf{I})\mathbf{F}_{\theta,\lambda}(\mathbf{X}_k) \tag{4.57}$$

$$\mathbf{V}_{k+1} = e^{\mathbf{D}\Delta}\mathbf{V}_k + \mathbf{L}\mathbf{F}_{\theta,\lambda}(\mathbf{X}_k) + \bar{\mathbf{w}}_k^{(d)}, \tag{4.58}$$

where $\bar{\mathbf{w}}_k^{(d)}$ is a $6N$ dimensional discrete-time Gauss Markov process.

Brownian Dynamics Simulation Algorithm In the BD simulation Algorithm 4.2 below, we use the following notation:

The algorithm runs for L iterations where L is user specified. Each iteration l, $l = 1, 2, \ldots, L$, runs for a random number of discrete-time steps until an ion crosses the channel. We denote these random times as $\hat{\tau}_{\mathcal{R}_1, \mathcal{R}_2}^{(l)}$ if the ion has crossed from $\mathcal{R}_1$ to R_2 and $\hat{\tau}_{\mathcal{R}_2, \mathcal{R}_1}^{(l)}$ if the ion has crossed from $\mathcal{R}_2$ to R_1. Thus

$$\hat{\tau}_{\mathcal{R}_1, \mathcal{R}_2}^{(l)} = \min\{k : \zeta_k^{(d)} \in \mathcal{P}_2\}, \quad \hat{\tau}_{\mathcal{R}_2, \mathcal{R}_1}^{(l)} = \min\{k : \zeta_k^{(d)} \in \mathcal{P}_1\}.$$

The positive ions $\{1, 2, \ldots, N/2\}$ are in $\mathcal{R}_1$ at steady state $\pi_\infty^{(\theta, \lambda)}$, and the positive ions $\{N/2 + 1, \ldots, 2N\}$ are in $\mathcal{R}_2$ at steady state.

$L_{\mathcal{R}_1, \mathcal{R}_2}$ is a counter that counts how many Na$^+$ions have crossed from $\mathcal{R}_1$ to $\mathcal{R}_2$, and $L_{\mathcal{R}_2, \mathcal{R}_1}$ counts how many Na$^+$ions have crossed from $\mathcal{R}_2$ to $\mathcal{R}_1$. Note that $L_{\mathcal{R}_1, \mathcal{R}_2} + L_{\mathcal{R}_2, \mathcal{R}_1} = L$. We only consider passage of positive Na$^+$ions $i = 1, \ldots, N$ across the ion channel since in a gramicidin-Achannel the ion channel current is caused only by Na$^+$ions.

Algorithm 4.2
```
Brownian Dynamics Simulation Algorithm (for Fixed θ and λ)
```

Input parameters θ for PMF and λ for applied external potential.

For $l = 1$ to L iterations:

    Step 1: Initialize all $2N$ ions according to stationary distribution $\pi_\infty^{(\theta, \lambda)}$ defined in equation 4.36.
    Open ion channel at discrete time $k = 0$ and set $k = 1$.

    Step 2: Propagate all $2N$ ions according to the time-discretized Brownian dynamical system 4.56 until time k^* at which an ion crosses the channel.

        $*$ If ion crossed ion channel from $\mathcal{R}_1$ to $\mathcal{R}_2$, i.e., for any ion $i^* \in \{1, 2, \ldots, N/2\}$, $z_{k^*}^{(i^*)} \geq \beta$, then set $\hat{\tau}_{\mathcal{R}_1, \mathcal{R}_2}^{(l)} = k^*$.
        Update number of crossings from $\mathcal{R}_1$ to $\mathcal{R}_2$: $L_{\mathcal{R}_1, \mathcal{R}_2} = L_{\mathcal{R}_1, \mathcal{R}_2} + 1$.

        $*$ If ion crossed ion channel from $\mathcal{R}_2$ to $\mathcal{R}_1$, i.e., for any ion $i^* \in \{N/2 + 1, \ldots, N\}$, $z_{k^*}^{(i)} \leq \alpha$ then set $\hat{\tau}_{\mathcal{R}_2, \mathcal{R}_1}^{(l)} = k^*$.
        Update number of crossings from $\mathcal{R}_2$ to $\mathcal{R}_1$: $L_{\mathcal{R}_2, \mathcal{R}_1} = L_{\mathcal{R}_2, \mathcal{R}_1} + 1$.

    Step 3: End for loop.

Compute the mean first passage time and mean current estimate after L iterations as

$$\hat{\tau}_{\mathcal{R}_1, \mathcal{R}_2}^{(\theta, \lambda)}(L) = \frac{1}{L_{\mathcal{R}_1, \mathcal{R}_2}} \sum_{l=1}^{L_{\mathcal{R}_1, \mathcal{R}_2}} \hat{\tau}_{\mathcal{R}_1, \mathcal{R}_2}^{(l)}, \quad \hat{\tau}_{\mathcal{R}_2, \mathcal{R}_1}^{(\theta, \lambda)}(L) = \frac{1}{L_{\mathcal{R}_2, \mathcal{R}_1}} \sum_{l=1}^{L_{\mathcal{R}_2, \mathcal{R}_1}} \hat{\tau}_{\mathcal{R}_2, \mathcal{R}_1}^{(l)}, \quad (4.59)$$

$$\hat{I}^{(\theta, \lambda)}(L) = q^+ \left(\frac{1}{\hat{\tau}_{\mathcal{R}_1, \mathcal{R}_2}^{(\theta, \lambda)}(L)} - \frac{1}{\hat{\tau}_{\mathcal{R}_2, \mathcal{R}_1}^{(\theta, \lambda)}(L)} \right). \quad (4.60)$$

The following result shows that the estimated current $\hat{I}^{(\theta, \lambda)}(L)$ obtained from a BD simulation run over L iterations is strongly consistent.

Theorem 4.5
For fixed PMF $\theta \in \Theta$ and applied external potential $\lambda \in \Lambda$, the ion channel current estimate $\hat{I}^{(\theta,\lambda)}(L)$ obtained from the BD simulation algorithm 4.2 over L iterations is strongly consistent, i.e.,

$$\lim_{L \to \infty} \hat{I}^{(\theta,\lambda)}(L) = I^{(\theta,\lambda)} \quad \text{w.p.1} \tag{4.61}$$

where $I^{(\theta,\lambda)}$ is the mean current defined in equation 4.42.

Proof Since by construction in algorithm 4.2, each of the L iterations is statistically independent, and $\mathbf{E}\left\{\hat{\tau}^{(l)}_{\mathcal{R}_1,\mathcal{R}_2}\right\}$, $\mathbf{E}\left\{\hat{\tau}^{(l)}_{\mathcal{R}_2,\mathcal{R}_1}\right\}$ are finite (see theorem 4.4), by Kolmogorov's strong law of large numbers

$$\lim_{L \to \infty} \hat{\tau}^{(\theta,\lambda)}_{\mathcal{R}_1,\mathcal{R}_2}(L) = \tau^{(\theta,\lambda)}_{\mathcal{R}_1,\mathcal{R}_2}, \quad \lim_{L \to \infty} \hat{\tau}^{(\theta,\lambda)}_{\mathcal{R}_2,R_1}(L) = \tau^{(\theta,\lambda)}_{\mathcal{R}_2,R_1} \quad \text{w.p.1.}$$

Thus $q^+ \left(\dfrac{1}{\hat{\tau}^{(\theta,\lambda)}_{\mathcal{R}_1,\mathcal{R}_2}(L)} - \dfrac{1}{\hat{\tau}^{(\theta,\lambda)}_{\mathcal{R}_2,R_1}(L)} \right) \to I^{(\theta,\lambda)}$ w.p.1 as $L \to \infty$. ∎

Remark: Instead of equation 4.42, if the mean rate definition in equation 4.48 is used for the mean current $\tilde{I}(\theta,\lambda)$, then the following minor modification of algorithm 4.2 yields consistent estimates of $\tilde{I}(\theta,\lambda)$. Instead of equations 4.59 and 4.60, use

$$\hat{\mu}^{(\theta,\lambda)}_{\mathcal{R}_1,\mathcal{R}_2}(L) = \frac{1}{L_{\mathcal{R}_1,\mathcal{R}_2}} \sum_{l=1}^{L_{\mathcal{R}_1,\mathcal{R}_2}} \frac{1}{\hat{\tau}^{(l)}_{\mathcal{R}_1,\mathcal{R}_2}}, \quad \hat{\mu}^{(\theta,\lambda)}_{\mathcal{R}_2,R_1}(L) = \frac{1}{L_{\mathcal{R}_2,R_1}} \sum_{l=1}^{L_{\mathcal{R}_2,R_1}} \frac{1}{\hat{\tau}^{(l)}_{\mathcal{R}_2,R_1}}, \tag{4.62}$$

$$\hat{I}^{(\theta,\lambda)}(L) = q^+ \left(\hat{\mu}^{(\theta,\lambda)}_{\mathcal{R}_1,\mathcal{R}_2}(L) - \hat{\mu}^{(\theta,\lambda)}_{\mathcal{R}_2,R_1}(L) \right). \tag{4.63}$$

Then a virtually identical proof to theorem 4.5 yields that $\hat{I}^{(\theta,\lambda)}(L) \to \tilde{I}(\theta,\lambda)$ w.p.1, as $L \to \infty$.

Implementation Details and Variations of Algorithm 4.2 In algorithm 4.2, the procedure of resetting all ions to $\pi^{(\theta,\lambda)}_\infty$ in step 1 when any ion crosses the channel can be expressed mathematically as

$$\zeta^{(d)}_{k+1} = \mathbf{1}_{\zeta^{(d)}_k \notin \mathcal{P}_2 \cup \mathcal{P}_1} [\mathbf{f}^{(d)}_{\theta,\lambda}(\zeta^{(d)}_k) + \mathbf{w}^{(d)}_k] + \mathbf{1}_{\zeta^{(d)}_k \in \mathcal{P}_2 \cup \mathcal{P}_1} \zeta^{(d)}_0, \quad \zeta^{(d)}_0 \sim \pi^{(\theta,\lambda)}_\infty, \tag{4.64}$$

where $\mathcal{P}_1$, $\mathcal{P}_2$ are defined in equation 4.40. The following approximations of algorithm 4.2 can be used in actual numerical simulations.

■ Instead of steps 2a and 2b, only remove the crossed ion denoted as i^* and put it back in its reservoir with probability $\pi^{(\theta,\lambda),\mathcal{R}_1}_\infty$ or $\pi^{(\theta,\lambda),\mathcal{R}_2}_\infty$ (eqs. 4.3 and 4.38) depending on whether it originated from $\mathcal{R}_1$ or $\mathcal{R}_2$. The other particles are not reset. With $\zeta^{(d)i^*}_k$ denoting the position and velocity of the crossed ion and $\zeta^{(\bar{d})}_k$ denoting the positions and velocities of the remaining $2N - 1$ ions, mathematically this is equivalent to replacing equation 4.64 by

$$\zeta_{k+1}^{(d)} = \mathbf{1}_{\zeta_k^{(d)} \notin \mathcal{P}_2 \cup \mathcal{P}_1}[\mathbf{f}_{\theta,\lambda}^{(d)}(\zeta_k^{(d)}) + \mathbf{w}_k^{(d)}] + \mathbf{1}_{\zeta_k^{(d)} \in \mathcal{P}_2 \cup \mathcal{P}_1}[\mathbf{f}_{\theta,\lambda}^{(d)}(\zeta^{\bar{(d)}}{}_k, \zeta^{(d)}{}_k^{i^*}) + \mathbf{w}_k^{(d)}],$$

$$(4.65)$$

where $\zeta^{(d)}{}_k^{i^*} \sim \pi_\infty^{(\theta,\lambda),\mathcal{R}_1}$ if $i \in \{1,\dots,N/2\}$, and $\zeta^{(d)}{}_k^{i^*} \sim \pi_\infty^{(\theta,\lambda),\mathcal{R}_2}$ if $i \in \{N+1,\dots,3N/2\}$.

- As in the above approximation (eq. 4.65), except that $\zeta^{(d)}{}_k^{i^*}$ is replaced according to an uniform distribution.

The above approximations are justified for three reasons:

1. Only one ion can be inside the gramicidin-Achannel $\mathcal{C}$ at any time instant. When this happens the ion channel behaves as though it is closed. Then the probabilistic construction of step 1 in section 4.5.1 applies.

2. The probability density functions of the remaining $2N-1$ ions converge rapidly to their stationary distribution and forget their initial distribution exponentially fast. This is due to the exponential ergodicity theorem 4.2. In comparison the time taken for an ion to cross the channel is significantly larger. As a result the removal of crossed particles and their replacement in the reservoir happens extremely infrequently. Between such events the probability density functions of the ions rapidly converge to their stationary distribution.

3. If an ion enters the channel $\mathcal{C}$, then the change in concentration of ions in the reservoir is of magnitude $1/N$. This is negligible if N is chosen sufficiently large.

4.6   Adaptive Brownian Dynamics Mesoscopic Simulation of Ion Channel

Having given a complete description of the dynamics of individual ions in section 4.4 and the Brownian dynamics algorithm for estimating the ion channel current, in this section we describe how the Brownian dynamics algorithms can be adaptively controlled to determine the molecular structure of the ion channel.

We will estimate the PMF profile U_θ parameterized by θ, by computing the θ that optimizes the fit between the mean current $I^{(\theta,\lambda)}$ (defined above in eq. 4.42) and the experimentally observed current $y(\lambda)$ defined below. Unfortunately, it is impossible to explicitly compute $I^{(\theta,\lambda)}$ from equation 4.42. For this reason we resort to a *stochastic optimization problem formulation* below, where consistent estimates of $I^{(\theta,\lambda)}$ are obtained via the Brownian dynamics simulation algorithm 4.2. The main algorithm presented in this section is the adaptive Brownian dynamics simulation algorithm (algorithm 4.3) which solves the stochastic optimization problem and yields the optimal PMF. We have showed the effective surface charge density along the protein of the inside surface of the ion channel from the PMF (Krishnamurthy and Chung, b).

4.6.1 Formulation of PMF Estimation as Stochastic Optimization Problem

The stochastic optimization problem formulation for determining the optimal PMF estimate comprises the following four ingredients:

Experimentally Observed Ion Channel Current $y(\lambda)$ Neurobiologists use the patch-clamp experimental setup to obtain experimental measurements of the current flowing through a single ion channel. Typically the measured discrete-time (digitized) current from a patch-clamp experiment is obtained by sampling the continuous-time observed current at 10 kHz (i.e., 0.1 millisecond intervals). Note that this is at a much slower timescale than the dynamics of individual ions which move around at a femtosecond timescale. Such patch clamping was widely regarded as a breakthrough in the 1970s for understanding the dynamics of ion channels at a millisecond timescale.

From patch-clamp experimental data, neurobiologists can obtain an accurate measurement of the actual current $y(\lambda)$ flowing through a gramicidin-Aion channel for various external applied potentials $\lambda \in \Lambda$. For example, as shown in Chung et al. (1991), the resulting discrete time series can be modeled as HMM. Then by using a HMM maximum likelihood estimator (Chung et al., 1991; James et al., 1996), accurate estimates of the open current level $y(\lambda)$ of the ion channel can be computed. Neurobiologists typically plot the relationship between the experimentally determined current $y(\lambda)$ vs. applied voltage λ on an IV curve—such curves provide a unique signature for an ion channel. For our purposes $y(\lambda)$ denotes the *true* (real-world) channel current.

Loss Function Let $n = 1, 2, \ldots$ denote the batch member. For fixed applied field $\lambda \in \Lambda$, consider at batch n, running the BD simulation algorithm 4.2, resulting in the simulated current $I_n^{(\theta, \lambda)}$. Define the mean square error loss function equation as

$$Q(\theta, \lambda) = \mathbf{E}\left\{ |I_n^{(\theta, \lambda)} - y(\lambda)|^2 \right\}, \tag{4.66}$$

where $Q(\theta, \lambda)_n = \left(I_n^{(\theta, \lambda)} - y(\lambda) \right)^2$. Define the total loss function obtained by adding the mean square error over all the applied fields $\lambda \in \Lambda$ on the IV curve as

$$Q(\theta) = \sum_{\lambda \in \Lambda} Q(\theta, \lambda). \tag{4.67}$$

The optimal PMF U_{θ^*} is determined by the parameter θ^* that best fits the mean current $I^{(\theta, \lambda)}$ to the experimentally determined IV curve of a gramicidin-Achannel, i.e..

$$\theta^* = \arg \min_{\theta \in \Theta} Q(\theta). \tag{4.68}$$

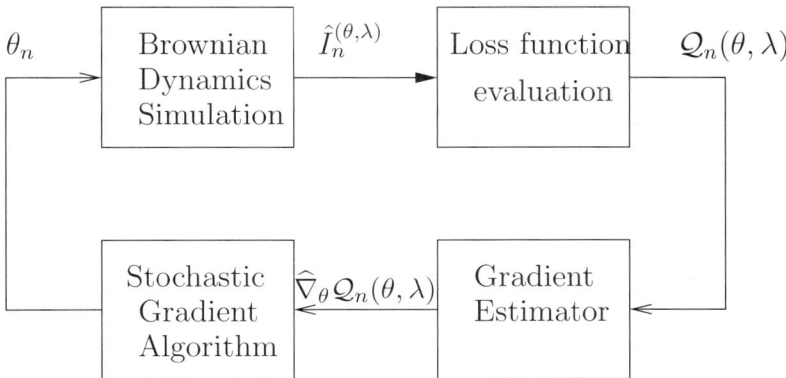

Figure 4.5   Adaptive Brownian dynamics simulation for estimating PMF.

Let Θ^* denote the set of local minima whose elements θ^* satisfy the second-order sufficient conditions for being a local minimum:

$$\widehat{\nabla}_\theta \mathcal{Q}(\theta) = 0, \quad \widehat{\nabla}_\theta^2 \mathcal{Q}(\theta) > 0, \tag{4.69}$$

where the notation $\nabla^2 \mathcal{Q}(\theta) > 0$ means that it is a positive definite matrix.

However, the deterministic optimization (eqs. 4.66,4.68) cannot be directly carried out since it is not possible to obtain explicit closed-form expressions for 4.66—this is because the partial differential equation 4.47 for the mean first passage times $\tau_{\mathcal{R}_2, R_1}^{(\theta, \lambda)}$ and $\tau_{\mathcal{R}_2, R_1}^{(\theta, \lambda)}$ cannot be solved explicitly. This motivates us to formulate the estimation of the PMF as a stochastic optimization problem.

4.6.2   Stochastic Gradient Algorithms for Estimating Potential of Mean Force (PMF) and the Need for Gradient Estimation

We now give a complete description of the adaptive Brownian dynamics simulation algorithm for computing the optimal PMF estimate U_{θ^*}. The algorithm is schematically depicted in fig. 4.5. Recall that $n = 0, 1, \cdots$, denotes batch number.

Algorithm 4.3
Adaptive Brownian Dynamics Simulation Algorithm for Estimating PMF

Step 0: Set batch index $n = 0$, and initialize $\theta_0 \in \Theta$.

Step 1 Evaluation of loss function: At batch n, evaluate loss function $\mathcal{Q}_n(\theta_n, \lambda)$ for each external potential $\lambda \in \Lambda$ according to equation 4.66. This uses one independent BD simulation (algorithm 4.2) for each λ.

Step 2 Gradient estimation: Compute gradient estimate $\widehat{\nabla}_\theta \mathcal{Q}_n(\theta_n, \lambda)$ either as a finite difference (see eq. 4.72 below), or according to the SPSA algorithm (eq. 4.73) below.

Step 3 Stochastic approximation algorithm: Update PMF estimate:

$$\theta_{n+1} = \theta_n - \epsilon_{n+1} \sum_{\lambda \in \Lambda} \widehat{\nabla}_\theta \mathcal{Q}_n(\theta_n, \lambda) \tag{4.70}$$

where ϵ_n denotes a decreasing step size (see discussion below for choice of step size).

Step 4: Set n to $n + 1$ and go to step 1.

A crucial aspect of the above algorithm is the gradient estimation step 2. In this step, an estimate $\widehat{\nabla}_\theta \mathcal{Q}_n(\theta, \lambda)$ of the gradient $\nabla_\theta \mathcal{Q}_n(\theta, \lambda)$ is computed. This gradient estimate is then fed to the stochastic gradient algorithm (step 3) which updates the PMF. Note that since the explicit dependence of $\mathcal{Q}_n(\theta, \lambda)$ on θ is not known, it is not possible to compute $\nabla_\theta \mathcal{Q}_n(\theta, \lambda)$. Thus we have to resort to gradient estimation, e.g., the finite difference estimators described below or a more sophisticated algorithm such as IPA (infinitesimal perturbation analysis).

The step size ϵ_n is typically chosen as

$$\epsilon_n = \epsilon / (n + 1 + R)^\kappa, \tag{4.71}$$

where $0.5 < \kappa \leq 1$ and R is some positive constant. Note that this choice of step size automatically satisfies the condition $\sum_{n=1}^{\infty} \epsilon_n = \infty$, which is required for convergence of algorithm 4.3.

Kiefer-Wolfowitz Finite Difference Gradient Estimator An obvious gradient estimator is obtained by finite differences as follows: Suppose θ is a p dimensional vector. Let $\mathbf{e}_1, \mathbf{e}_2, \ldots, \mathbf{e}_p$ denote p-dimensional unit vectors, where e_i is a unit vector with 1 in the ith position and zeros elsewhere. Then the two-sided finite difference gradient estimator is

$$\widehat{\nabla}_\theta \mathcal{Q}_n(\theta, \lambda) = \begin{bmatrix} \dfrac{\mathcal{Q}_n(\theta_n + \mu_n \mathbf{e}_1, \lambda) - \mathcal{Q}_n(\theta_n - \mu_n \mathbf{e}_1, \lambda)}{2\mu_n} \\ \dfrac{\mathcal{Q}_n(\theta_n + \mu_n \mathbf{e}_2, \lambda) - \mathcal{Q}_n(\theta_n - \mu_n \mathbf{e}_2, \lambda)}{2\mu_n} \\ \vdots \\ \dfrac{\mathcal{Q}_n(\theta_n + \mu_n \mathbf{e}_p, \lambda) - \mathcal{Q}_n(\theta_n - \mu_n \mathbf{e}_p, \lambda)}{2\mu_n} \end{bmatrix}. \tag{4.72}$$

Using equation 4.72 in algorithm 4.3 yields the so-called *Finite difference stochastic gradient algorithm*.

In the above gradient estimator, $\mu_k = \mu / (k+1)^\gamma$, where typically $\gamma < \kappa$ (where κ is defined in eq. 4.71), e.g., $\gamma = 0.101$ and $\kappa = 0.602$.

The main disadvantages of the above finite gradient estimator are twofold. First, the bias of the gradient estimate is $O(\mu_n^2)$, i.e.,

$$\mathbf{E}\left\{\widehat{\nabla}_\theta \mathcal{Q}_n(\theta, \lambda) | \theta_1, \ldots, \theta_n\right\} = O(\mu_n^2).$$

Second, the simulation cost of implementing the above estimator is large. It requires $2p$ BD simulations, since one BD simulation is required to evaluate $\mathcal{Q}_n(\theta_n + \mu_n \mathbf{e}_i, \lambda)$ and one BD simulation is required to evaluate $\mathcal{Q}_n(\theta_n - \mu_n \mathbf{e}_i, \lambda)$ for each $i = 1, 2, \ldots, p$.

Simultaneous Perturbation Stochastic Approximation (SPSA) Algorithm Unlike the Kiefer-Wolfowitz algorithm, the SPSA algorithm (Spall, 2003) is a novel method that picks a single random direction $\mathbf{d}_n$ along which direction the derivative is evaluated at each batch n. Thus the main advantage of SPSA compared to finite difference is that evaluating the gradient estimate $\widehat{\nabla}_\theta \mathcal{Q}_n(\theta, \lambda)$ in SPSA requires only two BD simulations, i.e., the number of evaluations is independent of the dimension p of the parameter vector θ. We refer the reader to Spall (2003) and to the Web site www.jhuapl.edu/SPSA/ for details, variations, and applications of the SPSA algorithm. The SPSA algorithm proceeds as follows: Generate the p-dimensional vector $\mathbf{d}_n$ with random elements $\mathbf{d}_n(i)$, $i = 1, \ldots, p$ simulated as follows:

$$\mathbf{d}_n(i) = \begin{cases} -1 & \text{with probability } 0.5 \\ +1 & \text{with probability } 0.5. \end{cases}$$

Then the SPSA algorithm uses the following gradient estimator together with the stochastic gradient algorithm (eq. 4.70):

$$\widehat{\nabla}_\theta \mathcal{Q}_n(\theta, \lambda) = \begin{bmatrix} \dfrac{\mathcal{Q}_n(\theta_n + \mu_n \mathbf{d}_n, \lambda) - \mathcal{Q}_n(\theta_n - \mu_n \mathbf{d}_n, \lambda)}{2\mu_n \mathbf{d}_n(1)} \\ \dfrac{\mathcal{Q}_n(\theta_n + \mu_n \mathbf{d}_n, \lambda) - \mathcal{Q}_n(\theta_n - \mu_n \mathbf{d}_n, \lambda)}{2\mu_n \mathbf{d}_n(2)} \\ \vdots \\ \dfrac{\mathcal{Q}_n(\theta_n + \mu_n \mathbf{d}_n, \lambda) - \mathcal{Q}_n(\theta_n - \mu_n \mathbf{d}_n, \lambda)}{2\mu_n \mathbf{d}_n(p)} \end{bmatrix}. \tag{4.73}$$

Here μ_k is chosen by a process similar to that of the Kiefer-Wolfowitz algorithm. Despite the substantial computational efficiency of SPSA compared to Kiefer-Wolfowitz, the asymptotic efficiency of SPSA is identical to the Kiefer-Wolfowitz algorithm. Thus SPSA can be viewed as a novel application of randomization in gradient estimation to break the curse of dimensionality. It can be proved that, like the finite gradient scheme, SPSA also has a bias $O(\mu_n^2)$ (Spall, 2003).

Remarks: In the SPSA algorithm above, the elements of $\mathbf{d}_n$ were chosen according to a Bernoulli distribution. In general, it is possible to generate the elements of $\mathbf{d}$ according to other distributions, as long as these distributions are symmetric, zero mean, and have bounded inverse moments; see Spall (2003) for a complete exposition of SPSA.

Convergence of Adaptive Brownian Dynamics Simulation Algorithm 4.3
Here we show that the estimates θ_n generated by algorithm 4.3 (whether using the Kiefer-Wolfowitz or SPSA algorithm) converge to a local minimum of the loss function.

Theorem 4.6
For batch size $L \to \infty$ in algorithm 4.2, the sequence of estimates $\{\theta_n\}$ generated by the controlled Brownian dynamics simulation algorithm 4.3, converge at $n \to \infty$ to a the locally optimal PMF estimate θ^* (defined in eq. 4.68) with probability 1.

Outline of Proof Since by construction of the BD algorithm, for fixed θ, $\mathcal{Q}_n(\theta, \lambda)$ are independent and identically distributed random variables, the proof of the above theorem involves showing strong convergence of a stochastic gradient algorithm with i.i.d. observations—which is quite straightforward. In Kushner and Yin (1997), almost sure convergence of stochastic gradient algorithms for state dependent Markovian noise under general conditions is presented. For the independent and identically distributed case we only need to verify the following condition for convergence.

Condition A.4.11 in section 8.4 of Kushner and Yin (1997) requires uniform integrability of $\widehat{\nabla}_\theta \mathcal{Q}_n(\theta_n, \lambda)$ in equation 4.70. This holds since the discretized version of the passage time $\hat{\tau}_{\mathcal{R}_1, \mathcal{R}_2}^{(\theta,\lambda)}(L) \geq 1$, implying that the estimate $\hat{I}_n^{(\theta,\lambda)}$ from equation 4.60 in algorithm 4.2 is uniformly bounded. Thus the evaluated loss $\mathcal{Q}_n(\theta, \lambda)$ (eq. 4.66) is uniformly bounded. This in turn implies that the finite difference estimate (eq. 4.72) for the Kiefer-Wolfowitz algorithm or equation 4.73 of the SPSA algorithm are uniformly bounded, which implies uniform integrability.

Then theorem 4.3 of Kushner and Yin (1997) implies that the sequence $\{\theta_n\}$ generated by the above controlled BD simulation algorithm 4.3 converges with probability 1 to the fixed points of the following ordinary differential equation:

$$\frac{d\theta}{dt} = -\mathbf{E}\left\{ \sum_{\lambda \in \Lambda} \widehat{\nabla}_\theta \mathcal{Q}_n(\theta, \lambda) \right\} = -\widehat{\nabla}_\theta Q(\theta),$$

namely the set Θ^* of local minima that satisfy the second-order sufficient conditions in equation 4.69.

4.7   Conclusions

This chapter has presented novel signal-processing and stochastic optimization (control) algorithms for two significant problems in ion channels—namely, the gating problem and the permeation problem. For the gating problem we presented novel discrete stochastic optimization algorithms and also a multiarmed bandit formulation for activating ion channels on a biological chip. For the permeation problem, we presented an adaptive controlled Brownian dynamics simulation algorithm for estimating the structure of the ion channel. We refer the reader to Krishnamurthy and Chung (a,b) for further details of the adaptive Brownian dynamics algorithm and convergence proofs.

The underlying theme of this chapter is the idea of *sensor adaptive signal processing* that transcends standard statistical signal processing (which deals with extracting signals from noisy measurements) to address the deeper issue of how to dynamically minimize sensor costs while simultaneously extracting a signal from noisy measurements. As we have seen in this chapter, the resulting problem is a dynamic stochastic optimization problem—whereas all traditional statistical signal-processing problems (such as optimal and adaptive filtering, parameter estimation) are merely static stochastic optimization problems. Furthermore, we have demonstrated the use of sophisticated new algorithms such as stochastic discrete optimization, partially observed Markov decision processes (POMDP), bandit processes, multiparticle Brownian dynamics simulations, and gradient estimation-based stochastic approximation algorithms. These novel methods provide a powerful set of algorithms that will supersede conventional signal-processing tools such as elementary stochastic approximation algorithms (e.g., the LMS algorithm), subspace methods, etc.

In current work we are examining several extensions of the ideas in this chapter, including estimating the shape of ion channels. Finally, it is hoped that this paper will motivate more researchers in the areas of statistical signal processing and stochastic control to apply their expertise to the exciting area of ion channels.

Acknowledgments

The author thanks his collaborator Dr. Shin-Ho Chung of the Australian National University, Canberra, for several useful discussions and for preparing fig. 4.4. This chapter is dedicated to the memory of the author's mother, Mrs. Bani Krishnamurthy, who passed away on November 26, 2004.

5   Spin Diffusion: A New Perspective in Magnetic Resonance Imaging

Timothy R. Field

5.1   Context

In this chapter we outline some emerging ideas relating to the detection of spin populations arising in magnetic resonance imaging (MRI). MRI techniques are already at a mature stage of development, widely used as a research tool and practised in clinical medicine, and provide the primary noninvasive method for studying internal brain structure and activity, with excellent spatial resolution. A lot of attention is typically paid to detecting spatial anomalies in brain tissue, e.g., brain tumors, and in localizing certain areas of the brain that correspond to particular stimuli, e.g., within the motor, visual, and auditory cortices. More recently, functional magnetic resonance imaging (fMRI) techniques have been successfully applied in psychological studies analyzing the temporal response of the brain to simple known stimuli. The possibility of enhancing these techniques to deal with more sophisticated neurological processes, characterized by physiological changes occurring on shorter timescales, provides the motivation for developing real-time imaging techniques, where there is much insight to be gained, e.g., in tracking the auditory system.

The concept of MRI is straightforward and can be briefly summarized as follows (e.g., Brown and Semelka, 2003). The physical phenomenon of magnetic resonance is due to the Zeeman effect, which accounts for the splitting of energy levels in atomic nuclei due to an applied magnetic field. In the presence of a background magnetic field B_0, the majority of protons (hydrogen nuclei) N_+ in a material tend toward their minimum energy spin configuration, with their spin vectors aligned with B_0, in the "spin-up" state (and thus in the minimum energy eigenstate of the quantum mechanical spin Hamiltonian). A smaller number N_- take up the excited state with spin antiparallel to B_0, the "spin-down" state. According to statistical mechanics arguments, the ratio of the spin-up to spin-down populations N_+/N_- is governed by the Bose-Einstein distribution. Thus, the majority of protons are able to absorb available energy and make a transition to an excited state, provided the energy is applied in a way that matches the resonance properties of the protons in the material. The details of this energy absorption stem from the design of the MR experiment. A pulse of energy is applied to the material inside the background field, which is absorbed (deterministically) and subsequently radiated away in a random-type fashion. Although the intrinsic nuclear atomic energies are very large in comparison, the *differences* between the spin-up/down energy levels, as predicted by the Zeeman interaction, lie in the radio frequency (RF) band of the electromagnetic spectrum. It is these energy differences (quantum gaps, if you like) that give rise to the received signal in MR experiments, through subsequent radiation of the absorbed energy. Although it is a quantum effect, many ideas from classical physics can be drawn into play in terms of a qualitative understanding of the origin of the MR signal. Conservation of energy is a key component, as also are the precessional dynamics of the net magnetization and Faraday's law of electromagnetic induction, described in section 5.2. It turns out that molecular environment affects the precise values of splitting in proton spin energy levels, so that, e.g., a proton in a fat molecule has a different absorption spectrum from a proton in water. The determination of the values of these "chemical shifts" is the basis of magnetic resonance spectroscopy (MRS). Once these "fingerprint" resonance absorption values are known, one can design spectrally selective RF pulses in MR experiments so that only protons in certain chemical environments, with magnetic resonance absorption properties matching the frequency (photon energy) spectrum of the pulse, are excited into states of higher energy.

The point of view taken here is that the spin population itself is the object of primary significance in constructing an image, especially when it comes to the study of neural brain activity. The reasons for this emphasis on the spin population size, as opposed to certain notions of its time average behavior, are twofold. First, it is the spin density of protons, in a given molecular environment, that is the fundamental object of the detection. Second, the spin population is something that can in principle be studied in real time. In contrast, the more familiar techniques known as $T1$- and $T2$-weighted imaging involve a statistical time average, measuring the respective "spin-lattice" and "spin-spin" relaxation times: $T1$ is the average restoration time for the longitudinal component and $T2$ the average decay time for the transverse component of the local magnetic field in the medium. Thereby, information concerning the short timescale properties of the population is necessarily lost. It is argued here that in principle one can infer the (local) size of a resonant spin population, from a large population whose dynamics is arbitrary, through a novel type of signal-processing technique which exploits certain ingredients in the physics of the spin dephasing process occurring in $T2$ relaxation.

The emphasis in this chapter is on the ideas and novel concepts, their significance in drawing together ideas from physics and signal processing, and the implications and new perspectives they provide in the context of magnetic resonance imaging. The detailed underlying mathematics, although an essential part of the sustained theoretical development, is highly technical and so reported separately, in elsewhere Field (2005). In section 5.2 we give the background to the description of an electromagnetic field interacting with a random population, in terms of the mathematics of stochastic differential equations (SDEs). Section 5.3 illustrates how this dynamics can be applied to an arbitrary spin population, in the context of constructing an image in magnetic resonance experiments. The possible implications of these ideas for future MRI systems are described in section 5.4, where we identify certain domains of validity of the proposed model and the appropriate corresponding choice of some design parameters that would be necessary for successful implementation. We provide, without proof, two key mathematical results behind this line of development, concerning the dynamics of spin-spin relaxation in proposition 5.1 and the observability of the spin population through statistical analysis of the phase fluctuations in theorem 5.1. The reader is referred to Field (2005) for their detailed mathematical derivation.

5.2   Conceptual Framework

Our purpose is to identify the dynamics of spin-spin ($T2$) relaxation using a geometrical description of the transverse spin population, and the mathematics of SDEs (e.g., Oksendal, 1998) to derive the continuous time statistical properties of the net transverse magnetization. In doing so we are led to an *exact* expression for the "hidden" state (the spin population level) in terms of the additional phase degrees of freedom in the MR signal, described in section 5.3. In our discussion we shall not confine ourselves to any specific choice of population model, and thus encompass the possibility of describing the highly nonstationary behavior characteristic of brain signals that encode information in real time in response to stimuli, such as occur in the auditory system.

Let us assume that the RF pulse is applied at a "pulse flip" angle of 90° to the longitudinal B_0 direction. As a result, RF energy ΔE is absorbed, and the net local magnetization is rotated into the transverse plane. Each component spin vector then rotates about the longitudinal axis at (approximately) the Larmor precessional frequency ω_0, governed by the following relations:

$$\Delta E = \tilde{}\,\omega_0 = h\gamma B_0/2\pi, \tag{5.1}$$

where γ is the gyromagnetic ratio (which varies throughout space depending on the details of the molecular environment). The resulting motion of the net local magnetization vector $\mathcal{M}_t$ can be understood by analogy with the Eulerian top in classical mechanics. As time progresses following the pulse, energy is transferred from the proton spins to the surroundings during the process of "spin-lattice" relaxation, and the longitudinal component of the net magnetization is gradually restored to the equilibrium value prior to the pulse being applied. Likewise, random exchange of energy between neighboring spins and small inhomogeneities in the total magnetic field cause perturbations in the phases of the transverse spin components and "dephasing" occurs, so that the net transverse component of magnetization decays to zero. The motion of $\mathcal{M}_t$ can thus be visualized as a

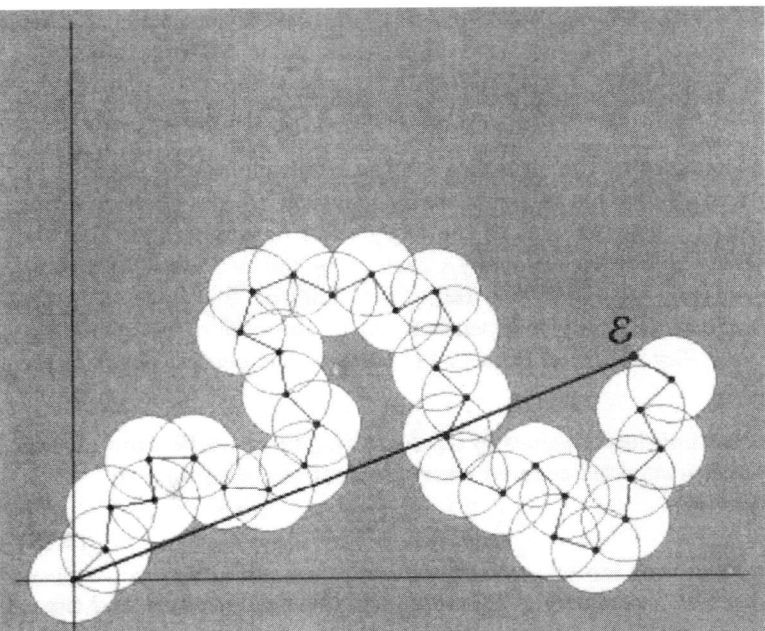

Figure 5.1 Geometry of transverse spin population—each point represents a constituent proton, with respect to which each connecting vector (in the direction of the random walk away from the origin) represents the transverse component of the associated spin vector.

precession about the longitudinal axis, over the surface of a cone whose opening angle tends from π to 0 as equilibrium is restored. It is convenient for visualization purposes to work in a rotating frame of reference, rotating at the Larmor frequency ω_0 about the longitudinal axis. (It is worth remarking that in the corresponding situation for radar scattering, ω_0 is the Doppler frequency arising from bulk wave motion in the scattering surface.) This brings each transverse spin vector to rest, for a perfect homogeneous B_0. Nevertheless it is the local *in*homogeneities in the total (background plus internal) magnetic field, due to the local magnetic properties of the medium, that give rise to spin-spin (or $T2$) relaxation constituting the (random) exchange of energy between spins.

The local perturbations in the total magnetic field can reasonably be considered as independent for each component spin, so that the resultant (transverse) spin vector can be modeled as a sum of independent random phasors. Thus, the pertinent expression for the resultant net transverse magnetization is

$$\mathcal{M}_t^{(N_t)} = \sum_{j=1}^{N_t} a_j \overbrace{\exp\left[i\varphi_t^{(j)}\right]}^{s^{(j)}}, \tag{5.2}$$

with (fluctuating) spin population size N_t, random phasor step $s^{(j)}$, and component amplitudes a_j. Observe that this random walk type model is directly analogous to what has been used in the dynamical description of Rayleigh scattering (Field and Tough, 2003b) introduced in Jakeman (1980). Thus, the geometrical spin structure, transverse to the background longitudinal magnetic field, lies in correspondence with the plane-polarized (perpendicular to the propagation vector) components of the electromagnetic field arising in (radar) scattering and (optical) propagation (Field and Tough, 2003a), where the same types of ideas, albeit for specific types of (stationary) populations, have been experimentally verified (Field and Tough, 2003b, 2005). The geometry of fig. 5.1 illustrates the isomorphism between the transverse spin structure of atomic nuclei and photon spin for a plane-polarized state, the latter familiar from radio theory as the (complex valued) electromagnetic amplitude perpendicular to the direction of propagation. Indeed, this duality between photon spin (EM wave polarization) and nuclear spin is a key conceptual ingredient in this development.

The dynamical structure of equation 5.2 is supplied by a (phase) diffusion model (Field and Tough, 2003b) which takes the component phases $\{\varphi_t^{(j)}\}$ to be a collection of (displaced) Wiener processes evolving on a suitable timescale. Thus $\varphi_t^{(j)} = \Delta^{(j)} + \mathcal{B}^{\frac{1}{2}} W_t^{(j)}$, with initialization $\Delta^{(j)}$. In magnetic resonance the 90° pulse explained above causes the spin phasors $s^{(j)}$ to be aligned initially; thus $\Delta^{(j)}$ are identical for all j. Let T be the *phase decoherence* or spin-spin relaxation time ($T2$) such that $\{\varphi_t^{(j)} \,|\, t \geq T\}$ have negligible correlation. Then for $t \geq T$, defining the (normalized) resultant by $m_t = \lim_{N \to \infty} \left[\mathcal{M}_t^{(N)} / N^{\frac{1}{2}} \right]$, we obtain the resultant spin dynamics or net magnetization in the transverse plane (cf. Field, 2005).

Proposition 5.1

For sufficiently large times $t \geq T$ the spin dynamics, for a constant population, is given by the complex Ornstein-Uhlenbeck equation

$$\mathrm{d}m_t = -\frac{1}{2}\mathcal{B}m_t\mathrm{d}t + \mathcal{B}^{\frac{1}{2}}\langle a^2 \rangle^{\frac{1}{2}}\mathrm{d}\xi_t, \tag{5.3}$$

where ξ_t is a complex-valued Wiener process (satisfying $|\mathrm{d}\xi_t|^2 = \mathrm{d}t$, $\mathrm{d}\xi_t^2 = 0$).

As the collection of spins radiates the absorbed energy, this gives rise to the received MR signal ψ_t (the free induction decay or FID), which is detected through the generation of electromotive force in a coil apparatus, due to the time-varying local magnetic field. This effect is the result of Faraday's law, i.e., Maxwell's vector equation $\nabla \times \mathbf{E} = -\partial \mathbf{B}/\partial t$ integrated around a current loop. The receiver coil is placed perpendicular to the transverse plane, and so only the transverse components of the change in magnetic flux contribute to the FID. The MR signal thus corresponds to an amplitude process that represents the (time derivative of the) net transverse magnetization, and has the usual in-phase (I) and quadrature-phase (Q) components familiar from radio theory, so $\psi = I + iQ$. Moreover it can be spatially localized using standard gradient field techniques (e.g., Brown and Semelka, 2003). The constant $\mathbf{B}$ in equation 5.3, which has dimensions of frequency, is (proportional to) the reciprocal of the spin-spin relaxation time $T2$.

In the case that the population size fluctuates over the timescale of interest, we introduce the continuum population variate x_t, and the receiver amplitude then has the compound representation

$$\psi_t = x_t^{\frac{1}{2}} m_t, \tag{5.4}$$

in which x_t and m_t are independent processes.

5.3   Image Extraction

An essential desired ingredient is the ability to handle real-time nonstationary behavior in the spin population. Moreover we do not wish to make any prior assumptions concerning the statistical behavior of this population, for it is this unknown state that we are trying to estimate from the observed MR signal. Indeed, its value depends on the external stimuli in a way that is not well understood, and it is precisely our purpose to uncover the nature of this dependence by processing the observable data in an intelligent way. But in doing so, we do not wish to prejudice our notions of spin population behavior, beyond some very generic assumptions set in place for the purpose of mathematical tractability.

Accordingly we shall assume that the population process x_t is an Ito process, i.e., that it satisfies the SDE (e.g., Oksendal, 1998):

$$\mathrm{d}x_t = \mathcal{A} b_t \mathrm{d}t + (2\mathcal{A}\Sigma_t)^{\frac{1}{2}} \mathrm{d}W_t^{(x)}, \tag{5.5}$$

in which the respective drift and diffusion parameters b_t, Σ_t are (real-valued) stochastic processes (not necessarily Ito), and in general include the effects of nonstationary and non-Gaussian behavior. Thus, we make no prior assumption concerning the nature of these parameters, and wish to estimate the values of $\{x_t\}$ from our observations of the MR signal.

The SDE for the resultant phase can be derived from equations 5.3, 5.4 and 5.5. Intriguingly, the behavior for a general population is *functionally* independent of the parameters b_t and Σ_t, the effect of these parameters coming through in the resulting *evolutionary* structure of the processes x_t, ψ_t. Calculation of the resulting squared phase fluctuations leads to the key noteworthy result of this chapter.

Theorem 5.1
The spin population is observable through the intensity-weighted squared phase fluctuations of the (FID) signal according to

$$x_t = \frac{2}{\mathcal{B}} z_t \mathrm{d}\theta_t^2 / \mathrm{d}t \tag{5.6}$$

throughout space and time, where $z_t = |\psi_t|^2$ and the field m_t is scaled so that $\langle a^2 \rangle = 1$.

The significance of this result is that the relation 5.6 is exact, instantaneous in nature, and moreover independent of the dynamics of the spin population. It is straightforward to illustrate this result with data sampled in discrete time (Field, 2005). The result approaches exactness as the pulse repetition rate tends to infinity. More precisely, for discretely sampled data theorem 5.1 implies that

$$z_i \delta\theta_i^2 \propto x_i n_i^2, \tag{5.7}$$

where i is a discrete time index and $\{n_i\}$ are an independent collection of $\mathcal{N}(0,1)$ distributed random variables. This can be used to estimate the state x_t via local time averaging. Applying a smoothing operation $\langle\cdot\rangle_\Delta$ to the left-hand side (the "observations") of 5.7 with window $\Delta = [t_0 - \Delta, t_0 + \Delta]$ yields an approximation to x_{t_0}, with an error that tends to zero as the number of pulses inside Δ tends to infinity and $\Delta \to 0$ (Field, 2005). In this respect we observe as a consequence that in order to achieve an improved signal-to-noise ratio (SNR), it is *sufficient merely to increase the pulse rate*, without (necessarily) requiring a high-amplitude signal. The term SNR is used here in the sense of extracting the signal x_t from ψ_t, thus overcoming the noise in m_t (cf. eq. 5.4).

This inference capability has been demonstrated in analysis of synthetic data, with population statistics chosen deliberately not to conform to the types of model usually encountered (Field, 2005). Instead of "filtering out" the noise to obtain the signal, we have exploited useful information contained in the random fluctuations in the phase. Indeed the phase noise is so strongly colored that, provided it is sampled at high enough frequency, it enables us to extract the precise values of the underlying state, in this case the population size of spins that have absorbed the RF energy. A comparison of time series of the state inferred from the observations alone, with the exact values recorded in generation of the synthetic data, shows a very high degree of correlation (Field, 2005).

5.4   Future Systems

For MR imaging purposes, we have proposed focusing on the local size of the transverse spin population that results from the applied RF pulse, which is assumed to be spectrally selective. Our results demonstrate, at a theoretical/simulation level, how this "signal" can be extracted through statistical analysis of the (phase) fluctuations in the received (complex) amplitude signal. In the MR context, the spin population, which assumes a Bose-Einstein distribution in equilibrium, responds in a dynamical nonstationary fashion to applied RF radiation, and our results suggest means for detecting this dynamical behavior using a combination of physical modeling and novel statistical signal-processing techniques based on the mathematics of SDEs.

The idea of focusing attention on the spin population size appears to be more intuitive than measurements of the single point statistic $T2$, the average transverse decay time, that is commonly used in $T2$-weighted imaging (Brown and Semelka, 2003). Primarily here one is concerned with estimating the population of spins that absorb energy from the RF pulse at different locations, since this implies the spin density of (protons in) the molecular environments of interest whose energy resonances (predicted through the Zeeman interaction) match those of the designed pulse. Our discussion demonstrates how the error in the estimate of a random population interacting with the electromagnetic field can be reduced to an arbitrarily small amount. In the MR context this suggests that a moderate/low background field strength B_0 and short pulse repetition time TR could be sufficient for generating real-time images. A complication posed by the high specific absorption rate SAR $\propto B_0^2/TR$ (which measures the deposition of energy in the medium in the form of heat) for short TR can presumably be overcome by using short-duration bursts of RF energy (just as in radar systems) to detect short timescale properties.

In summary, the results suggest means for real-time image construction in fMRI experiments at moderate to low magnetic field strength, and identify the choice of parameters necessary in the design of fMRI systems for the technique to be valid. There are indications that, for appropriate parameter values, the technique should succeed in extracting the real-time spin population behavior in the context of MR, without prior assumptions concerning the nature of this population needing to be made. Indeed, at the level of simulated data, this result has been verified (see fig. 2 in Field, 2005). The ability to track the spin population from the observed amplitude in local time suggests the possibility of detecting spatiotemporal changes in neural activity in the brain, e.g., in the localization and tracking of evoked responses.

Acknowledgments

The author is grateful to John Bienenstock and Michael Noseworthy of the Brain Body Institute, Simon Haykin, and José Príncipe.

6 What Makes a Dynamical System Computationally Powerful?

Robert Legenstein and *Wolfgang Maass*

We review methods for estimating the computational capability of a complex dynamical system. The main examples that we discuss are models for cortical neural microcircuits with varying degrees of biological accuracy, in the context of online computations on complex input streams. We address in particular the question to what extent earlier results about the relationship between the edge of chaos and the computational power of dynamical systems in discrete time for off-line computing also apply to this case.

6.1 Introduction

Most work in the theory of computations in circuits focuses on computations in feedforward circuits, probably because computations in feedforward circuits are much easier to analyze. But biological neural circuits are obviously recurrent; in fact the existence of feedback connections on several spatial scales is a characteristic property of the brain. Therefore an alternative computational theory had to be developed for this case. One neuroscientist who emphasized the need to analyze information processing in the brain in the context of dynamical systems theory was Walter Freeman, who started to write a number of influential papers on this topic in the 1960s; see Freeman (2000, 1975) for references and more recent accounts. The theoretical investigation of computational properties of recurrent neural circuits started shortly afterward. Earlier work focused on the engraving of attractors into such systems in order to restrict the dynamics to achieve well-defined properties. One stream of work in this direction (see, e.g., Amari, 1972; Cowan, 1968; Grossberg, 1967; Little, 1974) culminated in the influential studies of Hopfield regarding networks with stable memory, called Hopfield networks (Hopfield, 1982, 1984), and the work of Hopfield and Tank on networks which are able to find approximate solutions of hard combinatorial problems like the traveling salesman problem (Hopfield and Tank, 1985, 1986). The Hopfield network is a fully connected neural network of threshold or threshold-like elements. Such networks exhibit rich dynamics and are chaotic in general. However, Hopfield assumed symmetric weights, which strongly constrains the dynamics of the system. Specifically, one can show that only point attractors can emerge in the dynamics of the system, i.e., the activity of the elements always evolves to one of a set of stable states which is then kept forever.

Somewhat later the alternative idea arose to use the rich dynamics of neural systems that can be observed in cortical circuits rather than to restrict them (Buonomano and Merzenich, 1995). In addition one realized that one needs to look at online computations (rather than off-line or batch computing) in dynamical systems in order to capture the biologically relevant case (see Maass and Markram, 2005, for definitions of such basic concepts of computation theory). These efforts resulted in the "liquid state machine" model by Maass et al. (2002) and the "echo state network" by Jaeger (2002), which were introduced independently. The basic idea of these models is to use a recurrent network to hold and nonlinearly transform information about the past input stream in the high-dimensional transient state of the network. This information can then be used to produce in real time various desired online outputs by simple linear readout elements. These readouts can be trained to recognize common information in dynamical changing network states because of the high dimensionality of these states. It has been shown that these models exhibit high computational power (Jaeger and Haas, 2004; Joshi and Maass, 2005; Legenstein et al., 2003). However, the analytical study of such networks with rich dynamics is a hard job. Fortunately, there exists a vast body of literature on related questions in the context of many different scientific disciplines in the more general framework of dynamical systems theory. Specifically, a stream of research is concerned with system dynamics located at the boundary region between ordered and chaotic behavior, which was termed the "edge of chaos." This research is of special interest for the study of neural systems because it was shown that the behavior of dynamical systems is most interesting in this region. Furthermore, a link between computational power of dynamical systems and the edge of chaos was conjectured.

It is therefore a promising goal to use concepts and methods from dynamical systems theory to analyze neural circuits with rich dynamics and to get in this way better tools for understanding computation in the brain. In this chapter, we will take a tour, visiting research concerned with the edge of chaos and eventually arrive at a first step toward this goal. The aim of this chapter is to guide the reader through a stream of ideas which we believe are inspiring for research in neuroscience and molecular biology, as well as for the design of novel computational devices in engineering.

After a brief introduction of fundamental principles of dynamical systems theory and chaos in section 6.2, we will start our journey in section 6.3 in the field of theoretical biology. There, Kauffman studied questions of evolution and emerging order in organisms. We will see that depending on the connectivity structure,

Table 6.1   General Properties of Various Types of Dynamical Systems

	Cellular Automata	Iterative Maps	Boolean Circuits	Cortical Microcircuits and Gene Regulation networks
Analog states?	no	yes	no	yes
Continuous time?	no	no	no	yes
High dimensional?	yes	no	yes	yes
With noise?	no	no	no	yes
With online input?	no	no	usually no	yes

networks may operate either in an ordered or chaotic regime. Furthermore, we will encounter the edge of chaos as a transition between these dynamic regimes. In section 6.4, our tour will visit the field of statistical physics, where Derrida and others studied related questions and provided new methods for their mathematical analysis. In section 6.5 the reader will see how these ideas can be applied in the theory of computation. The study of cellular automata by Wolfram, Langton, Packard, and others led to the conjecture that complex computations are best performed in systems at the edge of chaos. The next stops of our journey in sections 6.6 and 6.7 will bring us close to our goal. We will review work by Bertschinger and Natschläger, who analyzed real-time computations on the edge of chaos in threshold circuits. In section 6.8, we will briefly examine self-organized criticality, i.e., how a system can adapt its own dynamics toward the edge of chaos. Finally, section 6.9 presents the efforts of the authors of this chapter to apply these ideas to computational questions in the context of biologically realistic neural microcircuit models. In this section we will analyze the edge of chaos in networks of spiking neurons and ask the following question: In what dynamical regimes are neural microcircuits computationally most powerful? Table 6.1 shows that neural microcircuits (as well as gene regulation networks) differ in several essential aspects from those examples for dynamical systems that are commonly studied in dynamical systems theory.

6.2   Chaos in Dynamical Systems

In this section we briefly introduce ideas from dynamical systems theory and chaos. A few slightly different definitions of chaos are given in the literature. Although we will mostly deal here with systems in discrete time and discrete state space, we start out with the well-established definition of chaos in continuous systems and return to discrete systems later in this section.

The subject known as *dynamics* deals with systems that evolve in time (Strogatz, 1994). The system in question may settle down to an equilibrium, may enter a periodic trajectory (limit cycle), or do something more complicated. In Kaplan and Glass (1995) the dynamics of a deterministic system is defined as being chaotic if it is aperiodic and bounded with sensitive dependence on initial conditions.

The phase space for an N-dimensional system is the space with coordinates $x_1, \ldots, x_N$. The state of an N-dimensional dynamical system at time t is represented by the state vector $\mathbf{x}(t) = (x_1(t), \ldots, x_N(t))$. If a system starts in some initial condition $\mathbf{x}(0)$, it will evolve according to its dynamics and describe a trajectory in state space. A steady state of the system is a state $\mathbf{x}_s$ such that if the system evolves with $\mathbf{x}_s$ as its initial state, it will remain in this state for all future times. Steady states may or may not be stable to small outside perturbations. For a stable steady state, small perturbations die out and the trajectory converges back to the steady state. For an unstable steady state, trajectories do not converge back to the steady state after arbitrarily small perturbations.

The general definition of an *attractor* is a set of points or states in state space to which trajectories within some volume of state space converge asymptotically over time. This set itself is invariant under the dynamic evolution of the system.[1] Therefore, a stable steady state is a zero-dimensional, or point, attractor. The set of initial conditions that evolve to an attractor A is called the *basin of attraction* of A. A *limit cycle* is an isolated closed trajectory. Isolated means that neighboring trajectories are not closed. If released in some point of the limit cycle, the system flows on the cycle repeatedly. The limit cycle is stable if all neighboring trajectories approach the limit cycle. A stable limit cycle is a simple type of attractor. Higher-dimensional and more complex types of attractors exist.

In addition, there exist also so-called *strange*, or *chaotic* attractors. For example all trajectories in a high-dimensional state space might be brought onto a two-dimensional surface of some manifold. The interesting property of such attractors is that, if the system is released in two different experiments from two points on the attractor which are arbitrarily close to each other, the subsequent trajectories remain on the attractor surface but diverge away from each other. After a sufficient time, the two trajectories can be arbitrarily far apart from each other. This extreme sensitivity to initial conditions is characteristic of chaotic behavior. In fact, exponentially fast divergence of trajectories (characterized by a positive Lyapunov exponent) is often used as a definition of chaotic dynamics (see, e.g., Kantz and Schreiber, 1997). There might be a lot of structure in chaotic dynamics since the trajectory of a high-dimensional system might be projected merely onto a two-dimensional surface. However, since the trajectory on the attractor is chaotic, the exact trajectory is practically not predictable (even if the system is deterministic).

Systems in discrete time and with a finite discrete state space differ from continuous systems in several aspects. First, since state variables are discrete, trajectories can merge, whereas in a continuous system they may merely approximate each other. Second, since there is a finite number of states, the system must eventually reenter a state previously encountered and will thereafter cycle repeatedly through this state cycle. These state cycles are the dynamical attractors of the discrete system. The set of states flowing into such a state cycle or lying on it constitutes the basin of attraction of that state cycle. The length of a state cycle is the number of states on the cycle. For example, the memory states in a Hopfield network (a network of artificial neurons with symmetric weights) are the stable states of the system. A Hopfield network does not have state cycles of length larger than one. The basins of attraction of memory states are used to drive the system from related initial conditions to the same memory state, hence constituting an associative memory device.

Characteristic properties of chaotic behavior in discrete systems are a large length of state cycles and high sensitivity to initial conditions. Ordered networks have short state cycles and their sensitivity to initial conditions is low, i.e., the state cycles are quite stable. We note that state cycles can be stable with respect to some small perturbations but unstable to others. Therefore, "quite stable" means in this context that the state cycle is stable to a high percentage of small perturbations. These general definitions are not very precise and will be made more specific for each of the subsequent concrete examples.

6.3 Randomly Connected Boolean Networks

The study of complex systems is obviously important in many scientific areas. In genetic regulatory networks, thousands or millions of coupled variables orchestrate developmental programs of an organism. In 1969, Kauffman started to study such systems in the simplified model of Boolean networks (Kauffman, 1969, 1993). He discovered some surprising results which will be discussed in this section. We will encounter systems in the ordered and in the chaotic phase. The specific phase depends on some simple structural feature of the system, and a phase transition will occur when this feature changes.

A Boolean network consists of N elements and connections between them. The state of its elements is described by binary variables $x_1, \ldots, x_N$. The dynamical behavior of each variable, whether it will be active (1) or inactive (-1) at the next time step, is governed by a Boolean function.[2] The (directed) connections between the elements describe possible interactions. If there is a connection from element i to element j, then the state of element i influences the state of element j in the next time step. We say that i is an input of element j.

An initial condition is given by a value for each variable $\mathbf{x}(0)$. Thereafter, the state of each element evolves according to the Boolean function assigned to it. We can describe the dynamics of the system by a set of iterated maps

$$x_1(t+1) = f_1(x_1(t), \ldots, x_N(t))$$
$$\vdots$$
$$x_N(t+1) = f_N(x_1(t), \ldots, x_N(t)),$$

where $f_1, \ldots, f_N$ are Boolean functions.[3] Here, all state variables are updated in parallel at each time step.

The stability of attractors in Boolean networks can be studied with respect to minimal perturbations. A minimal perturbation is just the flip of the activity of a single variable to the opposite state.

Kauffman studied the dynamics of Boolean networks as a function of the number of elements in the network N, and the average number K of inputs to each element in the net. Since he was not interested in the behavior of particular nets but rather in the expected dynamics of nets with some given N and K, he sampled at random from the ensemble of all such networks. Thus the K inputs to each element were first chosen at random and then fixed, and the Boolean function assigned to each element was also chosen at random and then fixed. For each such member of the ensemble, Kauffman performed computer simulations and examined the accumulated statistics.

The case $K = N$ is especially easy to analyze. Since the Boolean function of each element was chosen randomly from a uniform distribution, the successor to each circuit state is drawn randomly from a uniform distribution among the 2^N possible states. This leads to long state cycles. The median state cycle length is $0.5 \cdot 2^{N/2}$. Kauffman called such exponentially long state cycles *chaotic*.[4] These state cycles are unstable to most perturbations, hence there is a strong dependence on initial conditions. However, only a few different state cycles exit in this case: the expected number of state cycles is N/e. Therefore, there is some characteristic structure in the chaotic behavior in the sense that the system will end up in one of only a few long-term behaviors.

As long as K is not too small, say $K \geq 5$, the main features of the case $K = N$ persist. The dynamics is still governed by relatively few state cycles of exponential length, whose expected number is at most linear in N. For $K \geq 5$, these results can be derived analytically by a rough mean field approximation. For smaller K, the approximation becomes inaccurate. However, simulations confirm that exponential state cycle length and a linear number of state cycles are characteristic for random Boolean networks with $K \geq 3$. Furthermore, these systems show high sensitivity to initial conditions (Kauffman, 1993).

The case $K = 2$ is of special interest. There, a phase transition from ordered to chaotic dynamics occurs. Numerical simulations of these systems have revealed the following characteristic features of random Boolean networks with $K = 2$ (Kauffman, 1969). The expected median state cycle length is about $\sqrt{N}$. Thus, random Boolean networks with $K = 2$ often confine their dynamical behavior to tiny subvolumes of their state space, a strong sign of order. A more detailed analysis shows that most state cycles are short, whereas there are a few long ones. The number of state cycles is about $\sqrt{N}$ and they are inherently stable to about 80% to 90% of all minimal transient perturbations. Hence, the state cycles of the system have large basins of attraction and the sensitivity to initial conditions is low. In addition to these characteristics which stand in stark contrast to networks of larger K, we want to emphasize three further features. First, typically at least 70% of the N elements have some fixed active or inactive state which is identical for all the existing state cycles of the Boolean network. This behavior establishes a *frozen core* of elements. The frozen core creates walls of constancy which break the system into functionally isolated islands of unfrozen elements. Thus, these islands are prevented from influencing one another. The boundary regime where the frozen core is just breaking up and interaction between the unfrozen islands becomes possible is the phase transition between order and chaos. Second, altering transiently the activity of a single element typically propagates but causes only alterations in the activity of a small fraction of the elements in the system. And third, deleting any single element or altering its Boolean function typically causes only modest changes in state cycles and transients. The latter two points ensure that "damage" of the system is small. We will further discuss this interesting case in the next section.

Networks with $K = 1$ operate in an ordered regime and are of little interest for us here.

6.4   The Annealed Approximation by Derrida and Pomeau

The phase transition from order to chaos is of special interest. As we shall see in the sections below, there are reasons to believe that this dynamical regime is particularly well suited for computations. There were several attempts to understand the emerging order in random Boolean networks. In this section, we will review the approach of Derrida and Pomeau (1986). Their beautiful analysis gives an analytical answer to the question of where such a transition occurs.

In the original model, the connectivity structure and the Boolean functions f_i of the elements i were chosen randomly but were then fixed. The dynamics of the network evolved according to this fixed network. In this case the randomness is *quenched* because the functions f_i and the connectivity do not change with time. Derrida and Pomeau presented a simple *annealed approximation* to this model which explains why there is a critical value K_c of K where the transition from order to chaos appears. This approximation also allowed the calculation of many properties of the model. In contrast to the quenched model, the annealed approximation randomly reassigns the connectivity and the Boolean functions of the elements *at each time step*. Although the assumption of the annealed approximation is quite drastic, it turns out that its agreement with observations in simulations of the quenched model is surprisingly good. The benefits of the annealed model will become clear below.

It was already pointed out that exponential state cycle length is an indicator of chaos. In the annealed approximation, however, there are no fixed state cycles because the network is changed at every time step. Therefore, the calculations are based on the dependence on initial conditions. Consider two network states $\mathcal{C}_1, \mathcal{C}_2 \in \{-1, 1\}^N$. We define the *Hamming distance* $d(\mathcal{C}_1, \mathcal{C}_2)$ as the number of positions in which the two states are different. The question is whether two randomly chosen different initial network states eventually converge to the same pattern of activity over time. Or, stated in other words, given an initial state $\mathcal{C}_1$ which leads to a state $\mathcal{C}_1^{(t)}$ at time t and a different initial state $\mathcal{C}_2$ which leads to a state $\mathcal{C}_2^{(t)}$ at time t, will the Hamming distance $d(\mathcal{C}_1^{(t)}, \mathcal{C}_2^{(t)})$ converge to zero for

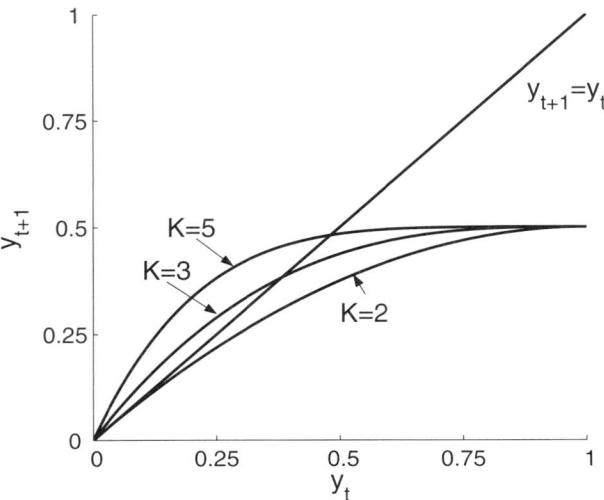

Figure 6.1 Expected distance between two states at time $t+1$ as a function of the state distance y_t between two states at time t, based on the annealed approximation. Points on the diagonal $y_t = y_{t+1}$ are fixed points of the map. The curves for $K \geq 3$ all have fixed points for a state distance larger than zero. The curve for $K = 2$ stays close to the diagonal for small state distances but does not cross it. Hence, for $K = 2$ state distances converge to zero for iterated applications of the map.

large t? Derrida and Pomeau found that this is indeed the case for $K \leq 2$. For $K \geq 3$, the trajectories will diverge.

To be more precise, one wants to know the probability $P_1(m, n)$ that the distance $d(\mathcal{C}_1', \mathcal{C}_2')$ between the states at time $t = 1$ is m given that the distance $d(\mathcal{C}_1, \mathcal{C}_2)$ at time $t = 0$ was n. More generally, one wants to estimate the probability $P_t(m, n)$ that the network states $\mathcal{C}_1^{(t)}, \mathcal{C}_2^{(t)}$ obtained at time t are at distance m, given that $d(\mathcal{C}_1, \mathcal{C}_2) = n$ at time $t = 0$. It now becomes apparent why the annealed approximation is useful. In the annealed approximation, the state transition probabilities at different time steps are independent, which is not the case in the quenched model. For large N, one can introduce continuous variables $\frac{n}{N} = x$. Derrida and Pomeau (1986) show that $P_1^{annealed}(m, n)$ for the annealed network has a peak around a value $m = N y_1$ where y_1 is given by

$$y_1 = \frac{1 - (1 - x)^K}{2} \,. \tag{6.1}$$

Similarly, the probability $P_t^{annealed}(m, n)$ has a peak at $m = N y_t$ with y_t given by

$$y_t = \frac{1 - (1 - y_{t-1})^K}{2} \tag{6.2}$$

for $t > 1$. The behavior of this iterative map can be visualized in the so called *Derrida plot*; see fig. 6.1. The plot shows the state distance at time $t + 1$ as a function of the state distance at time t. Points on the diagonal $y_t = y_{t+1}$ are fixed points of the map.

For $K \leq 2$, the fixed point $y = 0$ is the only fixed point of the map and it is stable. In fact, for any starting value y_1, we have $y_t \to 0$ for $t \to \infty$ in the limit of $N \to \infty$. For $K > 2$, the fixed point $y = 0$ becomes unstable and a new stable fixed point y^* appears. Therefore, the state distance need no longer always converge to zero. Hence there is a phase transition of the system at $K = 2$. The theoretical work of Derrida and Pomeau was important because before there was only empirical evidence for this phase transition.

We conclude that there exists an interesting transition region from order to chaos in these dynamical systems. For simplified models, this region can be determined analytically. In the following section we will find evidence that such phase transitions are of great interest for the computational properties of dynamical systems.

6.5 Computation at the Edge of Chaos in Cellular Automata

Evidence that systems exhibit superior computational properties near a phase transition came from the study of cellular automata. Cellular automata are quite similar to Boolean networks. The main differences are that connections between elements are local, and that an element may assume one out of k possible states at each time step (instead of merely two states as in Boolean networks). The former difference implies that there is a notion of space in a cellular automaton. More precisely, a d-dimensional space is divided into *cells* (the elements of the network). The state of a cell at time $t+1$ is a function only of its own state and the states of its immediate neighbors at time t. The latter difference is made explicit by defining a finite set Σ of cell states. The transition function Δ is a mapping from neighborhood states (including the cell itself) to the set of cell states. If the neighborhood is of size L, we have $\Delta : \Sigma^L \to \Sigma$.

What do we mean by "computation" in the context of cellular automata? In one common meaning, the transition function is interpreted as the program and the input is given by the initial state of the cellular automaton. Then, the system evolves for some specified number of time steps, or until some "goal pattern"— possibly a stable state—is reached. The final pattern is interpreted as the output of the automaton (Mitchell et al., 1993). In analogy to universal Turing machines, it has been shown that cellular automata are capable of universal computation (see, e.g., Codd, 1968; Smith, 1971; von Neumann, 1966). That is, there exist cellular automata which, by getting the algorithm to be applied as part of their initial configuration, can perform any computation which is computable by any Turing machine.

In 1984, Wolfram conjectured that such powerful automata are located in a special dynamical regime. Later, Langton identified this regime to lie on a phase transition between order and chaos (see below), i.e., in the regime which corresponds

Figure 6.2 Evolution of one-dimensional cellular automata. Each horizontal line represents one automaton state. Successive time steps are shown as successive horizontal lines. Sites with value 1 are represented by black squares; sites with value 0 by white squares. One example each for an automaton of class 1 (left), class 4 (middle), and class 3 (right) is given.

to random Boolean networks with $K = 2$.

Wolfram presented a qualitative characterization of one-dimensional cellular automaton behavior where the individual automata differed by their transfer function.[5] He found evidence that all one-dimensional cellular automata fall into four distinct classes (Wolfram, 1984). The dynamics for three of these classes are shown in fig. 6.2.

Class 1 automata evolve to a homogeneous state, i.e., a state where all cells are in the same state. Hence these systems evolve to a simple steady state.

Class 2 automata evolve to a set of separated simple stable states or separated periodic structures of small length. These systems have short state cycles.

Both of these classes operate in the ordered regime in the sense that state cycles are short.

Class 3 automata evolve to chaotic patterns.

Class 4 automata have long transients, and evolve "to complex localized structures" (Wolfram, 1984).

Class 3 automata are operating in the chaotic regime. By *chaotic*, Wolfram refers to the unpredictability of the exact automaton state after a few time steps. Successor states look more or less random. He also talks about nonperiodic patterns. Of course these patterns are periodic if the automaton is of finite size. But in analogy with the results presented above, one can say that state cycles are very long.

Transients are the states that emerge before the dynamics reaches a stable long-lasting behavior. They appear at the beginning of the state evolution. Once the system is on a state cycle, it will never revisit such transient states. The transients of class 4 automata can be identified with large basins of attraction or high stability of state cycles. Wolfram conjectured that class 4 automata are capable of universal computations.

In 1990, Langton systematically studied the space of cellular automata considered by Wolfram with respect to an order parameter λ (Langton, 1990). This parameter λ determines a crucial property of the transfer function Δ: the fraction of entries in Δ which do not map to some prespecified quiescent state s_q. Hence, for $\lambda = 0$, all local configurations map to s_q, and the automaton state moves to a homogeneous state after one time step for every initial condition. More generally, low λ values lead to ordered behavior. Rules with large λ tend to produce a completely different behavior.

Langton (1990) stated the following question: "Under what conditions will physical systems support the basic operations of information transmission, storage, and modification constituting the capacity to support computation?" When Langton went through different λ values in his simulations, he found that all automaton classes of Wolfram appeared in this parameterization. Moreover, he found that the interesting class 4 automata can be found at the phase transition between ordered and chaotic behavior for λ values between about 0.45 and 0.5, values of intermediate heterogeneity. Information-theoretic analysis supported the conjectures of Wolfram, indicating that the edge of chaos is the dominant region of computationally powerful systems.

Further evidence for Wolfram's hypothesis came from Packard (1988). Packard used genetic algorithms to genetically evolve one-dimensional cellular automata for a simple computational task. The goal was to develop in this way cellular automata which behave as follows: The state of the automaton should converge to the all-one state (i.e., the state where every cell is in state 1), if the fraction of one-states in the initial configuration is larger than 0.5. If the fraction of one-states in the initial configuration is below 0.5, it should evolve to the all-zero state.

Mutations were accomplished by changes in the transfer function (point mutations which changed only a single entry in the rule table, and crossover which merged two rule tables into a single one). After applying a standard genetic algorithm procedure to an initial set of cellular automaton rules, he examined the rule tables of the genetically evolved automata. The majority of the evolved rule tables had λ values either around 0.23 or around 0.83. These are the two λ values where the transition from order to chaos appears for cellular automata with two states per cell.[6] "Thus, the population appears to evolve toward that part of the space of rules that marks the transition to chaos" (Packard, 1988).

These results have later been criticized (Mitchell et al., 1993). Mitchell and collaborators reexamined the ideas of Packard and performed similar simulations with a genetic algorithm. The results of these investigations differed from Packard's results. The density of automata after evolution was symmetrically peaked around $\lambda = 0.5$, but much closer to 0.5 and definitely not in the transition region. They argued that the optimal λ value for a task should strongly depend on the task. Specifically, in the task considered by Packard one would expect a λ value close to 0.5 for a well-performing rule, because the task is symmetric with respect to the exchange of ones and zeros. A rule with $\lambda < 0.5$ tends to decrease the number of ones in the state vector because more entries in the rule table map the state to zero. This can lead to errors if the number of ones in the initial state is slightly larger than 0.5. Indeed, a rule which performs very well on this task, the Gacs-

Kurdyumov-Levin (GKL) rule, has $\lambda = 0.5$. It was suggested that artifacts in the genetic algorithm could account for the different results.

We want to return here to the notion of computation. Wolfram and Langton were interested in universal computations. Although universality results for automata are mathematically interesting, they do not contribute much to the goal of understanding computations in biological neural systems. Biological organisms usually face computational tasks which are quite different from the off-line computations on discrete batch inputs for which Turing machines are designed.

Packard was interested in automata which perform a specific kind of computation with the transition function being the "program." Mitchell at al. showed that there are complex tasks for which the best systems are not located at the edge of chaos. In Mitchell et al. (1993), a third meaning of computation in cellular automata—a kind of "intrinsic" computation—is mentioned: "Here, computation is not interpreted as the performance of a 'useful' transformation of the input to produce the output. Rather, it is measured in terms of generic, structural computational elements such as memory, information production, information transfer, logical operations, and so on. It is important to emphasize that the measurement of such intrinsic computational elements does not rely on a semantics of utility as do the preceding computational types" (Mitchell et al., 1993). It is worthwhile to note that this "intrinsic" computation in dynamical systems can be used by a readout unit which maps system states to desired outputs. This is the basic idea of the liquid state machine and echo state networks, and it is the basis of the considerations in the following sections.

To summarize, systems at the edge of chaos are believed to be computationally powerful. However, the type of computations considered so far are considerably different from computations in organisms. In the following section, we will consider a model of computation better suited for our purposes.

6.6   The Edge of Chaos in Systems with Online Input Streams

All previously considered computations were off-line computations where some initial state (the input) is transformed by the dynamics into a terminal state or state cycle (the output). However, computation in biological neural networks is quite different from computations in Turing machines or other traditional computational models. The input to an organism is a continuous stream of data and the organism reacts in real time (i.e., within a given time interval) to information contained in this input. Hence, as opposed to batch processing, the input to a biological system is a time varying signal which is mapped to a time-varying output signal. Such mappings are also called filters. In this section, we will have a look at recent work on real-time computations in threshold networks by Bertschinger and Natschläger (2004); see also Natschläger et al. (2005)). Results of experiments with closely related hardware models are reported in Schuermann et al. (2005).

Threshold networks are special cases of Boolean networks consisting of N elements (units) with states $x_i \in \{-1, 1\}, i = 1, \ldots, N$. In networks with online input, the state of each element depends on the state of exactly K randomly chosen other units, and in addition on an external input signal $u(\cdot)$ (the online input). At each time step, $u(t)$ assumes the value $\bar{u} + 1$ with probability r and the value $\bar{u} - 1$ with probability $1 - r$. Here, $\bar{u}$ is a constant input bias. The transfer function of the elements is not an arbitrary Boolean function but a randomly chosen threshold function of the form

$$x_i(t+1) = \Theta \left(\sum_{j=1}^{N} w_{ij} x_j(t) + u(t+1) \right), \tag{6.3}$$

where $w_{ij} \in \mathrm{R}$ is the weight of the connection from element j to element i and $\Theta(h) = +1$ if $h \geq 0$ and $\Theta(h) = -1$ otherwise. For each element, exactly K of its incoming weights are nonzero and chosen from a Gaussian distribution with zero mean and variance σ^2. Different dynamical regimes of such circuits are shown in fig. 6.3. The top row shows the online input, and below, typical activity patterns of networks with ordered, critical, and chaotic dynamics. The system parameters for each of these circuits are indicated in the phase plot below. The variance σ^2 of nonzero weights was varied to achieve the different dynamics. The transition from the ordered to the chaotic regime is referred to as the *critical line*.

Bertschinger and Natschläger used the approach of Derrida to determine the dynamical regime of these systems. They analyzed the change in Hamming distance between two (initial) states and their successor states provided that the same input is applied in both situations. Using Derrida's annealed approximation, one can calculate the Hamming distance $d(t + 1)$ given the Hamming distance $d(t)$ of the states at time t. If arbitrarily small distances tend to increase, the network operates in the chaotic phase. If arbitrarily small distances tend to decrease, the network operates in the ordered phase. This can also be expressed by the stability of the fixed point $d^* = 0$. In the ordered phase, this fixed point is the only fixed point and it is stable. In the chaotic phase, another fixed point appears and $d^* = 0$ becomes unstable. The fixed point 0 is stable if the absolute value of the slope of the map at $d(t) = 0$,

$$\alpha = \frac{\partial d(t+1)}{\partial d(t)} \bigg|_{d(t)=0},$$

is smaller than 1. Therefore, the transition from order to chaos (the critical line) is given by the line $|\alpha| = 1$. This line can be characterized by the equation

$$r P_{BF}(\bar{u} + 1) + (1 - r) P_{BF}(\bar{u} - 1) = \frac{1}{K}, \tag{6.4}$$

where the bit-flip probability $P_{BF}(v)$ is the probability that a single changed state component in the K inputs to a unit that receives the current online input v leads to a change of the output of that unit. This result has a nice interpretation. Consider

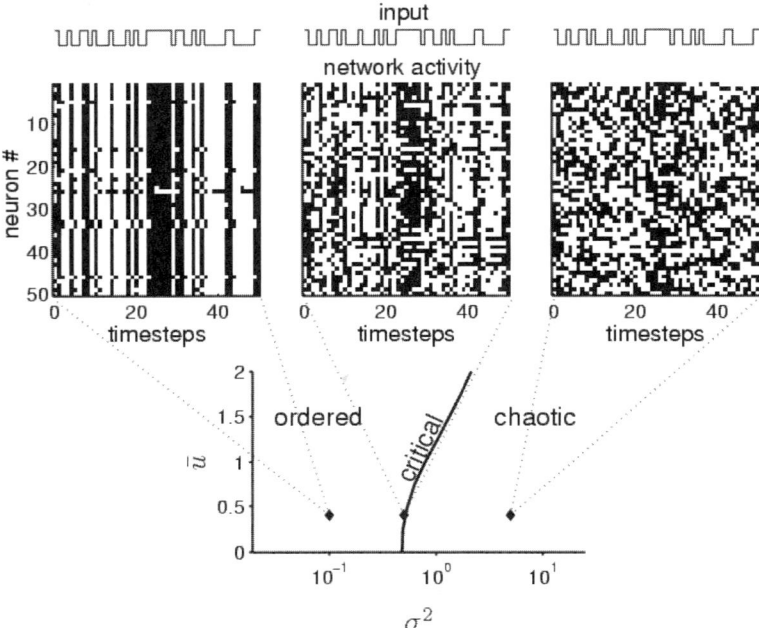

Figure 6.3 Threshold networks with online input streams in different dynamical regimes. The top row shows activity patterns for ordered (left), critical (middle), and chaotic behavior (right). Each vertical line represents the activity in one time step. Black (white) squares represent sites with value 1 (−1). Successive vertical lines represent successive circuit states. The input to the network is shown above the plots. The parameters σ^2 and $\bar{u}$ of these networks are indicated in the phase plot below. Further parameters: number of input connections $K = 4$, number of elements $N = 250$.

a value of $r = 1$, i.e., the input to the network is constant. Consider two network states $\mathcal{C}_1$, $\mathcal{C}_2$ which differ only in one state component. This different component is on average mapped to K elements (because each gate receives K inputs, hence there are altogether $N \cdot K$ connections). If the bit-flip probability in each of these units is larger than $1/K$, then more than one of these units will differ on average in the successor states $\mathcal{C}_1'$, $\mathcal{C}_2'$. Hence, differences are amplified. If the bit-flip probability of each element is smaller than $1/K$, the differences will die out on average.

6.7 Real-Time Computation in Dynamical Systems

In the previous section we were interested in the dynamical properties of systems with online input. The work we discussed there was influenced by recent ideas concerning computation in neural circuits that we will sketch in this section.

The idea to use the rich dynamics of neural systems which can be observed in cortical circuits, rather than to restrict them, resulted in the *liquid state machine* model by Maass et al. (2002) and the *echo state network* by Jaeger (2002).[7] They assume time series as inputs and outputs of the system. A recurrent network is used to hold nonlinearly transformed information about the past input stream in the state of the network. It is followed by a memoryless readout unit which simply looks at the current state of the circuit. The readout can then learn to map the current state of the system onto some target output. Superior performance of echo state networks for various engineering applications is suggested by the results of Jaeger and Haas (2004).

The requirement that the network is operating in the ordered phase is important in these models, although it is usually described with a different terminology. The ordered phase can be described by using the notion of *fading memory* (Boyd and Chua, 1985). Time-invariant fading memory filters are exactly those filters which can be represented by Volterra series. Informally speaking, a network has fading memory if its state at time t depends (up to some finite precision) only on the values (up to some finite precision) of its input from some finite time window $[t - T, t]$ into the past (Maass et al., 2002). This is essentially equivalent to the requirement that if there are no longer any differences in the online inputs then the state differences converge to 0, which is called *echo state property* in Jaeger (2002).

Besides the fading memory property, another property of the network is important for computations on time series: the pairwise separation property (Maass et al., 2002). Roughly speaking, a network has the pairwise separation property if for any two input time series which differed in the past, the network assumes at subsequent time points different states.

Chaotic networks have such separation property, but they do not have fading memory since differences in the initial state are amplified. On the other hand, very ordered systems have fading memory but provide weak separation. Hence, the separation property and the fading memory property are antagonistic. Ideally, one would like to have high separation on salient differences in the input stream but still keep the fading memory property (especially for variances in the input stream that do not contribute salient information). It is therefore of great interest to analyze these properties in models for neural circuits.

A first step in this direction was made by Bertschinger and Natschläger (2004) in the context of threshold circuits. Similar to section 6.6, one can analyze the evolution of the state separation resulting from two input streams u_1 and u_2 which differ at time t with some probability. The authors defined the *network-mediated separation* (short: NM-separation) of a network. Informally speaking, the NM-separation is roughly the amount of state distance in a network which results from differences in the input stream minus the amount of state difference resulting from different initial states. Hence, the NM-separation has a small value in the ordered regime, where both terms are small, but also in the chaotic regime, where both terms are large. Indeed, it was shown that the NM-separation peaks at the critical line, which is shown in fig. 6.4a. Hence, Bertschinger and Natschläger (2004) offer a new interpretation for the critical line and provide a more direct link between the edge of chaos and computational power.

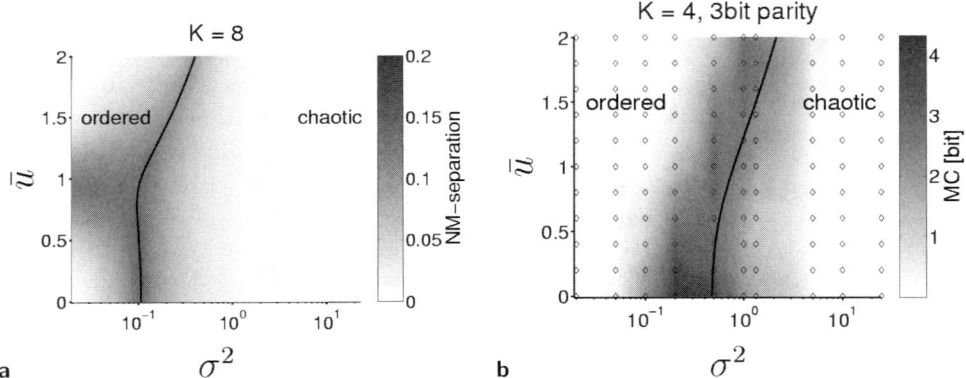

Figure 6.4   The network-mediated separation and computational performance for a 3-bit parity task with different settings of parameters σ^2 and $\bar{u}$. (a) The NM-separation peaks at the critical line. (b) High performance is achieved near the critical line. The performance is measured in terms of the memory capacity MC (Jaeger, 2002). The memory capacity is defined as the mutual information MI between the network output and the target function summed over all delays $\tau > 0$ on a test set. More formally, $MC = \sum_{\tau=0}^{\infty} MI(v_\tau, y_\tau)$, where $v_\tau(\cdot)$ denotes the network output and $y_\tau(t) = PARITY(u(t-\tau), u(t-\tau-1), u(t-\tau-2))$ is the target output.

Since the separation property is important for the computational properties of the network, one would expect that the computational performance peaks near the critical line. This was confirmed with simulations where the computational task was to compute the delayed 3-bit parity[8] of the input signal. The readout neuron was implemented by a simple linear classifier $C(\mathbf{x}(t)) = \Theta(\mathbf{w} \cdot \mathbf{x}(t) + w_0)$ which was trained with linear regression. Note that the parity task is quite complex since it partitions the set of all inputs into two classes which are not linearly separable (and can therefore not be represented by the linear readout alone), and it requires memory. Figure 6.4b shows that the highest performance is achieved for parameter values close to the critical line, although it is not clear why the performance drops for increasing values of $\bar{u}$. In contrast to preceding work (Langton, 1990; Packard, 1988), the networks used were not optimized for a specific task. Only the linear readout was trained to extract the specific information from the state of the system. This is important since it decouples the dynamics of the network from a specific task.

6.8   Self-Organized Criticality

Are there systems in nature with dynamics located at the edge of chaos? Since the edge of chaos is a small boundary region in the space of possible dynamics, only a vanishingly small fraction of systems should operate in this dynamical regime. However, it was argued that such "critical" systems are abundant in nature (see, e.g., Bak et al., 1988). How is this possible if critical dynamics may occur only accidentally in nature? Bak and collaborators argue that a class of dissipative coupled systems naturally evolve toward critical dynamics (Bak et al., 1988). This phenomenon was termed self-organized criticality (SOC) and it was demonstrated with a model of a sand pile. Imagine building up a sand pile by randomly adding sand to the pile, a grain at a time. As sand is added, the slope will increase. Eventually, the slope will reach a critical value. Whenever the local slope of the pile is too steep, sand will slide off, therefore reducing the slope locally. On the other hand, if one starts with a very steep pile it will collapse and reach the critical slope from the other direction.

In neural systems, the topology of the network and the synaptic weights strongly influence the dynamics. Since the amount of genetically determined connections between neurons is limited, self-organizing processes during brain development as well as learning processes are assumed to play a key role in regulating the dynamics of biological neural networks (Bornholdt and Röhl, 2003). Although the dynamics is a global property of the network, biologically plausible learning rules try to estimate the global dynamics from information available at the local synaptic level and they only change local parameters. Several SOC rules have been suggested (Bornholdt and Röhl, 2003, 2000; Christensen et al., 1998; Natschläger et al., 2005). In Bornholdt and Röhl (2003), the degree of connectivity was regulated in a locally connected network (i.e., only neighboring neurons are connected) with stochastic state update dynamics. A local rewiring rule was used which is related to Hebbian learning. The main idea of this rule is that the average correlation between the activities of two neurons contains information about the global dynamics. This rule only relies on information available on the local synaptic level.

Self-organized criticality in systems with online input streams (as discussed in section 6.6) was considered in Natschläger et al. (2005). According to section 6.6, the dynamics of a threshold network is at the critical line if the bit-flip probability P_{BF} (averaged over the external and internal input statistics) is equal to $1/K$, where K is the number of inputs to a unit. The idea is to estimate the bit-flip probability of a unit by the mean distance of the internal activation of that unit from the firing threshold. This distance is called the *margin*. Intuitively, a node with an activation much higher or lower than its firing threshold is rather unlikely to change its output if a single bit in its inputs is flipped. Each node i then applies synaptic scaling to its weights w_{ij} in order to adjust itself toward the critical line:

$$w_{ij}(t+1) = \begin{cases} \frac{1}{1+\nu} \cdot w_{ij}(t) & \text{if } P_{BF}^{est_i}(t) > \frac{1}{K} \\ (1+\nu) \cdot w_{ij}(t) & \text{if } P_{BF}^{est_i}(t) < \frac{1}{K} \end{cases}, \qquad (6.5)$$

where $0 < \nu \ll 1$ is the learning rate and $P_{BF}^{est_i}(t)$ is an estimate of the bit-flip probability P_{BF}^i of unit i. It was shown by simulations that this rule keeps the dynamics in the critical regime, even if the input statistics change. The computational capabilities of randomly chosen circuits with this synaptic scaling rule acting online during computation were tested in a setup similar to that discussed in section 6.7.

The performance of these networks was as high as for circuits where the parameters were a priori chosen in the critical regime, and they stayed in this region. This shows that systems can perform specific computations while still being able to react to changing input statistics in a flexible way.

6.9   Toward the Analysis of Biological Neural Systems

Do cortical microcircuits operate at the edge of chaos? If biology makes extensive use of the rich internal dynamics of cortical circuits, then the previous considerations would suggest this idea. However, the neural elements in the brain are quite different from the elements discussed so far. Most important, biological neurons communicate with spikes, discrete events in continuous time. In this section, we will investigate the dynamics of spiking circuits and ask: In what dynamical regimes are neural microcircuits computationally powerful? We propose in this section a conceptual framework and new quantitative measures for the investigation of this question (see also Maass et al., 2005).

In order to make this approach feasible, in spite of numerous unknowns regarding synaptic plasticity and the distribution of electrical and biochemical signals impinging on a cortical microcircuit, we make in the present first step of this approach the following simplifying assumptions:

1. Particular neurons ("readout neurons") learn via synaptic plasticity to extract specific information encoded in the spiking activity of neurons in the circuit.

2. We assume that the cortical microcircuit itself is highly recurrent, but that the impact of feedback that a readout neuron might send back into this circuit can be neglected.[9]

3. We assume that synaptic plasticity of readout neurons enables them to learn arbitrary linear transformations. More precisely, we assume that the input to such readout neurons can be approximated by a term $\sum_{i=1}^{n-1} w_i x_i(t)$, where $n-1$ is the number of presynaptic neurons, $x_i(t)$ results from the output spike train of the ith presynaptic neuron by filtering it according to the low-pass filtering property of the membrane of the readout neuron,[10] and w_i is the efficacy of the synaptic connection. Thus $w_i x_i(t)$ models the time course of the contribution of previous spikes from the ith presynaptic neuron to the membrane potential at the soma of this readout neuron. We will refer to the vector $\mathbf{x}(t)$ as the "circuit state at time t" (although it is really only that part of the circuit state which is directly observable by readout neurons).

All microcircuit models that we consider are based on biological data for generic cortical microcircuits (as described in section 6.9.1), but have different settings of their parameters.

6.9.1   Models for Generic Cortical Microcircuits

Our empirical studies were performed on a large variety of models for generic cortical microcircuits (we refer to Maass et al., 2004, , for more detailed definitions and explanations). All circuit models consisted of leaky integrate-and-fire neurons[11] and biologically quite realistic models for dynamic synapses.[12] Neurons (20% of which were randomly chosen to be inhibitory) were located on the grid points of a 3D grid of dimensions $6 \times 6 \times 15$ with edges of unit length. The probability of a synaptic connection from neuron a to neuron b was proportional to $exp(-D^2(a,b)/\lambda^2)$, where $D(a,b)$ is the Euclidean distance between a and b, and λ is a spatial connectivity constant (not to be confused with the λ parameter used by Langton). Synaptic efficiencies w were chosen randomly from distributions that reflect biological data (as in Maass et al., 2002), with a common scaling factor W_{scale}.

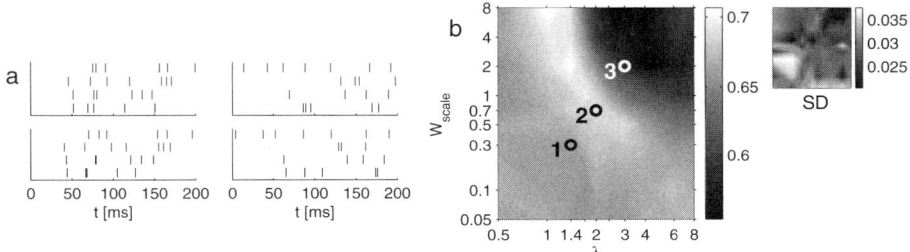

Figure 6.5   Performance of different types of neural microcircuit models for classification of spike patterns. (a) In the top row are two examples of the 80 spike patterns that were used (each consisting of 4 Poisson spike trains at 20 Hz over 200 ms), and in the bottom row are examples of noisy variations (Gaussian jitter with SD 10 ms) of these spike patterns which were used as circuit inputs. (b) Fraction of examples (for 200 test examples) that were correctly classified by a linear readout (trained by linear regression with 500 training examples). Results are shown for 90 different types of neural microcircuits C with λ varying on the x-axis and W_{scale} on the y-axis (20 randomly drawn circuits and 20 target classification functions randomly drawn from the set of 2^{80} possible classification functions were tested for each of the 90 different circuit types, and resulting correctness rates were averaged). Circles mark three specific choices of λ, W_{scale} pairs for comparison with other figures; see fig. 6.6. The standard deviation of the result is shown in the inset on the upper right.

Linear readouts from circuits with $n-1$ neurons were assumed to compute a weighted sum $\sum_{i=1}^{n-1} w_i x_i(t) + w_0$ (see section 6.9). In order to simplify notation we assume that the vector $\mathbf{x}(t)$ contains an additional constant component $x_0(t) = 1$, so that one can write $\mathbf{w} \cdot \mathbf{x}(t)$ instead of $\sum_{i=1}^{n-1} w_i x_i(t) + w_0$. In the case of classification tasks we assume that the readout outputs 1 if $\mathbf{w} \cdot \mathbf{x}(t) \geq 0$, and 0 otherwise.

In order to investigate the influence of synaptic connectivity on computational performance, neural microcircuits were drawn from this distribution for 10 different values of λ (which scales the number and average distance of synaptically connected neurons) and 9 different values of W_{scale} (which scales the efficacy of all synaptic connections). Twenty microcircuit models C were drawn for each of these 90 different assignments of values to λ and W_{scale}. For each circuit a linear readout was trained to perform one (randomly chosen) out of 2^{80} possible classification tasks on noisy variations u of 80 fixed spike patterns as circuit inputs u. See fig. 6.5 for two examples of such spike patterns. The target performance of any such circuit was to output at time $t = 200$ ms the class (0 or 1) of the spike pattern from which the preceding circuit input had been generated (for some arbitrary partition of the 80 fixed spike patterns into two classes). Each spike pattern u consisted of four Poisson spike trains over 200 ms. Performance results are shown in fig. 6.5b for 90 different types of neural microcircuit models.

6.9.2   Locating the Edge of Chaos in Neural Microcircuit Models

It turns out that the previously considered characterizations of the edge of chaos are not too successful in identifying those parameter values in the map of fig. 6.5b that yield circuits with large computational power (Maass et al., 2005). The reason is that large initial state differences (as they are typically caused by different spike input patterns) tend to yield for most values of the circuit parameters nonzero state differences not only while the online spike inputs are different, but also long afterward when the online inputs agree during subsequent seconds (even if the random internal noise is identical in both trials). But if one applies the definition of the edge of chaos via Lyapunov exponents (see Kantz and Schreiber, 1997), the resulting edge of chaos lies for the previously introduced type of computations (classification of noisy spike templates by a trained linear readout) in the region of the best computational performance (see the map in fig. 6.5b, which is repeated for easier comparison in fig. 6.6d). For this definition one looks for the exponent $\mu \in \mathbb{R}$ that provides through the formula

$$\delta_{\Delta T} \approx \delta_0 \cdot e^{\mu \Delta T}$$

the best estimate of the state separation $\delta_{\Delta T}$ at time ΔT after the computation was started in two trials with an initial state difference δ_0. We generalize this analysis to the case with online input by choosing exactly the same online input (and the same random noise) during the intervening time interval of length ΔT, and by averaging the resulting state differences $\delta_{\Delta T}$ over many random choices of such online inputs (and internal noise). As in the classical case with off-line input it turns out to be essential to apply this estimate for $\delta_0 \to 0$, since $\delta_{\Delta T}$ tends to saturate for each fixed value δ_0. This can be seen in fig. 6.6a, which shows results of this experiment for a δ_0 that results from moving a single spike that occurs in the online input at time $t = 1$s by 0.5 ms. This experiment was repeated for three

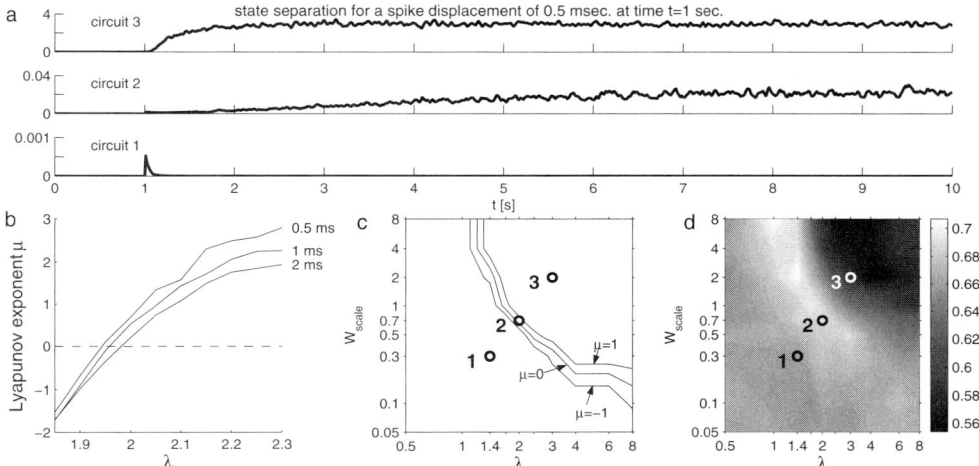

Figure 6.6   Analysis of small input differences for different types of neural micro-circuit models as specified in section 6.9.1. Each circuit C was tested for two arrays u and v of 4 input spike trains at 20 Hz over 10 s that differed only in the timing of a single spike at time $t = 1$ s. (a) A spike at time $t = 1$ s was delayed by 0.5 ms. Temporal evolution of Euclidean differences between resulting circuit states $\mathbf{x}_u(t)$ and $\mathbf{x}_v(t)$ with 3 different values of λ, W_{scale} according to the three points marked in panel c. For each parameter pair, the average state difference of 40 randomly drawn circuits is plotted. (b) Lyapunov exponents μ along a straight line between the points marked in panel c with different delays of the delayed spike. The delay is denoted on the right of each line. The exponents were determined for the average state difference of 40 randomly drawn circuits. (c) Lyapunov exponents μ for 90 different types of neural microcircuits C with λ varying on the x-axis and W_{scale} on the y-axis (the exponents were determined for the average state difference of 20 randomly drawn circuits for each parameter pair). A spike in u at time $t = 1$ s was delayed by 0.5 ms. The contour lines indicate where μ crosses the values -1, 0, and 1. (d) Computational performance of these circuits (same as fig. 6.5b), shown for comparison with panel c.

different circuits with parameters chosen from the 3 locations marked on the map in fig. 6.6c. By determining the best-fitting μ for $\Delta T = 1.5$s for three different values of δ_0 (resulting from moving a spike at time $t = 1$s by 0.5, 1, 2 ms) one gets the dependence of this Lyapunov exponent on the circuit parameter λ shown in fig. 6.6b (for values of λ and W_{scale} on a straight line between the points marked in the map of fig. 6.6c). The middle curve in fig. 6.6c shows for which values of λ and W_{scale} the Lyapunov exponent is estimated to have the value 0. By comparing it with those regions on this parameter map where the circuits have the largest computational power (for the classification of noisy spike patterns, see fig. 6.6d), one sees that this line runs through those regions which yield the largest computational power for these computations. We refer to Mayor and Gerstner (2005) for other recent work on studies of the relationship between the edge of chaos and the computational power of spiking neural circuit models.

Although this estimated edge of chaos coincides quite well with points of best computational performance, it remains an unsatisfactory tool for predicting parameter regions with large computational power for three reasons:

1. Since the edge of chaos is a lower-dimensional manifold in a parameter map (in this case a curve in a 2D map), it cannot predict the (full dimensional) regions of a parameter map with high computational performance (e.g., the regions with light shading in fig. 6.5b).

2. The edge of chaos does not provide intrinsic reasons *why* points of the parameter map yield small or large computational power.

3. It turns out that in some parameter maps *different* regions provide circuits with large computational power for *different* classes of computational tasks (as shown in Maass et al. (2005), for computations on spike patterns and for computations with firing rates). But the edge of chaos can at best single out peaks for *one* of these regions. Hence it cannot possibly be used as a universal predictor of maximal computational power for all types of computational tasks.

These three deficiencies suggest that one has to think about different strategies to approach the central question of this chapter. The strategy we will pursue in the following is based on the assumption that the computational function of cortical microcircuits is not fully genetically encoded, but rather emerges through various forms of plasticity ("learning") in response to the actual distribution of signals that the neural microcircuit receives from its environment. From this perspective the question about the computational function of cortical microcircuits C turns into the following questions:

- What functions (i.e., maps from circuit inputs to circuit outputs) can the circuit C *learn* to compute?

- How well can the circuit C generalize a specific learned computational function to new inputs?

In the following, we propose quantitative criteria based on rigorous mathematical principles for evaluating a neural microcircuit C with regard to these two questions. We will compare in section 6.9.5 the predictions of these quantitative measures with the actual computational performance achieved by neural microcircuit models as discussed in section 6.9.1.

6.9.3   A Measure for the Kernel-Quality

One expects from a powerful computational system that significantly different input streams cause significantly different internal states and hence may lead to different outputs. Most real-world computational tasks require that the circuit give a desired output not just for two, but for a fairly large number m of significantly different inputs. One could of course test whether a circuit C can separate each of the $\binom{m}{2}$ pairs of such inputs. But even if the circuit can do this, we do not know whether a neural readout from such circuit would be able to produce given target outputs for these m inputs.

Therefore we propose here the *linear separation property* as a more suitable quantitative measure for evaluating the computational power of a neural microcircuit (or more precisely, the kernel quality of a circuit; see below). To evaluate the linear separation property of a circuit C for m different inputs $u_1, \ldots, u_m$ (which are in the following always functions of time, i.e., input streams such as, for example, multiple spike trains) we compute the rank of the $n \times m$ matrix M whose columns are the circuit states $\mathbf{x}_{u_i}(t_0)$ that result at some fixed time t_0 for the preceding input stream u_i. If this matrix has rank m, then it is *guaranteed* that *any* given assignment of target outputs $y_i \in \mathrm{R}$ at time t_0 for the inputs u_i can be implemented by this circuit C (in combination with a linear readout). In particular, each of the 2^m possible binary classifications of these m inputs can then be carried out by a *linear* readout from this fixed circuit C. Obviously such insight is much more informative than a demonstration that some *particular* classification task can be carried out by such circuit C. If the rank of this matrix M has a value $r < m$, then this value r can still be viewed as a measure for the computational power of this circuit C, since r is the number of "degrees of freedom" that a linear readout has in assigning target outputs y_i to these inputs u_i (in a way which can be made mathematically precise with concepts of linear algebra). Note that this rank measure for the linear separation property of a circuit C may be viewed as an empirical measure for its *kernel quality*, i.e., for the complexity and diversity of nonlinear operations carried out by C on its input stream in order to boost the classification power of a subsequent *linear* decision hyperplane (see Vapnik, 1998).

6.9.4   A Measure for the Generalization-Capability

Obviously the preceding measure addresses only one component of the computational performance of a neural circuit C. Another component is its capability to *generalize* a learned computational function to *new* inputs. Mathematical criteria for generalization capability are derived by Vapnik (1998) (see ch. 4 in Cherkassky and Mulier, 1998, for a compact account of results relevant for our arguments). According to this mathematical theory one can quantify the generalization capability of any learning device in terms of the VC-dimension of the class $\mathcal{H}$ of hypotheses that are potentially used by that learning device.[13] More precisely: if VC-dimension ($\mathcal{H}$) is substantially smaller than the size of the training set S_{train}, one can prove that this learning device generalizes well, in the sense that the hypothesis (or input-output map) produced by this learning device is likely to have for new examples an error rate which is not much higher than its error rate on S_{train}, provided that the new examples are drawn from the same distribution as the training examples (see eq. 4.22 in Cherkassky and Mulier, 1998).

We apply this mathematical framework to the class $\mathcal{H}_C$ of all maps from a set

S_{univ} of inputs u into $\{0, 1\}$ that can be implemented by a circuit C. More precisely: $\mathcal{H}_C$ consists of all maps from S_{univ} into $\{0, 1\}$ that could possibly be implemented by a linear readout from circuit C with fixed internal parameters (weights etc.) but arbitrary weights $\mathbf{w} \in \mathbb{R}^n$ of the readout (which classifies the circuit input u as belonging to class 1 if $\mathbf{w} \cdot \mathbf{x}_u(t_0) \geq 0$, and to class 0 if $\mathbf{w} \cdot \mathbf{x}_u(t_0) < 0$).

Whereas it is very difficult to achieve tight theoretical bounds for the VC-dimension of even much simpler neural circuits (see Bartlett and Maass, 2003), one can efficiently estimate the VC-dimension of the class $\mathcal{H}_C$ that arises in our context for some finite ensemble S_{univ} of inputs (that contains all examples used for training or testing) by using the following mathematical result (which can be proved with the help of Radon's theorem):

Theorem 6.1

Let r be the rank of the $n \times s$ matrix consisting of the s vectors $\mathbf{x}_u(t_0)$ for all inputs u in S_{univ} (we assume that S_{univ} is finite and contains s inputs). Then $r \leq$ VC-dimension$(\mathcal{H}_C) \leq r + 1$.

Proof Idea. Fix some inputs $u_1, \ldots, u_r$ in S_{univ} so that the resulting r circuit states $\mathbf{x}_{u_i}(t_0)$ are linearly independent. The first inequality is obvious since this set of r linearly independent vectors can be shattered by linear readouts from the circuit C. To prove the second inequality one assumes for a contradiction that there exists a set $v_1, \ldots, v_{r+2}$ of $r+2$ inputs in S_{univ} so that the corresponding set of $r+2$ circuit states $\mathbf{x}_{v_i}(t_0)$ can be shattered by linear readouts. This set M of $r+2$ vectors is contained in the r-dimensional space spanned by the linearly independent vectors $\mathbf{x}_{u_1}(t_0), \ldots, \mathbf{x}_{u_r}(t_0)$. Therefore Radon's theorem implies that M can be partitioned into disjoint subsets M_1, M_2 whose convex hulls intersect. Since these sets M_1, M_2 cannot be separated by a hyperplane, it is clear that no linear readout exists that assigns value 1 to points in M_1 and value 0 to points in M_2. Hence $M = M_1 \cup M_2$ is not shattered by linear readouts, a contradiction to our assumption. ∎

We propose to use the rank r defined in theorem 6.1 as an estimate of VC-dimension$(\mathcal{H}_C)$, and hence as a measure that informs us about the generalization capability of a neural microcircuit C. It is assumed here that the set S_{univ} contains many noisy variations of the same input signal, since otherwise learning with a randomly drawn training set $S_{train} \subseteq S_{univ}$ has no chance to generalize to new noisy variations. Note that each family of computational tasks induces a particular notion of what aspects of the input are viewed as noise, and what input features are viewed as signals that carry information which is relevant for the target output for at least one of these computational tasks. For example, for computations on spike patterns some small jitter in the spike timing is viewed as noise. For computations on firing rates even the sequence of interspike intervals and the temporal relations between spikes that arrive from different input sources are viewed as noise, as long as these input spike trains represent the same firing rates.

An example for the former computational task was discussed in section 6.9.1. This task was to output at time $t = 200$ ms the class (0 or 1) of the spike pattern

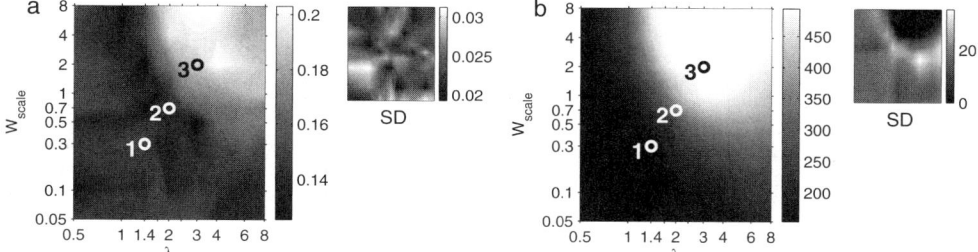

Figure 6.7   Measuring the generalization capability of neural microcircuit models. (a) Test error minus train error (error was measured as the fraction of examples that were misclassified) in the spike pattern classification task discussed in section 6.9.1 for 90 different types of neural microcircuits (as in fig. 6.5b). The standard deviation is shown in the inset on the upper right. (b) Generalization capability for spike patterns: estimated VC-dimension of $\mathcal{H}_C$ (for a set S_{univ} of inputs u consisting of 500 jittered versions of 4 spike patterns), for 90 different circuit types (average over 20 circuits; for each circuit, the average over 5 different sets of spike patterns was used). The standard deviation is shown in the inset on the upper right. See section 6.9.5 for details.

from which the preceding circuit input had been generated (for some arbitrary partition of the 80 fixed spike patterns into two classes; see section 6.9.1). For a poorly generalizing network, the difference between train and test error is large. One would suppose that this difference becomes large as the network dynamics become more and more chaotic. This is indeed the case; see fig. 6.7a. The transition is is pretty well predicted by the estimated VC-dimension of $\mathcal{H}_C$; see fig. 6.7b.

6.9.5   Evaluating the Influence of Synaptic Connectivity on Computational Performance

We now test the predictive quality of the two proposed measures for the computational power of a microcircuit on spike patterns. One should keep in mind that the proposed measures do not attempt to test the computational capability of a circuit for one particular computational task, but rather for *any* distribution on S_{univ} and for a very large (in general, infinitely large) family of computational tasks that have in common only a particular bias regarding which aspects of the incoming spike trains may carry information that is relevant for the target output of computations, and which aspects should be viewed as noise. fig. 6.8a explains why the lower left part of the parameter map in fig. 6.5b is less suitable for any such computation, since there the kernel quality of the circuits is too low.[14] Figure 6.8b explains why the upper right part of the parameter map in fig. 6.5b is less suitable, since a higher VC-dimension (for a training set of fixed size) entails poorer generalization capability. We are not aware of a theoretically founded way of combining both measures into a single value that predicts overall computational

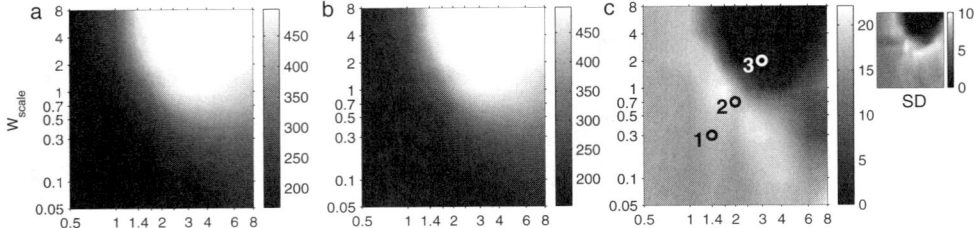

Figure 6.8 Values of the proposed measures for computations on spike patterns. (a) Kernel quality for spike patterns of 90 different circuit types (average over 20 circuits, mean $SD = 13$). (b) Generalization capability for spike patterns: estimated VC-dimension of $\mathcal{H}_C$ (for a set S_{univ} of inputs u consisting of 500 jittered versions of 4 spike patterns), for 90 different circuit types (same as fig. 6.7b). (c) Difference of both measures (the standard deviation is shown in the inset on the upper right). This should be compared with actual computational performance plotted in fig. 6.5b.

performance. But if one just takes the difference of both measures (after scaling each linearly into a common range $[0,1]$), then the resulting number (see fig. 6.8c) predicts quite well which types of neural microcircuit models perform well for the particular computational tasks considered in Figure 6.5b.[15]

Results of further tests of the predictive power of these measures are reported in Maass et al. (2005). These tests have been applied there to a completely different parameter map, and to diverse classes of computational tasks.

6.10   Conclusions

The need to understand computational properties of complex dynamical systems is becoming more urgent. New experimental methods provide substantial insight into the inherent dynamics of the computationally most powerful classes of dynamical systems that are known: neural systems and gene regulation networks of biological organisms. More recent experimental data show that simplistic models for computations in such systems are not adequate, and that new concepts and methods have to be developed in order to understand their computational function. This short review has shown that several old ideas regarding computations in dynamical systems receive new relevance in this context, once they are transposed into a more realistic conceptual framework that allows us to analyze also online computations on continuous input streams. Another new ingredient is the investigation of the temporal evolution of information in a dynamical system from the perspective of models for the (biological) user of such information, i.e., from the perspective of neurons that receive inputs from several thousand presynaptic neurons in a neural circuit, and from the perspective of gene regulation mechanisms that involve thousands of transcription factors. Empirical evidence from the area of machine learning supports the hypothesis that readouts of this type, which are able to sample not just two or three, but thousands of coordinates of the state vector of a dynamical system, impose different (and in general, less obvious) constraints on the dynamics of a high-dimensional dynamical system in order to employ such system for complex computations on continuous input streams. One might conjecture that unsupervised learning and regulation processes in neural systems adapt the system dynamics in such a way that these constraints are met. Hence, suitable variations of the idea of self-organized criticality may help us to gain a system-level perspective of synaptic plasticity and other adaptive processes in neural systems.

Notes

[1] For the sake of completeness, we give here the definition of an attractor according to Strogatz (1994): He defines an attractor to be a closed set A with the following properties: (1) A is an *invariant set*: any trajectory $\mathbf{x}(t)$ that starts in A stays in A for all time. (2) A attracts an open set of initial conditions: there is an open set U containing A such that if $\mathbf{x}(t) \in U$, then the distance from $\mathbf{x}(t)$ to A tends to zero as $t \to \infty$. The largest such U is called the *basin of attraction* of A. (3) A is *minimal*: there is no proper subset of A that satisfies conditions 1 and 2.

[2] In Kauffman (1969), the inactive state of a variable is denoted by 0. We use -1 here for reasons of notational consistency.

[3] Here, x_i potentially depends on all other variables $x_1, \ldots, x_N$. The function f_i can always be restricted such that x_i is determined by the inputs to elements i only.

[4] In Kauffman (1993), a state cycle is also called an attractor. Because such state cycles can be unstable to most minimal perturbations, we will avoid the term *attractor* here.

[5] Wolfram considered automata with a neighborhood of five cells in total and two possible cell states. Since he considered "totalistic" transfer functions only (i.e., the function depends on the *sum* of the neighborhood states only), the number of possible transfer functions was small. Hence, the behavior of all such automata could be studied.

[6] In the case of two-state cellular automata, high λ values imply that most state transitions map to the single nonquiescent state that leads to ordered dynamics. The most heterogeneous rules are found at $\lambda = 0.5$.

[7] The model in Maass et al. (2002) was introduced in the context of biologically inspired neural microcircuits. The network consisted of spiking neurons. In Jaeger (2002), the network consisted of sigmoidal neurons.

[8] The delayed 3-bit parity of an input signal $u(\cdot)$ is given by $PARITY(u(t-\tau), u(t-\tau-1), u(t-\tau-2))$ for delays $\tau > 0$. The function $PARITY$ outputs 1 if the number of inputs which assume the value $\bar{u} + 1$ is odd and -1 otherwise.

[9] This assumption is best justified if such readout neuron is located, for example, in another brain area that receives massive input from many neurons in this microcircuit and only has diffuse backward projection. But it is certainly problematic and should be addressed in future elaborations of the present approach.

[10] One can be even more realistic and filter it also by a model for the short-term dynamics of the synapse into the readout neuron, but this turns out to make no difference for the analysis proposed in this chapter.

[11] Membrane voltage V_m modeled by $\tau_m \frac{dV_m}{dt} = -(V_m - V_{resting}) + R_m \cdot (I_{syn}(t) + I_{background} + I_{noise})$, where $\tau_m = 30$ ms is the membrane time constant, I_{syn} models synaptic inputs from other neurons in the circuits, $I_{background}$ models a constant unspecific background input, and I_{noise} models noise in the input. The membrane resistance R_m was chosen as $1 M\Omega$.

[12] Short-term synaptic dynamics was modeled according to Markram et al. (1998), with distributions of synaptic parameters U (initial release probability), D (time constant for depression), F (time constant for facilitation) chosen to reflect empirical data (see Maass et al., 2002, for details).

[13] The VC-dimension (of a class $\mathcal{H}$ of maps H from some universe S_{univ} of inputs into $\{0, 1\}$) is defined as the size of the largest subset $S \subseteq S_{univ}$ which can be *shattered* by $\mathcal{H}$. One says that $S \subseteq S_{univ}$ is shattered by $\mathcal{H}$ if for *every* map $f : S \to \{0, 1\}$ there exists a map H in $\mathcal{H}$ such that

$H(u) = f(u)$ for all $u \in S$ (this means that *every* possible binary classification of the inputs $u \in S$ can be carried out by some hypothesis H in $\mathcal{H}$).

[14]The rank of the matrix consisting of 500 circuit states $\mathbf{x}_u(t)$ for $t = 200$ ms was computed for 500 spike patterns over 200 ms as described in section 6.9.3; see fig. 6.5a. For each circuit, the average over five different sets of spike patterns was used.

[15]Similar results arise if one records the analog values of the circuit states with a limited precision of, say, 1%.

7   A Variational Principle for Graphical Models

Martin J. Wainwright and *Michael I. Jordan*

Graphical models bring together graph theory and probability theory in a powerful formalism for multivariate statistical modeling. In statistical signal processing—as well as in related fields such as communication theory, control theory, and bioinformatics—statistical models have long been formulated in terms of graphs, and algorithms for computing basic statistical quantities such as likelihoods and marginal probabilities have often been expressed in terms of recursions operating on these graphs. Examples include hidden Markov models, Markov random fields, the forward-backward algorithm, and Kalman filtering (Kailath et al., 2000; Pearl, 1988; Rabiner and Juang, 1993). These ideas can be understood, unified, and generalized within the formalism of graphical models. Indeed, graphical models provide a natural framework for formulating variations on these classical architectures, and for exploring entirely new families of statistical models.

The recursive algorithms cited above are all instances of a general recursive algorithm known as the *junction tree algorithm* (Lauritzen and Spiegelhalter, 1988). The junction tree algorithm takes advantage of factorization properties of the joint probability distribution that are encoded by the pattern of missing edges in a graphical model. For suitably sparse graphs, the junction tree algorithm provides a systematic and practical solution to the general problem of computing likelihoods and other statistical quantities associated with a graphical model. Unfortunately, many graphical models of practical interest are not "suitably sparse," so that the junction tree algorithm no longer provides a viable computational solution to the problem of computing marginal probabilities and other expectations. One popular source of methods for attempting to cope with such cases is the *Markov chain Monte Carlo* (MCMC) framework, and indeed there is a significant literature on the application of MCMC methods to graphical models (Besag and Green, 1993; Gilks et al., 1996). However, MCMC methods can be overly slow for practical applications in fields such as signal processing, and there has been significant interest in developing faster approximation techniques.

The class of *variational methods* provides an alternative approach to computing approximate marginal probabilities and expectations in graphical models. Roughly speaking, a variational method is based on casting a quantity of interest (e.g., a likelihood) as the solution to an optimization problem, and then solving a perturbed version of this optimization problem. Examples of variational methods for computing approximate marginal probabilities and expectations include the "loopy" form of the *belief propagation* or *sum-product* algorithm (McEliece et al., 1998; Yedidia et al., 2001) as well as a variety of so-called *mean field* algorithms (Jordan et al., 1999; Zhang, 1996).

Our principal goal in this chapter is to give a mathematically precise and computationally oriented meaning to the term *variational* in the setting of graphical models—a meaning that reposes on basic concepts in the field of convex analysis (Rockafellar, 1970). Compared to the somewhat loose definition of *variational* that is often encountered in the graphical models literature, our characterization has certain advantages, both in clarifying the relationships among existing algorithms, and in permitting fuller exploitation of the general tools of convex optimization in the design and analysis of new algorithms. Briefly, the core issues can be summarized as follows. In order to define an optimization problem, it is necessary to specify both a cost function to be optimized, and a constraint set over which the optimization takes place. Reflecting the origins of most existing variational methods in statistical physics, developers of variational methods generally express the function to be optimized as a "free energy," meaning a functional on probability distributions. The set to be optimized over is often left implicit, but it is generally taken to be the set of all probability distributions. A basic exercise in constrained optimization yields the Boltzmann distribution as the general form of the solution. While useful, this derivation has two shortcomings. First, the optimizing argument is a joint probability distribution, not a set of marginal probabilities or expectations. Thus, the derivation leaves us short of our goal of a variational representation for computing marginal probabilities. Second, the set of all probability distributions is a very large set, and formulating the optimization problem in terms of such a set provides little guidance in the design of computationally efficient approximations.

Our approach addresses both of these issues. The key insight is to formulate the optimization problem not over the set of all probability distributions, but rather over a finite-dimensional set $\mathcal{M}$ of *realizable mean parameters*. This set is convex in general, and it is a polytope in the case of discrete random variables. There are several natural ways to approximate this convex set, and a broad range of extant algorithms turn out to involve particular choices of approximations. In particular, as we will show, the "loopy" form of the sum-product or belief propagation algorithm involves an *outer approximation* to $\mathcal{M}$, whereas the more classical mean field algorithms, on the other hand, involve an *inner approximation* to the set $\mathcal{M}$. The characterization of belief propagation as an optimization over an outer approximation of a certain convex set does not arise readily within the standard formulation of variational methods. Indeed, given an optimization over all possible probability distributions, it is difficult to see how to move "outside" of such a set. Similarly, while the standard formulation does provide some insight into the differences between belief propagation and mean field methods (in that they optimize different "free energies"), the standard formulation does not involve the set $\mathcal{M}$, and hence does not reveal the fundamental difference in terms of outer versus inner approximations.

The core of the chapter is a variational characterization of the problem solved by the junction tree algorithm—that of computing exact marginal probabilities and expectations associated with subsets of nodes in a graphical model. These probabilities are obtained as the maximizing arguments of an optimization over the set $\mathcal{M}$. Perhaps surprisingly, this problem is a convex optimization problem for a broad class of graphical models. With this characterization in hand, we show how variational methods arise as "relaxations"—that is, simplified optimization problems that involve some approximation of the constraint set, the cost function, or both. We show how a variety of standard variational methods, ranging from classical mean-field to cluster variational methods, fit within this framework. We also discuss new methods that emerge from this framework, including a relaxation based on semidefinite constraints and a link between reweighted forms of the max-product algorithm and linear programming.

The remainder of the chapter is organized as follows. The first two sections are devoted to basics: section 7.1 provides an overview of graphical models and section 7.2 is devoted to a brief discussion of exponential families. In section 7.3, we develop a general variational representation for computing marginal probabilities and expectations in exponential families. section 7.4 illustrates how various exact methods can be understood from this perspective. The rest of the chapter— sections 7.5 through 7.7—is devoted to the exploration of various relaxations of this exact variational principle, which in turn yield various algorithms for computing approximations to marginal probabilities and other expectations.

7.1   Background

7.1.1   Graphical Models

A graphical model consists of a collection of probability distributions that factorize according to the structure of an underlying graph. A graph $G = (V, E)$ is formed by a collection of vertices V and a collection of edges E. An edge consists of a pair of vertices, and may either be directed or undirected. Associated with each vertex $s \in V$ is a random variable x_s taking values in some set $\mathcal{X}_s$, which may either be continuous (e.g., $\mathcal{X}_s = \mathrm{R}$) or discrete (e.g., $\mathcal{X}_s = \{0, 1, \dots, m-1\}$). For any subset A of the vertex set V, we define $x_A := \{x_s \mid s \in A\}$.

Directed Graphical Models In the directed case, each edge is directed from parent to child. We let $\pi(s)$ denote the set of all parents of given node $s \in V$. (If s has no parents, then the set $\pi(s)$ should be understood to be empty.) With this notation, a *directed graphical model* consists of a collection of probability distributions that factorize in the following way:

$$p(\mathbf{x}) = \prod_{s \in V} p(x_s \mid x_{\pi(s)}). \tag{7.1}$$

It can be verified that our use of notation is consistent, in that $p(x_s \mid x_{\pi(s)})$ is, in fact, the conditional distribution for the global distribution $p(\mathbf{x})$ thus defined.

Undirected Graphical Models In the undirected case, the probability distribution factorizes according to functions defined on the *cliques* of the graph (i.e., fully connected subsets of V). In particular, associated with each clique C is a *compatibility function* $\psi_C : \mathcal{X}^n \to R_+$ that depends only on the subvector x_C. With this notation, an *undirected graphical model* (also known as a *Markov random field*) consists of a collection of distributions that factorize as

$$p(\mathbf{x}) = \frac{1}{Z} \prod_C \psi_C(x_C), \tag{7.2}$$

where the product is taken over all cliques of the graph. The quantity Z is a constant chosen to ensure that the distribution is normalized. In contrast to the directed case 7.1, in general the compatibility functions ψ_C need not have any obvious or direct relation to local marginal distributions.

Families of probability distributions as defined as in equation 7.1 or 7.2 also have a characterization in terms of conditional independencies among subsets of random variables. We will not use this characterization in this chapter, but refer the interested reader to Lauritzen (1996) for a full treatment.

7.1.2   Inference Problems and Exact Algorithms

Given a probability distribution $p(\cdot)$ defined by a graphical model, our focus will be solving one or more of the following *inference problems*:

1. computing the likelihood;

2. computing the marginal distribution $p(x_A)$ over a particular subset $A \subset V$ of nodes;

3. computing the conditional distribution $p(x_A \mid x_B)$, for disjoint subsets A and B, where $A \cup B$ is in general a proper subset of V;

4. computing a mode of the density (i.e., an element $\widehat{\mathbf{x}}$ in the set $\arg\max_{\mathbf{x} \in \mathcal{X}^n} p(\mathbf{x})$).

Problem 1 is a special case of problem 2, because the likelihood is the marginal probability of the observed data. The computation of a conditional probability in problem 3 is similar in that it also requires marginalization steps, an initial one to obtain the numerator $p(x_A, x_B)$, and a further step to obtain the denominator $p(x_B)$. In contrast, the problem of computing modes stated in problem 4 is fundamentally different, since it entails maximization rather than integration. Although problem 4 is not the main focus of this chapter, there are important connections between the problem of computing marginals and that of computing modes; these are discussed in section 7.7.2.

To understand the challenges inherent in these inference problems, consider the case of a discrete random vector $\mathbf{x} \in \mathcal{X}^n$, where $\mathcal{X}_s = \{0, 1, \ldots, m-1\}$ for each vertex $s \in V$. A naive approach to computing a marginal at a single node—say $p(x_s)$—entails summing over all configurations of the form $\{\mathbf{x}' \mid x_s' = x_s\}$. Since this set has m^{n-1} elements, it is clear that a brute-force approach will rapidly become intractable as n grows. Similarly, computing a mode entails solving an integer programming problem over an exponential number of configurations. For continuous random vectors, the problems are no easier[1] and typically harder, since they require computing a large number of integrals.

Both directed and undirected graphical models involve factorized expressions for joint probabilities, and it should come as no surprise that exact inference algorithms treat them in an essentially identical manner. Indeed, to permit a simple unified treatment of inference algorithms, it is convenient to convert directed models to undirected models and to work exclusively within the undirected formalism. Any directed graph can be converted, via a process known as moralization (Lauritzen and Spiegelhalter, 1988), to an undirected graph that—at least for the purposes of solving inference problems—is equivalent. Throughout the rest of the chapter, we assume that this transformation has been carried out.

Message-passing on trees   For graphs without cycles—also known as *trees*—these inference problems can be solved exactly by recursive "message-passing" algorithms of a dynamic programming nature, with a computational complexity that scales only linearly in the number of nodes. In particular, for the case of computing marginals, the dynamic programming solution takes the form of a general algorithm known as the *sum-product algorithm*, whereas for the problem of computing modes it takes the form of an analogous algorithm known as the *max-product algorithm*. Here we provide a brief description of these algorithms; further details can be found in various sources (Aji and McEliece, 2000; Kschischang and Frey, 1998; Lauritzen and Spiegelhalter, 1988; Loeliger, 2004).

We begin by observing that the cliques of a tree-structured graph $T = (V, E(T))$ are simply the individual nodes and edges. As a consequence, any tree-structured graphical model has the following factorization:

$$p(\mathbf{x}) = \frac{1}{Z} \prod_{s \in V} \psi_s(x_s) \prod_{(s,t) \in E(T)} \psi_{st}(x_s, x_t). \tag{7.3}$$

Here we describe how the sum-product algorithm computes the marginal distribution $\mu_s(x_s) := \sum_{\{\mathbf{x}' \mid x_s' = x_s\}} p(\mathbf{x})$ for every node of a tree-structured graph. We will focus in detail on the case of discrete random variables, with the understanding that the computations carry over (at least in principle) to the continuous case by replacing sums with integrals.

Sum-product algorithm   The essential principle underlying the sum-product algorithm on trees is divide and conquer: we solve a large problem by breaking it down into a sequence of simpler problems. The tree itself provides a natural way to break down the problem as follows. For an arbitrary $s \in V$, consider the set of its neighbors $\mathcal{N}(s) = \{u \in V \mid (s, u) \in E\}$. For each $u \in \mathcal{N}(s)$, let $T_u = (V_u, E_u)$ be the subgraph formed by the set of nodes (and edges joining them) that can be reached from u by paths that *do not* pass through node s. The key property of a tree is that each such subgraph T_u is again a tree, and T_u and T_v are disjoint for $u \neq v$. In this way, each vertex $u \in \mathcal{N}(s)$ can be viewed as the root of a subtree T_u, as illustrated in fig. 7.1a. For each subtree T_t, we define $x_{V_t} := \{x_u \mid u \in V_t\}$. Now consider the collection of terms in equation 7.3 associated with vertices or edges in T_t: collecting all of these terms yields a subproblem $p(x_{V_t}; T_t)$ for this subtree.

Now the conditional independence properties of a tree allow the computation of the marginal at node μ_s to be broken down into a product of the form

$$\mu_s(x_s) \;\propto\; \psi_s(x_s) \prod_{t \in \mathcal{N}(s)} M_{ts}^*(x_s). \tag{7.4}$$

Each term $M_{ts}^*(x_s)$ in this product is the result of performing a partial summation for the subproblem $p(x_{V_t}; T_t)$ in the following way:

$$M_{ts}^*(x_s) \;=\; \sum_{\{x_{T_t}' \mid x_s' = x_s\}} \psi_{st}(x_s, x_t') \; p(x_{T_t}'; T_t). \tag{7.5}$$

For fixed x_s, the subproblem defining $M_{ts}^*(x_s)$ is again a tree-structured summation, albeit involving a subtree T_t *smaller* than the original tree T. Therefore, it too can be broken down recursively in a similar fashion. In this way, the marginal at node s can be computed by a series of recursive updates.

Rather than applying the procedure described above to each node separately, the *sum-product algorithm* computes the marginals for all nodes simultaneously and in parallel. At each iteration, each node t passes a "message" to each of its neighbors $u \in \mathcal{N}(t)$. This message, which we denote by $M_{tu}(x_u)$, is a function of the possible states $x_u \in \mathcal{X}_u$ (i.e., a vector of length $|\mathcal{X}_u|$ for discrete random variables). On the full graph, there are a total of $2|E|$ messages, one for each direction of each edge. This full collection of messages is updated, typically in parallel, according to the following recursion:

$$M_{ts}(x_s) \;\leftarrow\; \kappa \sum_{x_t'} \left\{ \psi_{st}(x_s, x_t') \, \psi_t(x_t') \prod_{u \in \mathcal{N}(t)/s} M_{ut}(x_t') \right\}, \tag{7.6}$$

where $\kappa > 0$ is a normalization constant. It can be shown (Pearl, 1988) that for tree-structured graphs, iterates generated by the update 7.6 will converge to a unique fixed point $M^* = \{M_{st}^*, M_{ts}^*, \ (s, t) \in E\}$ after a finite number of iterations. Moreover, component M_{ts}^* of this fixed point is precisely equal, up to a normalization constant, to the subproblem defined in equation 7.5, which justifies our abuse of notation post hoc. Since the fixed point M^* specifies the solution to all of the subproblems, the marginal μ_s at every node $s \in V$ can be computed easily via equation 7.4.

Max-product algorithm Suppose that the summation in the update 7.6 is replaced by a maximization. The resulting *max-product algorithm* solves the problem of finding a mode of a tree-structured distribution $p(\mathbf{x})$. In this sense, it represents a generalization of the Viterbi algorithm (Forney, 1973) from chains to arbitrary tree-structured graphs. More specifically, the max-product updates will converge to another unique fixed point M^*—distinct, of course, from the sum-product fixed point. This fixed point can be used to compute the *max-marginal* $\nu_s(x_s) := \max_{\{\mathbf{x}' \mid x_s' = x_s\}} p(\mathbf{x}')$ at each node of the graph, in an analogous way to the computation of ordinary sum-marginals. Given these max-marginals, it is straightforward to compute a mode $\widehat{\mathbf{x}} \in \arg\max_{\mathbf{x}} p(\mathbf{x})$ of the distribution (Dawid, 1992; Wainwright et al., 2004). More generally, updates of this form apply to arbitrary *commutative semirings* on tree-structured graphs (Aji and McEliece, 2000; Dawid, 1992). The pairs "sum-product" and "max-product" are two particular examples of such an algebraic structure.

Junction Tree Representation We have seen that inference problems on trees can be solved exactly by recursive message-passing algorithms. Given a graph with cycles, a natural idea is to cluster its nodes so as to form a *clique tree*—that is, an acyclic graph whose nodes are formed by cliques of G. Having done so, it is tempting to simply apply a standard algorithm for inference on trees. However, the clique tree must satisfy an additional restriction so as to ensure consistency of these computations. In particular, since a given vertex $s \in V$ may appear in multiple cliques (say C_1 and C_2), what is required is a mechanism for enforcing consistency among the different appearances of the random variable x_s.

In order to enforce consistency, it turns out to be necessary to restrict attention to those clique trees that satisfy a particular graph-theoretic property. In particular, we say that a clique tree satisfies the *running intersection property* if for any two clique nodes C_1 and C_2, all nodes on the unique path joining them contain the intersection $C_1 \cap C_2$. Any clique tree with this property is known as a *junction tree*.

For what type of graphs can one build junction trees? An important result in graph theory asserts that a graph G has a junction tree if and only if it is *triangulated*.[2] This result underlies the *junction tree algorithm* (Lauritzen and Spiegelhalter, 1988) for exact inference on arbitrary graphs, which consists of the following three steps:

Step 1: Given a graph with cycles G, triangulate it by adding edges as necessary.

Step 2: Form a junction tree associated with the triangulated graph.

Step 3: Run a tree inference algorithm on the junction tree.

We illustrate these basic steps with an example.

Example 7.1

Consider the 3×3 grid shown in the top panel of fig. 7.1b. The first step is to form a triangulated version, as shown in the bottom panel of fig. 7.1b. Note that the graph would *not* be triangulated if the additional edge joining nodes 2 and 8

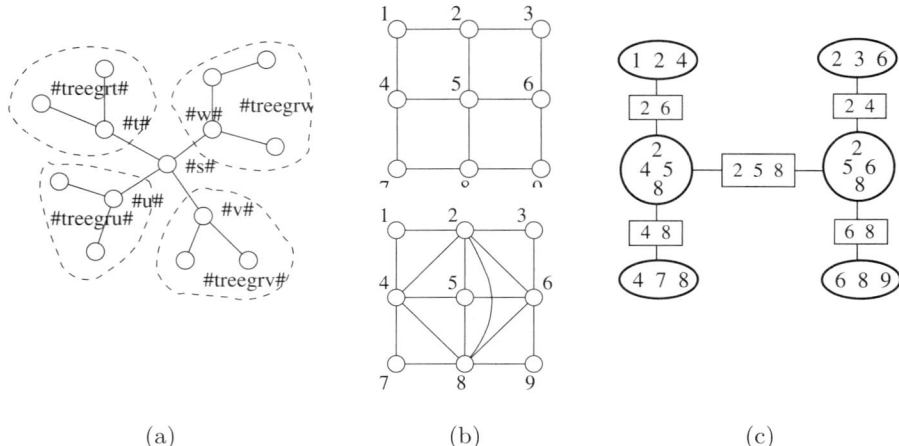

(a)   (b)   (c)

Figure 7.1   (a): Decomposition of a tree, rooted at node s, into subtrees. Each neighbor (e.g., u) of node s is the root of a subtree (e.g., T_u). Subtrees T_u and T_v, for $t \neq u$, are disconnected when node s is removed from the graph. (b), (c) Illustration of junction tree construction. Top panel in (b) shows original graph: a 3×3 grid. Bottom panel in (b) shows triangulated version of original graph. Note the two 4-cliques in the middle. (c) Corresponding junction tree for triangulated graph in (b), with maximal cliques depicted within ellipses. The rectangles are separator sets; these are intersections of neighboring cliques.

were not present. Without this edge, the 4-cycle $(2 - 4 - 8 - 6 - 2)$ would lack a chord. Figure 7.1c shows a junction tree associated with this triangulated graph, in which circles represent maximal cliques (i.e., fully connected subsets of nodes that cannot be augmented with an additional node and remain fully connected), and boxes represent *separator sets* (intersections of cliques adjacent in the junction tree). ◇

An important by-product of the junction tree construction is an alternative representation of the probability distribution defined by a graphical model. Let $\mathcal{C}$ denote the set of all maximal cliques in the triangulated graph, and define $\mathcal{S}$ as the set of all separator sets in the junction tree. For each separator set $S \in \mathcal{S}$, let $d(S)$ denote the number of maximal cliques to which it is adjacent. The junction tree framework guarantees that the distribution $p(\cdot)$ factorizes in the form

$$p(\mathbf{x}) = \frac{\prod_{C \in \mathcal{C}} \mu_C(x_C)}{\prod_{S \in \mathcal{S}} [\mu_S(x_S)]^{d(S)-1}}, \tag{7.7}$$

where μ_C and μ_S are the marginal distributions over the cliques and separator sets respectively. Observe that unlike the representation of equation 7.2, the decomposition of equation 7.7 is directly in terms of marginal distributions, and does not require a normalization constant (i.e., $Z = 1$).

Example 7.2 Markov Chain

Consider the Markov chain $p(x_1, x_2, x_3) = p(x_1)\, p(x_2\,|\,x_1)\, p(x_3\,|\,x_2)$. The cliques in a graphical model representation are $\{1, 2\}$ and $\{2, 3\}$, with separator $\{2\}$. Clearly the distribution cannot be written as the product of marginals involving only the cliques. However, if we include the separator, it can be factorized in terms of its marginals—viz., $p(x_1, x_2, x_3) = \frac{p(x_1, x_2) p(x_2, x_3)}{p(x_2)}$. $\diamond$

To anticipate the development in the sequel, it is helpful to consider the following "inverse" perspective on the junction tree representation. Suppose that we are given a set of functions $\tau_C(x_C)$ and $\tau_S(x_S)$ associated with the cliques and separator sets in the junction tree. What conditions are necessary to ensure that these functions are valid marginals for some distribution? Suppose that the functions $\{\tau_S, \tau_C\}$ are *locally consistent* in the following sense:

$$\sum_{x_S} \tau_S(x_S) = 1 \qquad\qquad \text{normalization} \qquad\qquad (7.8\text{a})$$

$$\sum_{\{\mathbf{x}'_C \,|\, \mathbf{x}'_S = x_S\}} \tau_C(x'_C) = \tau_S(x_S) \qquad\qquad \text{marginalization}. \qquad\qquad (7.8\text{b})$$

The essence of the junction tree theory described above is that such local consistency is both necessary and sufficient to ensure that these functions are valid marginals for some distribution.

Finally, turning to the computational complexity of the junction tree algorithm, the computational cost grows exponentially in the size of the maximal clique in the junction tree. The size of the maximal clique over all possible triangulations of a graph defines an important graph-theoretic quantity known as the *treewidth* of the graph. Thus, the complexity of the junction tree algorithm is exponential in the treewidth. For certain classes of graphs, including chains and trees, the treewidth is small and the junction tree algorithm provides an effective solution to inference problems. Such families include many well-known graphical model architectures, and the junction tree algorithm subsumes many classical recursive algorithms, including the forward-backward algorithms for hidden Markov models (Rabiner and Juang, 1993), the Kalman filtering-smoothing algorithms for state-space models (Kailath et al., 2000), and the pruning and peeling algorithms from computational genetics (Felsenstein, 1981). On the other hand, there are many graphical models (e.g., grids) for which the treewidth is infeasibly large. Coping with such models requires leaving behind the junction tree framework, and turning to approximate inference algorithms.

7.1.3   Message-Passing Algorithms for Approximate Inference

In the remainder of the chapter, we present a general variational principle for graphical models that can be used to derive a class of techniques known as *variational inference algorithms*. To motivate our later development, we pause to give a high-level description of two variational inference algorithms, with the goal of highlighting their simple and intuitive nature.

The first variational algorithm that we consider is a so-called "loopy" form of the sum-product algorithm (also referred to as the *belief propagation* algorithm). Recall that the sum-product algorithm is designed as an exact method for trees; from a purely algorithmic point of view, however, there is nothing to prevent one from running the procedure on a graph with cycles. More specifically, the message updates 7.6 can be applied at a given node while ignoring the presence of cycles— essentially pretending that any given node is embedded in a tree. Intuitively, such an algorithm might be expected to work well if the graph is suitably "tree like," such that the effect of messages propagating around cycles is appropriately diminished. This algorithm is in fact widely used in various applications that involve signal processing, including image processing, computer vision, computational biology, and error-control coding.

A second variational algorithm is the so-called *naive mean field* algorithm. For concreteness, we describe it in application to a very special type of graphical model, known as the Ising model. The Ising model is a Markov random field involving a binary random vector $\mathbf{x} \in \{0, 1\}^n$, in which pairs of adjacent nodes are coupled with a weight θ_{st}, and each node has an observation weight θ_s. (See examples 7.4 and 7.11 for a more detailed description of this model.) To motivate the mean field updates, we consider the Gibbs sampler for this model, in which the basic update step is to choose a node $s \in V$ randomly, and then to update the state of the associated random variable according to the conditional probability with neighboring states fixed. More precisely, denoting by $\mathcal{N}(s)$ the neighbors of a node $s \in V$, and letting $x_{\mathcal{N}(s)}^{(p)}$ denote the state of the neighbors of s at iteration p, the Gibbs update for x_s takes the following form:

$$x_s^{(p+1)} = \begin{cases} 1 & \text{if } u \leq \{1 + \exp[-(\theta_s + \sum_{t \in \mathcal{N}(s)} \theta_{st} x_t^{(p)})]\}^{-1} \\ 0 & \text{otherwise} \end{cases}, \qquad (7.9)$$

where u is a sample from a uniform distribution $\mathcal{U}(0, 1)$. It is well known that this procedure generates a sequence of configurations that converge (in a stochastic sense) to a sample from the Ising model distribution.

In a dense graph, such that the cardinality of $\mathcal{N}(s)$ is large, we might attempt to invoke a law of large numbers or some other concentration result for $\sum_{t \in \mathcal{N}(s)} \theta_{st} x_t^{(p)}$. To the extent that such sums are concentrated, it might make sense to replace sample values with expectations, which motivates the following averaged version of equation 7.9:

$$\mu_s \leftarrow \left\{ 1 + \exp\left[-(\theta_s + \sum_{t \in \mathcal{N}(s)} \theta_{st} \mu_t) \right] \right\}^{-1}, \qquad (7.10)$$

in which μ_s denotes an estimate of the marginal probability $p(x_s = 1)$. Thus, rather than flipping the random variable x_s with a probability that depends on the state of its neighbors, we update a parameter μ_s using a deterministic function of the corresponding parameters $\{\mu_t \mid t \in \mathcal{N}(s)\}$ at its neighbors. Equation 7.10

defines the naive mean field algorithm for the Ising model, which can be viewed as a message-passing algorithm on the graph.

At first sight, message-passing algorithms of this nature might seem rather mysterious, and do raise some questions. Do the updates have fixed points? Do the updates converge? What is the relation between the fixed points and the exact quantities? The goal of the remainder of this chapter is to shed some light on such issues. Ultimately, we will see that a broad class of message-passing algorithms, including the mean field updates, the sum-product and max-product algorithms, as well as various extensions of these methods can all be understood as solving either exact or approximate versions of a certain variational principle for graphical models.

7.2   Graphical Models in Exponential Form

We begin by describing how many graphical models can be viewed as particular types of exponential families. Further background can be found in the books by Efron (1978) and Brown (1986). This exponential family representation is the foundation of our later development of the variational principle.

7.2.1   Maximum Entropy

One way in which to motivate exponential family representations of graphical models is through the principle of maximum entropy. The set up for this principle is as follows: given a collection of functions $\phi_\alpha : \mathcal{X}^n \to \mathrm{R}$, suppose that we have observed their expected values—that is, we have

$$\mathrm{E}[\phi_\alpha(\mathbf{x})] = \mu_\alpha \qquad \text{for all } \alpha \in \mathcal{I}, \tag{7.11}$$

where $\mu = \{\mu_\alpha \mid \alpha \in \mathcal{I}\}$ is a real vector, $\mathcal{I}$ is an index set, and $d := |\mathcal{I}|$ is the length of the vectors μ and $\phi := \{\phi_\alpha \mid \alpha \in \mathcal{I}\}$.

Our goal is use the observations to infer a full probability distribution. Let $\mathcal{P}$ denote the set of all probability distributions p over the random vector $\mathbf{x}$. Since there are (in general) many distributions $p \in \mathcal{P}$ that are consistent with the observations 7.11, we need a principled method for choosing among them. The principle of maximum entropy is to choose the distribution p_{ME} such that its *entropy*, defined as $H(p) := -\sum_{\mathbf{x} \in \mathcal{X}^n} p(\mathbf{x}) \log p(\mathbf{x})$, is maximized. More formally, the maximum entropy solution p_{ME} is given by the following constrained optimization problem:

$$p_{ME} := \underset{p \in \mathcal{P}}{\arg\max}\, H(p) \qquad \text{subject to constraints 7.11.} \tag{7.12}$$

One interpretation of this principle is as choosing the distribution with maximal uncertainty while remaining faithful to the data.

Presuming that problem 7.12 is feasible, it is straightforward to show using a

Lagrangian formulation that its optimal solution takes the form

$$p(\mathbf{x}; \theta) \quad \propto \quad \exp\Big\{ \sum_{\alpha \in \mathcal{I}} \theta_\theta \phi_\alpha(\mathbf{x}) \Big\}, \tag{7.13}$$

which corresponds to a distribution in exponential form. Note that the exponential decomposition 7.13 is analogous to the product decomposition 7.2 considered earlier.

In the language of exponential families, the vector $\theta \in \mathrm{R}^d$ is known as the *canonical parameter*, and the collection of functions $\phi = \{\phi_\alpha \mid \alpha \in \mathcal{I}\}$ are known as *sufficient statistics*. In the context of our current presentation, each canonical parameter θ_α has a very concrete interpretation as the Lagrange multiplier associated with the constraint $\mathrm{E}[\phi_\alpha(\mathbf{x})] = \mu_\alpha$.

7.2.2   Exponential Families

We now define exponential families in more generality. Any exponential family consists of a particular class of densities taken with respect to a fixed base measure $\boldsymbol{\nu}$. The base measure is typically counting measure (as in our discrete example above), or Lebesgue measure (e.g., for Gaussian families). Throughout this chapter, we use $\langle a, b \rangle$ to denote the ordinary Euclidean inner product between two vectors a and b of the same dimension. Thus, for each fixed $\mathbf{x} \in \mathcal{X}^n$, the quantity $\langle \theta, \phi(\mathbf{x}) \rangle$ is the Euclidean inner product in R^d of the two vectors $\theta \in \mathrm{R}^d$ and $\phi(\mathbf{x}) = \{\phi_\alpha(\mathbf{x}) \mid \alpha \in \mathcal{I}\}$.

With this notation, the *exponential family* associated with ϕ consists of the following parameterized collection of density functions:

$$p(\mathbf{x}; \theta) \;=\; \exp\big\{ \langle \theta, \phi(\mathbf{x}) \rangle \,-\, A(\theta) \big\}. \tag{7.14}$$

The quantity A, known as the *log partition function* or *cumulant generating function*, is defined by the integral:

$$A(\theta) \;=\; \log \int_{\mathcal{X}^n} \exp\langle \theta, \phi(\mathbf{x}) \rangle \, \boldsymbol{\nu}(d\mathbf{x}). \tag{7.15}$$

Presuming that the integral is finite, this definition ensures that $p(\mathbf{x}; \theta)$ is properly normalized (i.e., $\int_{\mathcal{X}^n} p(\mathbf{x}; \theta)\boldsymbol{\nu}(d\mathbf{x}) = 1$). With the set of potentials ϕ fixed, each parameter vector θ indexes a particular member $p(\mathbf{x}; \theta)$ of the family. The canonical parameters θ of interest belong to the set

$$\Theta \;:=\; \{\theta \in \mathrm{R}^d \mid A(\theta) < \infty\}. \tag{7.16}$$

Throughout this chapter, we deal exclusively with *regular* exponential families, for which the set Θ is assumed to be open.

We summarize for future reference some well-known properties of A:

Lemma 7.1

The cumulant generating function A is convex in terms of θ. Moreover, it is infinitely differentiable on Θ, and its derivatives correspond to cumulants.

As an important special case, the first derivatives of A take the form

$$\frac{\partial A}{\partial \theta_\alpha} = \int_{\mathcal{X}^n} \phi_\alpha(\mathbf{x}) p(\mathbf{x};\theta)\boldsymbol{\nu}(d\mathbf{x}) = \mathrm{E}_\theta[\phi_\alpha(\mathbf{x})], \tag{7.17}$$

and define a vector $\mu := \mathrm{E}_\theta[\boldsymbol{\phi}(\mathbf{x})]$ of *mean parameters* associated with the exponential family. There are important relations between the canonical and mean parameters, and many inference problems can be formulated in terms of the mean parameters. These correspondences and other properties of the cumulant generating function are fundamental to our development of a variational principle for solving inference problems.

7.2.3   Illustrative Examples

In order to illustrate these definitions, we now discuss some particular classes of graphical models that commonly arise in signal and image processing problems, and how they can be represented in exponential form. In particular, we will see that graphical structure is reflected in the choice of sufficient statistics, or equivalently in terms of constraints on the canonical parameter vector.

We begin with an important case—the Gaussian Markov random field (MRF)— which is widely used for modeling various types of imagery and spatial data (Luettgen et al., 1994; Szeliski, 1990).

Example 7.3 Gaussian Markov Random Field

Consider a graph $G = (V, E)$, such as that illustrated in fig. 7.2(a), and suppose that each vertex $s \in V$ has an associated Gaussian random variable x_s. Any such scalar Gaussian is a (2-dimensional) exponential family specified by sufficient statistics x_s and x_s^2. Turning to the Gaussian random vector $\mathbf{x} := \{x_s \mid s \in V\}$, it has an exponential family representation in terms of the sufficient statistics $\{x_s, x_s^2 \mid s \in V\} \cup \{x_s x_t \mid (s,t) \in E\}$, with associated canonical parameters $\{\theta_s, \theta_{ss} \mid s \in V\} \cup \{\theta_{st} \mid (s,t) \in E\}$. Here the additional cross-terms $x_s x_t$ allow for possible correlation between components x_s and x_t of the Gaussian random vector. Note that there are a total of $d = 2n + |E|$ sufficient statistics.

The sufficient statistics and parameters can be represented compactly as $(n+1) \times (n+1)$ symmetric matrices:

$$\mathbf{X} = \begin{bmatrix} 1 \\ \mathbf{x} \end{bmatrix} \begin{bmatrix} 1 & \mathbf{x} \end{bmatrix} \qquad U(\theta) := \begin{bmatrix} 0 & \theta_1 & \theta_2 & \dots & \theta_n \\ \theta_1 & \theta_{11} & \theta_{12} & \dots & \theta_{1n} \\ \theta_2 & \theta_{21} & \theta_{22} & \dots & \theta_{2n} \\ \vdots & \vdots & \vdots & \vdots & \vdots \\ \theta_n & \theta_{n1} & \theta_{n2} & \dots & \theta_{nn} \end{bmatrix} \tag{7.18}$$

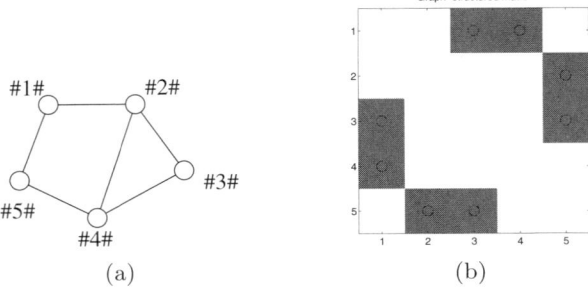

(a)   (b)

Figure 7.2   (a) A simple Gaussian model based on a graph G with 5 vertices. (b) The adjacency matrix of the graph G in (a), which specifies the sparsity pattern of the matrix $Z(\theta)$.

We use $Z(\theta)$ to denote the lower $n \times n$ block of $U(\theta)$; it is known as the *precision matrix*. We say that $\mathbf{x}$ forms a Gaussian Markov random field if its probability density function decomposes according to the graph $G = (V, E)$. In terms of our canonical parameterization, this condition translates to the requirement that $\theta_{st} = 0$ whenever $(s, t) \notin E$. Alternatively stated, the precision matrix $Z(\theta)$ must have the same zero-pattern as the adjacency matrix of the graph, as illustrated in fig. 7.2b.

For any two symmetric matrices C and D, it is convenient to define the inner product $\langle C, \, D \rangle := \mathrm{trace}(C\,D)$. Using this notation leads to a particularly compact representation of a Gaussian MRF:

$$p(\mathbf{x}; \theta) \; = \; \exp\big\{\langle U(\theta), \, \mathbf{X} \rangle - A(\theta)\big\}, \tag{7.19}$$

where $A(\theta) := \log \int_{\mathrm{R}^n} \exp\big[\langle U(\theta), \, \mathbf{X} \rangle\big] d\mathbf{x}$ is the log cumulant generating function. The integral defining $A(\theta)$ is finite only if the $n \times n$ precision matrix $Z(\theta)$ is negative definite, so that the domain of A has the form $\Theta = \{\theta \in \mathrm{R}^d \mid Z(\theta) \prec 0\}$.

Note that the mean parameters in the Gaussian model have a clear interpretation. The singleton elements $\mu_s = \mathrm{E}_\theta[x_s]$ are simply the Gaussian mean, whereas the elements $\mu_{ss} = \mathrm{E}_\theta[x_s^2]$ and $\mu_{st} = \mathrm{E}_\theta[x_s x_t]$ are second-order moments. $\Diamond$

Markov random fields involving *discrete* random variables also arise in many applications, including image processing, bioinformatics, and error-control coding (Durbin et al., 1998; Geman and Geman, 1984; Kschischang et al., 2001; Loeliger, 2004). As with the Gaussian case, this class of Markov random fields also has a natural exponential representation.

Example 7.4 Multinomial Markov Random Field

Suppose that each x_s is a multinomial random variable, taking values in the space $\mathcal{X}_s = \{0, 1, \dots, m_s - 1\}$. In order to represent a Markov random field over the vector $\mathbf{x} = \{x_s \mid s \in V\}$ in exponential form, we now introduce a particular set of sufficient statistics that will be useful in what follows. For each $j \in \mathcal{X}_s$, let $\mathrm{I}_j(x_s)$ be an indicator function for the event $\{x_s = j\}$. Similarly, for each pair $(j, k) \in \mathcal{X}_s \times \mathcal{X}_t$, let $\mathrm{I}_{jk}(x_s, x_t)$ be an indicator for the event $\{(x_s, x_t) = (j, k)\}$. These building blocks yield the following set of sufficient statistics:

$$\{\mathrm{I}_j(x_s) \mid s \in V, \ j \in \mathcal{X}_s\} \cup \{\mathrm{I}_j(x_s)\mathrm{I}_k(x_t) \mid (s, t) \in E, \ (j, k) \in \mathcal{X}_s \times \mathcal{X}_t\}. \quad (7.20)$$

The corresponding canonical parameter θ has elements of the form

$$\theta = \{\theta_{s;j} \mid s \in V, \ j \in \mathcal{X}_s\} \cup \{\theta_{st;jk} \mid (s, t) \in E, \ (j, k) \in \mathcal{X}_s \times \mathcal{X}_t\}. \quad (7.21)$$

It is convenient to combine the canonical parameters and indicator functions using the shorthand notation $\theta_s(x_s) := \sum_{j \in \mathcal{X}_s} \theta_{s;j} \mathrm{I}_j(x_s)$; the quantity $\theta_{st}(x_s, x_t)$ can be defined similarly.

With this notation, a multinomial MRF with pairwise interactions can be written in exponential form as

$$p(\mathbf{x}; \theta) \ = \ \exp\Big\{\sum_{s \in V} \theta_s(x_s) + \sum_{(s,t) \in E} \theta_{st}(x_s, x_t) - A(\theta)\Big\}, \quad (7.22)$$

where the cumulant generating function is given by the summation

$$A(\theta) := \log \sum_{\mathbf{x} \in \mathcal{X}^n} \exp\Big\{\sum_{s \in V} \theta_s(x_s) + \sum_{(s,t) \in E} \theta_{st}(x_s, x_t)\Big\}.$$

In signal-processing applications of these models, the random vector $\mathbf{x}$ is often viewed as hidden or partially observed (for instance, corresponding to the correct segmentation of an image). Thus, it is frequently the case that the functions θ_s are determined by noisy observations, whereas the terms θ_{st} control the coupling between variables x_s and x_t that are adjacent on the graph (e.g., reflecting spatial continuity assumptions). See fig. 7.3(a) for an illustration of such a multinomial MRF defined on a two-dimensional lattice, which is a widely used model in statistical image processing (Geman and Geman, 1984). In the special case that $\mathcal{X}_s = \{0, 1\}$ for all $s \in V$, the family represeted by equation 7.22 is known as the *Ising model*.

Note that the mean parameters associated with this model correspond to particular marginal probabilities. For instance, the mean parameters associated with vertex s have the form $\mu_{s;j} \ = \ \mathrm{E}_\theta[\mathrm{I}_j(x_s)] \ = \ p(x_s = j; \theta)$, and the mean parameters μ_{st} associated with edge (s, t) have an analogous interpretation as pairwise marginal values.

$\diamondsuit$

Example 7.5 Hidden Markov Model

A very important special case of the multinomial MRF is the hidden Markov model (HMM), which is a chain-structured graphical model widely used for the modeling of time series and other one-dimensional signals. It is conventional in the HMM literature to refer to the multinomial random variables $\mathbf{x} = \{x_s \mid s \in V\}$ as "state variables." As illustrated in fig. 7.3b, the edge set E defines a chain

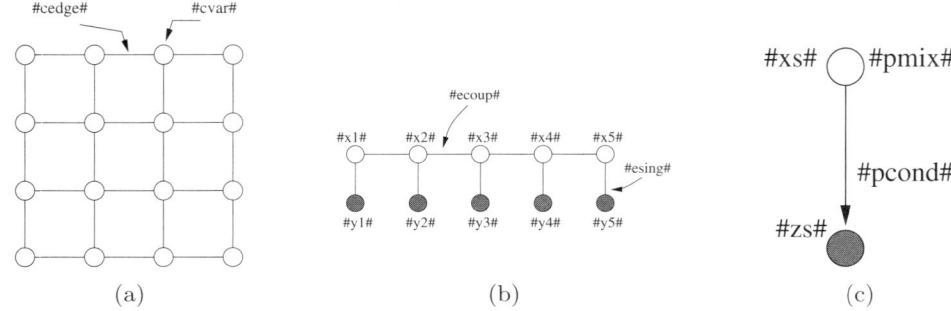

Figure 7.3 (a) A multinomial MRF on a 2D lattice model. (b) A hidden Markov model (HMM) is a special case of a multinomial MRF for a chain-structured graph. (c) The graphical representation of a scalar Gaussian mixture model: the multinomial x_s indexes components in the mixture, and y_s is conditionally Gaussian (with exponential parameters γ_s) given the mixture component x_s.

linking the state variables. The parameters $\theta_{st}(x_s, x_t)$ define the *state transition matrix*; if this transition matrix is the same for all pairs s and t, then we have a *homogeneous* Markov chain. Associated with each multinomial state variable x_s is a noisy *observation* y_s, defined by the conditional probability distribution $p(y_s | x_s)$. If we condition on the observed value of y_s, this conditional probability is simply a function of x_s, which we denote by $\theta_s(x_s)$. Given these definitions, equation 7.22 describes the conditional probability distribution $p(\mathbf{x} | \mathbf{y})$ for the HMM. In fig. 7.3b, this conditioning is captured by shading the corresponding nodes in the graph. Note that the cumulant generating function $A(\theta)$ is, in fact, equal to the log likelihood of the observed data. $\diamondsuit$

Graphical models are not limited to cases in which the random variables at each node belong to the same exponential family. More generally, we can consider heterogeneous combinations of exponential family members. A very natural example, which combines the two previous types of graphical model, is that of a Gaussian mixture model. Such mixture models are widely used in modeling various classes of data, including natural images, speech signals, and financial time series data; see the book by Titterington et al. (1986) for further background.

Example 7.6 Mixture Model
As shown in fig. 7.3c, a *scalar* mixture model has a very simple graphical interpretation. In particular, let x_s be a multinomial variable, taking values in $\mathcal{X}_s = \{0, 1, 2, \dots, m_s - 1\}$, specified in exponential parameter form with a function $\theta_s(x_s)$. The role of x_s is to specify the choice of mixture component in the mixture model, so that our mixture model has m_s components in total. We now let y_s be conditionally Gaussian given x_s, so that the conditional distribution $p(y_s | x_s; \gamma_s)$ can be written in exponential family form with canonical parameters γ_s that are a function of x_s. Overall, the pair (x_s, y_s) form a very simple graphical model in exponential form, as shown in fig. 7.3c.

The pair (x_s, y_s) serves a basic block for building more sophisticated graphical models. For example, one model is based on assuming that the mixture vector $\mathbf{x}$ is a multinomial MRF defined on an underlying graph $G = (V, E)$, whereas the components of $\mathbf{y}$ are conditionally independent given the mixture vector $\mathbf{x}$. These assumptions lead to an exponential family $p(\mathbf{y}, \mathbf{x}; \theta, \gamma)$ of the form

$$\prod_{s \in V} p(y_s \mid x_s; \gamma_s) \, \exp\left\{ \sum_{s \in V} \theta_s(x_s) + \sum_{(s,t) \in E} \theta_{st}(x_s, x_t)] \right\}. \qquad (7.23)$$

For tree-structured graphs, Crouse et al. (1998) have applied this type of mixture model to applications in wavelet-based signal processing. $\diamond$

This type of mixture model is a particular example of a broad class of graphical models that involve heterogeneous combinations of exponential family members (e.g., hierarchical Bayesian models).

7.3   An Exact Variational Principle for Inference

With this set up, we can now rephrase inference problems in the language of exponential families. In particular, this chapter focuses primarily on the following two problems:

- computing the cumulant generating function $A(\theta)$,
- computing the vector of mean parameters $\mu := \mathrm{E}_\theta[\phi(\mathbf{x})]$.

In Section 7.7.2 we discuss a closely related problem—namely, that of computing a mode of the distribution $p(\mathbf{x}; \theta)$.

The problem of computing the cumulant generating function arises in a variety of signal-processing problems, including likelihood ratio tests (for classification and detection problems) and parameter estimation. The computation of mean parameters is also fundamental, and takes different forms depending on the underlying graphical model. For instance, it corresponds to computing means and covariances in the Gaussian case, whereas for a multinomial MRF it corresponds to computing marginal distributions.

The goal of this section is to show how both of these inference problems can be represented *variationally*—as the solution of an optimization problem. The variational principle that we develop, though related to the classical "free energy" approach of statistical physics (Yedidia et al., 2001), also has important differences. The classical principle yields a variational formulation for the cumulant generating function (or log partition function) in terms of optimizing over the space of all distributions. In our approach, on the other hand, the optimization is not defined over all distributions—a very high or infinite-dimensional space—but rather over the much lower dimensional space of mean parameters. As an important consequence, solving this variational principle yields not only the cumulant generating function but also the full set of mean parameters $\mu = \{\mu_\alpha \mid \alpha \in \mathcal{I}\}$.

7.3.1 Conjugate Duality

The cornerstone of our variational principle is the notion of *conjugate duality*. In this section, we provide a brief introduction to this concept, and refer the interested reader to the standard texts (Hiriart-Urruty and Lemaréchal, 1993; Rockafellar, 1970) for further details. As is standard in convex analysis, we consider *extended real-valued* functions, meaning that they take values in the extended real line $R_* := R \cup \{+\infty\}$. Associated with any convex function $f : R^d \to R_*$ is a conjugate dual function $f^* : R^d \to R_*$, which is defined as follows:

$$f^*(y) := \sup_{x \in R^d} \big\{ \langle y,\, x \rangle - f(x) \big\}. \tag{7.24}$$

This definition illustrates the concept of a *variational definition*: the function value $f^*(y)$ is specified as the solution of an optimization problem parameterized by the vector $y \in R^d$.

As illustrated in fig. 7.4, the value $f^*(y)$ has a natural geometric interpretation as the (negative) intercept of the hyperplane with normal $(y, -1)$ that supports the epigraph of f. In particular, consider the family of hyperplanes of the form $\langle y,\, x \rangle - c$, where y is a fixed normal direction and $c \in R$ is the intercept to be adjusted. Our goal is to find the smallest c such that the resulting hyperplane supports the

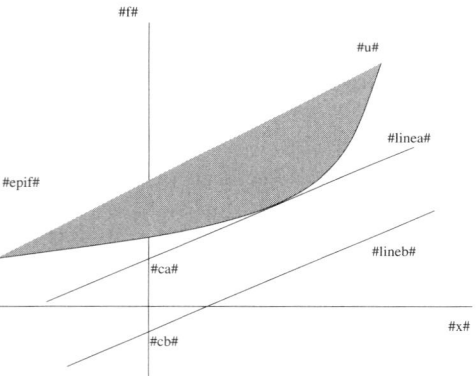

Figure 7.4 Interpretation of conjugate duality in terms of supporting hyperplanes to the epigraph of f, defined as $\mathrm{epi}(f) := \{(x, y) \in R^d \times R \mid f(x) \leq y\}$. The dual function is obtained by translating the family of hyperplane with normal y and intercept $-c$ until it just supports the epigraph of f (the shaded region).

epigraph of f. Note that the hyperplane $\langle y,\, x \rangle - c$ lies below the epigraph of f if and only if the inequality $\langle y,\, x \rangle - c \leq f(x)$ holds for all $x \in R^d$. Moreover, it can be seen that the smallest c for which this inequality is valid is given by $c^* = \sup_{x \in \mathrm{R}^d} \left\{ \langle y, x \rangle - f(x) \right\}$, which is precisely the value of the dual function. As illustrated in fig. 7.4, the geometric interpretation is that of moving the hyperplane (by adjusting the intercept c) until it is just tangent to the epigraph of f.

For convex functions meeting certain technical conditions, taking the dual *twice* recovers the original function. In analytical terms, this fact means that we can generate a variational representation for convex f in terms of its dual function as follows:

$$f(x) = \sup_{y \in \mathrm{R}^d} \left\{ \langle x, y \rangle - f^*(y) \right\}. \tag{7.25}$$

Our goal in the next few sections is to apply conjugacy to the cumulant generating function A associated with an exponential family, as defined in equation 7.15. More specifically, its dual function takes the form

$$A^*(\mu) := \sup_{\theta \in \Theta} \left\{ \langle \theta, \mu \rangle - A(\theta) \right\}, \tag{7.26}$$

where we have used the fact that, by definition, the function value $A(\theta)$ is finite only if $\theta \in \Theta$. Here $\mu \in \mathrm{R}^d$ is a vector of so-called dual variables of the same dimension as θ. Our choice of notation—using μ for the dual variables—is deliberately suggestive: as we will see momentarily, these dual variables turn out to be precisely the mean parameters defined in equation 7.17.

Example 7.7

To illustrate the computation of a dual function, consider a scalar Bernoulli random variable $x \in \{0, 1\}$, whose distribution can be written in the exponential family form as $p(x; \theta) = \exp\{\theta x - A(\theta)\}$. The cumulant generating function is given by $A(\theta) = \log[1 + \exp(\theta)]$, and there is a single dual variable $\mu = \mathrm{E}_\theta[x]$. Thus, the variational problem 7.26 defining A^* takes the form

$$A^*(\mu) = \sup_{\theta \in \mathrm{R}} \left\{ \theta \mu - \log[1 + \exp(\theta)] \right\}. \tag{7.27}$$

If $\mu \in (0, 1)$, then taking derivatives shows that the supremum is attained at the unique $\theta \in \mathrm{R}$ satisfying the well-known logistic relation $\theta = \log[\mu/(1 - \mu)]$. Substituting this logistic relation into equation 7.27 yields that for $\mu \in (0, 1)$, we have $A^*(\mu) = \mu \log \mu + (1 - \mu) \log(1 - \mu)$. By taking limits $\mu \to 1-$ and $\mu \to 0+$, it can be seen that this expression is valid for μ in the closed interval $[0, 1]$.

Figure 7.5 illustrates the behavior of the supremum in equation 7.27 for $\mu \notin [0, 1]$. From our geometric interpretation of the value $A^*(\mu)$ in terms of supporting hyperplanes, the dual value is $+\infty$ if no supporting hyperplane can be found. In this particular case, the log partition function $A(\theta) = \log[1 + \exp(\theta)]$ is bounded below by the line $\theta = 0$. Therefore, as illustrated in fig. 7.5a, any slope $\mu < 0$ cannot support epi A, which implies that $A^*(\mu) = +\infty$. A similar picture holds for the case $\mu > 1$, as shown in fig. 7.5b. Consequently, the dual function is equal to $+\infty$ for $\mu \notin [0, 1]$. $\diamondsuit$

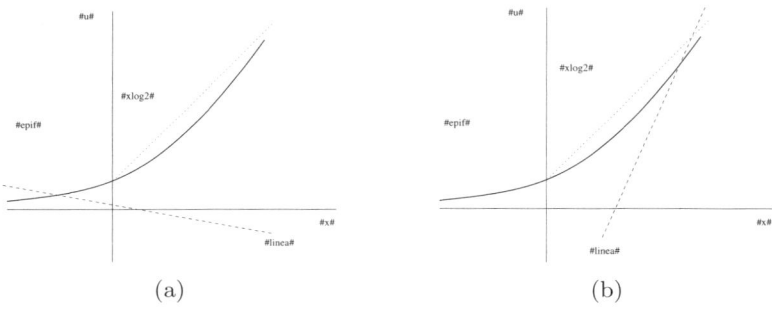

$$(a) \qquad\qquad\qquad\qquad (b)$$

Figure 7.5 Behavior of the supremum defining $A^*(\mu)$ for (a) $\mu < 0$ and (b) $\mu > 1$. The value of the dual function corresponds to the negative intercept of the supporting hyperplane to epi A with slope μ.

As the preceding example illustrates, there are two aspects to characterizing the dual function A^*: determining its domain (i.e., the set on which it takes a finite value); and specifying its precise functional form on the domain. In example 7.7, the domain of A^* is simply the closed interval $[0, 1]$, and its functional form on its domain is that of the binary entropy function. In the following two sections, we consider each of these aspects in more detail for general graphical models in exponential form.

7.3.2 Sets of Realizable Mean Parameters

For a given $\mu \in \mathbb{R}^d$, consider the optimization problem on the right-hand side of equation 7.26: since the cost function is differentiable, a first step in the solution is to take the derivative with respect to θ and set it equal to zero. Doing so yields the zero-gradient condition:

$$\mu = \nabla A(\theta) = \mathrm{E}_\theta[\boldsymbol{\phi}(\mathbf{x})], \tag{7.28}$$

where the second equality follows from the standard properties of A given in lemma 7.1.

We now need to determine the set of $\mu \in \mathbb{R}^d$ for which equation 7.28 has a solution. Observe that any $\mu \in \mathbb{R}^d$ satisfying this equation has a natural interpretation as a *globally realizable mean parameter*—i.e., a vector that can be realized by taking expectations of the sufficient statistic vector $\boldsymbol{\phi}$. This observation motivates defining the following set:

$$\mathcal{M} := \big\{ \mu \in \mathbb{R}^d \mid \exists \ p(\cdot) \ \text{such that} \ \int \boldsymbol{\phi}(\mathbf{x})p(\mathbf{x})\boldsymbol{\nu}(d\mathbf{x}) = \mu \big\}, \tag{7.29}$$

which corresponds to all realizable mean parameters associated with the set of sufficient statistics $\boldsymbol{\phi}$.

Example 7.8 Gaussian Mean Parameters

The Gaussian MRF, first introduced in example 7.3, provides a simple illustration of the set $\mathcal{M}$. Given the sufficient statistics that define a Gaussian, the associated mean parameters are either first-order moments (e.g., $\mu_s = \mathrm{E}[x_s]$), or second-order moments (e.g., $\mu_{ss} = \mathrm{E}[x_s^2]$ and $\mu_{st} = \mathrm{E}[x_s x_t]$). This full collection of mean parameters can be compactly represented in matrix form:

$$
W(\mu) := \mathrm{E}_\theta \begin{bmatrix} 1 \\ \mathbf{x} \end{bmatrix} \begin{bmatrix} 1 & \mathbf{x} \end{bmatrix} = \begin{bmatrix}
1 & \mu_1 & \mu_2 & \cdots & \mu_n \\
\mu_1 & \mu_{11} & \mu_{12} & \cdots & \mu_{1n} \\
\mu_2 & \mu_{21} & \mu_{22} & \cdots & \mu_{2n} \\
\vdots & \vdots & \vdots & \vdots & \vdots \\
\mu_n & \mu_{n1} & \mu_{n2} & \cdots & \mu_{nn}
\end{bmatrix}. \tag{7.30}
$$

The Schur product lemma (Horn and Johnson, 1985) implies that $\det W(\mu) = \det \mathrm{cov}(\mathbf{x})$, so that a mean parameter vector $\mu = \{\mu_s \mid s \in V\} \cup \{\mu_{st} \mid (s,t) \in E\}$ is globally realizable if and only if the matrix $W(\mu)$ is strictly positive definite. Thus, the set $\mathcal{M}$ is straightforward to characterize in the Gaussian case. $\Diamond$

Example 7.9 Marginal Polytopes

We now consider the case of a multinomial MRF, first introduced in example 7.4. With the choice of sufficient statistics (eq. 7.20), the associated mean parameters are simply local marginal probabilities, viz.,

$$
\mu_{s;j} := p(x_s = j; \theta) \ \forall \, s \in V, \quad \mu_{st;jk} := p((x_s, x_t) = (j,k); \theta) \ \forall \, (s,t) \in E. \tag{7.31}
$$

In analogy to our earlier definition of $\theta_s(x_s)$, we define functional versions of the mean parameters as follows:

$$
\mu_s(x_s) := \sum_{j \in \mathcal{X}_s} \mu_{s;j} \mathrm{I}_j(x_s), \quad \mu_{st}(x_s, x_t) := \sum_{(j,k) \in \mathcal{X}_s \times \mathcal{X}_t} \mu_{st;jk} \mathrm{I}_{jk}(x_s, x_t). \tag{7.32}
$$

With this notation, the set $\mathcal{M}$ consists of all singleton marginals μ_s (as s ranges over V) and pairwise marginals μ_{st} (for edges (s,t) in the edge set E) that can be realized by a distribution with support on $\mathcal{X}^n$. Since the space $\mathcal{X}^n$ has a finite number of elements, the set $\mathcal{M}$ is formed by taking the convex hull of a finite number of vectors. As a consequence, it must be a *polytope*, meaning that it can be described by a finite number of linear inequality constraints. In this discrete case, we refer to $\mathcal{M}$ as a *marginal polytope*, denoted by $\mathrm{MARG}(G)$; see fig. 7.6 for an idealized illustration.

As discussed in section 7.4.2, it is straightforward to specify a set of necessary conditions, expressed in terms of local constraints, that any element of $\mathrm{MARG}(G)$ must satisfy. However—and in sharp contrast to the Gaussian case—characterizing the marginal polytope exactly for a general graph is intractable, as it must require an exponential number of linear inequality constraints. Indeed, if it were possible to characterize $\mathrm{MARG}(G)$ with polynomial-sized set of constraints, then this would imply the polynomial-time solvability of various NP-complete problems (see section 7.7.2 for further discussion of this point). $\Diamond$

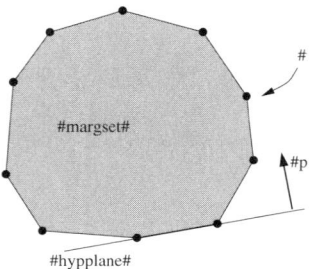

Figure 7.6 Geometrical illustration of a marginal polytope. Each vertex corresponds to the mean parameter $\mu_{\mathbf{e}} := \phi(\mathbf{e})$ realized by the distribution $\delta_{\mathbf{e}}(\mathbf{x})$ that puts all of its mass on the configuration $\mathbf{e} \in \mathcal{X}^n$. The faces of the marginal polytope are specified by hyperplane constraints $\langle a_j,\, \mu \rangle \leq b_j$.

7.3.3   Entropy in Terms of Mean Parameters

We now turn to the second aspect of the characterization of the conjugate dual function A^*—that of specifying its precise functional form on its domain $\mathcal{M}$. As might be expected from our discussion of maximum entropy in section 7.2.1, the form of the dual function A^* turns out to be closely related to entropy. Accordingly, we begin by defining the entropy in a bit more generality: Given a density function p taken with respect to base measure $\boldsymbol{\nu}$, its entropy is given by

$$H(p) = -\int_{\mathcal{X}^n} p(\mathbf{x}) \log\left[p(\mathbf{x})\right] \boldsymbol{\nu}(dx) \;=\; -\mathrm{E}_p[\log p(\mathbf{x})]. \tag{7.33}$$

With this set up, now suppose that μ belongs to the interior of $\mathcal{M}$. Under this assumption, it can be shown (Brown, 1986; Wainwright and Jordan, 2003b) that there exists a canonical parameter $\theta(\mu) \in \Theta$ such that

$$\mathrm{E}_{\theta(\mu)}[\phi(\mathbf{x})] = \mu. \tag{7.34}$$

Substituting this relation into the definition of the dual function (eq. 7.26) yields

$$A^*(\mu) \;=\; \langle \mu,\, \theta(\mu) \rangle - A(\theta(\mu)) \;=\; \mathrm{E}_{\theta(\mu)}\left[\log p(\mathbf{x}; \theta(\mu))\right],$$

which we recognize as the negative entropy $-H(p(\mathbf{x}; \theta(\mu)))$, where μ and $\theta(\mu)$ are dually coupled via equation 7.34.

Summarizing our development thus far, we have established that the dual function A^* has the following form:

$$A^*(\mu) = \begin{cases} -H(p(\mathbf{x}; \theta(\mu))) & \text{if } \mu \text{ belongs to the interior of } \mathcal{M} \\ +\infty & \text{if } \mu \text{ is outside the closure of } \mathcal{M}. \end{cases} \tag{7.35}$$

An alternative way to interpret this dual function A^* is by returning to the maximum entropy problem originally considered in section 7.2.1. More specifically, suppose that we consider the optimal value of the maximum entropy problem given in 7.12, considered parametrically as a function of the constraints μ. Essentially, what we have established is that the parametric form of this optimal value function is the dual function—that is:

$$A^*(\mu) \;=\; \max_{p \in \mathcal{P}} H(p) \quad \text{such that } \mathrm{E}_p[\phi_\alpha(\mathbf{x})] = \mu_\alpha \text{ for all } \alpha \in \mathcal{I}. \tag{7.36}$$

In this context, the property that $A^*(\mu) = +\infty$ for a constraint vector μ outside of $\mathcal{M}$ has a concrete interpretation: it corresponds to *infeasibility* of the maximum entropy problem (eq. 7.12).

Exact variational principle   Given the form of the dual function (eq. 7.35), we can now use the conjugate dual relation 7.25 to express A in terms of an optimization problem involving its dual function and the mean parameters:

$$A(\theta) \;=\; \sup_{\mu \in \mathcal{M}} \left\{ \langle \theta, \, \mu \rangle - A^*(\mu) \right\}. \tag{7.37}$$

Note that the optimization is restricted to the set $\mathcal{M}$ of globally realizable mean parameters, since the dual function A^* is infinite outside of this set. Thus, we have expressed the cumulant generating function as the solution of an optimization problem that is convex (since it entails maximizing a concave function over the convex set $\mathcal{M}$), and low dimensional (since it is expressed in terms of the mean parameters $\mu \in \mathrm{R}^d$).

In addition to representing the value $A(\theta)$ of the cumulant generating function, the variational principle 7.35 also has another important property. More specifically, the nature of our dual construction ensures that the optimum is always attained at the vector of mean parameters $\mu = \mathrm{E}_\theta[\phi(\mathbf{x})]$. Consequently, solving this optimization problem yields both the value of the cumulant generating function *as well as* the full set of mean parameters. In this way, the variational principle 7.37 based on exponential families differs fundamentally from the classical free energy principle from statistical physics.

7.4   Exact Inference in Variational Form

In order to illustrate the general variational principle given in equation 7.37, it is worthwhile considering important cases in which it can be solved exactly. Accordingly, this section treats in some detail the case of a Gaussian MRF on an arbitrary graph—for which we rederive the normal equations—as well as the case of a multinomial MRF on a tree, for which we sketch out a derivation of the sum-product algorithm from a variational perspective. In addition to providing a novel perspective on exact methods, the variational principle 7.37 also underlies a variety of methods for approximate inference, as we will see in section 7.5.

7.4.1   Exact Inference in Gaussian Markov Random Fields

We begin by considering the case of a Gaussian Markov random field (MRF) on an arbitrary graph, as discussed in examples 7.3 and 7.8. In particular, we showed in the latter example that the set $\mathcal{M}_{Gauss}$ of realizable Gaussian mean parameters μ is determined by a positive definiteness constraint on the matrix $W(\mu)$ of mean parameters defined in equation 7.30.

We now consider the form of the dual function $A^*(\mu)$. It is well known (Cover and Thomas, 1991) that the entropy of a multivariate Gaussian random vector can be written as

$$H(p) = \frac{1}{2}\log \det \operatorname{cov}(\mathbf{x}) + \frac{n}{2}\log 2\pi e,$$

where $\operatorname{cov}(\mathbf{x})$ is the $n\times n$ covariance matrix of $\mathbf{x}$. By recalling the definition (eq. 7.30) of $W(\mu)$ and applying the Schur complement formula (Horn and Johnson, 1985), we see that $\det \operatorname{cov}(\mathbf{x}) = \det W(\mu)$, which implies that the dual function for a Gaussian can be written in the form

$$A^*_{Gauss}(\mu) = -\frac{1}{2}\log \det W(\mu) - \frac{n}{2}\log 2\pi e, \tag{7.38}$$

valid for all $\mu \in \mathcal{M}_{Gauss}$. (To understand the negative signs, recall from equation 7.35 that A^* is equal to the negative entropy for $\mu \in \mathcal{M}_{Gauss}$.) Combining this exact expression for A^*_{Gauss} with our characterization of $\mathcal{M}_{Gauss}$ leads to

$$A_{Gauss}(\theta) = \sup_{W(\mu)\succ 0,\; W_{11}(\mu)=1} \left\{ \langle U(\theta),\, W(\mu)\rangle + \frac{1}{2}\log \det W(\mu) + \frac{n}{2}\log 2\pi e\right\}, \tag{7.39}$$

which corresponds to the variational principle 7.37 specialized to the Gaussian case.

We now show how solving the optimization problem 7.39 leads to the *normal equations* for Gaussian inference. In order to do so, it is convenient to introduce the following notation for different blocks of the matrices $W(\mu)$ and $U(\theta)$:

$$W(\mu) = \begin{bmatrix} 1 & z^T(\mu) \\ z(\mu) & Z(\mu) \end{bmatrix}, \qquad U(\theta) = \begin{bmatrix} 0 & z^T(\theta) \\ z(\theta) & Z(\theta) \end{bmatrix}. \tag{7.40}$$

In this definition, the submatrices $Z(\mu)$ and $Z(\theta)$ are $n \times n$, whereas $z(\mu)$ and $z(\theta)$ are $n \times 1$ vectors.

Now if $W(\mu) \succ 0$ were the only constraint in problem 7.39, then, using the fact that $\nabla \log \det W = W^{-1}$ for any symmetric positive matrix W, the optimal solution to problem 7.39 would simply be $W(\mu) = -2[U(\theta)]^{-1}$. Accordingly, if we enforce the constraint $[W(\mu)]_{11} = 1$ using a Lagrange multiplier λ, then it follows from the Karush-Kuhn-Tucker conditions (Bertsekas, 1995b) that the optimal solution will assume the form $W(\mu) = -2[U(\theta) + \lambda^* E_{11}]^{-1}$, where λ^* is the optimal setting of the Lagrange multiplier and E_{11} is an $(n + 1) \times (n + 1)$ matrix with a one in the upper left hand corner, and zero in all other entries. Using the standard formula for the inverse of a block-partitioned matrix (Horn and Johnson, 1985), it is straightforward to verify that the blocks in the optimal $W(\mu)$ are related to the blocks of $U(\theta)$ by the relations:

$$Z(\mu) - z(\mu)z^T(\mu) = -2[Z(\theta)]^{-1} \tag{7.41a}$$

$$z(\mu) = -[Z(\theta)]^{-1} z(\theta) \tag{7.41b}$$

(The multiplier λ^* turns out not to be involved in these particular blocks.) In order to interpret these relations, it is helpful to return to the definition of $U(\theta)$ given in equation 7.18, and the Gaussian density of equation 7.19. In this way, we see that the first part of equation 7.41 corresponds to the fact that the covariance matrix is the inverse of the precision matrix, whereas the second part corresponds to the normal equations for the mean $z(\mu)$ of a Gaussian. Thus, as a special case of the general variational principle 7.37, we have rederived the familiar equations for Gaussian inference.

It is worthwhile noting that the derivation did not exploit any particular features of the graph structure. The Gaussian case is remarkable in this regard, in that both the dual function A^* and the set $\mathcal{M}$ of realizable mean parameters can be characterized simply for an arbitrary graph. However, many methods for solving the normal equations 7.41 as efficiently as possible, including Kalman filtering on trees (Willsky, 2002), make heavy use of the underlying graphical structure.

7.4.2   Exact Inference on Trees

We now turn to the case of tree-structured Markov random fields, focusing for concreteness on the multinomial case, first introduced in example 7.4 and treated in more depth in example 7.9. Recall from the latter example that for a multinomial MRF, the set $\mathcal{M}$ of realizable mean parameters corresponds to a marginal polytope, which we denote by $\mathrm{MARG}(G)$.

There is an obvious set of local constraints that any member of $\mathrm{MARG}(G)$ must satisfy. For instance, given their interpretation as local marginal distributions, the vectors μ_s and μ_{st} must of course be nonnegative. In addition, they must satisfy normalization conditions (i.e., $\sum_{x_s} \mu_s(x_s) = 1$), and the pairwise marginalization conditions (i.e., $\sum_{x_t} \mu_{st}(x_s, x_t) = \mu_s(x_s)$). Accordingly, we define for any graph G the following constraint set:

$$\mathrm{LOCAL}(G) := \{ \mu \geq 0 \mid \sum_{x_s} \mu_s(x_s) = 1, \ \sum_{x_t} \mu_{st}(x_s, x_t) = \mu_s(x_s) \ \forall (s,t) \in E \}. \tag{7.42}$$

Since any set of singleton and pairwise marginals (regardless of the underlying graph structure) must satisfy these local consistency constraints, we are guaranteed that $\mathrm{MARG}(G) \subseteq \mathrm{LOCAL}(G)$ for *any* graph G. This fact plays a significant role in our later discussion in section 7.6 of the Bethe variational principle and sum-product on graphs with cycles. Of most importance to the current development is the following consequence of the junction tree theorem (see section 7.1.2, subsection on junction-tree representation): when the graph G is tree-structured, then LOCAL(T) = MARG(T). Thus, the marginal polytope MARG(T) for trees has a very simple description given in 7.42.

The second component of the exact variational principle 7.37 is the dual function A^*. Here the junction tree framework is useful again: in particular, specializing the representation in equation 7.7 to a tree yields the following factorization:

$$p(\mathbf{x}; \mu) = \prod_{s \in V} \mu_s(x_s) \prod_{(s,t) \in E} \frac{\mu_{st}(x_s, x_t)}{\mu_s(x_s)\mu_t(x_t)} \qquad (7.43)$$

for a tree-structured distribution in terms of its mean parameters μ_s and μ_{st}.

From this decomposition, it is straightforward to compute the entropy *purely* as a function of the mean parameters by taking the logarithm and expectations and simplifying. Doing so yields the expression

$$-A^*(\mu) = \sum_{s \in V} H_s(\mu_s) - \sum_{(s,t) \in E} I_{st}(\mu_{st}), \qquad (7.44)$$

where the singleton entropy H_s and mutual information I_{st} are given by

$$H_s(\mu_s) := -\sum_{x_s} \mu_s(x_s) \log \mu_s(x_s), \qquad I_{st}(\mu_{st}) := \sum_{x_s, x_t} \mu_{st}(x_s, x_t) \log \frac{\mu_{st}(x_s, x_t)}{\mu_s(x_s)\mu_t(x_t)},$$

respectively. Putting the pieces together, the general variational principle 7.37 takes the following particular form:

$$A(\theta) = \max_{\mu \in \text{LOCAL}(T)} \left\{ \langle \theta, \mu \rangle + \sum_{s \in V} H_s(\mu_s) - \sum_{(s,t) \in E} I_{st}(\mu_{st}) \right\}. \qquad (7.45)$$

There is an important link between this variational principle for multinomial MRFs on trees, and the sum-product updates (eq. 7.6). In particular, the sum-product updates can be derived as an iterative algorithm for solving a Lagrangian dual formulation of the problem 7.45. This will be clarified in our discussion of the Bethe variational principle in section 7.6.

7.5   Approximate Inference in Variational Form

Thus far, we have seen how well-known methods for exact inference—specifically, the computation of means and covariances in the Gaussian case and the computation of local marginal distributions by the sum-product algorithm for tree-structured problems—can be rederived from the general variational principle (eq. 7.37). It is worthwhile isolating the properties that permit an exact solution of the variational principle. First, for both of the preceding cases, it is possible to characterize the set $\mathcal{M}$ of globally realizable mean parameters in a straightforward manner. Second, the entropy can be expressed as a closed-form function of the mean parameters μ, so that the dual function $A^*(\mu)$ has an explicit form.

Neither of these two properties holds for a general graphical model in exponential form. As a consequence, there are significant challenges associated with exploiting the variational representation. More precisely, in contrast to the simple cases discussed thus far, many graphical models of interest have the following properties:

1. the constraint set $\mathcal{M}$ of realizable mean parameters is extremely difficult to characterize in an explicit manner.

2. the negative entropy function A^* is defined indirectly—in a variational manner—so that it too typically lacks an explicit form.

These difficulties motivate the use of approximations to $\mathcal{M}$ and A^*. Indeed, a broad class of methods for approximate inference—ranging from mean field theory to cluster variational methods—are based on this strategy. Accordingly, the remainder of the chapter is devoted to discussion of approximate methods based on relaxations of the exact variational principle.

7.5.1 Mean Field Theory

We begin our discussion of approximate algorithms with mean field methods, a set of algorithms with roots in statistical physics (Chandler, 1987). Working from the variational principle 7.37, we show that mean field methods can be understood as solving an approximation thereof, with the essential restriction that the optimization is limited to a subset of distributions for which the dual function A^* is relatively easy to characterize. Throughout this section, we will refer to a distribution with this property as a *tractable* distribution.

Tractable Families   Let H represent a subgraph of G over which it feasible to perform exact calculations (e.g., a graph with small treewidth); we refer to any such H as a *tractable subgraph*. In an exponential formulation, the set of all distributions that respect the structure of H can be represented by a linear subspace of canonical parameters. More specifically, letting $\mathcal{I}(H)$ denote the subset of indices associated with cliques in H, the set of canonical parameters corresponding to distributions structured according to H is given by:

$$\mathcal{E}(H) := \{\theta \in \Theta \mid \theta_\alpha = 0 \quad \forall \; \alpha \in \mathcal{I}\backslash\mathcal{I}(H)\}. \tag{7.46}$$

We consider some examples to illustrate:

Example 7.10 Tractable Subgraphs

The simplest instance of a tractable subgraph is the completely disconnected graph $H_0 = (V, \emptyset)$ (see fig. 7.7b). Permissible parameters belong to the subspace $\mathcal{E}(H_0) := \{\theta \in \Theta \mid \theta_{st} = 0 \; \forall \; (s, t) \in E\}$, where θ_{st} refers to the collection of canonical parameters associated with edge (s, t). The associated distributions are of the product form $p(\mathbf{x}; \theta) = \prod_{s \in V} p(x_s; \theta_s)$, where θ_s refers to the collection of canonical parameters associated with vertex s.

To obtain a more structured approximation, one could choose a spanning tree $T = (V, E(T))$, as illustrated in fig. 7.7c. In this case, we are free to choose the canonical parameters corresponding to vertices and edges in T, but we must set to zero any canonical parameters corresponding to edges not in the tree. Accordingly, the subspace of tree-structured distributions is given by $\mathcal{E}(T) = \{\theta \mid \theta_{st} = 0 \ \forall \ (s, t) \notin E(T)\}$. $\Diamond$

For a given subgraph H, consider the set of all possible mean parameters that are realizable by tractable distributions:

$$\mathcal{M}_{tract}(G; H) := \{\mu \in \mathrm{R}^d \mid \mu = \mathrm{E}_\theta[\phi(\mathbf{x})] \text{ for some } \theta \in \mathcal{E}(H)\}. \qquad (7.47)$$

The notation $\mathcal{M}_{tract}(G; H)$ indicates that mean parameters in this set arise from taking expectations of sufficient statistics associated with the graph G, but that they must be realizable by a tractable distribution—i.e., one that respects the structure of H. See example 7.11 for an explicit illustration of this set when the tractable subgraph H is the fully disconnected graph. Since any μ that arises from a tractable distribution is certainly a valid mean parameter, the inclusion $\mathcal{M}_{tract}(G; H) \subseteq \mathcal{M}(G)$ always holds. In this sense, $\mathcal{M}_{tract}$ is an *inner approximation* to the set $\mathcal{M}$ of realizable mean parameters.

Optimization and Lower Bounds   We now have the necessary ingredients to develop the mean field approach to approximate inference. Let $p(\mathbf{x}; \theta)$ denote the *target distribution* that we are interested in approximating. The basis of the mean field method is the following fact: any valid mean parameter specifies a lower bound on the cumulant generating function. Indeed, as an immediate consequence of the variational principle 7.37, we have:

$$A(\theta) \geq \langle \theta, \mu \rangle - A^*(\mu) \qquad (7.48)$$

for any $\mu \in \mathcal{M}$. This inequality can also be established by applying Jensen's inequality (Jordan et al., 1999).

Since the dual function A^* typically lacks an explicit form, it is not possible, at least in general, to compute the lower bound (eq. 7.48). The mean field approach circumvents this difficulty by restricting the choice of μ to the tractable subset $\mathcal{M}_{tract}(G; H)$, for which the dual function has an explicit form A_H^*. As long as μ belongs to $\mathcal{M}_{tract}(G; H)$, then the lower bound 7.48 will be computable.

Of course, for a nontrivial class of tractable distributions, there are many such bounds. The goal of the mean field method is the natural one: find the best approximation μ^{MF}, as measured in terms of the tightness of the bound. This optimal approximation is specified as the solution of the optimization problem

$$\sup_{\mu \in \mathcal{M}_{tract}(G; H)} \left\{ \langle \mu, \theta \rangle - A_H^*(\mu) \right\}, \qquad (7.49)$$

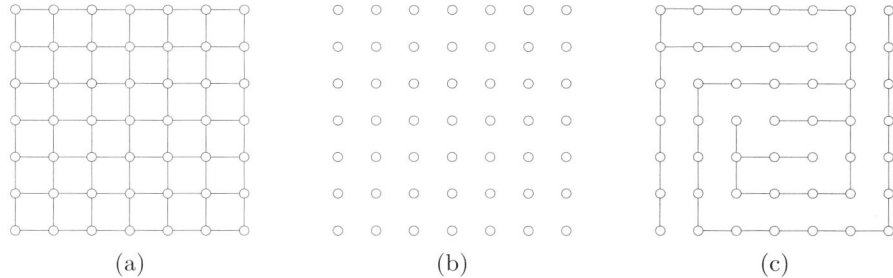

Figure 7.7 Graphical illustration of the mean field approximation. (a) Original graph is a 7×7 grid. (b) Fully disconnected graph, corresponding to a naive mean field approximation. (c) A more structured approximation based on a spanning tree.

which is a relaxation of the exact variational principle 7.37. The optimal value specifies a lower bound on $A(\theta)$, and it is (by definition) the best one that can be obtained by using a distribution from the tractable class.

An important alternative interpretation of the mean field approach is in terms of minimizing the Kullback-Leibler (KL) divergence between the approximating (tractable) distribution and the target distribution. Given two densities p and q, the KL divergence is given by

$$D(p \parallel q) = \int_{\mathcal{X}^n} \log \frac{p(\mathbf{x})}{q(\mathbf{x})} p(\mathbf{x}) \boldsymbol{\nu}(d\mathbf{x}). \tag{7.50}$$

To see the link to our derivation of mean field, consider for a given mean parameter $\mu \in \mathcal{M}_{tract}(G; H)$, the difference between the log partition function $A(\theta)$ and the quantity $\langle \mu, \theta \rangle - A_H^*(\mu)$:

$$D(\mu \parallel \theta) = A(\theta) + A_H^*(\mu) - \langle \mu, \theta \rangle.$$

A bit of algebra shows that this difference is equal to the KL divergence 7.50 with $q = p(\mathbf{x}; \theta)$ and $p = p(\mathbf{x}; \mu)$ (i.e., the exponential family member with mean parameter μ). Therefore, solving the mean field variational problem 7.49 is equivalent to minimizing the KL divergence subject to the constraint that μ belongs to tractable set of mean parameters, or equivalently that p is a tractable distribution.

7.5.2 Naive Mean Field Updates

The *naive mean field* (MF) approach corresponds to choosing a fully factorized or product distribution in order to approximate the original distribution. The naive mean field updates are a particular set of recursions for finding a stationary point of the resulting optimization problem.

Example 7.11

As an illustration, we derive the naive mean field updates for the *Ising model*, which is a special case of the multinomial MRF defined in example 7.4. It involves binary variables, so that $\mathcal{X}_s = \{0, 1\}$ for all vertices $s \in V$. Moreover, the canonical parameters are of the form $\theta_s(x_s) = \theta_s x_s$ and $\theta_{st}(x_s, x_t) = \theta_{st} x_s x_t$ for real numbers θ_s and θ_{st}. Consequently, the exponential representation of the Ising model has the form

$$p(\mathbf{x}; \theta) \propto \exp\Big\{ \sum_{s \in V} \theta_s x_s + \sum_{(s,t) \in E} \theta_{st} x_s x_t \Big\}.$$

Letting H_0 denote the fully disconnected graph (i.e., without any edges), the tractable set $\mathcal{M}_{tract}(G; H_0)$ consists of all mean parameters $\{\mu_s, \mu_{st}\}$ that arise from a product distribution. Explicitly, in this binary case, we have

$$\mathcal{M}_{tract}(G; H_0) := \{(\mu_s, \mu_{st}) \mid 0 \le \mu_s \le 1, \ \mu_{st} = \mu_s \, \mu_t \}.$$

Moreover, the negative entropy of a product distribution over binary random variables decomposes into the sum $A_{H_0}^*(\mu) = \sum_{s \in V} \big[\mu_s \log \mu_s + (1 - \mu_s) \log(1 - \mu_s) \big]$. Accordingly, the associated naive mean field problem takes the form

$$\max_{\mu \in \mathcal{M}_{tract}(G; H_0)} \big\{ \langle \mu, \, \theta \rangle - A_{H_0}^*(\mu) \big\}.$$

In this particular case, it is convenient to eliminate μ_{st} by replacing it by the product $\mu_s \mu_t$. Doing so leads to a reduced form of the problem:

$$\max_{\{\mu_s\} \in [0,1]^n} \Big\{ \sum_{s \in V} \theta_s \mu_s + \sum_{(s,t) \in E} \theta_{st} \mu_s \mu_t - \sum_{s \in V} \big[\mu_s \log \mu_s + (1 - \mu_s) \log(1 - \mu_s) \big] \Big\}. \tag{7.51}$$

Let F denote the function of μ within curly braces in equation 7.51. It can be seen that the function F is strictly concave in a given fixed coordinate μ_s when all the other coordinates are held fixed. Moreover, it is straightforward to show that the maximum over μ_s with $\mu_t, t \neq s$ fixed is attained in the interior $(0, 1)$, and can be found by taking the gradient and setting it equal to zero. Doing so yields the following update for μ_s:

$$\mu_s \ \leftarrow \ \sigma\Big(\theta_s + \sum_{t \in \mathcal{N}(s)} \theta_{st} \mu_t\Big), \tag{7.52}$$

where $\sigma(z) := [1 + \exp(-z)]^{-1}$ is the logistic function. Applying equation 7.52 iteratively to each node in succession amounts to performing coordinate ascent in the objective function for the mean field variational problem 7.51. Thus, we have derived the update equation presented earlier in equation 7.10. $\diamond$

Similarly, it is straightforward to apply the naive mean field approximation to other types of graphical models, as we illustrate for a multivariate Gaussian.

Example 7.12 Gaussian Mean Field

The mean parameters for a multivariate Gaussian are of the form $\mu_s = \mathrm{E}[x_s]$, $\mu_{ss} = \mathrm{E}[x_s^2]$ and $\mu_{st} = \mathrm{E}[x_s x_t]$ for $s \neq t$. Using only Gaussians in product form, the set of tractable mean parameters takes the form

$$\mathcal{M}_{tract}(G; H_0) = \{\mu \in \mathrm{R}^d \mid \mu_{st} = \mu_s \mu_t \; \forall s \neq t, \; \mu_{ss} - \mu_s^2 > 0 \}.$$

As with naive mean field on the Ising model, the constraints $\mu_{st} = \mu_s \mu_t$ for $s \neq t$ can be imposed directly, thereby leaving only the inequality $\mu_{ss} - \mu_s^2 > 0$ for each node. The negative entropy of a Gaussian in product form can be written as $A_{Gauss}^*(\mu) = -\sum_{s=1}^n \frac{1}{2} \log(\mu_{ss} - \mu_s^2) - \frac{n}{2} \log 2\pi e$. Combining A_{Gauss}^* with the constraints leads to the naive MF problem for a multivariate Gaussian:

$$\sup_{\{(\mu_s, \mu_{ss}) \mid \mu_{ss} - \mu_s^2 > 0\}} \left\{ \langle U(\theta), \, W(\mu) \rangle + \sum_{s=1}^n \frac{1}{2} \log(\mu_{ss} - \mu_s^2) + \frac{n}{2} \log 2\pi e \right\},$$

where the matrices $U(\theta)$ and $W(\mu)$ are defined in equation 7.40. Here it should be understood that any terms $\mu_{st}, s \neq t$ contained in $W(\mu)$ are replaced with the product $\mu_s \mu_t$.

Taking derivatives with respect to μ_{ss} and μ_s and rearranging yields the stationary conditions $\frac{1}{2(\mu_{ss} - \mu_s^2)} = -\theta_{ss}$ and $\frac{\mu_s}{2(\mu_{ss} - \mu_s^2)} = \theta_s + \sum_{t \in \mathcal{N}(s)} \theta_{st} \mu_t$. Since $\theta_{ss} < 0$, we can combine both equations into the update $\mu_s \leftarrow -\frac{1}{\theta_{ss}} \{\theta_s + \sum_{t \in \mathcal{N}(s)} \theta_{st} \mu_t \}$. In fact, the resulting algorithm is equivalent to the Gauss-Jacobi method for solving the normal equations, and so is guaranteed to converge under suitable conditions (Demmel, 1997), in which case the algorithm computes the correct mean vector $[\mu_1 \ldots \mu_n]$. $\Diamond$

7.5.3 Structured Mean Field and Other Extensions

Of course, the essential principles underlying the mean field approach are not limited to fully factorized distributions. More generally, one can consider classes of tractable distributions that incorporate additional structure. This *structured mean field approach* was first proposed by Saul and Jordan (1996), and further developed by various researchers. In this section, we discuss only one particular example in order to illustrate the basic idea, and refer the interested reader elsewhere (Wainwright and Jordan, 2003b; Wiegerinck, 2000) for further details.

Example 7.13 Structured Mean Field for Factorial Hidden Markov Models

The factorial hidden Markov model, as described in Ghahramani and Jordan (1997), has the form shown in fig. 7.8a. It consists of a set of M Markov chains ($M = 3$ in this diagram), which share at each time a common observation (shaded nodes). Such models are useful, for example, in modeling the joint dependencies between speech and video signals over time.

Although the separate chains are independent a priori, the common observation induces an effective coupling between all nodes at each time (a coupling which is captured by the moralization process mentioned earlier). Thus, an equivalent model is shown in fig. 7.8b, where the dotted ellipses represent the induced coupling of each observation.

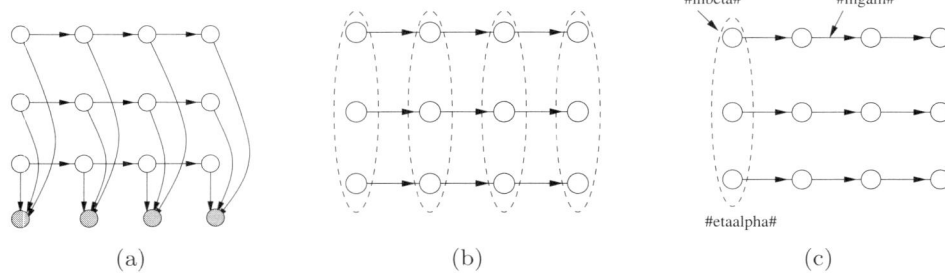

(a)   (b)   (c)

Figure 7.8   Structured mean field approximation for a factorial HMM. (a) Original model consists of a set of hidden Markov models (defined on chains), coupled at each time by a common observation. (b) An equivalent model, where the ellipses represent interactions among all nodes at a fixed time, induced by the common observation. (c) Approximating distribution formed by a product of chain-structured models. Here μ_α and μ_δ are the sets of mean parameters associated with the indicated vertex and edge respectively.

A natural choice of approximating distribution in this case is based on the subgraph H consisting of the decoupled set of M chains, as illustrated in fig. 7.8c. The decoupled nature of the approximation yields valuable savings on the computational side. In particular, it can be shown (Saul and Jordan, 1996; Wainwright and Jordan, 2003b) that all intermediate quantities necessary for implementing the structured mean field updates can be calculated by applying the forward-backward algorithm (i.e., the sum-product updates as an exact method) to each chain separately.   ◇

In addition to structured mean field, there are various other extensions to naive mean field, which we mention only in passing here. A large class of techniques, including linear response theory and the TAP method (Kappen and Rodriguez, 1998; Opper and Saad, 2001; Plefka, 1982), seek to improve the mean field approximation by introducing higher-order correction terms. Although the lower bound on the log partition function is not usually preserved by these higher-order methods, Leisink and Kappen (2001) demonstrated how to generate tighter lower bounds based on higher-order expansions.

7.5.4   Geometric View of Mean Field

An important fact about the mean field approach is that the variational problem (eq. 7.49) may be nonconvex, so that there may be local minima, and the mean field updates can have multiple solutions.

One way to understand this nonconvexity is in terms of the set of tractable mean parameters: under fairly mild conditions, it can be shown (Wainwright and Jordan, 2003b) that the set $\mathcal{M}_{tract}(G; H)$ is nonconvex. Figure 7.9 provides a geometric illustration for the case of a multinomial MRF, for which the set $\mathcal{M}$ is a marginal polytope.

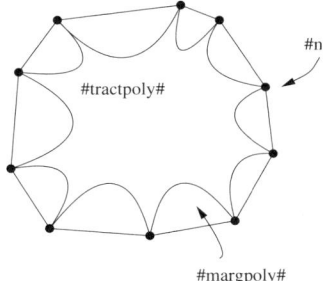

Figure 7.9 The set $\mathcal{M}_{tract}(G; H)$ of mean parameters that arise from tractable distributions is a nonconvex inner bound on $\mathcal{M}(G)$. Illustrated here is the multinomial case where $\mathcal{M}(G) \equiv \mathrm{MARG}(G)$ is a polytope. The circles correspond to mean parameters that arise from delta distributions with all their mass on a single configuration , and belong to both $\mathcal{M}(G)$ and $\mathcal{M}_{tract}(G; H)$.

A practical consequence of this nonconvexity is that the mean field updates are often sensitive to the initial conditions. Moreover, the mean field method can exhibit *spontaneous symmetry breaking*, wherein the mean field approximation is asymmetric even though the original problem is perfectly symmetric; see Jaakkola (2001) for an illustration of this phenomenon. Despite this nonconvexity, the mean field approximation becomes exact for certain types of models as the number of nodes n grows to infinity (Baxter, 1982).

7.5.5 Parameter Estimation and Variational Expectation Maximization

Mean field methods also play an important role in the problem of parameter estimation, in which the goal is to estimate model parameters on the basis of partial observations. The expectation-maximization (EM) algorithm (Dempster et al., 1977) provides a general approach to maximum likelihood parameter estimation in the case in which some subset of variables are observed whereas others are unobserved. Although the EM algorithm is often presented as an alternation between an expectation step (E step) and a maximization step (M step), it is also possible to take a variational perspective on EM, and view both steps as maximization steps (Csiszar and Tusn'ady, 1984; Neal and Hinton, 1999). More concretely, in the exponential family setting, the E step reduces to the computation of expected sufficient statistics—i.e., mean parameters. As we have seen, the variational framework provides a general class of methods for computing approximations of mean parameters. This observation suggests a general class of *variational EM algorithms*, in which the approximation provided by a variational inference algorithm is substituted for the mean parameters in the E step. In general, as a consequence of making such a substitution, one loses the guarantees that are associated with the EM algorithm. In the specific case of mean field algorithms, however, a convergence guarantee is retained: in particular, the algorithm will converge to a stationary point of a lower bound for the likelihood function (Wainwright and Jordan, 2003b).

7.6   The Bethe Entropy Approximation and the Sum-Product Algorithm

In this section, we turn to another important message-passing algorithm for approximate inference, known either as *belief propagation* or the *sum-product algorithm*. In section 7.4.2, we described the use of the sum-product algorithm for trees, in which context it is guaranteed to converge and perform exact inference. When the same message-passing updates are applied to graphs with cycles, in contrast, there are no such guarantees; nonetheless, this "loopy" form of the sum-product algorithm is widely used to compute approximate marginals in various signal-processing applications, including phase unwrapping (Frey et al., 2001), low-level vision (Freeman et al., 2000), and channel decoding (Richardson and Urbanke, 2001).

The main idea of this section is the connection between the sum-product updates and the Bethe variational principle. The presentation given here differs from the original work of Yedidia et al. (2001), in that we formulate the problem purely in terms of mean parameters and marginal polytopes. This perspective highlights a key point: mean field and sum-product, though similar as message-passing algorithms, are fundamentally different at the variational level. In particular, whereas the essence of mean field is to *restrict* optimization to a limited class of distributions for which the negative entropy and mean parameters can be characterized *exactly*, the the sum-product algorithm, in contrast, is based on *enlarging* the constraint set and *approximating* the entropy function.

The standard Bethe approximation applies to an undirected graphical model with potential functions involving at most pairs of variables, which we refer to as a *pairwise Markov random field*. In principle, by selectively introducing auxiliary variables, any undirected graphical model can be converted into an equivalent pairwise form to which the Bethe approximation can be applied; see Weiss and Freeman (2000) for a detailed description of this procedure. Moreover, although the Bethe approximation can be developed more generally, we also limit our discussion to a multinomial MRF, as discussed earlier in examples 7.4 and 7.9. We also make use of the local marginal functions $\mu_s(x_s)$ and $\mu_{st}(x_s, x_t)$, as defined in equation 7.32. As discussed in Example 7.9, the set $\mathcal{M}$ associated with a multinomial MRF is the marginal polytope MARG(G).

Recall that there are two components to the general variational principle 7.37: the set of realizable mean parameters (given by a marginal polytope in this case), and the dual function A^*. Developing an approximation to the general principle requires approximations to both of these components, which we discuss in turn in the following sections.

7.6.1   Bethe Entropy Approximation

From equation 7.35, recall that dual function A^* corresponds to the maximum entropy distribution consistent with a given set of mean parameters; as such, it typically lacks a closed-form expression. An important exception to this general rule is the case of a tree-structured distribution: as discussed in section 7.4.2, the function A^* for a tree-structured distribution has a closed-form expression that is straightforward to compute; see, in particular, equation 7.44.

Of course, the entropy of a distribution defined by a graph with cycles will not, in general, decompose additively like that of a tree. Nonetheless, one can imagine using the decomposition in equation 7.44 as an approximation to the entropy. Doing so yields an expression known as the *Bethe approximation* to the entropy on a graph with cycles:

$$H_{Bethe}(\mu) := \sum_{s \in V} H_s(\mu_s) - \sum_{(s,t) \in E} I_{st}(\mu_{st}). \tag{7.53}$$

To be clear, the quantity $H_{Bethe}(\mu)$ is an approximation to the negative dual function $-A^*(\mu)$. Moreover, our development in section 7.4.2 shows that this approximation is exact when the graph is tree-structured.

An alternative form of the Bethe entropy approximation can be derived by writing mutual information in terms of entropies as $I_{st}(\mu_{st}) = H_s(\mu_s) + H_t(\mu_t) - H_{st}(\mu_{st})$. In particular, expanding the mutual information terms in this way, and then collecting all the single-node entropy terms yields $H_{Bethe}(\mu) = \sum_{s \in V} (1 - d_s) H_s(\mu_s) + \sum_{(s,t) \in E} H_{st}(\mu_{st})$, where d_s denotes the number of neighbors of node s. This representation is the form of the Bethe entropy introduced by Yedidia et al. (2001); however, the form given in equation 7.53 turns out to be more convenient for our purposes.

7.6.2   Tree-Based Outer Bound

Note that the Bethe entropy approximation H_{Bethe} is certainly well defined for any $\mu \in \text{MARG}(G)$. However, as discussed earlier, characterizing this polytope of realizable marginals is a very challenging problem. Accordingly, a natural approach is to specify a subset of necessary constraints, which leads to an outer bound on $\text{MARG}(G)$. Let $\tau_s(x_s)$ and $\tau_{st}(x_s, x_t)$ be a set of candidate marginal distributions. In section 7.4.2, we considered the following constraint set:

$$\text{LOCAL}(G) = \left\{ \tau \geq 0 \mid \sum_{x_s} \tau_s(x_s) = 1, \quad \sum_{x_s} \tau_{st}(x_s, x_t) = \tau_t(x_t) \right\}. \tag{7.54}$$

Although LOCAL(G) is an exact description of the marginal polytope for a tree-structured graph, it is only an outer bound for graphs with cycles. (We demonstrate this fact more concretely in example 7.14.) For this reason, our change in notation— i.e., from μ to τ—is quite deliberate, with the goal of emphasizing that members τ of LOCAL(G) need not be realizable. We refer to members of LOCAL(G) as *pseudomarginals* (these are sometimes referred to as beliefs).

Example 7.14 Pseudomarginals

We illustrate using a binary random vector on the simplest possible graph for which LOCAL(G) is not an exact description of MARG(G)—namely, a single cycle with three nodes. Consider candidate marginal distributions $\{\tau_s, \tau_{st}\}$ of the form

$$\tau_s := \begin{bmatrix} 0.5 & 0.5 \end{bmatrix}, \qquad \tau_{st} := \begin{bmatrix} \beta_{st} & 0.5 - \beta_{st} \\ 0.5 - \beta_{st} & \beta_{st} \end{bmatrix}, \qquad (7.55)$$

where $\beta_{st} \in [0, 0.5]$ is a parameter to be specified independently for each edge (s, t). It is straightforward to verify that $\{\tau_s, \tau_{st}\}$ belong to LOCAL(G) for any choice of $\beta_{st} \in [0, 0.5]$.

First, consider the setting $\beta_{st} = 0.4$ for all edges (s, t), as illustrated in fig. 7.10a. It is not difficult to show that the resulting marginals thus defined are realizable; in fact, they can be obtained from the distribution that places probability 0.35 on each of the configurations $[0\ 0\ 0]$ and $[1\ 1\ 1]$, and probability 0.05 on each of the remaining six configurations. Now suppose that we perturb one of the pairwise marginals—say τ_{13}—by setting $\beta_{13} = 0.1$. The resulting problem is illustrated in fig. 7.10b. Observe that there are now strong (positive) dependencies between the pairs of variables (x_1, x_2) and (x_2, x_3): both pairs are quite likely to agree (with probability 0.8). In contrast, the pair (x_1, x_3) can only share the

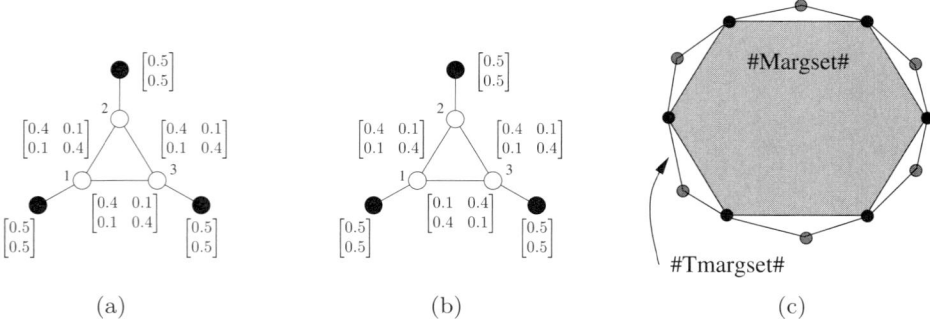

(a)   (b)   (c)

Figure 7.10   (a), (b): Illustration of the marginal polytope for a single cycle graph on three nodes. Setting $\beta_{st} = 0.4$ for all three edges gives a globally consistent set of marginals. (b) With β_{13} perturbed to 0.1, the marginals (though locally consistent) are no longer globally so. (c) For a more general graph, an idealized illustration of the tree-based constraint set LOCAL(G) as an outer bound on the marginal polytope MARG(G).

same value relatively infrequently (with probability 0.2). This arrangement should provoke some doubt. Indeed, it can be shown that $\tau \notin \mathrm{MARG}(G)$ by attempting but *failing* to construct a distribution that realizes τ, or alternatively and much more directly by using the idea of semidefinite constraints (see example 7.15). $\diamondsuit$

More generally, figure 7.10c provides an idealized illustration of the constraint set $\mathrm{LOCAL}(G)$, and its relation to the exact marginal polytope $\mathrm{MARG}(G)$. Observe that the set $\mathrm{LOCAL}(G)$ is another polytope that is a *convex outer approximation* to $\mathrm{MARG}(G)$. It is worthwhile contrasting with the *nonconvex inner approximation* used by a mean field approximation, as illustrated in fig. 7.9.

7.6.3   Bethe Variational Problem and Sum-Product

Note that the Bethe entropy is also well defined for any pseudomarginal in $\mathrm{LOCAL}(G)$. Therefore, it is valid to consider a constrained optimization problem over the set $\mathrm{LOCAL}(G)$ in which the cost function involves the Bethe entropy approximation H_{Bethe}. Indeed, doing so leads to the so-called *Bethe variational problem*:

$$\max_{\tau \in \mathrm{LOCAL}(G)} \left\{ \langle \theta, \, \tau \rangle + \sum_{s \in V} H_s(\tau_s) - \sum_{(s,t) \in E} I_{st}(\tau_{st}) \right\}. \tag{7.56}$$

Although ostensibly similar to a (structured) mean field approach, the Bethe variational problem (BVP) is fundamentally different in a number of ways. First, as discussed in section 7.5.1, a mean field method is based on an exact representation of the entropy, albeit over a limited class of distributions. In contrast, with the exception of tree-structured graphs, the Bethe entropy is a bona fide *approximation* to the entropy. For instance, it is not difficult to see that it can be negative, which of course can never happen for an exact entropy. Second, the mean field approach entails optimizing over an *inner bound* on the marginal polytope, which ensures that any mean field solution is always globally consistent with respect to at least one distribution, and that it yields a lower bound on the log partition function. In contrast, since $\mathrm{LOCAL}(G)$ is a strict outer bound on the set of realizable marginals $\mathrm{MARG}(G)$, the optimizing pseudomarginals τ^* of the BVP may not be globally consistent with any distribution.

7.6.4   Solving the Bethe Variational Problem

Having formulated the Bethe variational problem, we now consider iterative methods for solving it. Observe that the set $\mathrm{LOCAL}(G)$ is a polytope defined by $\mathcal{O}(n + |E|)$ constraints. A natural approach to solving the BVP, then, is to attach Lagrange multipliers to these constraints, and find stationary points of the Lagrangian. A remarkable fact, established by Yedidia et al. (2001), is that the sum-product updates of equation 7.6 can be rederived as a method for trying to find such Lagrangian stationary points.

A bit more formally, for each $x_s \in \mathcal{X}_s$, let $\lambda_{st}(x_s)$ be a Lagrange multiplier associated with the constraint $C_{ts}(x_s) = 0$, where $C_{ts}(x_s) := \tau_s(x_s) - \sum_{x_t} \tau_{st}(x_s, x_t)$. Our approach is to consider the following partial Lagrangian corresponding to the Bethe variational problem 7.56:

$$\mathcal{L}(\tau; \lambda) := \langle \theta, \tau \rangle + H_{Bethe}(\tau) + \sum_{(s,t) \in E} \Big[\sum_{x_s} \lambda_{ts}(x_s) C_{ts}(x_s) + \sum_{x_t} \lambda_{st}(x_t) C_{st}(x_t) \Big].$$

The key insight of Yedidia et al. (2001) is that any fixed point of the sum-product updates specifies a pair (τ^*, λ^*) such that

$$\nabla_\tau \mathcal{L}(\tau^*; \lambda^*) = 0, \qquad \nabla_\lambda \mathcal{L}(\tau^*; \lambda^*) = 0 \qquad (7.57)$$

In particular, the Lagrange multipliers can be used to specify messages of the form $M_{ts}(x_s) = \exp(\lambda_{ts}(x_s))$. After taking derivatives of the Lagrangian and equating them to zero, some algebra then yields the familiar message-update rule:

$$M_{ts}(x_s) = \kappa \sum_{x_t} \exp \big\{ \theta_{st}(x_s, x_t) + \theta_t(x_t) \big\} \prod_{u \in \mathcal{N}(t) \setminus s} M_{ut}(x_t). \qquad (7.58)$$

We refer the reader to Yedidia et al. (2001) or Wainwright and Jordan (2003b) for further details of this derivation. By construction, any fixed point M^* of these updates specifies a pair (τ^*, λ^*) that satisfies the stationary[3] conditions given in equation 7.57.

This variational formulation of the sum-product updates—namely, as an algorithm for solving a constrained optimization problem—has a number of important consequences. First of all, it can be used to guarantee the existence of sum-product fixed points. Observe that the cost function in the Bethe variational problem 7.56 is continuous and bounded above, and the constraint set $\text{LOCAL}(G)$ is nonempty and compact; therefore, at least some (possibly local) maximum is attained. Moreover, since the constraints are linear, there will always be a set of Lagrange multipliers associated with any local maximum (Bertsekas, 1995b). For any optimum in the relative interior of $\text{LOCAL}(G)$, these Lagrange multipliers can be used to construct a fixed point of the sum-product updates.

For graphs with cycles, this Lagrangian formulation provides no guarantees on the convergence of the sum-product updates; indeed, whether or not the algorithm converges depends both on the potential strengths and the topology of the graph. Several researchers (Heskes et al.; Welling and Teh, 2001; Yuille, 2002) have proposed alternatives to sum-product that are guaranteed to converge, albeit at the price of increased computational cost. It should also be noted that with the exception of trees and other special cases (McEliece and Yildirim, 2002; Pakzad and Anantharam, 2002), the BVP is usually a nonconvex problem, in that H_{Bethe} fails to be concave. As a consequence, there may be multiple local optima to the BVP, and there are no guarantees that sum-product (or other iterative algorithms) will find a global optimum.

As illustrated in fig. 7.10c, the constraint set $\text{LOCAL}(G)$ of the Bethe variational problem is a strict outer bound on the marginal polytope $\text{MARG}(G)$. Since the exact marginals of $p(\mathbf{x}; \theta)$ must always lie in the marginal polytope, a natural question is whether solutions to the Bethe variational problem ever fall into the region $\text{LOCAL}(G) \setminus \text{MARG}(G)$. There turns out be a straightforward answer to this question, stemming from an alternative reparameterization-based characterization of sum-product fixed points (Wainwright et al., 2003b). One consequence of this characterization is that for any vector τ of pseudomarginals in the interior of $\text{LOCAL}(G)$, it is possible to specify a distribution for which τ is a sum-product fixed point. As a particular example, it is possible to construct a distribution $p(\mathbf{x}; \theta)$ such that the pseudomarginal τ discussed in example 7.14 is a fixed point of the sum-product updates.

7.6.5 Extensions Based on Clustering And Hypertrees

From our development in the previous section, it is clear that there are two *distinct* components to the Bethe variational principle: (1) the entropy approximation H_{Bethe}, and (2) the approximation $\text{LOCAL}(G)$ to the set of realizable marginal parameters. In principle, the BVP could be strengthened by improving either one, or both, of these components. One natural generalization of the BVP, first proposed by Yedidia et al. (2002) and further explored by various researchers (Heskes et al.; McEliece and Yildirim, 2002; Minka, 2001), is based on working with clusters of variables. The approximations in the Bethe approach are based on trees, which are special cases of junction trees based on cliques of size two. A natural strategy, then, is to strengthen the approximations by exploiting more complex junction trees, also known as hypertrees. Our description of this procedure is very brief, but further details can be found in various sources (Wainwright and Jordan, 2003b; Yedidia et al., 2002).

Recall that the essential ingredients in Bethe variational principle are local (pseudo)marginal distributions on nodes and edges (i.e., pairs of nodes). These distributions, subject to edgewise marginalization conditions, are used to specify the Bethe entropy approximation. One way to improve the Bethe approach, which is based on pairs of nodes, is to build entropy approximations and impose marginalization constraints on *larger* clusters of nodes. To illustrate, suppose that the original graph is simply the 3×3 grid shown in fig. 7.11a. A particular grouping of the nodes, which is known as Kikuchi four-plaque clustering in statistical physics (Yedidia et al., 2002), is illustrated in fig. 7.11b. This operation creates four new "supernodes" or clusters, each consisting of four nodes from the original graph. These clusters, as well as their overlaps—which turn out to be critical to track for certain technical reasons (Yedidia et al., 2002)—are illustrated in fig. 7.11c.

Given a clustering of this type, we now consider a set of marginal distributions τ_h, where h ranges over the clusters. As with the singleton τ_s and pairwise τ_{st} that define the Bethe approximation, we require that these higher-order cluster marginals be suitably normalized (i.e., $\sum_{x'_h} \tau_h(x'_h) = 1$), and be consistent with one

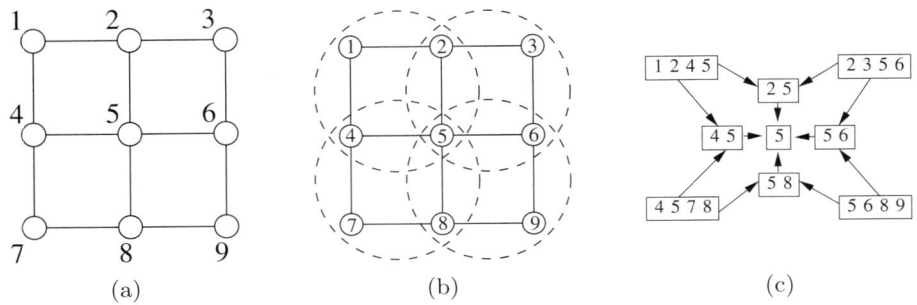

Figure 7.11   (a) Ordinary 3×3 grid. (b) Clustering of the vertices into groups of 4, known as Kikuchi four-plaque clustering. (c) Poset diagram of the clusters as well as their overlaps. Pseudomarginals on these subsets must satisfy certain local consistency conditions, and are used to define a higher-order entropy approximation.

another whenever they overlap. More precisely, for any pair $g \subseteq h$, the following *marginalization* condition $\sum_{\{x'_h \mid x'_g = x_g\}} \tau_h(x'_h) = \tau_g(x_g)$ must hold. Imposing these normalization and marginalization conditions leads to a higher-order analog of the constraint $\mathrm{LOCAL}(G)$ previously defined in equation 7.54.

In analogy to the Bethe entropy approximation, we can also consider a hypertree-based approximation to the entropy. There are certain technical aspects to specifying such entropy approximations, in that it turns out to be critical to ensure that the local entropies are weighted with certain "overcounting" numbers (Wainwright and Jordan, 2003b; Yedidia et al., 2002). Without going into these details here, the outcome is another relaxed variational principle, which can be understood as a higher-level analog of the Bethe variational principle.

7.7   From the Exact Principle to New Approximations

The preceding sections have illustrated how a variety of known methods—both exact and approximate—can be understood in an unified manner on the basis of the general variational principle given in equation 7.37. In this final section, we turn to a brief discussion of several new approximate methods that also emerge from this same variational principle. Given space constraints, our discussion in this chapter is necessarily brief, but we refer to reader to the papers of (Wainwright and Jordan, 2003a,b; Wainwright et al., 2002, 2003a) for further details.

7.7.1   Exploiting Semidefinite Constraints for Approximate Inference

As discussed in section 7.5, one key component in any relaxation of the exact variational principle is an approximation of the set $\mathcal{M}$ of realizable mean parameters. Recall that for graphical models that involve discrete random variables, we refer to this set as a *marginal polytope*. Since any polytope is specified by a finite collection of halfspace constraints (see fig. 7.6), one very natural way in which to generate an outer approximation is by including only a *subset* of these halfspace constraints. Indeed, as we have seen in section 7.6, it is precisely this route that the Bethe approximation and its clustering-based extensions follow.

However, such *polyhedral relaxations* are not the only way in which to generate outer approximations to marginal polytopes. Recognizing that elements of the marginal polytope are essentially *moments* leads very naturally to the idea of a *semidefinite relaxation*. Indeed, the use of semidefinite constraints for characterizing moments has a very rich history, both with classical work (Karlin and Studden, 1966) on scalar random variables, and more recent work (Lasserre, 2001; Parrilo, 2003) on the multivariate case.

Semidefinite Outer Bounds on Marginal Polytopes We use the case of a multinomial MRF defined by a graph $G = (V, E)$, as discussed in example 7.4, in order to illustrate the use of semidefinite constraints. Although the basic idea is quite generally applicable (Wainwright and Jordan, 2003b), herein we restrict ourselves to binary variables (i.e., $\mathcal{X}_s = \{0, 1\}$) so as to simplify the exposition. Recall that the sufficient statistics in a binary MRF take the form of certain indicator functions, as defined in equation 7.20. In fact, this representation is overcomplete (in that there are linear dependencies among the indicator functions); in the binary case, it suffices to consider only the sufficient statistics $x_s = \mathbb{I}_1(x_s)$ and $x_s x_t = \mathbb{I}_{11}(x_s, x_t)$. Our goal, then, is to characterize the set of all first- and second-order moments, defined by $\mu_s = \mathrm{E}[x_s]$ and $\mu_{st} = \mathrm{E}[x_s x_t]$ respectively, that arise from taking expectations with respect to a distribution with its support restricted to $\{0, 1\}^n$. Rather than focusing on just the pairs μ_{st} for edges $(s, t) \in E$, it is convenient to consider the full collection of pairwise moments $\{\mu_{st} \mid s, t \in V\}$.

Suppose that we are given a vector $\mu \in \mathbb{R}^d$ (where $d = n + \binom{n}{2}$), and wish to assess whether or not it is a globally realizable moment vector (i.e., whether there exists some distribution $p(\mathbf{x})$ such that $\mu_s = \sum_{\mathbf{x}} p(\mathbf{x})\, x_s$ and $\mu_{st} = \sum_{\mathbf{x}} p(\mathbf{x})\, x_s x_t$). In order to derive a *necessary* condition, we suppose that such a distribution p exists, and then consider the following $(n + 1) \times (n + 1)$ moment matrix:

$$
\mathrm{E}_p\left\{ \begin{bmatrix} 1 \\ \mathbf{x} \end{bmatrix} \begin{bmatrix} 1 & \mathbf{x} \end{bmatrix} \right\} =
\begin{bmatrix}
1 & \mu_1 & \mu_2 & \cdots & \mu_{n-1} & \mu_n \\
\mu_1 & \mu_1 & \mu_{12} & \cdots & & \mu_{1n} \\
\mu_2 & \mu_{21} & \mu_2 & \cdots & & \mu_{2n} \\
\vdots & \vdots & \vdots & \vdots & \vdots & \vdots \\
\mu_{n-1} & \vdots & \vdots & \vdots & \vdots & \mu_{n,(n-1)} \\
\mu_n & \mu_{n1} & \mu_{n2} & \cdots & \mu_{(n-1),n} & \mu_n
\end{bmatrix}, \quad (7.59)
$$

which we denote by $M_1[\mu]$. Note that in calculating the form of this moment matrix, we have made use of the relation $\mu_s = \mu_{ss}$, which holds because $x_s = x_s^2$ for any binary-valued quantity.

We now observe that any such moment matrix is necessarily positive semidefinite, which we denote by $M_1[\mu] \succeq 0$. (This positive semidefiniteness can be verified as follows: letting $\mathbf{y} := (1, \ \mathbf{x})$, then for any vector $a \in \mathrm{R}^{n+1}$, we have $a^T M_1[\mu]a = a^T \mathrm{E}[\mathbf{y}\mathbf{y}^T]a = \mathrm{E}[\|a^T\mathbf{y}\|^2]$, which is certainly nonnegative). Therefore, we conclude that the semidefinite constraint set $\mathrm{SDEF}_1 := \{\mu \in \mathrm{R}^d \mid M_1[\mu] \succeq 0\}$ is an outer bound on the exact marginal polytope.

Example 7.15

To illustrate the use of the outer bound SDEF_1, recall the pseudomarginal vector τ that we constructed in example 7.14 for the single cycle on three nodes. In terms of our reduced representation (involving only expectations of the singletons x_s and pairwise functions $x_s x_t$), this pseudomarginal can be written as follows:

$$\tau_s = 0.5 \text{ for } s = 1, 2, 3, \qquad \tau_{12} = \tau_{23} = 0.4, \qquad \tau_{13} = 0.1.$$

Suppose that we now construct the matrix M_1 for this trial set of mean parameters; it takes the following form:

$$M_1[\tau] = \begin{bmatrix} 1 & 0.5 & 0.5 & 0.5 \\ 0.5 & 0.5 & 0.4 & 0.1 \\ 0.5 & 0.4 & 0.5 & 0.4 \\ 0.5 & 0.1 & 0.4 & 0.5 \end{bmatrix}.$$

A simple calculation shows that it is not positive definite, so that $\tau \notin \mathrm{SDEF}_1$. Since SDEF_1 is an outer bound on the marginal polytope, this reasoning shows—in a very quick and direct manner— that τ is not a globally valid moment vector. In fact, the semidefinite constraint set SDEF_1 can be viewed as the first in a sequence of progressively tighter relaxations on the marginal polytope.

Log-Determinant Relaxation   We now show how to use such semidefinite constraints in approximate inference. Our approach is based on combining the first-order semidefinite outer bound SDEF_1 with Gaussian-based entropy approximation. The end result is a log-determinant problem that represents another relaxation of the exact variational principle (Wainwright and Jordan, 2003a). In contrast to the Bethe and Kikuchi approaches, this relaxation is convex (and hence has a unique optimum), and moreover provides an upper bound on the cumulant generating function.

Our starting point is the familiar interpretation of the Gaussian as the maximum entropy distribution subject to covariance constraints (Cover and Thomas, 1991). In particular, given a continuous random vector $\widetilde{\mathbf{x}}$, its differential entropy $h(\widetilde{\mathbf{x}})$ is always upper bounded by the entropy of a Gaussian with matched covariance, or in analytical terms

$$h(\widetilde{\mathbf{x}}) \le \frac{1}{2} \log \det \mathrm{cov}(\widetilde{\mathbf{x}}) + \frac{n}{2} \log(2\pi e), \tag{7.60}$$

where $\mathrm{cov}(\widetilde{\mathbf{x}})$ is the covariance matrix of $\widetilde{\mathbf{x}}$. The upper bound 7.60 is not directly applicable to a random vector taking values in a discrete space (since differential entropy in this case diverges to minus infinity). However, a straightforward discretization argument shows that for any discrete random vector $\mathbf{x} \in \{0, 1\}^n$, its (ordinary) discrete entropy can be upper bounded in terms of the matrix $M_1[\mu]$ of mean parameters as

$$H(\mathbf{x}) = -A^*(\mu) \leq \frac{1}{2} \log \det \left\{ M_1[\mu] + \frac{1}{12} \operatorname{blkdiag}[0, I_n] \right\} + \frac{n}{2} \log(2\pi e), \quad (7.61)$$

where $\operatorname{blkdiag}[0, I_n]$ is a $(n+1) \times (n+1)$ block-diagonal matrix with a 1×1 zero block, and an $n \times n$ identity block.

Finally, putting all the pieces together leads to the following result (Wainwright and Jordan, 2003a): the cumulant generating function $A(\theta)$ is upper bounded by the solution of the following *log-determinant optimization problem*:

$$A(\theta) \leq \max_{\tau \in \text{SDEF}_1} \left\{ \langle \theta, \mu \rangle + \frac{1}{2} \log \det \left[M_1(\tau) + \frac{1}{12} \operatorname{blkdiag}[0, I_n] \right] \right\} + \frac{n}{2} \log(2\pi e).$$

$$(7.62)$$

Note that the constraint $\tau \in \text{SDEF}_1$ ensures that $M_1(\tau) \succeq 0$, and hence a fortiori that $M_1(\tau) + \frac{1}{12} \operatorname{blkdiag}[0, I_n]$ is positive definite. Moreover, an important fact is that the optimization problem in equation 7.62 is a determinant maximization problem, for which efficient interior point methods have been developed (Vandenberghe et al., 1998).

Just as the Bethe variational principle (eq. 7.56) is a tree-based approximation, the log-determinant relaxation (eq. 7.62) is a Gaussian-based approximation. In particular, it is worthwhile comparing the structure of the log-determinant relaxation (eq. 7.62) to the exact variational principle for a multivariate Gaussian, as described in section 7.4.1. In contrast to the Bethe variational principle, in which all of the constraints defining the relaxation are local, this new principle (eq. 7.62) imposes some quite *global* constraints on the mean parameters. Empirically, these global constraints are important for strongly coupled problems, in which the performance log-determinant relaxation appears much more robust than the sum-product algorithm (Wainwright and Jordan, 2003a). In summary, starting from the exact variational principle (eq. 7.37), we have derived a new relaxation, whose properties are rather different than the Bethe and Kikuchi variational principles.

7.7.2 Relaxations for Computing Modes

Recall from our introductory comments in section 7.1.2 that, in addition to the problem of computing expectations and likelihoods, it is also frequently of interest to compute the mode of a distribution. This section is devoted to a brief discussion of mode computation, and more concretely how the exact variational principle (eq. 7.37), as well as relaxations thereof, again turns out to play an important role.

Zero-Temperature Limits   In order to understand the role of the exact variational principle 7.37 in computing modes, consider a multinomial MRF of the form $p(\mathbf{x}; \theta)$, as discussed in example 7.4. Of interest to us is the one-parameter family of distributions $\{p(\mathbf{x}; \beta\theta) \mid \beta > 0\}$, where β is the real number to be varied. At one extreme, if $\beta = 0$, then there is no coupling, and the distribution is simply uniform over all possible configurations. The other extreme, as $\beta \to +\infty$, is more interesting: in this limit, the distribution concentrates all of its mass on the configuration (or subset of configurations) that are modes of the distribution. Taking this limit $\beta \to +\infty$ is known as *zero-temperature* limit, since the parameter β is typically viewed as inverse temperature in statistical physics. This argument suggests that there should be a link between computing modes and the limiting behavior of the marginalization problem as $\beta \to +\infty$.

In order to develop this idea a bit more formally, we begin by observing that the exact variational principle 7.37 holds for the distribution $p(\mathbf{x}; \beta\theta)$ for any value of $\beta \geq 0$. It can be shown (Wainwright and Jordan, 2003b) that if we actually take a suitably scaled limit of this exact variational principle as $\beta \to +\infty$, then we recover the following variational principle for computing modes:

$$\max_{\mathbf{x} \in \mathcal{X}^n} \langle \theta, \, \phi(\mathbf{x}) \rangle \;=\; \max_{\mu \in \mathrm{MARG}(G)} \langle \theta, \, \mu \rangle \tag{7.63}$$

Since the log probability $\log p(\mathbf{x}; \theta)$ is equal to $\langle \theta, \, \phi(\mathbf{x}) \rangle$ (up to an additive constant), the left-hand side is simply the problem of computing the mode of the distribution $p(\mathbf{x}; \theta)$. On the right-hand side, we simply have a linear program, since the constraint set $\mathrm{MARG}(G)$ is a polytope, and the cost function $\langle \theta, \, \mu \rangle$ is linear in μ (with θ fixed). This equivalence means that, at least in principle, we can compute a mode of the distribution by solving a linear program (LP) over the marginal polytope. The geometric interpretation is also clear: as illustrated in fig. 7.6, vertices of the marginal polytope are in one-to-one correspondence with configurations $\mathbf{x}$. Since any LP achieves its optimum at a vertex (Bertsimas and Tsitsiklis, 1997), solving the LP is equivalent to finding the mode.

Linear Programming and Tree-Reweighted Max-Product   Of course, the LP-based reformulation in equation 7.63 is not practically useful for precisely the same reasons as before—it is extremely challenging to characterize the marginal polytope $\mathrm{MARG}(G)$ for a general graph. Many computationally intractable optimization problems (e.g., MAX-CUT) can be reformulated as LPs over the marginal polytope, as in equation 7.63, which underscores the inherent complexity of characterizing marginal polytopes. Nonetheless, this variational formulation motivates the idea of forming relaxations using outer bounds on the marginal polytope. For various classes of problems in combinatorial optimization, both linear programming and semidefinite relaxations of this flavor have been studied extensively.

Here we briefly describe an LP relaxation that is very natural given our development of the Bethe variational principle in section 7.6. In particular, we consider using the local constraint set $\mathrm{LOCAL}(G)$, as defined in equation 7.54, as an outer bound of the marginal polytope MARG(G). Doing so leads to the following LP relaxation for the problem of computing the mode of a multinomial MRF:

$$\max_{\mathbf{x}\in\mathcal{X}^n} \langle \theta,\, \phi(\mathbf{x})\rangle \;=\; \max_{\mu\in\mathrm{MARG}(G)} \langle \theta,\, \mu\rangle \;\leq\; \max_{\tau\in\mathrm{LOCAL}(G)} \langle \theta,\, \mu\rangle. \qquad (7.64)$$

Since the relaxed constraint set LOCAL(G)—like the original set MARG(G)—is a polytope, the relaxation on the right-hand side of equation 7.64 is a linear program. Consequently, the optimum of the relaxed problem must be attained at a vertex (possibly more than one) of the polytope LOCAL(G).

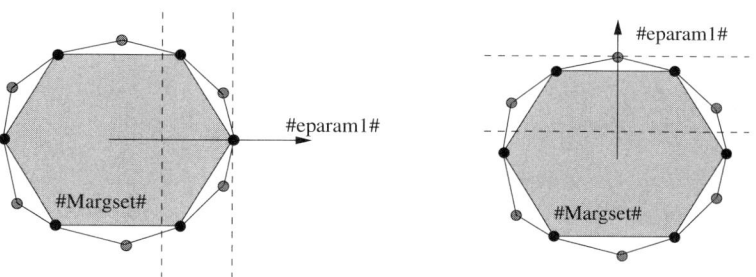

Figure 7.12   The constraint set LOCAL(G) is an outer bound on the exact marginal polytope. Its vertex set includes all the vertices of MARG(G), which are in one-to-one correspondence with optimal solutions of the integer program. It also includes additional fractional vertices, which are *not* vertices of MARG(G).

We say that a vertex of LOCAL(G) is *integral* if all of its components are zero or one, and *fractional* otherwise. The distinction between fractional and integral vertices is crucial, because it determines whether or not the LP relaxation 7.64 specified by LOCAL(G) is tight. In particular, there are only two possible outcomes to solving the relaxation:

1. The optimum is attained at a vertex of MARG(G), in which case the upper bound in equation 7.64 is tight, and a mode can be obtained.

2. The optimum is attained only at one or more fractional vertices of LOCAL(G), which lie strictly outside MARG(G). In this case, the upper bound of equation 7.64 is loose, and the relaxation does not output the optimal configuration.

Figure 7.12 illustrates both of these possibilities. The vector θ^1 corresponds to case 1, in which the optimum is attained at a vertex of MARG(G). The vector θ^2 represents a less fortunate setting, in which the optimum is attained only at a fractional vertex of LOCAL(G). In simple cases, one can explicitly demonstrate a fractional vertex of the polytope LOCAL(G).

Given the link between the sum-product algorithm and the Bethe variational principle, it would be natural to conjecture that the max-product algorithm can be derived as an algorithm for solving the LP relaxation 7.64. For trees (in which case the LP 7.64 is exact), this conjecture is true: more precisely, it can be shown (Wainwright et al., 2003a) that the max-product algorithm (or the Viterbi algorithm) is an iterative method for solving the dual problem of the LP 7.64. However, this statement is false for graphs with cycles, since it is straightforward to construct problems (on graphs with cycles) for which the max-product algorithm will output a nonoptimal configuration. Consequently, the max-product algorithm does not specify solutions to the dual problem, since any LP relaxation will output either a configuration with a guarantee of correctness, or a fractional vertex. However, Wainwright et al. (2003a) derive a tree-reweighted analog of the max-product algorithm, which does have provable connections to dual optimal solutions of the tree-based relaxation 7.64.

7.8   Conclusion

A fundamental problem that arises in applications of graphical models—whether in signal processing, machine learning, bioinformatics, communication theory, or other fields—is that of computing likelihoods, marginal probabilities, and other expectations. We have presented a variational characterization of the problem of computing likelihoods and expectations in general exponential-family graphical models. Our characterization focuses attention on both the constraint set and the objective function. In particular, for exponential-family graphical models, the constraint set $\mathcal{M}$ is a convex subset in a finite-dimensional space, consisting of all realizable mean parameters. The objective function is the sum of a linear function and an entropy function. The latter is a concave function, and thus the overall problem—that of maximizing the objective function over $\mathcal{M}$—is a convex problem. In this chapter, we discussed how the junction tree algorithm and other exact inference algorithms can be understood as particular methods for solving this convex optimization problem. In addition, we showed that a variety of approximate inference algorithms—including loopy belief propagation, general cluster variational methods and mean field methods—can be understood as methods for solving particular relaxations of the general variational principle. More concretely, we saw that belief propagation involves an outer approximation of $\mathcal{M}$ whereas mean field methods involve an inner approximation of $\mathcal{M}$. In addition, this variational principle suggests a number of new inference algorithms, as we briefly discussed.

It is worth noting certain limitations inherent to the variational framework as presented in this chapter. In particular, we have not discussed curved exponential families, but instead limited our treatment to regular families. Curved exponential families are useful in the context of directed graphical models, and further research is required to develop a general variational treatment of such models. Similarly, we have dealt exclusively with exponential family models, and not treated nonparametric models. One approach to exploiting variational ideas for nonparametric models is through exponential family approximations of nonparametric distributions; for example, Blei and Jordan (2004) have presented inference methods for Dirichlet process mixtures that are based on the variational framework presented here.

Notes

[1] The Gaussian case is an important exception to this statement.

[2] That graph is triangulated means that every cycle of length four or longer has a chord.

[3] Some care is required in dealing with the boundary conditions $\tau_s(x_s) \geq 0$ and $\tau_{st}(x_s, x_t) \geq 0$; see Yedidia et al. (2001) for further discussion.

8   Modeling Large Dynamical Systems with Dynamical Consistent Neural Networks

Hans-Georg Zimmermann, Ralph Grothmann,
Anton Maximilian Schäfer, and Christoph Tietz

Recurrent neural networks are typically considered to be relatively simple architectures, which come along with complicated learning algorithms. Most researchers focus on improving these algorithms. Our approach is different: Rather than focusing on learning and optimization algorithms, we concentrate on the network architecture. Unfolding in time is a well-known example of this modeling philosophy. Here, a temporal algorithm is transferred into an architectural framework such that the learning can be done using an extension of standard error backpropagation.

As we will show, many difficulties in the modeling of dynamical systems can be solved with neural network architectures. We exemplify architectural solutions for the modeling of open systems and the problem of unknown external influences.

Another research area is the modeling of high-dimensional systems with large neural networks. Instead of modeling, e.g., a financial market as small sets of time series, we try to integrate the information from several markets into an integrated model. Standard neural networks tend to overfit, like other statistical learning systems. We will introduce a new recurrent neural network architecture in which overfitting and the associated loss of generalization abilities is not a major problem. In this context we will point to different sources of uncertainty which have to be handled when dealing with recurrent neural networks. Furthermore, we will show that sparseness of the network's transition matrix is not only important to dampen overfitting but also provides new features such as an optimal memory design.

8.1   Introduction

Recurrent neural networks (RNNs) allow the identification of dynamical systems in the form of high-dimensional, nonlinear state space models. They offer an explicit modeling of time and memory and allow us, in principle, to model any type of dynamical systems (Elman, 1990; Haykin, 1994; Kolen and Kremer, 2001; Medsker and Jain, 1999). The basic concept is as old as the theory of artificial neural networks, so, e.g., unfolding in time of neural networks and related modifications of the backpropagation algorithm can be found in Werbos (1974) and Rumelhart et al. (1986). Different types of learning algorithms are summarized by Pearlmutter (1995). Nevertheless, over the last 15 years most time series problems have been approached with feedforward neural networks. The appeal of modeling time and memory in recurrent networks is opposed to the apparently better numerical tractability of a pattern-recognition approach as represented by feedforward neural networks. Still, some researchers did enhance the theory of recurrent neural networks. Recent developments are summarized in the books of Haykin (1994), Kolen and Kremer (2001), Soofi and Cao (2002), and Medsker and Jain (1999).

Our approach differs from the outlined research directions in a significant but, at first sight nonobvious, way. Instead of focusing on algorithms, we put network architectures in the foreground. We show that a network architecture automatically implies using an adjoint solution algorithm for the parameter identification problem. This correspondence between architecture and equations holds for simple as well as complex network architectures. The underlying assumption is that the associated parameter optimization problem is solved by error backpropagation through time, i.e., a shared weights extension of the standard error backpropagation algorithm.

recurrent neural networks   In technical and economical applications virtually all systems of interest are open dynamical systems (see section 8.2). This means that the dynamics of the system is determined partly by an autonomous development and partly by external drivers of the system environment. The measured data always reflect a superposition of both parts. If we are interested in forecasting the development of the system, extracting the autonomous subsystem is the most relevant task. It is the only part of the open system that can be predicted (see subsec. 8.2.3). A related question is the sequence length of the unfolding in time which is necessary to approximate the recurrent system (see subsec. 8.2.2).

error correction neural networks   The outlined concepts are only applicable if we have a perfectly specified open dynamical system, where all external drivers are known. Unfortunately, this assumption is virtually never fulfilled in real-world applications. Even if we knew all the external system drivers, it would be questionable whether an appropriate amount of training data would be available. As a consequence, the task of identifying the open system is misspecified right from the beginning. On this problem, we introduce error correction neural networks (ECNN) (Zimmermann et al., 2002b) (see section 8.2.4).

dynamical consistent neural networks   Another weakness of our modeling framework is the implicit assumption that we only have to analyze a small number of time series. This is also uncommon in real-world applications. For instance, in economics we face coherent markets and not a single interest or foreign exchange rate. A market or a complex technical plant is intrinsically high dimensional. Now the major problem is that all our neural networks tend to overfit if we increase the model dimensionality in order to approach the true high-dimensional system dynamics. We therefore present recurrent network architectures, which work even for very large state spaces (see section 8.3). These networks also combine different operations of small neural networks (e.g., processing of input information) into one shared state transition matrix. Our experiments indicate that this stabilizes the model behavior to a large extent (see section 8.3.1).

If one iterates an open system into the future, the standard assumption is that the system environment remains constant. As this is not true for most real-world applications, we introduce dynamical consistent recurrent neural networks, which try to forecast also the external influences (see section 8.3.2). We then combine the concepts of large networks and dynamic consistency with error correction. We show that ECNNs can be extended in a slightly different way than basic recurrent networks (see section 8.3.3).

We also demonstrate that some types of dynamical systems can more easily be analyzed with a (dynamical consistent) recurrent network, while others are more appropriate for ECNNs. Our intention is to merge the different aspects of the competing network architectures within a single recurrent neural network. We call it DCNN, for *dynamical consistent neural network*. We found that DCNNs allow us to model even small deviations in the dynamics without losing the generalization abilities of the model. We point out that the networks presented so far create state trajectories of the dynamics that are close to the observed ones, whereas the DCNN evolves exactly on the observed trajectories (see section 8.3.4). Finally, we introduce a DCNN architecture for partially known observables to generalize our models from a differentiation between past and future to the (time-independent) availability of information (see section 8.3.5).

uncertainties

The identification and forecasting of dynamical systems has to cope with a number of uncertainties in the underlying data as well as in the development of the dynamics (see section 8.4). Cleaning noise is a technique which allows the model itself—within the training process—to correct corrupted or noisy data (see section 8.4.1). Working with finite unfolding in time brings up the problem of initializing the internal state at the first time step. We present different approaches to achieve a desensitization of the model's behavior from the initial state and simultaneously improve the generalization abilities (see section 8.4.2). To stabilize the network against uncertainties of the environment's future development we further apply noise to the inputs in the future part of the network (see section 8.4.3).

function & structure

Working with (high-dimensional) recurrent networks raises the question of how a desired network function can be supported by a certain structure of the transition matrix (see section 8.5). As we will point out, sparseness alone is not sufficient to optimize the network functions regarding conservation and superposition of information. Only with an inflation of the internal dimension of the recurrent neural network can we implement an optimal balance between memory and computation effects (see sections 8.5.1 and 8.5.2). In this context we work out that sparseness of the transition matrix is actually a necessary condition for large neural networks (see section 8.5.3). Furthermore we analyze the information flow in sparse networks and present an architectural solution which speeds up the distribution of information (see section 8.5.4).

Finally, section 8.6 summarizes our contributions to the research field of recurrent neural networks.

8.2   Recurrent Neural Networks (RNN)

Figure 8.1) illustrates a dynamical system (Zimmermann and Neuneier, 2001, p. 321).

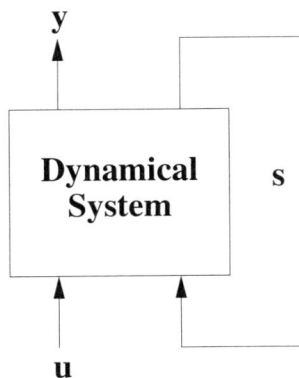

Figure 8.1   Identification of a dynamical system using a discrete time description: input u, hidden states s, and output y.

The dynamical system (fig. 8.1) can be described for discrete time grids as a set of equations (eq. 8.1), consisting of a state transition and an output equation (Haykin, 1994; Kolen, 2001):

$$
\begin{aligned}
s_{t+1} &= f(s_t, u_t) & \text{state transition} \\
y_t &= g(s_t) & \text{output equation}
\end{aligned}
\tag{8.1}
$$

The state transition is a mapping from the present internal hidden state of the system s_t and the influence of external inputs u_t to the new state s_{t+1}. The output equation computes the observable output y_t.

The system can be viewed as a partially observable autoregressive dynamic state transition $s_t \rightarrow s_{t+1}$ which is also driven by external forces u_t. Without the external inputs the system is called an autonomous system (Haykin, 1994; Mandic and Chambers, 2001). However, in reality most systems are driven by a superposition of an autonomous development and external influences.

system identification

The task of identifying the dynamical system of equation 8.1 can be stated as the problem of finding (parameterized) functions f and g such that a distance measurement (eq. 8.2) between the observed data y_t^d and the computed data y_t of the model is minimal:[1]

$$\sum_{t=1}^{T} \left(y_t - y_t^d\right)^2 \rightarrow \min_{f,g} \tag{8.2}$$

If we assume that the state transition does not depend on s_t, i.e., $y_t = g(s_t) = g(f(u_{t-1}))$, we are back in the framework of feedforward neural networks (Neuneier and Zimmermann, 1998). However, the inclusion of the internal hidden dynamics makes the modeling task much harder, because it allows varying intertemporal dependencies. Theoretically, in the recurrent framework an event s_{t+1} is explained by a superposition of external inputs $u_t, u_{t-1}, \ldots$ from all previous time steps (Haykin, 1994; Mandic and Chambers, 2001).

8.2.1   Representing Dynamic Systems by Recurrent Neural Networks

basic RNN
The identification task of equations 8.1 and 8.2 can be easily modeled by a *recurrent neural network* (Haykin, 1994; Zimmermann and Neuneier, 2001)

$$
\begin{aligned}
s_{t+1} &= \tanh(As_t + c + Bu_t) & \text{state transition} \\
y_t &= Cs_t & \text{output equation}
\end{aligned}
\tag{8.3}
$$

where A, B, and C are weight matrices of appropriate dimensions and c is a bias, which handles offsets in the input variables u_t.

Note that the output equation $y_t = Cs_t$ is implemented as a linear function. It is straightforward to show that this is not a functional restriction by using an augmented inner state vector (Zimmermann and Neuneier, 2001, pp. 322–323).

By specifying the functions f and g as a neural network with weight matrices A, B and C and a bias vector c, we have transformed the system identification task of equation 8.2 into a parameter optimization problem:

$$\sum_{t=1}^{T} \left(y_t - y_t^d\right)^2 \rightarrow \min_{A,B,C,c} \tag{8.4}$$

As Hornik et al. (1992) proved for feedforward neural networks, it can be shown that recurrent neural networks (eq. 8.3) are universal approximators, as they can approximate any arbitrary dynamical system (eq. 8.1) with a continuous output function g.

8.2.2   Finite Unfolding in Time

In this section we discuss an architectural representation of recurrent neural networks that enables us to solve the parameter optimization problem of equation 8.4 by an extended version of standard backpropagation (Haykin, 1994; Rumelhart et al., 1986).[2] Figure 8.2 unfolds the network of equation 8.3 (fig. 8.2, left) over time using *shared weight matrices* A, B, and C (fig. 8.2, right). Shared weights share the same memory for storing their weights, i.e., the weight values are the same at each time step of the unfolding and for every pattern $t \in \{1, \ldots, T\}$ (Haykin, 1994; Rumelhart et al., 1986). This guarantees that we have in every time step the same dynamics.

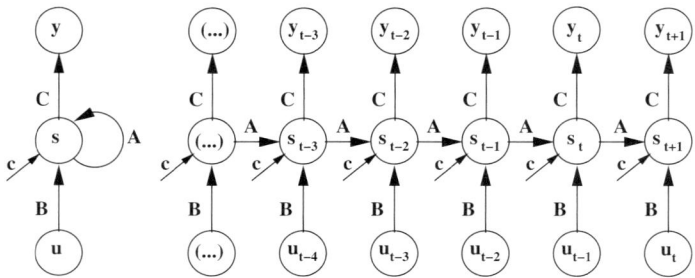

Figure 8.2   Finite unfolding using shared weight matrices A, B, and C.

We approximate the recurrence of the system with a finite unfolding which truncates after a certain number of time steps $m \in \mathbb{N}$. The important question to solve is the determination of the correct amount of past information needed to predict y_{t+1}. Since the outputs are explained by more and more external information, the error of the outputs is decreasing with each additional time step from left to right until a minimum error is achieved. This saturation level indicates the maximum number of time steps m which contribute relevant information for modeling the present time state. A more detailed description is given in Zimmermann and Neuneier (2001).

backpropagation through time

We train the unfolded recurrent neural network shown in fig. 8.2 (right) with error backpropagation through time, which is a shared weights extension of standard backpropagation (Haykin, 1994; Rumelhart et al., 1986). Error backpropagation is an efficient way of calculating the partial derivatives of the network error function. Thus, all parts of the network are provided with error information.

advantages of the RNN

In contrast to typical feedforward neural networks, RNNs are able to explicitly model memory. This allows the identification of intertemporal dependencies. Furthermore, recurrent networks contain less free parameters. In a feedforward neural network an expansion of the delay structure automatically increases the number of weights (left panel of fig. 8.3). In the recurrent formulation, the shared matrices A, B, and C are reused when more delayed input information from the past is needed (right panel of fig. 8.3).

Additionally, if weights are shared more often, more gradient information is available for learning. As a consequence, potential overfitting is not as dangerous in recurrent as in feedforward networks. Due to the inclusion of temporal structure in the network architecture, our approach is applicable to tasks where only a small training set is available (Zimmermann and Neuneier, 2001, p. 325).

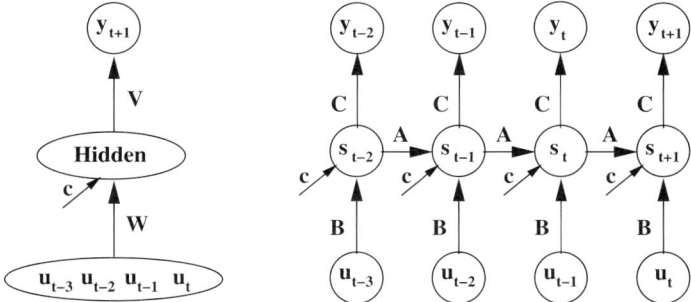

Figure 8.3 An additional time step leads in the feedforward framework (left) with $y_{t+1} = V \tanh(Wu + c)$ to a higher dimension of the input vector u, whereas the number of free parameters remains constant in recurrent networks (right), due to the use of shared weights.

8.2.3 Overshooting

An obvious generalization of the network in fig. 8.2 is the extension of the autonomous recurrence (matrix A) in future direction $t + 2, t + 3, \ldots$ (see fig. 8.4) (Zimmermann and Neuneier, 2001, pp. 326–327). If this so-called *overshooting* leads to good predictions, we get a whole sequence of forecasts as an output. This is especially interesting for decision support systems. The number of autonomous iterations into the future, which we define with $n \in N$, most often depends on the required forecast horizon of the application. Note that overshooting does not add new parameters, since the shared weight matrices A and C are reused.

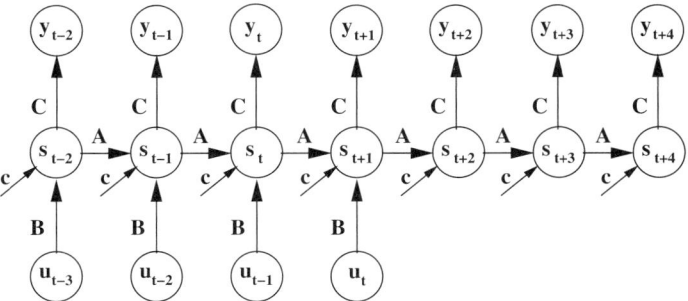

Figure 8.4 Overshooting extends the autonomous part of the dynamics.

The most important property of the overshooting network (fig. 8.4) is the concatenation of an input-driven system and an autonomous system. One may argue that the unfolding-in-time network (fig. 8.2) already consists of recurrent functions, and that this recurrent structure has the same modeling characteristics as the overshooting network. This is definitely not true, because the learning algorithm leads to different models for each of the architectures. Backpropagation learning usually tries to model the relationship between the most recent inputs and the output because the fastest adaptation takes place in the shortest path between input and output. Thus, learning mainly focuses on u_t. Only later in the training process may learning also extract useful information from input vectors u_τ $(t - m \leq \tau < t)$ which are more distant from the output. As a consequence, the unfolding-in-time network (fig. 8.2, right) tries to rely as much as possible on the part of the dynamics which is driven by the most recent inputs $u_t, \ldots, u_{t-k}$ with $k < m$. In contrast, the overshooting network (fig. 8.4) forces the learning through additional future outputs $y_{t+2}, \ldots, y_{t+n}$ to focus on modeling an internal autonomous dynamics (Zimmermann and Neuneier, 2001).

In summary, overshooting generates additional valuable forecast information about the analyzed dynamical system and stabilizes learning.

8.2.4   Error Correction Neural Networks (ECNN)

If we have a complete description of all external influences, recurrent neural networks (eq. 8.3) allow us to identify the intertemporal relationships (Haykin, 1994). Unfortunately, our knowledge about the external forces is typically incomplete and our observations might be noisy. Under such conditions, learning with finite data sets leads to the construction of incorrect causalities due to learning by heart (over-fitting). The generalization properties of such a model are questionable (Neuneier and Zimmermann, 1998).

If we are unable to identify the underlying system dynamics due to insufficient input information or unknown influences, we can refer to the actual model error $y_t - y_t^d$, which can be interpreted as an indicator that our model is misleading. Handling this error information as an additional input, we extend equation 8.1, obtaining:

$$
\begin{aligned}
s_{t+1} &= f(s_t, u_t, y_t - y_t^d), \\
y_t &= g(s_t).
\end{aligned}
\tag{8.5}
$$

The state transition s_{t+1} is a mapping from the previous state s_t, external influences u_t, and a comparison between model output y_t and observed data y_t^d. If the model error $(y_t - y_t^d)$ is zero, we have a perfect description of the dynamics. However, due to unknown external influences or noise, our knowledge about the dynamics is often incomplete. Under such conditions, the model error $(y_t - y_t^d)$ quantifies the model's misfit and serves as an indicator of short-term effects or external shocks (Zimmermann et al., 2002b).

error   correction
network

Using weight matrices A, B, C and D of appropriate dimensions corresponding to s_t, u_t, and $(y_t - y_t^d)$ and a bias c, a neural network approach to 8.5 can be written as

$$s_{t+1} = \tanh(As_t + c + Bu_t + D\tanh(Cs_t - y_t^d)),$$
$$y_t = Cs_t.$$

(8.6)

In 8.6 the output y_t is computed by Cs_t and compared to the observation y_t^d. The matrix D adjusts a possible difference in the dimension between the error correction term and s_t. The system identification is now a parameter optimization task of appropriately sized weight matrices A, B, C, D, and the bias c (Zimmermann et al., 2002b):

system identifica-
tion

$$\sum_{t=1}^{T}\left(y_t - y_t^d\right)^2 \rightarrow \min_{A,B,C,D,c}$$

(8.7)

finite unfolding

We solve the system identification task of 8.7 by finite unfolding in time using shared weights (see section 8.2.2). Figure 8.5 depicts the resulting neural network solution of 8.6.

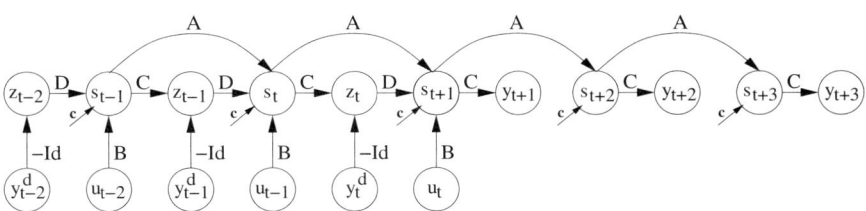

Figure 8.5   Error correction neural network (ECNN) using unfolding in time and overshooting. Note that $-Id$ is the fixed negative of an appropriate-sized identity matrix, while z_τ with $t - m \leq \tau \leq t$ are output clusters with target values of zero in order to optimize the error correction mechanism.

The ECNN (eq. 8.6) is best understood by analyzing the dependencies of s_t, u_t, $z_t = Cs_t - y_t^d$, and s_{t+1}. The ECNN has two different inputs: the externals u_t directly influencing the state transition; and the targets y_t^d. Only the difference between y_t and y_t^d has an impact on s_{t+1} (Zimmermann et al., 2002b). At all future time steps $t < \tau \leq t + n$, we have no compensation of the internal expectations y_τ, and thus the system offers forecasts $y_\tau = Cs_\tau$.

overshooting

The autonomous part of the ECNN is—analogous to the RNN case (see section 8.2.3)—extended into the future by overshooting. Besides all advantages described in section 8.2.3, overshooting influences the learning of the ECNN in an extended way. A forecast provided by the ECNN is in general, based on a modeling of the recursive structure of a dynamical system (coded in the matrix A) and on the error correction mechanism which acts as an external input (coded in C and D). Now, the overshooting enforces an autoregressive substructure allowing long-term forecasts. Of course, we have to supply target values for the additional output clusters y_τ, $t < \tau \leq t + n$. Due to the shared weights, there is no change in the number of model parameters (Zimmermann et al., 2002b).

8.3   Dynamical Consistent Neural Networks (DCNN)

The neural networks described in section 8.2 not only learn from data, but also integrate prior knowledge and first principles into the modeling in the form of architectural concepts.

market dynamics

However, the question arises if the outlined neural networks are a sufficient framework for the modeling of complex nonlinear dynamical systems, which can only be understood by analyzing the interrelationship of different subdynamics. Consider the following economic example: The dynamics of the US dollar–euro foreign exchange market is clearly influenced by the development of other major foreign exchange, stock or commodity markets (Murphy, 1999). In other words, movements of the US dollar–euro foreign exchange rate can only be comprehended by a combined analysis of the behavior of other coherent markets. This means that a model of the US dollar and euro foreign exchange market must also learn the dynamics of related markets and intermarket dependencies. Now it is important to note that, due to their computational power (in the sense of modeling high-dimensional nonlinear dynamics), the described medium-sized recurrent neural networks are only capable of modeling a single market's dynamics. From this point of view an integrated approach of market modeling is hardly possible within the framework of those networks. Hence, we need *large* neural networks.

A simple scaling up of the presented neural networks would be misleading. Our experiments indicate that scaling up the networks by increasing the dimension of the internal state results in overfitting due to the large number of free parameters.

overfitting

Overfitting is a critical issue, because the neural network does not only learn the underlying dynamics, but also the noise included in the data. Especially in economic applications, overfitting poses a serious problem.

In this section we deal with architectures which are feasible for large recurrent neural networks. These architectures are based on a redesign of the recurrent neural networks introduced in section 8.2. Most of the resulting networks cannot even be designed with a low-dimensional internal state (see section 8.3.1). In addition, we focus on a consistency problem of traditional statistical modeling: Typically one assumes that the environment of the system remains unchanged when the dynamics is iterated into the future. We show that this is a questionable statistical assumption, and solve the problem with a dynamical consistent recurrent neural network (see section 8.3.2). Thereafter, we deal with large error correction networks and integrate dynamical consistency into this framework (see section 8.3.3). Finally, we point out that large RNNs and large ECNNs are appropriate for different types of dynamical systems. Our intention is to merge the different characteristics of the two models in a unified neural network architecture. We call it DCNN for *dynamical consistent neural network* (see section 8.3.4). Finally we discuss the problem of partially known observables (see section 8.3.5).

8.3.1   Normalization of Recurrent Networks

Let us revisit the basic time-delay recurrent neural network of 8.3. The state transition equation s_t is a nonlinear combination of the previous state s_{t-1} and external influences u_t using matrices A and B. The network output y_t is computed from the present state s_t employing matrix C. The network output is therefore a nonlinear composition applying the transformations A, B, and C.

In preparation for the development of large networks we first separate the state equation of the recurrent network (eq. 8.3) into a past and a future part. In this framework s_t is always regarded as the present time state. That means that for this pattern t all states s_τ with $\tau \leq t$ belong to the past part and those with $\tau > t$ to the future part. The parameter τ is hereby always bounded by the length of the unfolding m and the length of the overshooting n (see sections 8.2.2 and 8.2.3), such that we have $\tau \in \{t - m, \dots, t + n\}$ for all $t \in \{m, \dots, T - n\}$ with T as the available number of data patterns. The present time ($\tau = t$) is included in the past part, as these state transitions share the same characteristics. We get the following representation of the optimization problem:

$$
\begin{aligned}
\tau \leq t: \quad & s_{\tau+1} \;=\; \tanh(As_\tau + c + Bu_\tau) \\
\tau > t: \quad & s_{\tau+1} \;=\; \tanh(As_\tau + c) \\
& y_\tau \;=\; Cs_\tau, \qquad \sum_{t=m}^{T-n} \sum_{\tau=t-m}^{t+n} (y_\tau - y_\tau^d)^2 \;\to\; \min_{A,B,C,c}
\end{aligned}
\qquad (8.8)
$$

As shown in section 8.2, these equations can be easily transformed into a neural network architecture (see fig. 8.4).

In this model, past and future iterations are consistent under the assumption of a constant future environment. The difficulty with this kind of recurrent neural network is the training with backpropagation through time, because a sequence of different connectors has to be balanced. The gradient computation is not regular, i.e., we do not have the same learning behavior for the weight matrices in different time steps. In our experiments we found that this problem becomes more important for training large recurrent neural networks. Even the training itself is unstable due to the concatenated matrices A, B, and C. As training changes weights in all of these matrices, different effects or tendencies—even opposing ones—can influence them and may superpose. This implies that no clear learning direction or weight changes result from a certain backpropagated error.

The question arises of how to redesign the basic recurrent architecture (eq. 8.8) to improve learning behavior and stability especially for large networks.

normalized recur-
rent networks

As a solution, we propose the neural network of 8.9, which incorporates besides the bias c only one connector, the matrix A. The corresponding architecture is depicted in fig. 8.6. Note that from now on we change the formulation of the system

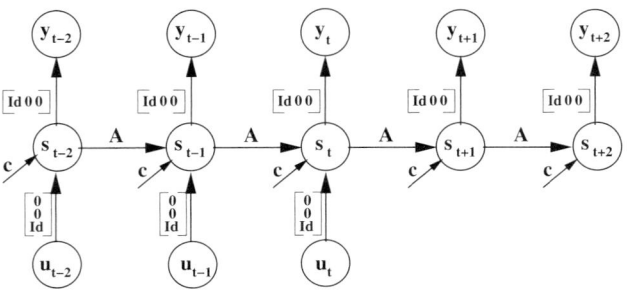

Figure 8.6   Normalized recurrent neural network.

equations (e.g., eq. 8.8) from a forward ($s_{t+1} = f(s_t, u_t)$) to a backward formulation ($s_t = f(s_{t-1}, u_t)$). As we will see, the backward formulation is internally equivalent to a forward model.

$$
\begin{aligned}
\tau \leq t: \quad s_\tau &= \tanh\left(As_{\tau-1} + c + \begin{bmatrix} 0 \\ 0 \\ Id \end{bmatrix} u_\tau\right) \\
\tau > t: \quad s_\tau &= \tanh(As_{\tau-1} + c)
\end{aligned}
$$

(8.9)

$$
y_\tau = [Id \ 0 \ 0]s_\tau, \qquad \sum_{t=m}^{T-n}\sum_{\tau=t-m}^{t+n} (y_\tau - y_\tau^d)^2 \to \min_{A,c}
$$

We call this model a normalized recurrent neural network (NRNN). It avoids the stability and learning problems resulting from the concatenation of the three matrices A, B, and C. The modeling is now focused solely on the transition matrix A. The matrices between input and hidden as well as between hidden and output layers are fixed and therefore not learned during the training process. This implies that all free parameters—as they are combined in one matrix—are now treated the same way by backpropagation.

It is important to note that the normalization or concentration on only one single matrix is paid for with an oversized (high-dimensional) internal state. At first view it seems that in this network architecture (fig. 8.6) the external input u_τ is directly connected to the corresponding output y_τ. This is not the case, though, because we increase the dimension of the internal state s_τ, such that the input u_τ has no direct influence on the output y_τ. Assuming that we have a number p of network outputs, q computational hidden neurons, and r external inputs, the dimension of the internal state would be $\dim(s) \geq p + q + r$.

With the matrix $[Id \ 0 \ 0]$ we connect only the first p neurons of the internal state s_τ to the output layer y_τ. This connector is a fixed identity matrix of appropriate size. Consequently, the neural network is forced to generate the p outputs of the neural network at the first p components of the state vector s_τ.

Let us now focus on the last r state neurons, which are used for the processing of the external inputs u_τ. The connector $[0 \ 0 \ Id]^T$ between the externals u_τ and the internal state s_τ is an appropriately sized fixed identity matrix. More precisely, the connector is designed such that the input u_τ is connected to the last state neurons. Recalling that the network outputs are located at the first p internal states, this composition avoids a direct connection between input and output. It delays the impact of the externals u_τ on the outputs y_τ by at least one time step.

To additionally support the internal processing and to increase the network's computational power, we add a number q of hidden neurons between the first p and the last r state neurons. This composition ensures that input and output processing of the network are separate.

Besides the bias vector c the state transition matrix A holds the only tunable parameters of the system. Matrix A does not only code the autonomous and the externally driven parts of the dynamics, but also the processing of the external inputs u_τ and the computation of the network outputs y_τ. The bias added to the internal state handles offsets in the input variables u_τ.

large networks   Remarkably, the normalized recurrent network of 8.9 can only be designed as a large neural network. If the internal network state is too small, the inputs and outputs cannot be separated, as the external inputs would at least partially cover the internal states at which the outputs are read out. Thus, the identification of the network outputs at the first p internal states would become impossible.

Our experiments indicate that recurrent neural networks in which the only tunable parameters are located in a single state transition matrix (e.g., eq. 8.9) show a more stable training process, even if the dimension of the internal state is very large. Having trained the large network to convergence, many weights of the state transition matrix will be dispensable without derogating the functioning of the network. Unneeded weights can be singled out by using a weight decay penalty and standard pruning techniques (Haykin, 1994; Neuneier and Zimmermann, 1998).

modeling observables   In the normalized recurrent neural network (eq. 8.9) we consider inputs and outputs independently. This distinction between externals u_τ and the network output y_τ is arbitrary and mainly depends on the application or the view of the model builder instead of the real underlying dynamical system. Therefore, for the following model we take a different point of view. We merge inputs and targets into one group of variables, which we call *observables*. So we now look at the model as a high-dimensional dynamical system where input and output represent the observable variables of the environment. The hidden units stand for the unobservable part of the environment, which nevertheless can be reconstructed from the observations. This is an integrated view of the dynamical system.

We implement this approach by replacing the externals u_τ with the (observable) targets y_τ^d in the normalized recurrent network. Consequently, the output y_τ

and the external input y_τ^d have now identical dimensions.

$$\tau \leq t : \quad s_\tau = \tanh \left(As_{\tau-1} + c + \begin{bmatrix} 0 \\ 0 \\ Id \end{bmatrix} y_\tau^d \right)$$

$$\tau > t : \quad s_\tau = \tanh(As_{\tau-1} + c) \tag{8.10}$$

$$y_\tau = [Id \ 0 \ 0]s_\tau, \qquad \sum_{t=m}^{T-n} \sum_{\tau=t-m}^{t+n} (y_\tau - y_\tau^d)^2 \to \min_{A,c}$$

The corresponding model architecture is shown in fig. 8.7.

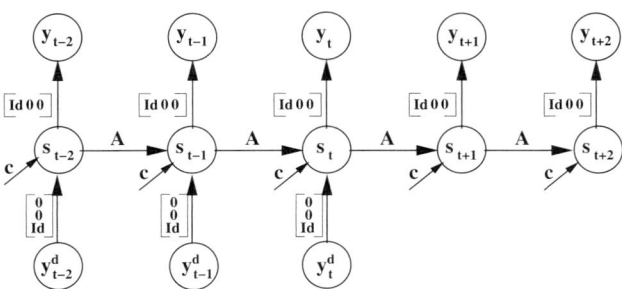

Figure 8.7   Normalized recurrent net modeling the dynamics of observables y_τ^d.

Note that, because of the one-step time delay between input and output, y_τ^d and y_τ are not directly connected. Furthermore, it is important to understand that we now take a totally different view of the dynamical system. In contrast to 8.9, this network (eq. 8.10) not only generates forecasts for the dynamics of interest but for all external observables y_τ^d. Consequently, the first r state neurons are used for the identification of the network outputs. They are followed by q computational hidden neurons, and r state neurons that read in the external inputs.

8.3.2   Dynamical Consistent Recurrent Neural Networks (DCRNN)

The models presented so far are all statistical but not dynamical consistent, as we assume that the environment stays constant for the future part of the network. In the following we improve our models with dynamical consistency.

An open dynamical system is partially driven by an autonomous development and partially by external influences. When the dynamics is iterated into the future, the development of the system environment is unknown. Now, one of the standard statistical paradigms is to assume that the external influences are not significantly changing in the future part. This means that the expected value of a shift in an external input y_τ^d with $\tau > t$ is 0 by definition. For that reason we have so far neglected the external inputs y_τ^d in the normalized recurrent neural network at all future unfolding time steps, $\tau > t$ (see eq. 8.10).

Especially when we consider fast-changing external variables with a high impact on the dynamics of interest, the above assumption is very questionable. In relation to 8.10 it even poses a contradiction, as the observables are assumed to be constant on the input and variable on the output side. Even in case of a slowly changing environment, long-term forecasts become doubtful. The longer the forecast horizon is, the more the statistical assumption is violated. A statistical model is therefore not consistent from a dynamical point of view. For a dynamical consistent approach, one has to integrate assumptions about the future development of the environment into the modeling of the dynamics.

For that reason we propose a network that uses its own predictions as replacements for the unknown future observables. This is expressed by an additional fixed matrix in the state equation. The resulting DCRNN is:

$$
\begin{aligned}
\tau \leq t: \quad s_\tau &= \begin{bmatrix} Id & 0 & 0 \\ 0 & Id & 0 \\ 0 & 0 & 0 \end{bmatrix} \tanh(As_{\tau-1} + c) + \begin{bmatrix} 0 \\ 0 \\ Id \end{bmatrix} y_\tau^d \\
\\
\tau > t: \quad s_\tau &= \begin{bmatrix} Id & 0 & 0 \\ 0 & Id & 0 \\ Id & 0 & 0 \end{bmatrix} \tanh(As_{\tau-1} + c)
\end{aligned}
\tag{8.11}
$$

$$
y_\tau = \begin{bmatrix} Id & 0 & 0 \end{bmatrix} s_\tau, \qquad \sum_{t=m}^{T-n} \sum_{\tau=t-m}^{t+n} (y_\tau - y_\tau^d)^2 \to \min_{A,c}
$$

state vector

Similarly to the end of section 8.3.1, we look at the state vector s_τ in a very structured way. The recursion of the state equations (eq. 8.11) acts in the past ($\tau \leq t$) and future ($\tau > t$) always on the same partitioning of that vector. For all $\tau \in \{t - m, \ldots, t + n\}$, s_τ can be described as

$$
s_\tau = \begin{bmatrix} y_\tau \\ h_\tau \\ \left\{ \begin{array}{l} \tau \leq t: \ y_\tau^d \\ \tau > t: \ y_\tau \end{array} \right\} \end{bmatrix} = \begin{bmatrix} \text{expectations} \\ \text{hidden states} \\ \left\{ \begin{array}{l} \tau \leq t: \ \text{observations} \\ \tau > t: \ \text{expectations} \end{array} \right\} \end{bmatrix}.
\tag{8.12}
$$

This means that in the first r components of the state vector we have the expectations y_τ, i.e., the predictions of the model. The q components in the middle of the vector represent the hidden units h_τ. They are actually responsible for the development of the dynamics. In the last r components of the vector we find in the past ($\tau \leq t$) the observables y_τ^d, which the model receives as external input. In the future ($\tau > t$) the model replaces these unknown future observables by its own expectations y_τ. This replacement is modeled with two consistency matrices:

consistency matrices

$$C_{\leq} = \begin{bmatrix} Id & 0 & 0 \\ 0 & Id & 0 \\ 0 & 0 & 0 \end{bmatrix} \text{ and } C_{>} = \begin{bmatrix} Id & 0 & 0 \\ 0 & Id & 0 \\ Id & 0 & 0 \end{bmatrix}. \tag{8.13}$$

Let us explain one recursion of the state equation (eq. 8.11) in detail: In the past ($\tau \leq t$) we start with a state vector $s_{\tau-1}$, which has the structure of 8.12. This vector is first multiplied with the transition matrix A. After adding the bias c, the vector is sent through the nonlinearity tanh. The consistency matrix then keeps the first $r+q$ components (expectations and hidden states) of the state vector but deletes (by multiplication with zero) the last r ones. These are finally replaced by the observables y_τ^d, such that s_τ again has the partitioning of 8.12. Note that in contrast to the normalized recurrent neural network (eq. 8.10) the observables are now added to the state vector after the nonlinearity. This is important for the consistency structure of the model.

The recursion in the future state transition ($\tau > t$) differs from the one in the past in terms of the structure of the consistency matrix and the missing external input. The latter is now replaced with an additional identity block in the future consistency matrix $C_{>}$, which maps the first r components of the state vector, the expectations y_τ, to its last r components. Thus we get the desired partitioning of s_τ (eq. 8.12) and the model becomes dynamical consistent.

Figure 8.8 illustrates this architecture. Note that the nonlinearity and the final calculation of the state vector are separate and hence modeled in two different layers. This follows from the dynamical consistent state equation (eq. 8.11), in which the observables are added separately from the nonlinear component.

transition matrix   Regarding the single transition matrix A, we want to point out that in a statistical consistent recurrent network (eq. 8.10) the matrix has to model the state transformation over time and the merging of the input information. However, the

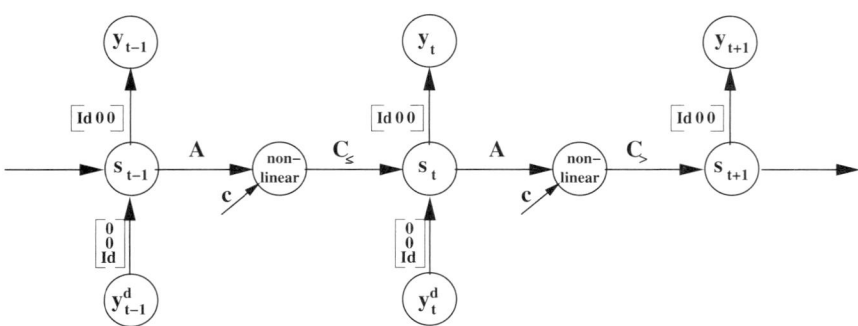

Figure 8.8 Dynamical consistent recurrent neural network (DCRNN). At all future time steps of the unfolding the network uses its own forecasts as substitutes for the unknown development of the environment.

network is only triggered by the external drivers up to the present time step t. In a dynamical consistent network we have forecasts of the external influences, which can be used as future inputs. Thus, the transition matrix A is always dedicated to the same task: modeling the dynamics.

8.3.3 Dynamical Consistent Error Correction NNs (DCECNN)

The ECNN is a nonlinear state space model employing the shared weight matrices A, B, C, and D (eq. 8.6). Matrix A computes the state transformation over time, B processes the external input information, C derives the network output, and D is responsible for the error correction mechanism. The latter composition of nonlinear transformations A, B, C, and D is difficult to handle when the network's internal state is high dimensional.

Therefore we developed a dynamical consistent error correction neural network (DCECNN) of the form of 8.14. It is an analogous approach to the DCRNN (eq. 8.11) and consequently the equations are very similar. The only changes concern the two consistency matrices $C_{\leq}$ and $C_{>}$.

$$
\tau \leq t : \quad s_\tau = \begin{bmatrix} Id & 0 & 0 \\ 0 & Id & 0 \\ -Id & 0 & 0 \end{bmatrix} \tanh(As_{\tau-1} + c) + \begin{bmatrix} 0 \\ 0 \\ Id \end{bmatrix} y_\tau^d
$$

$$
\tau > t : \quad s_\tau = \begin{bmatrix} Id & 0 & 0 \\ 0 & Id & 0 \\ 0 & 0 & 0 \end{bmatrix} \tanh(As_{\tau-1} + c)
$$

$$
y_\tau = \begin{bmatrix} Id & 0 & 0 \end{bmatrix} s_\tau, \qquad \sum_{t=m}^{T-n} \sum_{\tau=t-m}^{t+n} (y_\tau - y_\tau^d)^2 \to \min_{A,c} .
$$

$$(8.14)$$

state vector Due to the error correction, the definition or partitioning of the state vector differs in the last r components. We now have for all $\tau \in \{t - m, \dots, t + n\}$

$$
s_\tau = \begin{bmatrix} y_\tau \\ h_\tau \\ \begin{cases} \tau \leq t : e_\tau \\ \tau > t : 0 \end{cases} \end{bmatrix} = \begin{bmatrix} \text{expectations} \\ \text{hidden states} \\ \begin{cases} \tau \leq t : \text{error correction} \\ \tau > t : \quad 0 \end{cases} \end{bmatrix} . \qquad (8.15)
$$

In the past part ($\tau \leq t$) we get the error correction term in the state vector by subtracting the expectations y_τ from the observations y_τ^d. This is performed by the negative identity matrix $-Id$ within the consistency matrix $C_{\leq}$. In the future part ($\tau > t$) we expect that our model is correct. Therefore we replace the error correction by zero. The future consistency matrix $C_{>}$ simply overwrites the last r

components of the state vector with zero. Analogous to the DCRNN, the internal transition matrix A is only used for the modeling of the dynamics over time.

The graphical illustration of a dynamical consistent error correction neural network is identical to the recurrent one (fig. 8.8), but note, that the consistency matrices $C_{\leq}$ and $C_{>}$ have changed their structure.

8.3.4 Dynamical Consistent Neural Networks (DCNN)

Dynamical consistent neural networks(see section 8.3.2) are most appropriate if the observed dynamics is not hidden by noise and evolves smoothly over time, e.g., modeling of a sine curve. However, modeling can only be successful if we know all external drivers of the system and the dynamics is not influenced by external shocks. In many real-world applications, e.g. trading (see Zimmermann et al. (2002a)), this is simply not true. The dynamics of interest is often covered with noise. External shocks or unknown external influences disturb the system dynamics. In this case, one should apply DCECNNs (see section 8.3.3), which describe the dynamics with an internal expectation and its deviation from the observables. Now the question arises of whether and how we can merge the different model characteristics within a single dynamical consistent neural network (DCNN).

There are two different ways to set up this combination. In our first approach (eq. 8.18) we keep the framework of the DCRNN (eq. 8.11), whereas the second one (eq. 8.22) is based on the DCECNN (eq. 8.14).

The first approach is based on the DCRNN. Consequently, the state vector s_τ has, in the past ($\tau \leq t$) and future ($\tau \leq t$) for all $\tau \in \{t - m, \dots, t + n\}$, the partitioning of 8.16 (see also eq. 8.12).

$$
s_\tau =
\begin{bmatrix}
y_\tau \\
h_\tau \\
\left\{ \tau \leq t : y_\tau^d \right\} \\
\left\{ \tau > t : y_\tau \right\}
\end{bmatrix}
=
\begin{bmatrix}
\text{expectations} \\
\text{hidden states} \\
\left\{ \tau \leq t : \text{observations} \right\} \\
\left\{ \tau > t : \text{expectations} \right\}
\end{bmatrix}
\tag{8.16}
$$

In comparison to the DCRNN (eq. 8.11) the recursion of the new model (eq. 8.18) is extended by an additional consistency matrix

$$
C =
\begin{bmatrix}
0 & 0 & Id \\
0 & Id & 0 \\
-Id & 0 & Id
\end{bmatrix}
\tag{8.17}
$$

between the state vector and the transition matrix A. As we will see, this matrix ensures that the model is supplied with the information of the observables y_τ^d as well as the error corrections e_τ. We call this approach DCNN1 (eq. 8.18). The corresponding network architecture is depicted in fig. 8.9.

$$\tau \leq t: \; s_\tau = \begin{bmatrix} Id & 0 & 0 \\ 0 & Id & 0 \\ 0 & 0 & 0 \end{bmatrix} \tanh\left(A \begin{bmatrix} 0 & 0 & Id \\ 0 & Id & 0 \\ -Id & 0 & Id \end{bmatrix} s_{\tau-1} + c \right) + \begin{bmatrix} 0 \\ 0 \\ Id \end{bmatrix} y_\tau^d$$

$$\tau > t: \; s_\tau = \begin{bmatrix} Id & 0 & 0 \\ 0 & Id & 0 \\ Id & 0 & 0 \end{bmatrix} \tanh\left(A \begin{bmatrix} 0 & 0 & Id \\ 0 & Id & 0 \\ -Id & 0 & Id \end{bmatrix} s_{\tau-1} + c \right)$$

$$y_\tau = \begin{bmatrix} Id & 0 & 0 \end{bmatrix} s_\tau, \qquad \sum_{t=m}^{T-n} \sum_{\tau=t-m}^{t+n} (y_\tau - y_\tau^d)^2 \to \min_{A,c}$$

$$(8.18)$$

To describe how the model evolves, we explain the state equations step by step: We start with a state vector $s_{\tau-1}$ which has the structure of 8.16. Through the multiplication with the additional consistency matrix C the state vector is transformed into a vector with the partitioning

inner state vector

$$\tilde{s}_\tau = \begin{bmatrix} y_\tau^d \\ h_\tau \\ e_\tau \end{bmatrix} = \begin{bmatrix} \text{observations} \\ \text{hidden states} \\ \text{error correction} \end{bmatrix} \qquad (8.19)$$

for all $\tau \in \{t-m, \dots, t+n\}$. This inner state vector $\tilde{s}_\tau$ contains the observables and the error correction and combines the ideas of DCRNN and DCECNN. The rest of the recursion is identical with the DCRNN (eq. 8.11). As before, the only learnable parameters of the network are located in matrix A and the bias c.

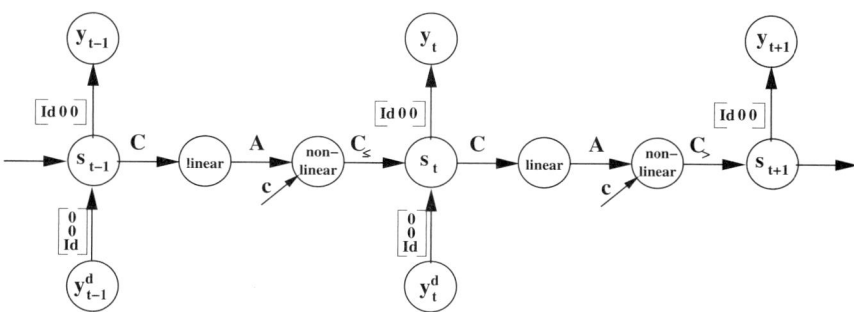

Figure 8.9   Dynamical consistent neural network (DCNN).

DCNN2

As already mentioned, the second approach to a dynamical consistent neural network (DCNN2) is based on the DCECNN model (dq. 8.14). The state vector s_τ

assumes the corresponding structure (see eq. 8.15):

$$s_\tau = \begin{bmatrix} y_\tau \\ h_\tau \\ \left\{ \begin{matrix} \tau \le t: & e_\tau \\ \tau > t : 0 \end{matrix} \right\} \end{bmatrix} = \begin{bmatrix} \text{expectations} \\ \text{hidden states} \\ \left\{ \begin{matrix} \tau \le t: & \text{error correction} \\ \tau > t: & 0 \end{matrix} \right\} \end{bmatrix} . \tag{8.20}$$

Analogous to the development of the DCNN1 (eq. 8.18) the DCECNN equation is extended by an additional consistency matrix C, which now has the structure

$$C = \begin{bmatrix} Id & 0 & Id \\ 0 & Id & 0 \\ 0 & 0 & Id \end{bmatrix} . \tag{8.21}$$

The resulting DCNN2 can be described with the following set of equations:

$$\tau \le t : s_\tau = \begin{bmatrix} Id & 0 & 0 \\ 0 & Id & 0 \\ -Id & 0 & 0 \end{bmatrix} \tanh \left(A \begin{bmatrix} Id & 0 & Id \\ 0 & Id & 0 \\ 0 & 0 & Id \end{bmatrix} s_{\tau-1} + c \right) + \begin{bmatrix} 0 \\ 0 \\ Id \end{bmatrix} y_\tau^d$$

$$\tau > t : s_\tau = \begin{bmatrix} Id & 0 & 0 \\ 0 & Id & 0 \\ 0 & 0 & 0 \end{bmatrix} \tanh \left(A \begin{bmatrix} Id & 0 & Id \\ 0 & Id & 0 \\ 0 & 0 & Id \end{bmatrix} s_{\tau-1} + c \right)$$

$$y_\tau = \begin{bmatrix} Id & 0 & 0 \end{bmatrix} s_\tau, \qquad \sum_{t=m}^{T-n} \sum_{\tau=t-m}^{t+n} (y_\tau - y_\tau^d)^2 \rightarrow \min_{A,c}$$

$$\tag{8.22}$$

Looking at the multiplication $C \cdot s_{\tau-1}$ we can easily confirm that—supposing that $s_{\tau-1}$ is structured as in 8.20—we once again get an inner state vector $\tilde{s}_\tau$ partitioned as in 8.19.

This implies that the transition matrix A is applied in both models to the same inner state vector $\tilde{s}_\tau$. Consequently, although the two models look quite different, they share an identical modeling of the dynamics. It may depend on additional modeling tools or a particular application which approach is preferable.

The network architecture for the alternative approach, DCNN2, is identical to DCNN1 (fig. 8.8), but note that the consistency matrices C, $C_\le$, and $C_>$ differ.

advantages of the DCNN   Opposite to the DCRNN (eq. 8.11) and DCECNN (eq. 8.14) the two approaches to the DCNN (eqs. 8.18 and 8.22) compute the state trajectory of the dynamics in the past exactly on the observed path. This follows from the partitioning of the inner state vector $\tilde{s}_\tau$ (eq. 8.19), which is responsible for the calculation of the dynamics. It contains the observables in the first r components, which are directly used to determine the prediction y_τ. The error corrections, which are now located in the last r components, act as additional inputs. Furthermore, the DCNN offers an interesting new insight into the observation of dynamical systems: Typically, small movements of the dynamics are treated as noise, and thus the modeling focuses on larger shifts in the dynamics. Our view is different. We believe that small system changes characterize the autonomous part of our open system, while the large swings originate at least partially from the external forces. If we neglect small system changes, we also suppress valuable substructure in our observations. We found that DCNNs allow us to model even small changes in the dynamics without losing the generalization abilities of the model. This introduces a new perspective on the structure/noise dilemma in modeling dynamical systems.

8.3.5   Partially Known Observables

So far our models have always distinguished between a past and a future development of the state equation. We assumed that in the past part ($\tau \leq t$) all the identified observables are available. In the future part ($\tau > t$) we accepted that we do not know anything about the observables and hence we replaced them by the model's own expectations.

In many practical applications we have observables which are not available for all time steps in the past. In contrast, one might have observables which are also available in the future, e.g., calendar data. In the following we therefore switch from a model differentiating between past and future to a modeling structure which distinguishes between available and missing external inputs.

The DCNN with partially known observables merges the two state equations of the DCNN (e.g., eq. 8.22) into one single equation that allows us to differentiate between available and unavailable observables. Consequently, it is a reformulation of the normal DCNN providing an easier and more general structure. The simplification in one equation makes the model also more tractable for further discussions (see sections 8.4 and 8.5). The following model (eq. 8.23) is based on the DCNN2 (eq. 8.22), but an analogous model can also easily be created for DCNN1 (eq. 8.18). For all $\tau \in \{t - m, \ldots, t + n\}$, we have

$$s_\tau = \begin{bmatrix} Id & 0 & 0 \\ 0 & Id & 0 \\ E & 0 & 0 \end{bmatrix} \tanh\left(A \begin{bmatrix} Id & 0 & Id \\ 0 & Id & 0 \\ 0 & 0 & Id \end{bmatrix} s_{\tau-1} + c \right) + \begin{bmatrix} 0 \\ 0 \\ Id \end{bmatrix} y_\tau^E \tag{8.23}$$

$$y_\tau = \begin{bmatrix} Id & 0 & 0 \end{bmatrix} s_\tau, \qquad \sum_{t=m}^{T-n} \sum_{\tau=t-m}^{t+n} (y_\tau - y_\tau^d)^2 \to \min_{A,c}.$$

In this model the external inputs y_τ^E and the included matrix E are defined as follows:

$$y_\tau^E := \left\{ \begin{array}{ll} 0 & \text{if} \quad \text{input missing} \\ y_\tau^d & \text{input available} \end{array} \right\} \tag{8.24}$$

and

$$E := \left\{ \begin{array}{cc} 0 & \text{input missing} \\ & \text{if} \\ -1 & \text{input available} \end{array} \right\}. \tag{8.25}$$

It is important to note that the inner consistency matrix C is independent of the input availability. We only adapt the consistency matrix

$$C_E = \begin{bmatrix} Id & 0 & 0 \\ 0 & Id & 0 \\ E & 0 & 0 \end{bmatrix}. \tag{8.26}$$

The structure guarantees that an error correction is calculated in the last r state components if external input is available. Thus, we have a time-independent combination of the former two state equations (eq. 8.22). The corresponding model architecture (fig. 8.10) does not change significantly in comparison to the former (time-oriented) DCNN (fig. 8.9).

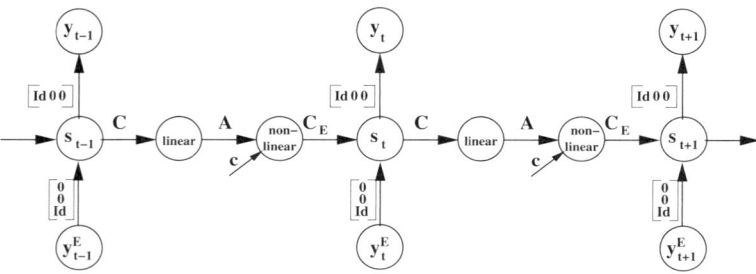

Figure 8.10   DCNN with partially known observables.

The DCNN with partially known observables is more general in the sense of observable availability and hence better applicable to real-world problems. The following discussions are mainly based on this model.

8.4   Handling Uncertainty

In practical applications our models have to cope with several forms of uncertainty. So far we have neglected their possible influence on generalization performance. Uncertainty can disturb the development of the internal dynamics and seriously harm the quality of our forecasts. In this section we present several methods which reduce the model's dependency on uncertain data.

There are actually three major sources of uncertainty. First, the input data itself might be corrupted or noisy. We deal with that problem in section 8.4.1. In the framework of finitely unfolded in time recurrent neural networks we also have the uncertainty of the initial state. We present different approaches to overcome that uncertainty and achieve a desensitization of the model from the unknown initialization (see section 8.4.2). Finally, we discuss the uncertainty of the future inputs and question once more the assumption of a constant environment (see section 8.4.3).

8.4.1   Handling Data Noise

So far we have always assumed our input data to be correct. In most practical applications this is not true. In the following we present an approach which tries to minimize input uncertainty.

cleaning noise   *Cleaning noise* is a method which improves the model's learning behavior by correcting corrupted or noisy input data. The method is an enhancement of the *cleaning* technique which is described in detail in (Neuneier and Zimmermann, 1998). In short, cleaning considers the inputs as corrupted and adds corrections to the inputs if necessary. However, we want to keep the cleaning correction as small as possible. This leads to an extended error function

$$E_t^{y,x} = \frac{1}{2}[(y_t - y_t^d)^2 + (x_t - x_t^d)^2] = E_t^y + E_t^x \quad \rightarrow \min_{x_t, w} \quad . \tag{8.27}$$

Note that this new error function does not change the usual weight adaption rule

$$w^+ = w - \eta \frac{\partial E^y}{\partial w} \quad , \tag{8.28}$$

where $\eta > 0$ is the so-called learning rate and w^+ stands for the adapted weight. To calculate the cleaned input

$$x_t = x_t^d + \rho_t \tag{8.29}$$

we need the correction vectors ρ_t for all input data of the training set. The update rule for these corrections, initialized with $\rho_t = 0$, can be derived from typical adaption sequences:

$$x_t^+ = x_t - \eta \frac{\partial E^{y,x}}{\partial x}, \tag{8.30}$$

leading to

$$\rho_t^+ = (1 - \eta)\rho_t - \eta \frac{\partial E^y}{\partial x} \quad . \tag{8.31}$$

This is a nonlinear version of the error-in-variables concept from statistics.

We derive all the information needed, especially the residual error $\frac{\partial E^{y,x}}{\partial x}$, from training the network with backpropagation (fig. 8.18), which makes the computational effort negligible. It is important to note that in this way the corrections are performed by the model itself and not by applying external knowledge (see "observer-observation dilemma" in Neuneier and Zimmermann, 1998).

We now assume that the data is not only corrupted but also noisy. For that reason we add an extra noise vector, $-\rho_\tau$, to the cleaned value:

$$x_t = x_t^d + \rho_t - \rho_\tau \quad .\tag{8.32}$$

The noise vector ρ_τ is a randomly chosen row vector $\{\rho_{i\tau}\}_{i=1,\dots,r}$ of the cleaning matrix

$$C_{Cleaning} := \begin{bmatrix} \rho_{11} & \cdots & \cdots & \cdots & \rho_{r1} \\ \rho_{12} & \ddots & & & \rho_{r2} \\ \vdots & & \rho_{it} & & \vdots \\ \vdots & & & \ddots & \vdots \\ \rho_{1T} & \cdots & \cdots & \cdots & \rho_{rT} \end{bmatrix} ,$$

which stores the input error corrections of all data patterns. The matrix has the same size as the pattern matrix, as the number of rows equals the number of patterns T and the number of columns equals the number of inputs r.

One might wonder why disturb the cleaned input $x_t = x_t^d + \rho_t$ with an additional noise-term $-\rho_\tau$. The reason for this is, that we want to benefit from representing the whole input distribution to the network instead of only using one particular realization (Zimmermann and Neuneier, 1998).

local cleaning noise A variation on the Cleaning Noise method is called *local cleaning noise*. Cleaning noise adds to every training pattern the same noise term $-\rho_\tau$ and therefore assumes that the noise of the different inputs is correlated. Especially in high-dimensional models it is improbable that all the components of the input vector follow an identical or at least correlated noise distribution. For these cases we propose a method which is able to differentiate component-wise:

$$x_{it} = x_t^d + \rho_t - \rho_{i\tau}.\tag{8.33}$$

In contrast to the normal cleaning technique, the local version is correcting each component of the input vector x_{it} individually by a cleaning correction and a randomly taken entry $\rho_{i\tau}$ of the corresponding column $\{\rho_{it}\}_{t=1,\dots,T}$ of the cleaning matrix $C_{Cleaning}$.

A further advantage of the local cleaning technique is that—with the increased number of (local) correction terms $(T \cdot r)$—we can cover higher dimensions. In contrast, with the normal cleaning technique the dimension is bounded by the number of training patterns T, which can be insufficient for high-dimensional problems.

8.4.2 Handling the Uncertainty of the Initial State

One of the difficulties with finite unfolding in time is to find a proper initialization for the first state vector of the recurrent neural network. An obvious solution is to set the first state s_0 to zero. We then implicitly assume that the unfolding includes enough (past) time steps such that the misspecification of the initialization phase is compensated along the state transitions. In other words, the network accumulates information over time, and thus can eliminate the impact of the arbitrary initial state on the network outputs.

state   initialization   The model can be improved if we make the unfolded recurrent network less sensitive to the unknown initial state s_0. For this purpose we look for an initialization, for which the interpretation of the state recursion is consistent over time. Since the initialization procedure is identical for all types of DCNNs, we demonstrate the approach on the DCNN with partially known observables (eq. 8.23):

$$s_\tau = C_E \tanh(A \cdot C \cdot s_{\tau-1} + c) + \begin{bmatrix} 0 \\ 0 \\ Id \end{bmatrix} y_\tau^E$$

$$y_\tau = [Id \ 0 \ 0]s_\tau, \qquad \sum_{t=m}^{T-n} \sum_{\tau=t-m}^{t+n} (y_\tau - y_\tau^d)^2 \to \min_{A,c}. \tag{8.34}$$

In a first step we explicitly integrate a first state vector s_0 (see fig. 8.11). This is no longer set to zero but receives the target information y_{t-1}^{target} of the first output vector y_{t-1}. The vector is then multiplied by the consistency matrix C_E, such that the first r components of the first state vector s_{t-1} coincide with the first expected output. This avoids the generation of an excessively large error for the first output.

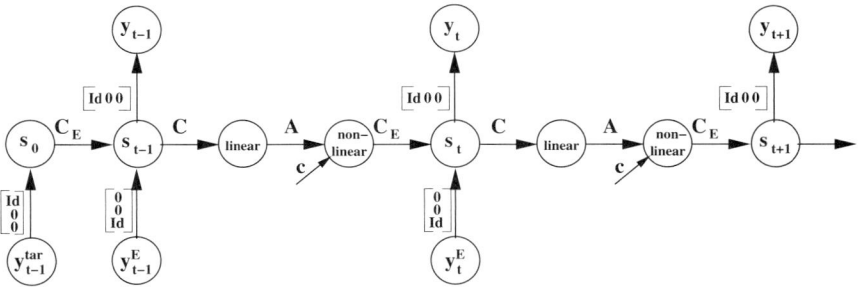

Figure 8.11   Time consistent initialization of a DCNN with an additional initial state s_0.

The hidden states of this model are arbitrarily initialized with zero. In a second step we add a noise term ε to the first state vector s_0 to stiffen the model against the uncertainty of the unknown initial state. A fixed noise term ε that is drawn from a predetermined noise distribution is clearly inadequate to handle the uncertainty of the initial state. Instead we apply—according to the cleaning noise method—an adaptive noise term, which fits best the volatility of the unknown initial state s_0. As explained in section 8.4.1, the characteristics of the adaptive noise term are automatically determined as a by-product of the error backpropagation algorithm.

residual error

The basic idea is as follows: The residual error ρ as measured at the initial state s_0 can be interpreted as the uncertainty stemming from missing information about the true initial state vector. If we disturb s_0 with a noise term which follows the distribution of the residual error of the network, we diminish the uncertainty about the unknown initial state during system identification. In addition, this allows a better fitting of the target values over the training set. A corresponding network architecture is depicted in fig. 8.12.

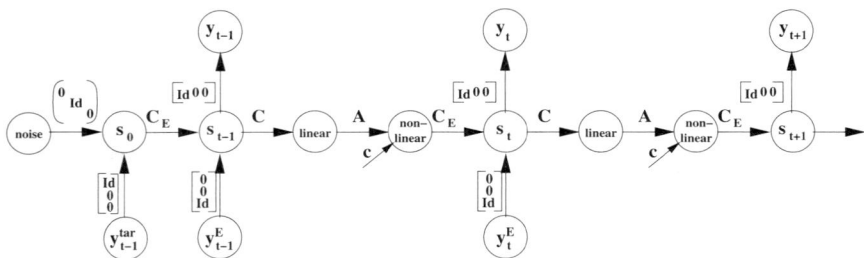

Figure 8.12   Desensitization of a DCNN from the unknown initial state s_0.

Technically, *noise* is introduced into the model via an additional input layer. The dimension of *noise* is equal to that of the internal state. The input values are fixed at zero over time. Due to the incomplete identity matrix between *noise* and the initial state the noise is only applied to the hidden values of the initial state, where no input information is available. The desensitization of the network from the initial state vector s_0 can therefore be seen as a self-scaling stabilizer of the modeling. Note that the noise term ρ is drawn randomly from the observed residual errors, without any prior assumption on the underlying noise distribution.

In general, a discrete-time state trajectory forms a sequence of points over time. Such a trajectory is comparable to a thread in the internal state space. The trajectory is very sensitive to the initial state vector s_0. If we apply noise to s_0, the space of all possible trajectories becomes a tube in the internal state space (fig. 8.13). Due to the characteristics of the adaptive noise term, which decreases over time, the tube contracts. This enforces the identification of a stable dynamical system. Consequently, the finite volume trajectories act as a regularization and stabilization of the dynamics.

initialization techniques

The question arises, what may be the best method to create an appropriate noise level. Table 8.1 gives an overview of several initialization techniques we have developed and examined so far. Remember that in all cases the corrections are only applied to the hidden variables of the initial state s_0.

We already explained the first three methods in section 8.4.1. The idea behind the initialization with *start noise* is, that we do not need a cleaning correction but solely focus on the noise term. *double start noise* tries to achieve a nearly symmetrical noise distribution, which is also double in comparison to normal *start*

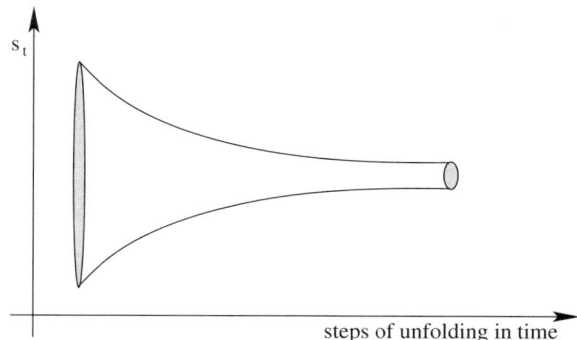

s_t

steps of unfolding in time

Figure 8.13   Creating a tube in the internal state space by applying noise to the initial state.

Table 8.1   Overview of Initialization Techniques

Cleaning:	s_0	$=$	$0 + \rho_t$
Cleaning noise:	s_0	$=$	$0 + \rho_t - \rho_\tau$
Local Cleaning noise:	s_{0_i}	$=$	$0 + \rho_t - \rho_{\tau_i}$
Start noise:	s_0	$=$	$0 + \rho_\tau$
Local start noise:	s_{0_i}	$=$	$0 + \rho_{\tau_i}$
Double start noise:	s_0	$=$	$0 + (\rho_\tau^1 - \rho_\tau^2)$
Double local start noise:	s_{0_i}	$=$	$0 + (\rho_{\tau_i}^1 - \rho_{\tau_i}^2)$

noise. In all cases *local* always corresponds to the individual application of a noise term to each component of the initial state s_0 (see *local cleaning noise* in section 8.4.1).

From top to bottom the methods listed in table 8.1 use less and less information about the training set. Hence *double start noise* emphasizes more the generalization abilities of the model. This is also confirmed by our experiments. Furthermore we could confirm that the local initialization techniques lead to better performance in high-dimensional models (see section 8.4.1).

8.4.3   Handling the Uncertainty of Unknown Future Inputs

In the past part of the network, the influence of the unknown externals is reflected in the error corrections as calculated by the backpropagation algorithm. In the future part we do not have any information about the correctness of our inputs. As explained in section 8.3, we either use our own forecasts as future inputs, or simply assume that the inputs in the future stay constant. The underlying assumption is that the observables evolve in the future like they did in the past. We cannot verify if this is correct. Anyway, for most practical applications it is a very questionable assumption.

To stabilize our model against these uncertainties of the future inputs we apply a Gaussian noise term $\varepsilon_{t+\tau}$ to the last r components of each future state vector $s_{t+\tau}$. The corresponding architecture is depicted in fig. 8.14.

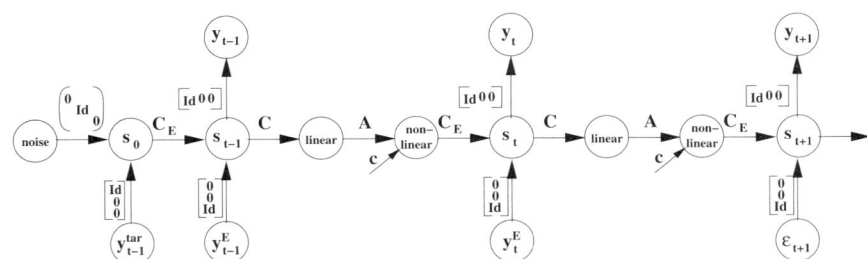

Figure 8.14   Handling the uncertainty of the future inputs by adding a noise term $\varepsilon_{t+\tau}$ to each future state vector $s_{t+\tau}$.

The additional noise is used during the training of the model to achieve a more stable output. For the actual deterministic forecast we either skip the application of noise to avoid a disturbance of the predictions or average our results over a sufficient number of different forecasts (Monte Carlo approach).

8.5   Function and Structure in Recurrent Neural Networks

Our discussion about function and structure in recurrent neural networks is focused on the autonomous part of the model, which is mapped by the internal state transition matrix A. So far the transition matrix A has always been assumed to be fully connected. In a fully connected matrix the information of a state vector s_t is processed using the weights in A to compute s_{t+1}. This implies that there is a high proportion of superposition (computation) but hardly any conservation of information (memory) from one state to a succeeding one (see the right panel of fig. 8.15).

For the identification of dynamical systems such memory can be essential, as information may be needed for computation in subsequent time steps. A shift register (see the left panel of fig. 8.15) is a simple example for the implementation of memory, as it only transports information within the state vector s. No superposition is performed in this transition matrix.

superposition and conservation of information   At first view we have two contradicting functions: superposition and conservation of information. Superposition of information is necessary to generate or adapt changes of the dynamics. In contrast, conservation of information causes memory effects by transporting information more or less unmodified to a subsequent state neuron. In this context, memory can be defined as the average number of state transitions necessary to transmit information from one state neuron to any other

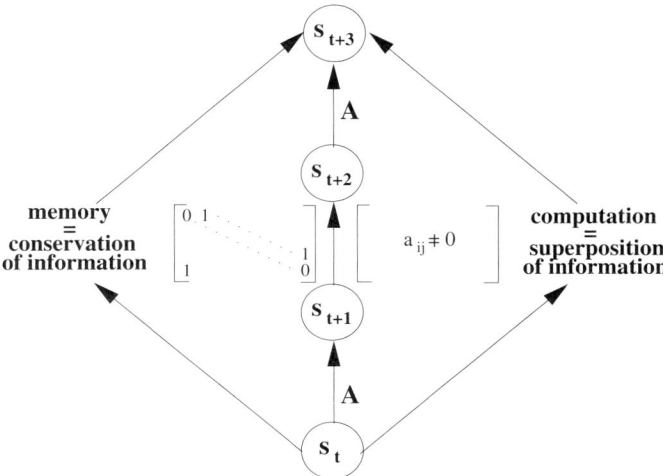

Figure 8.15 Function and structure in dynamical systems: computation versus memory in the transition matrix A.

one in a subsequent state. We call this number of necessary state transitions the *path length* of a neuron. To overcome the apparent dilemma between superposition and conservation of information the transition matrix A needs a structure which balances memory and computation effects. Sparseness of the transition matrix reduces the number of paths and the computation effect of the network but at the same time increases the average path length, and therefore allows for longer-lasting memory. A possible solution is an inflation of the recurrent network, i.e., of the transition matrix A.

We show that with such an inflation an optimal balance between memory and computation can be achieved (section 8.5.1). In this context we present conjectures about the optimal level of sparseness and the required minimum dimension. An experiment with artificial data underlines our results (section 8.5.2). In section 8.5.3 we conclude that sparseness is actually an essential condition for high-dimensional neural networks. Finally, we discuss in section 8.5.4 the information flow in sparse networks.

8.5.1   Inflation of Recurrent Neural Networks

Based on the length of the past unfolding m (see section 8.2.2) and the optimal state dimension $\dim(s)$ of a fully connected recurrent network, we can define a procedure for an optimal design of the neural network structure, which solves the dilemma between memory and computation.

The idea is to inflate the network to a higher dimensionality, while maintaining the computational complexity of the former lower-dimensional and fully connected network, and at the same time allowing for memory effects. With an inflated transi-

optimal inflation

tion matrix A we can optimize both superposition and conservation of information.

To determine the optimal dimension and the level of sparseness, we propose two conjectures, which we will empirically investigate in section 8.5.2. In a first step we calculate the new dimension of the internal state s by

$$\dim(s_{new}) := m \cdot \dim(s). \tag{8.35}$$

As the former dimension of s was supposed to be optimal, we have to ensure that the higher-dimensional network has the same superposition of information as the original one. This can be achieved by keeping the number of active weights constant. On average we want to have the same number of nonzero elements as in the former lower-dimensional network. Thus, the sparseness level of the new matrix A_{new} is given by

$$\text{initialize } A_{new} \text{ with } \quad Random\left(\frac{\dim(s)}{\dim(s_{new})}\right) = Random\left(\frac{1}{m}\right). \tag{8.36}$$

Hereby $Random(\cdot)$ represents the percentage of randomly initialized weights, whereas the remaining weights are set to zero. Proceeding this way, we replicate on average the computation effect of the former network. At the same time we increase the path lengths (memory) with the sparseness level of the new transition matrix A_{new}. Note that the sparseness level only depends on the length of the past unfolding m.

The conjecture (eq. 8.36) implies that the sparseness of A_{new} is generated randomly. In section 8.5.3 we present techniques which try to optimize the sparse structure and consequently the memory and computation abilities of the network.

training proce-
dure for RNNs

Based on our conjectures about inflation, a proper training procedure for recurrent neural networks should consist of four steps: First, one has to set up an appropriate network architecture (e.g., DCNN, eq. 8.23). Second, the length of the past unfolding m and the optimal internal state dimension $\dim(s)$ of the system have to be estimated by analyzing the network errors along the time steps of the unfolding (see section 8.2.2). Third, we use the estimated parameters m and $\dim(s)$ to determine the optimal dimensionality and sparseness (eqs. 8.35 and 8.36). Fourth, the inflated network is trained until convergence by backpropagation through time using, e.g., the *vario-eta* learning rule (Neuneier and Zimmermann, 1998).

8.5.2   Experiments: Testing Conjectures About Inflation

In the following experiments we want to evaluate our conjectures about optimal inflation of recurrent networks. To ensure a straight analysis of our proposed equations (eqs. 8.35 and 8.36), we modeled an artificial network which consists of an autonomous development only. We applied the network to forecast the development of the following artificial data generation process

$$s_t = \tanh(A \cdot s_{t-m}) + \epsilon_t \, , \tag{8.37}$$

where $\dim(s) = 5$, A is randomly initialized, $m = 3$, and ϵ is white noise with $\sigma_\epsilon = 0.2$, one time step ahead. The unfolding in time of the recurrent network includes five time steps from $t-3$ to $t+1$. As the data generation process is a closed dynamical system, there are no inputs, but time-delayed states s_{t-k} ($k = 1, \ldots, m$) are used as external influences.

Each of the following experiments is based on 100 Monte Carlo simulation runs. For each run we generated 1000 observations, 25% for training and 75% for testing purposes. The network was trained until convergence with error backpropagation through time using *vario-eta* learning (Neuneier and Zimmermann, 1998).

First we evaluated our conjecture about the sparse random initialization of the transition matrix A_{new}. For this purpose we randomly initialized matrix A_{new} with different levels of sparseness (100 test runs per sparseness degree). The dimension of the internal state was fixed according to 8.35 at $\dim(s_{new}) = 3 \cdot 5 = 15$ for all test runs. The mean square error of the network measured on the test set was used as a performance criterion.

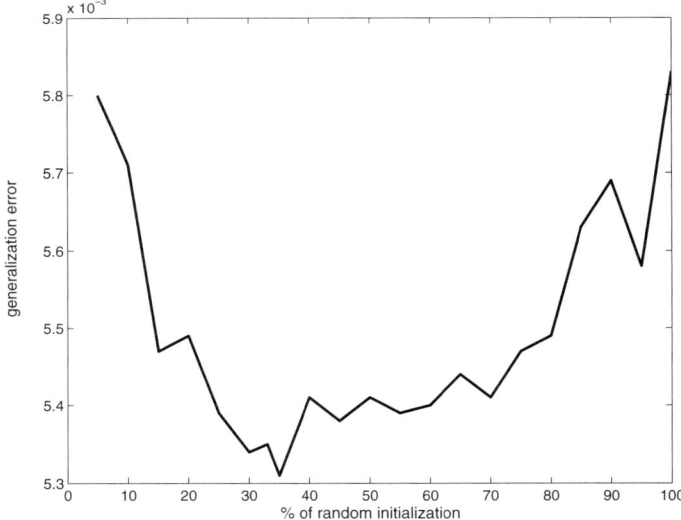

Figure 8.16 Effects of different degrees of sparseness on matrix A_{new}.

The results of the experiment (Figure 8.16) confirm our conjecture about an optimal sparseness level: If we initialize matrix A_{new} randomly 35% sparse, we observe the best performance (i.e., lowest average error on the test set). This corresponds to equation 8.36, where an optimal sparseness level computes to 33%.

The second series of experiments was connected with the optimal internal state dimension. During these experiments we kept the sparseness level constant at 33% (see eq. 8.36), whereas the dimension of the internal state was variable. We performed 100 runs for each dimension of the internal state with different random initializations of matrix A_{new}. Again we used the network error as an indicator of the model performance. The results are shown in fig. 8.17.

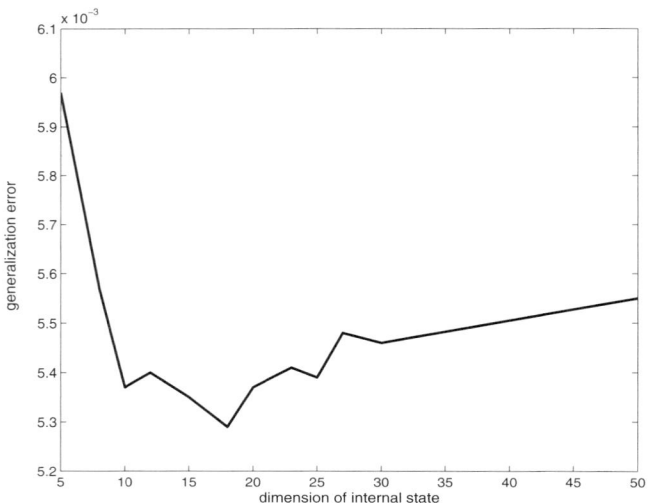

Figure 8.17   Impacts of different internal state dimensions $\dim(s_{new})$.

It turns out that the best performance is achieved if the dimension of the internal state is equal to $\dim(s_{new}) = 18$. Our conjecture of $\dim(s_{new}) = 3 \cdot 5 = 15$ (eq. 8.35) slightly underestimates the empirically measured optimal dimensionality. However, because of the noise term ϵ_t, we suppose that the optimal dimension of the system is larger than 5. This indicates that our conjecture in 8.35 is a helpful estimate of an optimal level of sparseness.

Both experiments show that mismatches between dimensionality and sparseness cause problems in the function (superposition and conservation) of the transition matrix. In other words, an unbalanced parameterization of the inflation leads to lower generalization performance of the network.

8.5.3   Sparseness as a Necessary Condition for Large Systems

One might come up with the idea of initializing a model with a fully connected transition matrix A, and then pruning it during the learning process until a desired degree of sparseness is reached. This approach is misleading, as sparseness is an essential condition for the performance of the backpropagation algorithm in large networks.

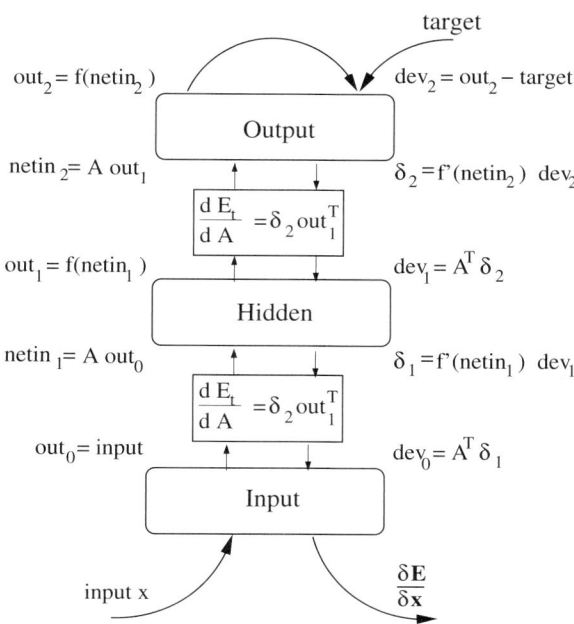

Figure 8.18 Forward and backward information flow in the backpropagation algorithm.

backpropagation

Figure 8.18 shows the forward and backward information flow in the backpropagation algorithm. [3] When we look at the calculations, it becomes obvious that if the transition matrix A is fully connected and $dim(s)$ is increasing, we get a growing number as well as a lengthening of the sums in the matrix times vector operations. Due to the law of large numbers the probability for large sum values also increases. This does not pose any problems in the forward flow of the algorithm. The hyperbolic tangent as the nonlinear activation function guarantees that the calculated values stay numerically tractable. In contrast, backward information flow is linear. In this part of the algorithm large values are spread all over a fully connected matrix A. They quickly sum up to values which cause numerical instabilities and may destroy the whole learning process. This can be avoided if we use a sparse transition matrix A. The number of summands is then smaller and therefore the probability of large sums is low.

sparse initialization of A

In the remainder of this section we want to discuss the question of how to choose a sparse transition matrix A that is still trainable to a stable model.

One intuitive answer is to initialize the model several times and then compare the different results. We performed 100 test runs of the neural network using different random initializations of matrix A. The prestructuring of the network followed our conjectures about inflation (eqs. 8.35 and 8.36).

An obvious approach to overcome the uncertainty of the random initialization is to pick the best-performing network out of the 100 test runs. However, it is not clear if 100 test runs are effectively required to find an appropriate model. To study how many test runs are needed to find a good solution with a minimum of computational effort, we picked all possible subsets of $k = 1, 2, \ldots, 100$ solutions out of the 100 test runs. From each subset we chose the solution with the lowest error on the test set and computed the average performance:

$$E_k = \frac{1}{\binom{100}{k}} \sum_{i_1 \leq \ldots \leq i_k} \min(e_{i_1}, e_{i_2}, \ldots, e_{i_k}). \tag{8.38}$$

The resulting average error curve for $k = 1, 2, \ldots, 100$ is depicted in fig. 8.19 (solid line).

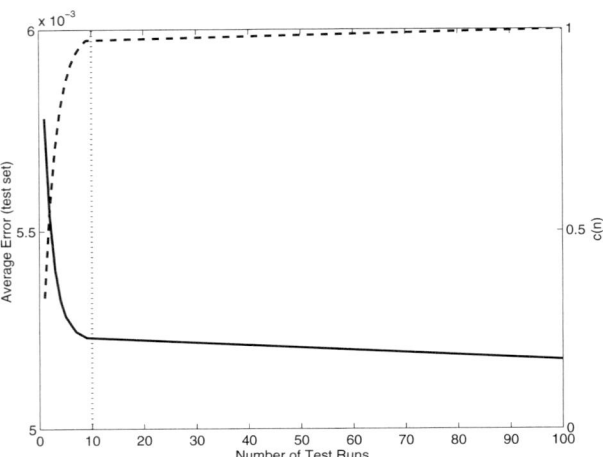

Figure 8.19 Estimating the number of random initializations for matrix A.

As can be seen from the error curve in fig. 8.19, an appropriate solution can be obtained on the average by choosing the best model out of a subset of 10 networks (vertical dotted line). Of course, the performance is worse than picking the best model out of 100 solutions. However, the additional computational effort does not justify the small improvement of performance.

coverage percent-
age
As an apparently easier guideline to determine the number of required test runs, we choose the number k such that the so-called coverage percentage,

$$c(k) = 1 - \left(1 - \frac{1}{m}\right)^k, \tag{8.39}$$

is close to 1. The idea behind the coverage percentage $c(k)$ in 8.39 is that the first inflated network covers $c(1) = 1/m$ active elements in the internal transition matrix A_{new}. Assuming that we have $c(k)$, the next initialization covers another percentage of the weights in the transition matrix, resulting in $c(k+1) = c(k) + (1/m) \cdot (1 - c(k))$. The coverage percentage $c(k)$ for the different numbers of initializations is also

pruning   &   re-
creation

reported in fig. 8.19 (dashed line). A number of $k = 10$ random initializations already leads to a coverage of $c(10) \approx 0.983$.

To further reduce the computational effort, we developed a more sophisticated approach, a process we call *pruning and re-creation of weights*. As described in section 8.5.1, we initialize matrix A with a sparseness level of $Random(1/m)$ (eq. 8.36). The idea is now to optimize the initial sparse structure by alternating weight pruning and re-creation. Using this method, matrix A is always sparse and the number of active weights stays constant. The network still gets the opportunity to replace active weights by initially inactive ones that it considers more important for the identification of the dynamics.

For the first step, the weight pruning, we use a test criterium similar to *optimal brain damage* (OBD) (LeCun et al., 1990):

$$test_w(w \neq 0) = \frac{\partial^2 E}{\partial w^2} w^2. \tag{8.40}$$

We prune a certain percentage (e.g., 5%) of the lowest values, as these weights w are assumed to be less important for the identification of the dynamics. To simplify our calculations we use

$$\frac{\partial^2 E}{\partial w^2} \approx \frac{1}{T} \sum_t g_t^2, \tag{8.41}$$

with $g_t := \frac{\partial E_t}{\partial w}$, as an approximation for the second derivative. Our simulations showed that this equivalence holds for a 95% level.

In the second step, the re-creation of inactive weights, we use the following test:

$$test_w(w = 0) \sim \frac{1}{T} \left| \sum_t g_t \right|. \tag{8.42}$$

We reactivate the weights w with the highest test values. This implies that we recover weights whose average of the absolute gradient information is high and which are therefore considered important for the identification of the dynamics. Note that we always re-create the same amount of weights we pruned in the first step to keep the sparseness level of the transition matrix A constant.

Our experiments showed that we can even prune and re-create weights simultaneously without losing modeling ability.

8.5.4   Information Flow in Sparse Recurrent Networks

In small networks with a full transition matrix A the information of a state neuron can reach every other one within one time step. This is different in (large) sparse networks, where state neurons have a longer path length on average.

As matrix A is sparse there is in most cases no direct connection between different state neurons. Hence, it can take several state transitions to transport information from one state neuron to another. As information might not reach a

undershooting

desired neuron in a limited number of time steps, this can be disadvantageous for the modeling ability of the network. The resulting question is, how we can speed up the information flow, i.e., shorten the path length?

In a simple recurrent network (e.g., eq. 8.9) the transition matrix A is applied once in every state transition. The idea is now to reduce the average path length with at least one additional undershooting step (Zimmermann and Neuneier, 2001). Undershooting means that we implement intermediate states $s_{\tau \pm \frac{1}{2}}$ which improve the computation of the network (fig. 8.20). These intermediate states have no external inputs. Like future unfolding time steps, they are only responsible for the development of the dynamics and therefore also improve numerical stability.

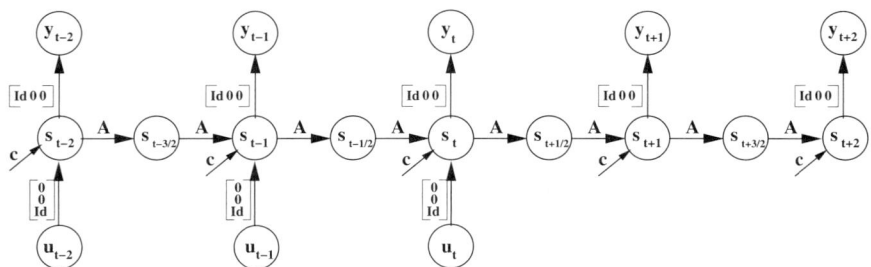

Figure 8.20 Undershooting improves the computation of a sparse matrix A and the numerical stability of the model.

The following formula gives a rough approximation of how many undershooting steps k are needed (eq. 8.43). As the state transition matrix A in the inflated network has, per definition (eq. 8.36), a sparseness of $1/m$, in each time step every state neuron only gets the information of approximately $1/m$ others. The equation now determines the number of undershooting steps which are needed to achieve a desired average path length (information flow) between all state neurons. The kth power of the product of the sparseness factor $1/m$ and the dimension of the state vector $\dim(s)$ must be higher than the number of state neurons ($\hat{=} \dim(s)$):

$$\left(\frac{1}{m} dim(s) \right)^k \geq dim(s)$$

$$\Rightarrow \qquad k \geq \frac{1}{1 - \frac{\log(m)}{\log(dim(s))}} \qquad (8.43)$$

$$\Rightarrow \qquad k \geq 1 + \frac{\log(m)}{\log(dim(s))}.$$

undershooting
with DCNN

Let us reconsider the equations of the DCNN with partially known observables (eq. 8.23):

$$s_\tau = C_E \tanh\left(A \cdot C \cdot s_{\tau-1} + c\right) + \begin{bmatrix} 0 \\ 0 \\ Id \end{bmatrix} y_\tau^E \tag{8.44}$$

$$y_\tau = [Id \ \ 0 \ \ 0]s_\tau, \qquad \sum_{t,\tau}(y_\tau - y_\tau^d)^2 \rightarrow \min_{A,c}.$$

Following the principle of undershooting, we add a state $s_{\tau-\frac{1}{2}}$ between the states $s_{\tau-1}$ and s_τ (fig. 8.21). Consequently, the matrix A is now applied twice between two consecutive time steps, which implies that the information flow is doubled. The consistency matrix C_E handles the lack of external inputs, such that the network stays dynamical consistent.

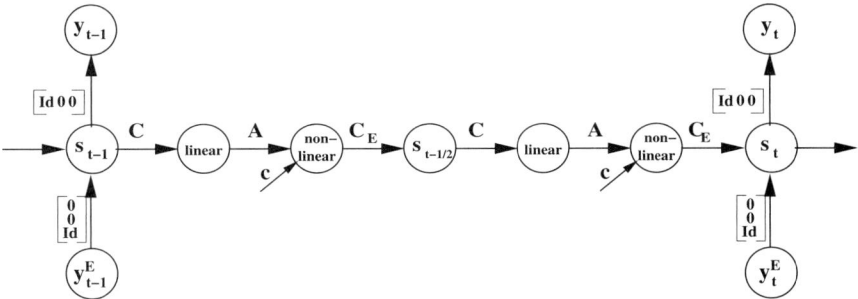

Figure 8.21   Undershooting doubles the information flow between two successive states by applying the transition matrix twice.

It is important to note that the solution is different from just decreasing the sparseness of the matrix A. The latter would not only cause numerical problems in the backpropagation algorithm but also disturb the balance between memory and computation.

8.6   Conclusion

In this chapter we focused on dynamical consistent neural networks (DCNN) for the modeling of open dynamical systems. After a short description of small recurrent networks, including error correction neural networks, we presented a new kind of dynamical consistent neural networks. These networks allow an integrated view on particular modeling problems and consequently show better generalization abilities. We concentrated the modeling of the dynamics on one single-transition matrix and also enhanced the model from a simple statistical to a dynamical consistent handling of missing input information in the future part. The networks are now able to map integrated system dynamics (e.g., financial markets) instead of only a small set of time series. The final DCNN combines the advantages of the former RNN and the ECNN.

Besides the new, more powerful architectures, the modeling involves a paradigm shift in the analysis of open systems (see fig. 8.1). In the beginning we looked at the description of a dynamical system from an *exterior* point of view. This means that we observed the information flow into and out of an open system and tried to reconstruct the interior. The long-term predictability of the model finally depended on the quality of the extracted autonomous subsystem.

world model In our new approach we describe dynamical systems from an *interior* viewpoint. Conceptually we start with a *world model* (fig. 8.22). Without loss of generalization, we assume that our variables of interest are all organized as the first elements in a large state vector. Identifying this first section of the state vector as our observables (y_t), we can reconstruct some more unobservable states (h_t) by their indirect influence on the observables. Nevertheless, there are an infinite number of variables which are unobservable and even unidentifiable. Their nearly infinite influence can be shrunk to a finite dimensional section in the state vector: the error correction part (e_t). If the error correction is equal to zero, knowledge about the unidentifiable variables is not necessary. Otherwise, it dispenses us from having to know the details of the unknown part of the world.

Clearly, the concept of a world model is a closed one from the beginning. As a consequence of dynamical consistency the closure concept even holds for finite-dimensional subsections of it. Therefore it models the dynamics as a closed system and is still able to keep model evolution exactly on the observed state trajectory (see DCNN2, eq. 8.22).

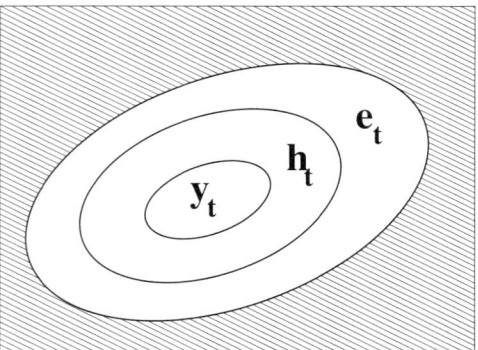

Figure 8.22 Variable space of the world model. y_t stands for the observables, h_t for the hidden variables which can be explained by the observables, and e_t for the error corrections, which close the gap between the observable and the unobservable part of the system.

We augmented the model-building process by incorporating prior knowledge. Learning from data is only one part of this process. The recurrent ECNN and the DCNN are two examples of this model-building philosophy. Remarkably, such a joint model-building framework does not only provide superior forecasts, but also a deeper understanding of the underlying dynamical system. On this basis it is also possible to analyze and to quantify the uncertainty of the predictions. This is especially important for the development of decision support systems.

Currently we test our models in several industrial applications. Further research is conducted concerning the optimal sparseness of the transition matrix A as well as the optimal initialization method for the state vector.

Acknowledgment

We thank J. Zwierz for the calculations and tests during our experiments in section 8.5. The extensive work performed by M. Pellegrino in proof reading this chapter is gratefully acknowledged. The computations were performed on our neural network modeling software SENN (Simulation Environment for Neural Networks), which is a product of Siemens AG.

Notes

[1] For other cost functions see Neuneier and Zimmermann (1998).

[2] For an overview of algorithmic methods see Pearlmutter (2001) and Medsker and Jain (1999).

[3] For further details, the reader is referred to Haykin (1994) and Bishop (1995).

9 Diversity in Communication: From Source Coding to Wireless Networks

Suhas Diggavi

Randomness is an inherent part of network communications. We broadly define diversity as creating multiple independent instantiations (conduits) of randomness for conveying information. In the past few years a trend is emerging in several areas of communications, where diversity is utilized for reliable transmission and efficiency. In this chapter, we give examples from three topics where diversity is beginning to play an important role.

9.1 Introduction

One of the main characteristics of network communication is the uncertainty (randomness): randomness in users' wireless transmission channels, randomness in users' geographical locations in a wireless network, and randomness in route failures and packet losses in networks. The randomness we study in this chapter can have timescales of variation that are comparable to the communication transmission times. This can result in complete failures in communication and therefore affect reliability. Such "nonergodic" losses can be combated if we somehow create independent instantiations of the randomness. We broadly define *diversity* as the method of conveying information through such multiple independent instantiations. The overarching theme of this chapter is how to create diversity and how we can use it as a tool to enhance performance. We study this idea through diversity in *multiple antennas*, *multiple users*, and *multiple routes*.

The functional modularities and abstractions of the network protocol known as *stack layering* (Keshav, 1997) contributed significantly to the success of the wired Internet infrastructure. The layering achieves a form of information hiding, providing only interface information to higher layers, and not the details of the implementation. The physical layer is dedicated to signal transmission, while the data-link layer implements functionalities of data framing, arbitrating access to transmission medium and some error control. The network layer abstracts the physical and data-link layers from the upper layers by providing an interface for end-to-end links. Hence, the task of routing and framing details of the link layer are hidden from the higher layers (transport and application layers). However, as we will see, the use of diversity necessarily causes cross-layer interactions. These cross-layer interactions form a subtext to the theme of this chapter.

Wireless communication hinges on transmitting information riding on radio (electromagnetic) waves, and hence the information undergoes attenuation effects (fading) of radio waves (see section 9.2 for more details). Such multipath fading is a source of randomness. Here diversity arises by utilizing independent realizations of fading in several domains, time (mobility), frequency (delay spread), and space (multiple antennas). Over the past decade research results have shown that multiple-antenna spatial diversity (space-time) communication can not only provide robustness, but also dramatically improve reliable data rates. These ideas are having a huge impact on the design of physical layer transmission techniques in next-generation wireless systems. *Multiple-antenna diversity* is the focus of section 9.3.

The wireless communication medium is naturally shared by several users using the same resources. Since the users' locations (and therefore their transmission conditions) are roughly independent, they experience independent randomness in local channel and interference conditions. Diversity in this case arises by utilizing the independent transmission conditions of the different users as conduits for transmitting information i.e., *multi-user diversity*. This can be utilized in two ways. One by allowing users access to resources when it is most advantageous to the overall network. This is a form of opportunistic scheduling and is examined in section 9.4.1. The other by using the users themselves as relays to transmit information from source to destination. This is a form of opportunistic relaying, and is studied in section 9.4.2. These multi-user diversity methods are the focus of section 9.4.

In transmission over networks, random route failures and packet losses degrade performance. Diversity here would be achieved by creating conduits with independent probability of route failures. For example, this can be done by transmission over multiple routes with no overlapping links. A fundamental question that arises is how we can best utilize the presence of such *route diversity*. In order to utilize these conduits, multiple description source coding generates multiple codeword streams to describe a source (such as images, voice, video, etc.). The design goal is to have a graceful degradation in performance (in terms of distortion) when only subsets of the transmitted streams are received. In section 9.5 we study fundamental bounds and design ideas for multiple description source coding.

Therefore, diversity not only plays a role in robustness, it can also result in remarkable gains in achievable performance over several disparate applications. The details of how diversity enhances performance are discussed in the sequel.

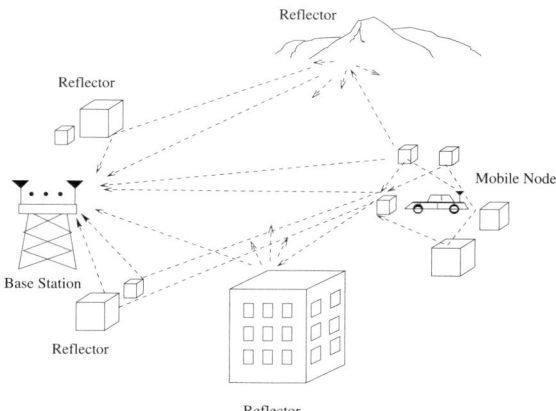

Figure 9.1   Radio propagation environment.

9.2   Transmission Models

Since a considerable part of this chapter is about wireless communication, it is essential to understand some of the rudiments of wireless channel characteristics. In this section, we focus on models for point-to-point wireless channels and also introduce some of the basic characteristics of transmission over (wireless) networks.

Wireless communication transmits information by riding (modulation) on electromagnetic (radio) waves with a carrier frequency varying from a few hundred megahertz to several gigahertz. Therefore, the behavior of the wireless channel is a function of the radio propagation effects of the environment.

A typical outdoor wireless propagation environment is illustrated in fig. 9.1, where the mobile wireless node is communicating with a wireless access point (base station). The signal transmitted from the mobile may reach the access point directly (line-of-sight) or through multiple reflections on local scatterers (buildings, mountains, etc.). As a result, the received signal is affected by multiple random attenuations and delays. Moreover, the mobility of either the nodes or the scattering environment may cause these random fluctuations to vary with time. Time variation results in the random waxing and waning of the transmitted signal strength over time. Finally, a shared wireless environment may incur interference (due to concurrent transmissions from other mobile nodes) to the transmitted signal.

The attenuation incurred by wireless propagation can be decomposed in three main factors: a signal attenuation due to the distance between communicating nodes (*path loss*), attenuation effects due to absorption in local structures such as buildings (*shadowing loss*), and rapid signal fluctuations due to constructive and destructive interference of multiple reflected radio wave paths (*fading loss*). Typically the path loss attenuation behaves as $1/d^\alpha$ as a function of distance d, with $\alpha \in [2, 6]$. More detailed models of wireless channels can be found in Jakes (1974) and Rappaport (1996).

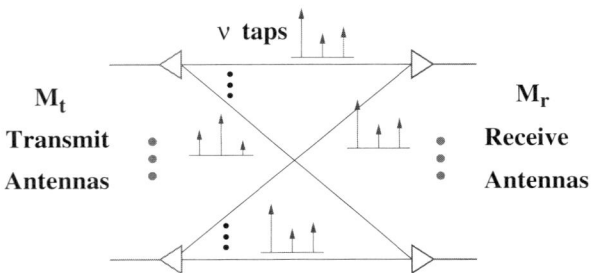

Figure 9.2   MIMO channel model.

9.2.1   Point-to-Point Model

For the purposes of this chapter we start with the following model:

$$y_c(t) = \int h_c(t; \tau)s(t - \tau)d\tau + z(t) \,, \tag{9.1}$$

where the transmitted signal $s(t) = g(t) * x(t)$ is the convolution of the information-bearing signal $x(t)$ with $g(t)$, the transmission shaping filter, $y_c(t)$ is the continuous time received signal, $h_c(t; \tau)$ is the response at time t of the time-varying channel if an impulse is sent at time $t - \tau$, and $z(t)$ is the additive Gaussian noise. The channel impulse response (CIR) depends on the combination of all three propagation effects and in addition contains the delay induced by the reflections.

To collect discrete-time sufficient statistics[1] of the information signal $x(t)$ we need to sample (9.1) faster than the Nyquist rate[2]. Therefore we focus on the following discrete-time model:

$$y(k) = y_c(kT_s) = \sum_{l=0}^{\nu} h(k; l)x(k - l) + z(k) \,, \tag{9.2}$$

where $y(k)$, $x(k)$, and $z(k)$ are the output, input, and noise samples at sampling instant k, respectively, and $h(k; l)$ represents the sampled time-varying channel impulse response of finite length ν. Modeling the channel as having a finite duration can be made arbitrarily accurate by appropriately choosing the channel memory ν.

Though the channel response $\{h(k; l)\}$ depends on all three radio propagation attenuation factors, in the timescales of interest the main variations come from the small-scale fading which is well modeled as a complex Gaussian random process.

Since we are interested in studying multiple-antenna diversity, we need to extend the model given in equation 9.2 to the multiple transmit (M_t) and receive (M_r) antenna case. The multi-input multi-output (MIMO) model is given by

$$\mathbf{y}(k) = \sum_{l=0}^{\nu} \mathbf{H}(k; l)\mathbf{x}(k - l) + \mathbf{z}(k) \,, \tag{9.3}$$

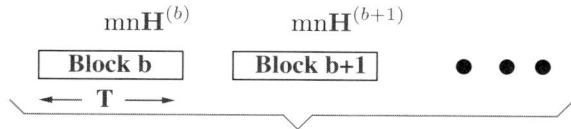

Block time–invariant transmission frames

Figure 9.3 Block time-invariant model.

where the $M_r \times M_t$ complex[3] matrix $\mathbf{H}(k; l)$ represents the lth tap of the channel matrix response with $\mathbf{x} \in \mathbb{C}^{M_t}$ as the input and $\mathbf{y} \in \mathbb{C}^{M_r}$ as the output (see fig. 9.2). The variations of the channel response between antennas arises due to variations in arrival directions of the reflected radio waves (Raleigh et al., 1994). The input vector may have independent entries to achieve high throughput (e.g., through spatial multiplexing) or correlated entries through coding or filtering to achieve high reliability (better distance properties, higher diversity, spectral shaping, or desirable spatial profile; see section 9.3). Throughout this chapter, the input is assumed to be zero mean and to satisfy an average power constraint, i.e., $\mathrm{E}[\|\mathbf{x}(k)\|^2] \leq P$. The vector $\mathbf{z} \in \mathbb{C}^{M_r}$ models the effects of noise and is assumed to be independent of the input and is modeled as a complex additive circularly symmetric Gaussian vector with $\mathbf{z} \sim \mathbb{CN}(0, \mathbf{R}_z)$, i.e., a complex Gaussian vector with mean $\mathbf{0}$ and covariance $\mathbf{R}_z$. In many cases we assume white noise, i.e., $\mathbf{R}_z = \sigma^2 \mathbf{I}$.

Finally, the basic point-to-point model given in equation 9.3 can be modified for an important special case. Many of the insights can be gained for the *flat fading* channel where we have $\nu = 0$ in equation 9.3. Unless otherwise mentioned, we will use this special case for illustration throughout this chapter. Also we examine the case where we transmit a block or frame of information. Here we encounter another important modeling assumption. If the transmission block is small enough so that the channel time variation within a transmission block can be neglected, we have a *block time-invariant* model. Such models are quite realistic for transmission blocks of lengths less than a millisecond and typical channel variation bandwidths. However, this does *not* imply that the channel remains constant during the entire transmission. Transmission blocks sent at various periods of time can experience different (independent) channel instantiations (see fig. 9.3). This can be utilized by coding across these different channel instantiations, as will be seen in section 9.3. Therefore, if the transmission block is of length T, for the *flat-fading* case, the specialization of equation 9.3 yields

$$\mathbf{Y}^{(b)} = \mathbf{H}^{(b)}\mathbf{X}^{(b)} + \mathbf{Z}^{(b)}, \tag{9.4}$$

where $\mathbf{Y}^{(b)} = [\mathbf{y}^{(b)}(0), \dots, \mathbf{y}^{(b)}(T-1)] \in \mathbb{C}^{M_r \times T}$ is the received sequence, $\mathbf{H}^{(b)} \in \mathbb{C}^{M_r \times M_t}$ is the block time-invariant channel fading matrix for transmission block b, $\mathbf{X}^{(b)} = [\mathbf{x}^{(b)}(0), \dots, \mathbf{x}^{(b)}(T-1)] \in \mathbb{C}^{M_t \times T}$ is the "space-time" information transmission sequence, and $\mathbf{Z}^{(b)} = [\mathbf{z}^{(b)}(0), \dots, \mathbf{z}^{(b)}(T-1)] \in \mathbb{C}^{M_r \times T}$.

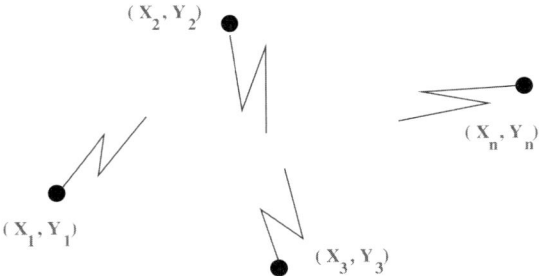

Figure 9.4   General multi-user wireless communication network.

9.2.2   Network Models

The wireless medium is inherently shared, and this directly motivates a study of multi-user communication techniques. Moreover, since we are also interested in multi-user diversity, we need to extend our model from the point-to-point scenario (eq. 9.2) to the network case. The general communication network (illustrated in fig. 9.4) consists of n nodes trying to communicate with each other. In the scalar flat-fading wireless channel, the received symbol $Y_i(t)$ at the ith node is given by

$$Y_i(t) = \sum_{\substack{j=1 \\ j \neq i}}^{n} h_{i,j} X_j(t) + Z_i(t), \tag{9.5}$$

where $h_{i,j}$ is determined by the channel attenuation between nodes i and j. Given this general model, one way of abstracting the multi-user communication problem is through embedding it in an underlying *communication graph* $\mathcal{G}_C$ where the n nodes are vertices of the graph and the edges of the graph represent a channel connecting the two nodes along with the interference from other nodes. The graph could be directed with constraints and channel transition probability depending on the directed graph. A general multi-user network is therefore a fully connected graph with the received symbol at each node described as a conditional distribution dependent on the messages transmitted by all other nodes. Such a graph is illustrated in fig. 9.5. We examine different communication topologies in section 9.4 and study the role of diversity in networks.

9.3   Multiple-Antenna Diversity

The first form of diversity that we examine in some detail is that of multiple-antenna diversity. A major development over the past decade has been the emergence of space-time (multiple-antenna) techniques that enable high-rate, reliable communication over fading wireless channels. In this section we highlight some of the theoretical underpinnings of this topic. More details about practical code construc-

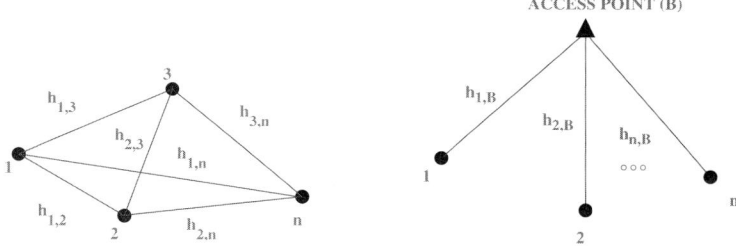

Figure 9.5 Graph representation of communication topologies. On the left is a general topology and on the right is a hierarchical topology.

tions can be found in Tarokh et al. (1998), Diggavi et al. (2004b), and references therein.

Reliable information transmission over fading channels has a long and rich history; see Ozarow et al. (1994) and references therein. The importance of multiple antenna diversity was recognized early; see, for example, Brennan (1959). However, most of the focus until the mid-1990s was on receive diversity, where multiple "looks" of the transmitted signal were obtained using many receive antennas (see equation 9.3 using $M_t = 1$). The use of multiple transmit antennas was restricted to sending the same signal over each antenna, which is a form of repetition coding (Wornell and Trott, 1997).

During the mid-1990s several researchers started to investigate the idea of coding across transmit antennas to obtain higher rate and reliability (Foschini, 1996; Tarokh et al., 1998; Telatar, 1999). One focus was on maximizing the reliable transmission rate, i.e., channel capacity, without requiring a bound on the rate at which error probability diminishes (Foschini, 1996; Telatar, 1999). However another point of view was explored where nondegenerate correlation was introduced between the information streams across the multiple transmit antennas in order to guarantee a certain bound on the rate at which the error probability diminishes (Tarokh et al., 1998). These approaches have led to the broad area of space-time codes, which is still an active research topic.

In section 9.3.1 we first start with an understanding of reliable transmission rate over multiple-antenna channels. In particular we examine the rate advantages of multiple transmit and receive antennas. Then in section 9.3.2 we introduce the notion of diversity order, which captures transmission reliability (error probability) in the high signal-to-noise ratio (SNR) regime. This allows us to develop criteria for space-time codes which guarantee a given reliability. section 9.3.3 examines the fundamental trade-off between maximizing rate and reliability.

9.3.1 Capacity of Multiple-Antenna Channels

The concept of capacity was first introduced by Shannon (1948), where it was shown that even in noisy channels, one can transmit information at positive rates with the error probability going to zero asymptotically in the coding block size. The seminal result was that for a noisy channel whose input at time k is $\{X_k\}$ and output is $\{Y_k\}$, there exists a number C such that

$$C = \lim_{T \to \infty} \left[\frac{1}{T} \sup_{p(x^T)} I(X^T; Y^T) \right], \qquad (9.6)$$

where the mutual information is given by $I(X^T; Y^T) = \mathrm{E}_{X^T, Y^T}[\log(\frac{p(x^T, y^T)}{p(x^T)p(y^T)})]$, $p(\cdot)$ is the probability density function, and for convenience we have denoted $X^T = \{X_1, \dots, X_T\}$ and similarly for Y^T (Cover and Thomas, 1991). In Shannon (1948) it was shown that asymptotically in block length T, there exist codes which can transmit information at all rates below C with arbitrarily small probability of error over the noisy channel. Perhaps the most famous illustration of this idea was the formula derived in Shannon (1948) for the capacity C of the additive white Gaussian noise channel with noise variance σ^2 and input power constraint P:

$$C = \frac{1}{2} \log(1 + \frac{P}{\sigma^2}). \qquad (9.7)$$

In this section we will focus mostly on the *flat-fading* channels where, in equation 9.3, we have $\nu = 0$. The generalizations of these ideas for *frequency-selective* channels (i.e., $\nu > 0$) can be easily carried out (see Biglieri et al., 1998; Diggavi et al., 2004b, and references therein). We begin with the case where we are allowed to develop transmit schemes which code across multiple (B) realizations of the channel matrix $\{\mathbf{H}^{(b)}\}_{b=1}^B$ (see fig. 9.3). In such a case, we can again define a notion of reliable transmission rate, where the error probability decays to zero when we develop codes across asymptotically large numbers of transmit blocks (i.e., $B \to \infty$). We examine this for a coherent receiver, where the receiver uses perfect channel state information $\{\mathbf{H}^{(b)}\}$ for each transmission block. But the transmitter is assumed not to have access to the channel realizations. To gain some intuition, consider first the case when each transmission block is large, i.e., $T \to \infty$. If we have one transmit antenna ($M_t = 1$), the channel vector response is a vector $\mathbf{h}^{(b)} \in \mathbb{C}^{M_r}$ (see equation 9.4 in section 9.2). Therefore the reliable transmission rate for any particular block can be generalized[4] $\{\mathbf{h}(k)\}$ from (9.7) as $\log(1 + \frac{\|\mathbf{h}^{(b)}\|^2 P}{\sigma^2})$. Note that when we are dealing with complex channels (as is usual in communication with in-phase and quadrature-phase transmissions), the factor of $1/2$ disappears (Neeser and Massey, 1993) when we adapt the expression from equation 9.7. Now, if one codes across a large number of transmission blocks ($B \to \infty$), for a stationary and ergodic sequence of $\{\mathbf{h}^{(b)}\}$ we would expect to get a reliable transmission rate that is the average of this quantity. This intuition has been made precise in Ozarow et al. (1994), and references therein, for flat-fading channels ($\nu = 0$), even when we do not have $T \to \infty$, but we have $B \to \infty$. Therefore when we have only receive diversity, i.e., $M_t = 1$, for a given M_r, it is shown (Ozarow et al., 1994) that the capacity is given by

$$C = \mathrm{E}\left[\log(1 + \frac{||\mathbf{h}||^2 P}{\sigma^2})\right], \tag{9.8}$$

where the expectation is taken over the fading channel $\{\mathbf{h}^{(b)}\}$ and the channel sequence is assumed to be stationary and ergodic. This is called the *ergodic channel capacity* (Ozarow et al., 1994). This is the rate at which information can be transmitted if there is *no feedback* of the channel state ($\{\mathbf{h}^{(b)}\}$) from the receiver to the transmitter. If there is feedback available about the channel state, one can do slightly better through optimizing the allocation of transmitted power by "waterfilling" over the fading channel states. The problem of studying the capacity of channels with causal transmitter-side information was introduced in Shannon (1958a), where a coding theorem for this problem was proved. Using ideas from there and *perfect* transmitter channel state information, capacity expressions that generalize equation 9.8 have been developed (Goldsmith and Varaiya, 1997). However, for fast time-varying channels the instantaneous feedback could be difficult, resulting in an outdated estimate of the channel being sent back (Caire and Shamai, 1999; Viswanathan, 1999). However, the basic question of impact of feedback on capacity of time-varying channels is still not completely understood, and for developing the basic ideas in this chapter, we will deal with the case where the transmitter does not have access to the channel state information. We refer the interested reader to Biglieri et al. (1998) for a more complete overview of such topics.

Now let us focus our attention on the multiple transmit and receive antenna channel where again as before we consider the coherent case, i.e., the receiver has perfect channel state information (CSI) $\mathbf{H}^{(b)}$. In the flat-fading case where $\nu = 0$, when we code across B transmission blocks, the mutual information for this case is

$$R^{(B)} = \frac{1}{BT} I(\{\mathbf{X}^{(b)}\}_{b=1}^B; \{\mathbf{Y}^{(b)}\}_{b=1}^B, \{\mathbf{H}^{(b)}\}_{b=1}^B),$$

since we assume that the receiver has access to CSI. Using the chain rule of mutual information (Cover and Thomas, 1991), this can be written as

$$R^{(B)} = \frac{1}{BT}\left[I(\{\mathbf{X}^{(b)}\}_{b=1}^B; \{\mathbf{H}^{(b)}\}_{b=1}^B) + I(\{\mathbf{X}^{(b)}\}_{b=1}^B; \{\mathbf{Y}^{(b)}\}_{b=1}^B | \{\mathbf{H}^{(b)}\}_{b=1}^B)\right]. \tag{9.9}$$

Using the the assumption that the input $\{\mathbf{x}(k)\}$ is independent of the fading process (as the transmitter does not have CSI), equation 9.9 is equal to

$$R^{(B)} = \frac{1}{BT}\mathrm{E}_{\mathcal{H}}\left[I\left(\{\mathbf{X}^{(b)}\}_{b=1}^B; \{\mathbf{Y}^{(b)}\}_{b=1}^B | \mathcal{H}^{(B)} = \{\mathbf{H}^{(b)}\}_{b=1}^B\right)\right]. \tag{9.10}$$

Now, if we use the memoryless property of the vector Gaussian channel obtained by conditioning on $\mathbf{H}^{(b)}$ and also due to the assumption[5] that $\{\mathbf{H}^{(b)}\}$ is i.i.d. over b, for when $B \to \infty$ we get that

$$\lim_{B \to \infty} \frac{1}{BT} I(\{\mathbf{X}^{(b)}\}_{b=1}^B; \{\mathbf{Y}^{(b)}\}_{b=1}^B, \{\mathbf{H}^{(b)}\}_{b=1}^B) = \mathrm{E}_{\mathcal{H}}[\log(\frac{|\mathbf{R}_z + \mathbf{H}\mathbf{R}_x\mathbf{H}^*|}{|\mathbf{R}_z|})], \tag{9.11}$$

where the expectation[6] is taken over the random channel realizations $\{\mathbf{H}^{(b)}\}$. An *operational* meaning to this expression can be given by showing that there exist codes which can transmit information at this rate with arbitrarily small probability of error (Telatar, 1999).

In general, it is difficult to evaluate equation 9.11 except for some special cases. If the random matrix $\mathbf{H}^{(b)}$ consists of zero-mean i.i.d. Gaussian elements, Telatar (1999) showed that

$$C = \mathrm{E}_{\mathcal{H}}[\log(|\mathbf{I} + \frac{P}{M_t\sigma^2}\mathbf{H}\mathbf{H}^*|)] \tag{9.12}$$

is the capacity of the fading matrix channel.[7] Therefore in this case, to achieve capacity the optimal codebook is generated from an i.i.d. Gaussian input $\{\mathbf{x}^{(b)}\}$ with $\mathbf{R}_x = \mathrm{E}[\mathbf{x}\mathbf{x}^*] = \frac{P}{M_t}\mathbf{I}$.

The expression in equation 9.12 shows that the capacity is dependent on the eigenvalue distribution of the random matrix $\mathbf{H}$ with Gaussian i.i.d. components. This important connection between capacity of multiple-antenna channels and the mathematics related to eigenvalues of random matrices (Edelman, 1989) was noticed in Telatar (1999), where it was shown that the capacity could be numerically computed using Laguerre polynomials (Edelman, 1989; Muirhead, 1982; Telatar, 1999).

Theorem 9.1

(Telatar, 1999) The capacity C of the channel with M_t transmitters and M_r receivers and average power constraint P is given by

$$C = \int_0^\infty \log(1 + \frac{P\lambda}{\sigma^2 M_t}) \sum_{k=0}^{T_{min}-1} \lambda^{T_{max}-T_{min}} [L_k^{T_{max}-T_{min}}(\lambda)]^2 \frac{k!}{k + T_{max} - T_{min}} e^{-\lambda} d\lambda \,,$$

where $T_{max} = \max(M_t, M_r)$, $T_{min} = \min(M_t, M_r)$, and $L_k^m(\cdot)$ is the generalized Laguerre polynomial of order k with parameter m (Gradshteyn and Ryzhik, 1994).

∎

In Foschini (1996) it was observed that when $M_t = M_r = M$ the capacity C grows linearly in M as $M \to \infty$.

Theorem 9.2

(Foschini, 1996) For $M_t = M_r = M$ the capacity C given by (9.12) grows asymptotically linearly in M, i.e.,

$$\lim_{M\to\infty} \frac{C}{M} = c^*(\mathrm{SNR}) \,, \tag{9.13}$$

where $c^*(\mathrm{SNR})$ is a constant depending on SNR.

∎

This quantifies the advantage of using multiple transmit and receive antennas and shows the promise of such architectures for high-rate reliable wireless communication.

To achieve the capacity given in equation 9.12, we require joint optimal (maximum-likelihood) decoding of all the receiver elements which could have large computational complexity. The channel model in equation 9.3 resembles a multi-user channel (Verdu, 1998) with user cooperation. A natural question to ask is whether the simpler decoding schemes proposed in multi-user detection would yield good performance on this channel. A motivation for this is seen by observing that for i.i.d. elements of the channel response matrix (flat-fading) the normalized cross-correlation matrix decouples (i.e., $\lim_{M_r \to \infty} \frac{1}{M_r} \mathbf{H}^* \mathbf{H} \to \mathbf{I}_{M_t}$). Therefore, since nature provides some decoupling, a simple "matched filter" receiver (Verdu, 1998) might perform quite well. In this context a matched filter for the flat-fading channel in equation 9.3 is given by $\tilde{\mathbf{y}}(k) = \mathbf{H}^*(k)\mathbf{y}(k)$. Therefore, component-wise this means that

$$\tilde{\mathbf{y}}_i(k) = ||\mathbf{h}_i(k)||^2 \mathbf{x}_i(k) + \sum_{\substack{j=1 \\ j \neq i}}^{M_t} \mathbf{h}_i^*(k)\mathbf{h}_j(k)\mathbf{x}_j(k) + \tilde{\mathbf{z}}_i(k), \quad i = 1, \ldots, M_t. \quad (9.14)$$

By ignoring the cross-coupling between the channels we decode $\hat{\mathbf{x}}_i$ by including the "interference" from $\{\mathbf{x}_j\}_{j \neq i}$ as part of the noise. However, a tension arises between the decoupling of the channels and the added "interference" $\sum_{\substack{j=1 \\ j \neq i}}^{M_t} \mathbf{h}_i^*(k)\mathbf{h}_j(k)\mathbf{x}_j(k)$ from the other antennas, which clearly grows with the number of antennas. It is shown in Diggavi (2001), that the two effects exactly cancel each other.

Proposition 9.1
If $\mathbf{H}(k) = [\mathbf{h}_1(k), \ldots, \mathbf{h}_{M_t}(k)] \in \mathbb{C}^{M_r \times M_t}$ and $\mathbf{h}_l(k) \sim \mathbb{CN}(0, \mathbf{I}_{M_r})$, $l = 1, \ldots, M_t$, are i.i.d., then

$$\lim_{\substack{M_r \to \infty \\ M_t = \lfloor \alpha M_r \rfloor}} \sum_{\substack{j=1 \\ j \neq i}}^{M_t} |\frac{\mathbf{h}_i^*(k)\mathbf{h}_j(k)}{M_r}|^2 = \alpha \text{ almost surely.}$$

Therefore, using this result it can be shown that the simple detector still retains the linear growth rate of the optimal decoding scheme (Diggavi, 2001). However, in the rate R_I achievable for this simple decoding scheme, we do pay a price in terms of rate growth with SNR.

Theorem 9.3
If $\mathbf{H}_{i,j} \sim \mathbb{CN}(0, 1)$, with i.i.d. elements, then

$$\lim_{\substack{M_t \to \infty \\ M_t = \lfloor \alpha M_r \rfloor}} \frac{1}{M_t} I(\mathbf{Y}, \mathbf{H}; \mathbf{X}) \geq \lim_{\substack{M_t \to \infty \\ M_t = \lfloor \alpha M_r \rfloor}} R_I / M_t = \log(1 + \frac{\frac{P}{\sigma^2 \alpha}}{1 + \frac{P}{\sigma^2}}).$$

$\blacksquare$

Multi-user detection (Verdu, 1998) is a good analogy to understand receiver structures in MIMO systems. The main difference is that unlike multiple access channels, the space-time encoder allows for cooperation between "users." Therefore, the encoder could introduce correlations that can simplify the job of the decoder.

Such encoding structures using space-time block codes are discussed further in Diggavi et al. (2004b), and references therein. An example of using the multi-user detection approach is the result in theorem 9.3 where a simple matched filter receiver is applied. Using more sophisticated linear detectors, such as the decorrelating receiver and the MMSE receiver (Verdu, 1998), one can improve performance while still maintaining the linear growth rate. The decision feedback structures also known as successive interference cancellation, or onion peeling (Cover, 1975; Patel and Holtzman, 1994; Wyner, 1974) can be shown to be optimal, i.e., to achieve the capacity, when an MMSE multi-user interference suppression is employed and the layers are peeled off (Cioffi et al., 1995; Varanasi and Guess, 1997). However, decision feedback structures inherently suffer from error propagation (which is not taken into account in the theoretical results) and could therefore have poor performance in practice, especially at low SNR. Thus, examining nondecision feedback structures is important in practice.

All of the above results illustrate that significant gains in information rate (capacity) are possible using multiple transmit and receive antennas. The intuition for the gains with multiple transmit and receive antennas is that there are a larger number of communication modes over which the information can be transmitted. This is formalized by the observation (Diggavi, 2001; Zheng and Tse, 2002) that the capacity as a function of SNR, $C(SNR)$, grows linearly in $\min(M_r, M_t)$, even for a finite number of antennas, asymptotically in the SNR.

Theorem 9.4

$$\lim_{SNR \to \infty} \frac{C(SNR)}{\log(SNR)} = \min(M_r, M_t).$$ (9.15)

∎

In the results above, the fundamental assumption was that the receiver had access to *perfect* channel state information, obtained through training or other methods. When the channel is slowly varying, the estimation error could be small since we can track the channel variations and one can quantify the effect of such estimation errors. As a rule of thumb, it is shown by Lapidoth and Shamai (2002) that if the estimation error is small compared to $\frac{1}{SNR}$, these results would hold. Another line of work assumes that the receiver does not have *any* channel state information. The question of the information rate that can be reliably transmitted over the multiple-antenna channel without channel state information was introduced in Hochwald and Marzetta (1999) and has also been examined in Zheng and Tse (2002). The main result from this line of work shows that the capacity growth is again (almost) linear in the number of transmit and receive antennas, as stated formally next.

Theorem 9.5

If the channel is block fading with block length T and we denote $K = \min(M_t, M_r)$, then for $T > K + M_t$, as $SNR \to \infty$, the capacity is[8]

$$C(SNR) = K \left(1 - \frac{K}{T}\right) \log(SNR) + c + o(1) \,,$$

where c is a constant depending only on M_r, M_t, T.

∎

In fact, Zheng and Tse (2002) go on to show that the rate achievable by using a training-based technique is only a constant factor away from the optimal, i.e., it attains the same capacity-SNR slope as in theorem 9.5. Further results on this topic can be found in Hassibi and Marzetta (2002). Therefore, even in the noncoherent block-fading case, there are significant advantages in using multiple antennas.

Most of the discussion above was for the flat-fading case where $\nu = 0$ in equation 9.3. However, these ideas can be easily extended for the block time-invariant frequency-selective channels where again the advantages of multiple-antenna channels can be established (Diggavi, 2001). However, when the channels are not block time-invariant, the characterization of the capacity of frequency-selective channels is an open question.

outage   **Outage** In all of the above results, the error probability goes to zero asymptotically in the number of coding blocks i.e., $B \to \infty$. Therefore, coding is assumed to take place *across* fading blocks, and hence it inherently uses the *ergodicity* of the channel variations. This approach would clearly entail large delays, and therefore Ozarow et al. (1994) introduced a notion of *outage*, where the coding is done (in the extreme case) just across one fading block, i.e., $B = 1$. Here the transmitter sees only one block of channel coefficients, and therefore the channel is *nonergodic*, and the strict Shannon-sense capacity is zero. However, one can define an outage probability that is the probability with which a certain rate R is possible. Therefore, for a block time-invariant channel with a single channel realization $\mathbf{H}^{(b)} = \mathbf{H}$ the outage probability can be defined as follows.

Definition 9.1

The outage probability for a transmission rate of R and a given transmission strategy $p(\mathbf{X})$ is defined as

$$P_{outage}(R, p(\mathbf{X})) = \mathrm{P}\left\{\mathbf{H} : I(\mathbf{X}; \mathbf{Y}|\mathbf{H}^{(b)} = \mathbf{H}) < R\right\}. \qquad (9.16)$$

Therefore, if one uses a white Gaussian codebook ($\mathbf{R}_x = \frac{P}{M_t}\mathbf{I}$) then (abusing notation by dropping the dependence on $p(\mathbf{X})$) we can write the outage probability at rate R as

$$P_{outage}(R) = \mathrm{P}\left\{\log(|\mathbf{I} + \frac{P}{M_t\sigma^2}\mathbf{H}\mathbf{H}^*|) < R\right\}. \qquad (9.17)$$

It has been shown (Zheng and Tse, 2003) that at high SNR the outage probability is the same as the frame-error probability in terms of the SNR exponent. Therefore, to evaluate the optimality of practical coding techniques, one can compare, for a given rate, how far the performance of the technique is from that predicted through an outage analysis. Moreover, the frame-error rates and outage capacity comparisons in Tarokh et al. (1998) can also be formally justified through this argument.

9.3.2   Diversity Order

In section 9.3.1 the focus was on achievable transmission rate. A more practical performance criterion is probability of error. This is particularly important when we are coding over a small number of blocks (low delay) where the Shannon capacity is zero (Ozarow et al., 1994) and we are in the outage regime as was seen above. By characterizing the error probability, we can also formulate design criteria for space-time codes.

Since we are allowed to transmit a coded sequence, we are interested in the probability that an erroneous codeword[9] $\mathbf{e}$ is mistaken for the transmitted codeword $\mathbf{x}$. This is called the *pairwise error probability* (PEP) and is used to bound the error probability. This analysis relies on the condition that the receiver has perfect channel state information. However, a similar analysis can be done when the receiver does not know the channel state information, but has statistical knowledge of the channel (Hochwald and Marzetta, 2000).

For simplicity, we shall again focus on a flat-fading channel (where $\nu = 0$) and when the channel matrix contains i.i.d. zero-mean Gaussian elements, i.e., $\mathbf{H}_{i,j} \sim \mathbb{C}\mathcal{N}(0,1)$. Many of these results can be easily generalized for $\nu > 0$ as well as for correlated fading and other fading distributions. Consider a codeword sequence $\mathbf{X} = [\mathbf{x}^t(0), \dots, \mathbf{x}^t(T-1)]^t$, where $\mathbf{x}(k) = [\mathbf{x}_1(k), \dots, \mathbf{x}_{M_t}(k)]^t$ (defined in eq. 9.4). In the case when the receiver has perfect channel state information, we can bound the PEP between two codeword sequences $\mathbf{x}$ and $\mathbf{e}$ (denoted by $P(\mathbf{x} \to \mathbf{e})$) as follows (Guey et al., 1999; Tarokh et al., 1998):

$$P(\mathbf{x} \to \mathbf{e}) \leq \left[\frac{1}{\prod_{n=1}^{M_t} (1 + \frac{E_s}{4N_0} \lambda_n)} \right]^{M_r}. \tag{9.18}$$

$E_s = \frac{P}{M_t}$ is the power per transmitted symbol, λ_n are the eigenvalues of the matrix $\mathbf{A}(\mathbf{x}, \mathbf{e}) = \mathbf{B}^*(\mathbf{x}, \mathbf{e})\mathbf{B}(\mathbf{x}, \mathbf{e})$, and

$$\mathbf{B}(\mathbf{x}, \mathbf{e}) = \begin{pmatrix} \mathbf{x}_1(0) - \mathbf{e}_1(0) & \cdots & \mathbf{x}_{M_t}(0) - \mathbf{e}_{M_t}(0) \\ \vdots & \vdots & \vdots \\ \mathbf{x}_1(N-1) - \mathbf{e}_1(N-1) & \cdots & \mathbf{x}_{M_t}(N-1) - \mathbf{e}_{M_t}(N-1) \end{pmatrix}. \tag{9.19}$$

If q denotes the rank of $\mathbf{A}(\mathbf{x}, \mathbf{e})$, (i.e., the number of nonzero eigenvalues) then we can bound equation 9.18 as

$$P(\mathbf{x} \to \mathbf{e}) \leq \left[\prod_{n=1}^{q} \lambda_n \right]^{-M_r} \left(\frac{E_s}{4N_0} \right)^{-qM_r}. \tag{9.20}$$

We define the notion of diversity order as follows.

Definition 9.2

A coding scheme which has an average error probability $\bar{P}_e(SNR)$ that behaves as

$$\lim_{SNR \to \infty} \frac{\log(\bar{P}_e(SNR))}{\log(SNR)} = -d \tag{9.21}$$

as a function of SNR is said to have a diversity order of d.

In words, a scheme with diversity order d has an error probability at high SNR behaving as $\bar{P}_e(SNR) \approx SNR^{-d}$ (see fig. 9.6). One reason to focus on such a behavior for the error probability can be seen from the following intuitive argument for a simple scalar fading channel ($M_t = 1 = M_r$). It is well known that for particular frame b, the error probability for binary transmission, conditioned on the channel realization $h^{(b)}$, is given by $P_e(h^{(b)}) = Q\left(\sqrt{2SNR} \, |h^{(b)}| \right)$ (Proakis, 1995). Hence if $|h^{(b)}|\sqrt{2SNR} \gg 1$, then $P_e(h^{(b)}) \approx 0$, and if $|h^{(b)}|\sqrt{2SNR} \ll 1$, then $P_e(h^{(b)}) \approx \frac{1}{2}$. Therefore a frame is in error with high probability when the channel gain $|h^{(b)}|^2 \ll \frac{1}{SNR}$, i.e., when the channel is in a "deep fade." Therefore the average error probability is well approximated by the probability that $|h^{(b)}|^2 \ll \frac{1}{SNR}$. For high SNR we can show that, for $h \sim \mathbb{C}\mathcal{N}(0,1)$, $P\left\{|h|^2 < \frac{1}{SNR}\right\} \approx \frac{1}{SNR}$, and this explains the behavior of the average error probability. Although this is a crude analysis, it brings out the most important difference between the additive white Gaussian noise (AWGN) channel and the fading channel. The typical way in which an error occurs in a fading channel is due to channel failure, i.e., when the channel gain $|h|$ is very small, less than $\frac{1}{SNR}$. On the other hand, in an AWGN channel errors occur when the noise is large, and since the noise is Gaussian it has an exponential tail, causing this to be very unlikely at high SNR.

Given the definition 9.2 of diversity order, we see that the diversity order in equation 9.20 is at most qM_r. Moreover, in inequality 9.20 we notice that we also obtain a coding gain of $(\prod_{n=1}^{q} \lambda_n)^{1/q}$.

Note that in order to obtain the average error probability, one can calculate a naive union bound using the pairwise error probability given in equation 9.20 but this may not be tight. A more careful upper bound for the error probability can be derived (Zheng and Tse, 2003). However, if we ensure that *every* pair of codewords satisfies the diversity order in equation 9.20, then clearly the average error probability satisfies it as well. This is true when the transmission rate is held constant with respect to SNR, i.e., a fixed-rate code. Therefore, in the case of fixed rate code design the simple pairwise error probability given in equation 9.20 is sufficient to obtain the correct diversity order.

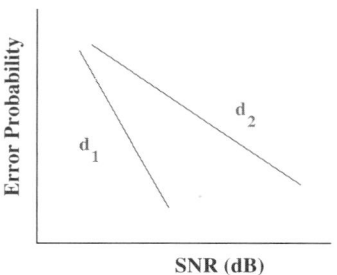

Figure 9.6   Relationship between error probability and diversity order.

In order to design practical codes that achieve a performance target we need to glean insights from the analysis to state design criteria. For example, in the flat-fading case of equation 9.20 we can state the following rank and determinant design criteria.

Design criteria for space-time codes over flat-fading channels (Tarokh et al., 1998):

Rank criterion: In order to achieve maximum diversity $M_t M_r$, the matrix $\mathbf{B}(\mathbf{x}, \mathbf{e})$ from equation 9.19 has to be full rank for any codewords $\mathbf{x}, \mathbf{e}$. If the minimum rank of $\mathbf{B}(\mathbf{x}, \mathbf{e})$ over all pairs of distinct codewords is q, then a diversity order of $q M_r$ is achieved.

Determinant criterion: For a given diversity order target of q, maximize $(\prod_{n=1}^{q} \lambda_n)^{1/q}$ over all pairs of distinct codewords.

Over the past few years, there have been significant developments in designing codes which can guarantee a given reliability (error probability). An exhaustive listing of all these developments is beyond the scope of this chapter, but we give a glimpse of the recent developments. The interested reader is referred to Diggavi et al. (2004b), and references therein.

Pioneering work on trellis codes for Gaussian channels was done in Ungerboeck (1982). In Tarokh et al. (1998), the first space-time trellis code constructions were presented. In this seminal work, trellis codes were carefully designed to meet the design criteria for minimizing error probability. In parallel a very simple coding idea for $M_t = 2$ was developed in Alamouti (1998). This code achieved maximal diversity order of $2M_r$ and had a very simple decoder associated with it. The elegance and simplicity of the Alamouti code has made it a candidate for next generation of wireless systems which are slated to utilize space-time codes. The basic idea of the Alamouti code was extended to orthogonal designs in Tarokh et al. (1999). The publication of Tarokh et al. (1998) and Alamouti (1998), created a significant community of researchers working on space-time code constructions. Over the past few years, there has been significant progress in the construction of space-time codes for coherent channels. The design of codes that are linear in the complex field was proposed in Hassibi and Hochwald (2002), and efficient decoders for such codes were given in Damen et al. (2000). Codes based on algebraic rotations and number-theoretic tools are developed in El-Gamal and Damen (2003) and Sethuraman et al. (2003). A common assumption in all these designs was that the receiver had perfect knowledge of the channel. Techniques based on channel estimation and the evaluation of the degradation in performance for space-time trellis codes was examined in Naguib et al. (1998). In another line of work, non-coherent space-time codes were proposed in Hochwald and Marzetta (2000). This also led to the design and analysis of differential space-time codes for flat fading channels (Hochwald and Sweldens, 2000; Hughes, 2000; Tarokh and Jafarkhani, 2000). This was also examined for frequency selective channels in Diggavi et al. (2002a).

As can be seen, the topic of space-time codes is still evolving and we just have a snapshot of the recent developments.

9.3.3   Rate-Diversity Tradeoff

A natural question that arises is how many codewords can we have which allow us to attain a certain diversity order. For a flat Rayleigh fading channel, this has been examined (Lu and Kumar, 2003; Tarokh et al., 1998) and the following result was obtained.[10]

Theorem 9.6
If we use a transmit signal with constellation of size $|\mathcal{S}|$ and the diversity order of the system is qM_r, then the rate R that can be achieved is bounded as

$$R \leq (M_t - q + 1) \log_2 |\mathcal{S}| \tag{9.22}$$

in bits per transmission.

■

One consequence of this result is that for maximum $(M_t M_r)$ diversity order we can transmit at most $\log_2 |\mathcal{S}|$ bits/sec/Hz. Note that the trade-off in theorem 9.6 is established with a constraint on the alphabet size of the transmit signal, which may not be fundamental from an information-theoretic point of view. An alternate viewpoint of the rate-diversity trade-off has been explored in Zheng and Tse (2003) from a Shannon-theoretic point of view. In that work the authors are interested in the multiplexing rate of a transmission scheme.

Definition 9.3
A coding scheme which has a transmission rate of $R(SNR)$ as a function of SNR is said to have a multiplexing rate r if

$$\lim_{SNR \to \infty} \frac{R(SNR)}{\log(SNR)} = r. \tag{9.23}$$

Therefore, the system has a rate of $r \log(SNR)$ at high SNR. One way to contrast this with the statement in theorem 9.6 is to note that the constellation size is also allowed to become larger with SNR. The naive union bound of the pairwise

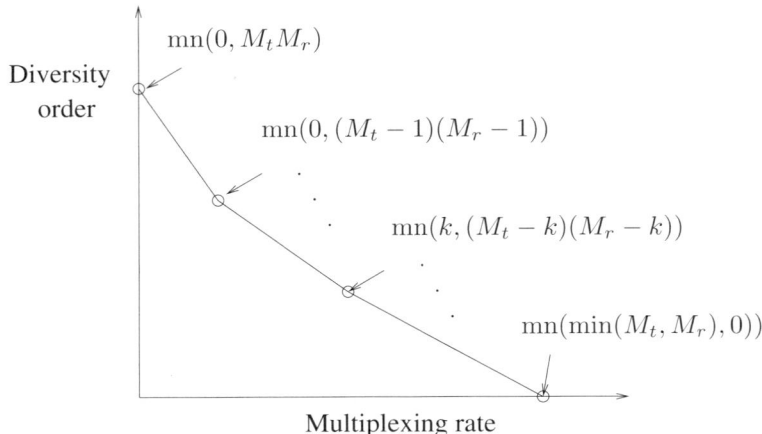

Figure 9.7 Rate-diversity trade-off curve.

error probability (eq. 9.18) has to be used with care if the constellation size is also increasing with SNR. There is a trade-off between the achievable diversity and the multiplexing gain, and $d^*(r)$ is defined as the supremum of the diversity gain achievable by *any* scheme with multiplexing gain r. The main result in Zheng and Tse (2003) states the following.

Theorem 9.7
For $T > M_t + M_r - 1$, and $K = \min(M_t, M_r)$, the optimal trade-off curve $d^*(r)$ is given by the piecewise linear function connecting points in $(k, d^*(k)), k = 0, \ldots, K$ where

$$d^*(k) = (M_r - k)(M_t - k). \tag{9.24}$$

∎

If $r = k$ is an integer, the result can be notionally interpreted as using $M_r - k$ receive antennas and $M_t - k$ transmit antennas to provide diversity while using k antennas to provide the multiplexing gain. However, this interpretation is not physical but really an intuitive explanation of the result in theorem 9.7. Clearly this result means that one can get large rates which grow with SNR if we reduce the diversity order from the maximum achievable. This diversity-multiplexing trade-off implies that a high multiplexing gain comes at the price of decreased diversity gain and is a manifestation of a corresponding trade-off between error probability and rate. This trade-off is depicted in fig. 9.7. Therefore, as illustrated in Theorems 9.6 and 9.7, the trade-off between diversity and rate is an important consideration both in terms of coding techniques (theorem 9.6) and in terms of Shannon theory (theorem 9.7).

A different question was proposed in Diggavi et al. (2003, 2004a), where it was asked whether there exists a strategy that combines high-rate communications with high reliability (diversity). Clearly the overall code will still be governed by the rate-diversity trade-off, but the idea is to ensure the reliability (diversity) of at least part of the total information. This allows a form of communication where the high-rate code opportunistically takes advantage of good channel realizations whereas the embedded high-diversity code ensures that at least part of the information is received reliably. In this case, the interest was not in a single pair of multiplexing rate and diversity order (r, d), but in a tuple (r_a, d_a, r_b, d_b) where rate r_a and diversity order d_a was ensured for part of the information with rate-diversity pair (r_b, d_b) guaranteed for the other part. A class of space-time codes with such desired characteristics have been constructed in Diggavi et al. (2003, 2004a).

From an information-theoretic point of view, Diggavi and Tse (2004) focused on the case when there is one degree of freedom (i.e., $\min(M_t, M_r) = 1$). In that case if we consider $d_a \geq d_b$ without loss of generality, the following result was established (Diggavi and Tse, 2004):

Theorem 9.8
When $\min(M_t, M_r) = 1$, then the diversity-multiplexing trade-off curve is successively refinable, i.e., for any multiplexing gains r_a and r_b such that $r_a + r_b \leq 1$, the diversity orders $d_a \geq d_b$,

$$d_a = d^*(r_a), \;\; d_b = d^*(r_a + r_b), \tag{9.25}$$

are achievable, where $d^*(r)$ is the optimal diversity order given in theorem 9.7.

∎

Since the overall code has to still be governed by the rate-diversity trade-off given in theorem 9.7, it is clear that the trivial outer bound to the problem is that $d_a \leq d^*(r_a)$ and $d_b \leq d^*(r_a + r_b)$. Hence theorem 9.3 shows that the best possible performance can be achieved. This means that for $\min(M_t, M_r) = 1$, we can design ideal *opportunistic* codes. This new direction of enquiry is being currently explored.

9.4   Multi-user Diversity

In section 9.3, we explored the importance of using many fading realizations through multiple antennas for reliable, high-rate, single-user wireless communication. In this section we explore another form of diversity where we can view different users as a form of *multi-user diversity*. This is because each user potentially has independent channel conditions and local interference environment. This implies that in fig. 9.5, the fading links between users are random and independent of each other. Therefore, this diversity in channel and interference conditions can be exploited by treating the independent links from *different* users as conduits for information transfer.

In order to explore this idea further we first digress to discuss communication topologies. As seen in section 9.2 (see fig. 9.5), we can view the n-user communication network through the underlying graph $\mathcal{G}_C$. One topology which is very commonly seen in practice is obtained by giving special status to one of the nodes as the base-station or access point. The other nodes can *only* communicate to the base station. We call such a topology the *hierarchical communication topology* (see fig. 9.5). An alternate topology that has emerged more recently is when the nodes organize themselves without a centralized base station. Such a topology is called an *ad hoc communication topology*, where the nodes relay information from source to destination, typically through multiple "nearest neighbor" communication hops (see also fig. 9.8). In both these topologies there is potential to utilize multi-user diversity, but the methods to do so are distinct. Therefore we explore them separately in Sections 9.4.1 and 9.4.2.

9.4.1 Opportunistic Scheduling

In the hierarchical topology, we distinguish between two types of problems; the first is the *uplink* channel where the nodes communicate to the access point (many-to-one communication or the *multiple access channel*), and the second is the *downlink* channel where the access point communicates to the nodes (one-to-many communication or the *broadcast channel*).

The idea of multi-user diversity can be further motivated by looking at the scalar fading multiple access channel. If the users are distributed across geographical areas, their channel responses will be different depending on their local environments. This is modeled by choosing the users' channels to vary according to channel distributions that are chosen to be independent and identical across users. The rate region for the uplink channel for this case was characterized in Knopp and Humblet (1995) where it was shown that in order to maximize the total information capacity (the sum rate), it is optimal to transmit *only* to the user with the best channel. For the scalar channel, the channel gain determines the best channel. The result (in Knopp and Humblet, 1995) when translated to rapidly fading channels results in a form of *time-division multiple access* (TDMA), where the users are not preassigned time slots, but are scheduled according to their respective channel conditions. Even if a particular user at the current time might be in a deep fade, there could be another user who has good channel conditions. Hence this strategy is a form of *multi-user diversity* where the diversity is viewed across users. Here the multi-user diversity (which arises through independent channel realizations across users) can be harnessed using an appropriate scheduling strategy. If the channels vary rapidly in time, the idea is to schedule users when their channel state is close to the peak rate that it can support. A similar result also holds for the scalar fading broadcast channel (Li and Goldsmith, 2001; Tse, 1997). Note that this requires feedback from the users to the base station about the channel conditions. The feedback could be just the received SNR. These results are proved on the basis of two assumptions. One is that all the users have identically distributed (i.e., symmetric) channels and the other is that we are interested in long-term rates. We focus on the first assumption, and later briefly return to the question about delay.

In wireless networks, the users' channel is almost never symmetric. Nodes that are closer to the base station experience much better channels on the average than nodes that are further away (due to path loss, see section 9.2). Therefore, using a TDMA technique that allows exclusive use of the channel to the best user would be inherently unfair to users who are further away. Suppose the long-term average rate $\{T_k\}$ is to be provided to the users. The criterion used in the result in Knopp and Humblet (1995) was the sum throughput of all the users, i.e., $\max \sum_k T_k$. This criterion can be maximized by only scheduling the nodes with strong channels, and this could be an unfair allocation of resources across users. In order to translate the intuition about multi-user diversity into practice, one would need to ensure fairness among users. The idea in Bender et al. (2000); Jalali et al. (2000) and Chaponniere et al., is to use a *proportionally fair* criterion for scheduling which maximizes $\sum_{k=1}^{K} \log(T_k)$. This idea is inherently used in the downlink scheduling algorithm used in IS-856 (Bender et al., 2000; Chaponniere et al.; Jalali et al., 2000) (also known as the *high data rate*—HDR 1xEV-DO system).

The scheduling algorithm implemented in the 1xEV-DO system keeps track of the average throughput $T_k(t)$ of user k in a past window of length t_c. Let the rate that can be supported to user k at time t be denoted by $R_k(t)$. At time t, the scheduling algorithm transmits to the user with the largest $\frac{R_k(t)}{T_k(t)}$ among the active users. The average throughputs are then updated given the current allocation. Since this idea ensures fairness while utilizing multi-user diversity, it is an instantiation of an *opportunistic scheduler*.

This scheduling algorithm described above relies on the rates supported by the users to vary rapidly in time. But this assumption can be violated when the channels are constant or are very slowly time-varying. In order to artificially induce time variations, Viswanath et al., 2002) propose to use multiple transmit antennas and introduce random phase rotations between the antennas to simulate fast fading. This idea of phase-sweeping for multiple antennas has been also proposed in Weerackody (1993) and Hiroike et al. (1992) in the context of creating time diversity in single-user systems. With such artificially induced fast channel variations, the same scheduling algorithm used in IS-856 (outlined above) inherently captures the multi-user spatial diversity of the network. In Viswanath et al. (2002), this technique is shown to achieve the maximal diversity order (see section 9.3.2) for each user, asymptotically in number of (uniformly distributed) users.

In a heavily loaded system (large number of users) and where there is a uniform distribution of users, the technique proposed in Viswanath et al. (2002) is attractive. However, for lightly loaded systems, *or* when delay is an important QoS criterion, its desirability is less clear. Given that the technique proposed in Viswanath et al. (2002) is based on a rate-based QoS criterion, it cannot provide delay guarantees for the jobs of different users. This motivates the discussion of scheduling algorithms for job-based QoS criteria.

In job-based criteria, the requests are assumed to come in at certain arrival times a_i, and we have information about the size s_i (say in bytes). *Response time* is defined to be $c_i - a_i$ where c_i is the time when a request was fully serviced and a_i is the arrival time of the request. This is a standard QoS criterion for a request.

Relative response is defined as $\frac{c_i - a_i}{s_i}$ (Bender et al., 1998). Relative response was proposed in the context of heterogeneous workloads, such as the Web, i.e., requests for data of different sizes (thus, different s_i). The above criteria relate to guarantees per request; we could also give guarantees only over all requests. For example, the overall performance criterion for a set of jobs could be the l_∞ norm, namely, $\max_i(c_i - a_i)$ (i.e., max response time) or $\max_i \frac{c_i - a_i}{s_i}$ (i.e., max relative response). Other criteria based on average instead of maximum are also studied.

The new generation of wireless networks can support multiple transmission rates depending on the channel conditions. Assuming an accurate communication-theoretic model for the physical layer achievable rates (as described in section 9.3), job-scheduling algorithms are proposed and analyzed for various QoS criteria in Becchetti et al. (2002). These algorithms utilize diverse job requirements of the users to provide provable guarantees in terms of the job-scheduling criteria.

These discussions just illustrate how multi-user diversity can be utilized in hierarchical networks. This form of opportunistic scheduling is an important part of the new generation of wireless data networks.

9.4.2 Mobile Ad Hoc Networks

In an ad hoc communication topology (network), one need not transmit information directly from source to destination, but instead can use other users which act as relays to help communication of information to its ultimate destination. Such multihop wireless networks have rich history (see, for example, Hou and Li, 1986, and references therein).

In an important step toward systematically understanding the capacity of wireless networks, Gupta and Kumar (2000) explored the behavior of wireless networks asymptotically in the number of users. In their setup, n nodes were placed independently and randomly at locations $\{S_i\}$ in a finite geographical area (a scaled unit disk). Also $m = \Theta(n)$ source and destination (S-D) pairs $\{(\mathcal{S}_i, \mathcal{T}_i)\}$ are randomly chosen as shown in fig. 9.8.[11] The model assumes that each source $\mathcal{S}_i$ has an infinite stream of (information) packets to send to its respective destination $\mathcal{T}_i$. The nodes are allowed to use any scheduling and relaying strategy through other nodes to send the packets from the sources to the destinations (see fig. 9.8). The goal is to analyze the best possible long-term throughput per S-D pair asymptotically in the number of nodes n.

In Gupta and Kumar (2000), a single-user communication model was used where each node transmitted information to its intended receiver (relay or destination node), and the receiver considered the interference from other nodes as part of the noise. Therefore in the communication model, a successful transmission of rate R occurred when the signal-to-interference-plus-noise ratio (SINR) was above a certain threshold β. Clearly, such a communication model can be improved by attempting to decode the "interference" from other nodes using sophisticated multi-user decoding (Verdu, 1998). But such a decoding strategy was not considered by Gupta and Kumar (2000) and therefore this need not be an information-

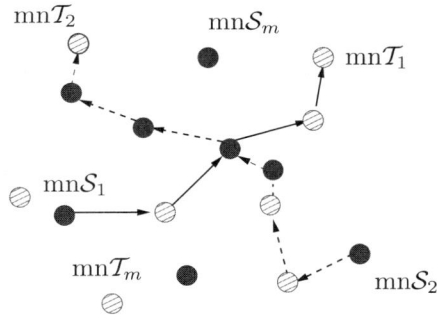

Figure 9.8   Routes from sources $\{\mathcal{S}_i\}$ denoted by filled circles to destinations $\{\mathcal{T}_i\}$ denoted by shaded circles.

theoretically optimal strategy. In order to represent wireless signal transmission, the signal strength variation was modeled only through path loss (see section 9.2) with exponent α. Therefore, if $\{P_i\}$ are the powers at which the various nodes transmitted, then the SINR from node i to node j is defined as

$$SINR = \frac{\frac{P_i}{|S_i - S_j|^\alpha}}{\sigma^2 + \sum_{\substack{k \in \mathcal{I} \\ k \neq i}} \frac{P_k}{|S_k - S_j|^\alpha}},$$ (9.26)

where $\mathcal{I}$ is the subset of users simultaneously transmitting at some time instant. Next, we need to define the notion of throughput per S-D pair more precisely.

Definition 9.4
For a scheduling and relay policy π, let $M_i^\pi(t)$ be the number of packets from source node $\mathcal{S}_i$ to its destination node $\mathcal{T}_i$ successfully delivered at time t. A long-term throughput $\tilde{\lambda}(n)$ is feasible if there exists a policy π such that for *every* source-destination pair

$$\lim \inf_{T \to \infty} \frac{1}{T} \sum_{t=1}^{T} M_i^\pi(t) \geq \tilde{\lambda}(n) \,.$$ (9.27)

We define the throughput $\lambda(n)$ as the highest achievable $\tilde{\lambda}(n)$.

∎

Note that $\lambda(n)$ is a random quantity which depends on the node locations of the users. Our interest is in the *scaling law* governing $\lambda(n)$, i.e., the behavior of $\lambda(n)$ asymptotically in n. One of the main results of Gupta and Kumar (2000) was the following.

Theorem 9.9
There exist constants c_1 and c_2 such that

$$\lim_{n \to \infty} \mathrm{P}\left\{\lambda(n) = \frac{c_1 R}{\sqrt{n \log n}} \text{ is feasible}\right\} = 1, \quad \lim_{n \to \infty} \mathrm{P}\left\{\lambda(n) = \frac{c_2 R}{\sqrt{n}} \text{ is feasible}\right\} = 0 \,.$$

∎

Therefore, the long-term per-user throughput decays as $O(\frac{1}{\sqrt{n}})$, showing that high per-user throughput may be difficult to attain in large-scale (fixed) wireless networks. This result has been recently strengthened: it was shown by Franceschetti et al. (2004) that $\lambda(n) = \Theta(\frac{1}{\sqrt{n}})$.

One way to interpret this result is the following. If n nodes are randomly placed in a unit disk, nearest neighbors (with high probability) are at a distance $O(\frac{1}{\sqrt{n}})$ apart. Gupta and Kumar (2000) show that it is important to schedule a large number of simultaneous short transmissions, i.e., between nearest-neighbors. If randomly chosen source-destination pairs are $O(1)$ distance apart and we can only schedule nearest neighbor transmissions, information has to travel $O(\sqrt{n})$ hops to reach its destination. Since there can be at most $O(n)$ simultaneous transmissions at a given time instant, this imposes a $O(\frac{1}{\sqrt{n}})$ upper bound on such a strategy. This is an intuitive argument, and a rigorous proof of theorem 9.9 is given in Gupta and Kumar (2000) among other interesting results.

Note that the coding strategy in theorem 9.9 was simple and the interference was treated as part of the noise. An open question concerns the throughput when we use sophisticated multi-user codes and decoding is used. Therefore, for such an information-theoretic characterization, understanding the rate region of the relay channel is an important component (Cover and Thomas, 1991). The relay channel was introduced in van der Meulen (1977), and the rate region for special cases was presented in Cover and El Gamal (1979). Recently Leveque and Telatar (2005); Xie and Kumar (2004), and Gupta and Kumar (2003) have established that even with network information-theoretic coding strategies, the per S-D pair throughput scaling law decays with the number of users n.

A natural question that arises is whether there is any mechanism by which one can improve the scaling law for throughput in wireless networks. Mobility was one such mechanism examined in Grossglauser and Tse (2002). In the model studied, random node mobility was allowed and the locations $\{S_i(t)\}$ vary in a uniform, stationary, and ergodic manner over the entire disk (see fig. 9.9).

In the presence of such symmetric (among users) and "space-filling" mobility patterns, the following surprising result was established in (Grossglauser and Tse, 2002).

Theorem 9.10

There exists a scheduling and relaying policy π and a constant $c > 0$ such that

$$\lim_{n \to \infty} \mathrm{P} \left\{ \lambda(n) = cR \text{ is feasible} \right\} = 1 . \tag{9.28}$$

∎

Therefore, node mobility allows us to achieve a per-user throughput of $\Theta(1)$. The main reason this was attainable was that packets are relayed only through a finite number of hops by utilizing node mobility. Thus, a node carries packets over $O(1)$ distance before relaying it, and therefore Grossglauser and Tse (2002) shows that, with high probability, if the mobility patterns are space-filling, the number of hops needed from source to destination is bounded instead of growing as $O(\sqrt{n})$ in

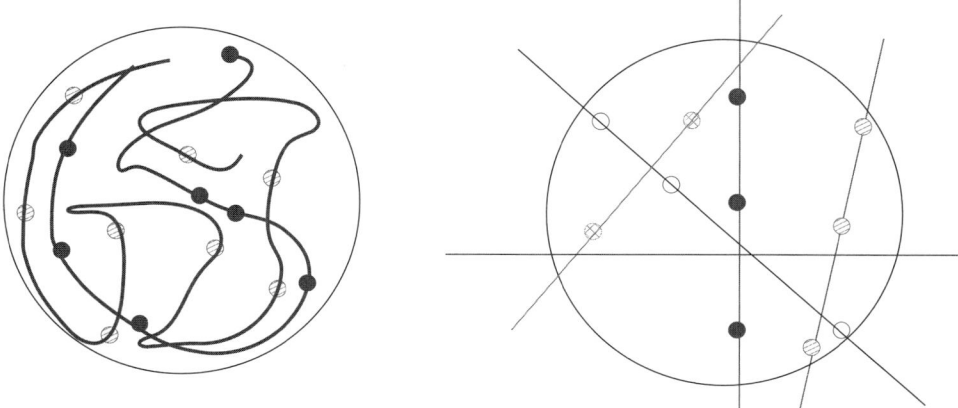

Figure 9.9 Mobility in ad hoc networks. The figure on the left shows a space-filling mobility model where the nodes uniformly cover the region. The figure on the right shows a limited one-dimensional mobility model where nodes move along *fixed* line segments.

the case of fixed (nonmobile) wireless networks (Gupta and Kumar, 2000). However, the above mobility model is a generous one, since (1) it is homogeneous, i.e., every node has the same mobility process, and (2) the sample path of each node "fills the space over time." This means that there is a nonzero probability that the node visits every part of the geographical region or area. A natural question is whether the throughput result in Grossglauser and Tse (2002) strongly depends on these two features of the mobility model.

In Diggavi et al. (2002b), a different mobility model is introduced which embodies two salient features that many real mobility processes seem to possess (e.g., cars traveling on roads, people walking in buildings or cities, trains, satellites circling earth), which are not captured by the model in Grossglauser and Tse (2002). First, an individual node typically visits only a small portion of the entire space, and rarely leaves this preferred region. Second, the nodes do move frequently within their preferred regions, and an individual region often covers a large distance. As an extreme abstraction of such mobility processes, Diggavi et al. (2002b) studied mobility patterns where nodes move along a given set of one-dimensional paths (see fig. 9.9). In particular, the mobility patterns were restricted to random line segments and once chosen, the configuration of line segments are fixed for all time. Therefore, given the configuration, the only randomness arose through user mobility along these line segments. In order to isolate the effects of one-dimensional mobility from edge effects, Diggavi et al. (2002b) studied a model in which the nodes are on a unit sphere but each node is constrained to move on a single-dimensional great circle. Therefore, a configuration in this case was a set of line segments (great circles) which were fixed throughout the communication period, and the nodes moved in randomly only on these one-dimensional paths. Thus, the homogeneity assumption in Grossglauser and Tse (2002) is now relaxed. In particular, there can be pairs of nodes that are far more likely to be in close proximity to each other than other pairs. For example, if two one-dimensional paths nearly overlap, the probability of close encounter between the nodes is significantly larger than for two paths that are "far apart." This lack of homogeneity implies, as shown in Diggavi et al. (2002b), that there are configurations where constant throughput is unattainable even with mobility.

Since the capacity of such a mobile ad hoc network then depends on the constellation of one-dimensional paths, the question becomes one of scaling laws for a *random* configuration. Therefore, the configurations themselves are chosen randomly with each one-dimensional path (great circle) chosen independently and with an identical uniform distribution. Given such a random configuration, the question then becomes whether "bad" configurations (where the per S-D pair throughput is not $\Theta(1)$) occur often. One of the key ideas in Diggavi et al. (2002b) was the identification and proof of *typical* ("good") configurations, on which the average long-term throughput per node is $\Theta(1)$. Intuitively the typical configurations defined in Diggavi et al. (2002b) are those where the fraction of one-dimensional paths intersecting any given area is *uniformly* close to its expected number. That is, the empirical probability counts are *uniformly* close to the underlying probability of a random one-dimensional path intersecting that area. Therefore, even for a particular deterministically chosen configuration which satisfies the typicality condition, the per S-D pair throughput is $\Theta(1)$. One of the main results in Diggavi et al. (2002b) is that if the one-dimensional paths are chosen (uniformly) randomly and independently, then for almost all constellations of such paths, the throughput per S-D pair is $\Theta(1)$. Therefore, for random configurations the probability of an *atypical* configuration is shown to go to zero asymptotically in network size n. Thus, although each node is restricted to move in a one-dimensional space, the same asymptotic performance is achieved as in the case when they can move in the entire two-dimensional region.

Theorem 9.11
Given a configuration $\mathcal{C}$, there exists a scheduling and relaying policy π and a constant $c > 0$ such that

$$\lim_{n \to \infty} \mathrm{P}\left\{\lambda(n) = cR \text{ is feasible } | \mathcal{C}\right\} = 1 \qquad (9.29)$$

for almost all configurations $\mathcal{C}$ as $n \to \infty$, i.e., the probability of the set of configurations for which the policy achieves a throughput of λ goes to 1 as $n \to \infty$.

∎

Next we give a flavor of the proof techniques used to prove theorem 9.11. First, we examine a relaying strategy where at each time, every node carries *source packets*, which originate from that node, and *relay packets*, which originated from other nodes and are to be forwarded to their final destinations. In phase I, each sender attempts to transmit a source packet to its nearest receiver, who

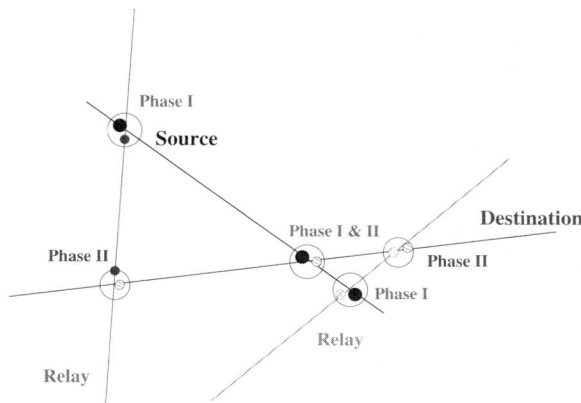

Figure 9.10 The relaying strategy for mobile nodes. In phase I, the source attempts to transfer packets to relays. During phase II, the relays attempt to transfer packets to the destination.

will serve as a relay for that packet. In phase II, each sender identifies its nearest receiver and attempts to transmit a relay packet destined for it, if the sender has one (see fig. 9.10). As in equation 9.26, a successful transmission of rate R occurs when the signal-to-interference-plus-noise ratio (SINR) is above a certain threshold β. Note that it can be shown that if the source nodes attempt to "wait" till it encounters its destination, the per S-D pair throughput cannot be $\Theta(1)$. Therefore every source spreads its traffic to random intermediate nodes depending on the mobility. Moreover, each packet is forwarded successfully to *only one* relay, i.e., there is no duplication. Mobility allows source-destination pairs to be able to relay information through several independent relay paths, since nodes have changing nearest neighbors due to mobility. This method of relaying information through independent attenuation links which vary over time is also a form of *multi-user diversity*. One can see this by observing that the transmission occurs over several realizations of the communication graph $\mathcal{G}_C$. The relaying strategy which utilizes mobility schedules transmissions over appropriate realizations of the graph. Conceptually, this use of independent relays to transmit information from source to destination is illustrated in fig. 9.11, where the strategy of Theorems 9.10 and 9.11 is used.

Intuitively, if the source is able to *uniformly* spread its traffic through each of its relays (see fig. 9.11) then we can expect to obtain $\Theta(1)$ throughput per S-D pair. In order for this to occur, we need to show two properties:

1. Every node spends the same order of time as the nearest neighbor to $\Theta(n)$ other nodes. This ensures that each source can spread its packets uniformly across $\Theta(n)$ other nodes, all acting as relays, and these packets can in turn be merged back into their respective final destinations.

2. When communicating with the nearest neighbor receiver, the capture probability is not vanishingly small even in a large system, even though there are $\Theta(n)$ interfering nodes transmitting simultaneously.

However, with one-dimensional mobility, it is shown in Diggavi et al. (2002b) that there exist configurations where these properties cannot be satisfied. This is where the identification of *typical configurations* becomes important. For typical configurations through a detailed technical argument it is shown in Diggavi et al. (2002b) that these properties hold. Moreover, for randomly chosen configurations, it is shown that such typical configurations occur with probability going to 1 asymptotically in n. Therefore, using these components, the proof of theorem 9.11 is completed.

throughput-delay   There is a dramatic gain in the per S-D pair throughput in theorems 9.10 and
trade-off   9.11 over theorem 9.9 from $O(\frac{1}{\sqrt{n}})$ to $\Theta(1)$. A natural question to ask is whether there is a cost to this improvement. The results in theorems 9.10 and 9.11 utilized node mobility to deliver the information from source to destination. Therefore, the timescale over which this is effective is dependent on the velocity of the nodes, which determines the rate of change of the topology. Hence we can expect there to be significantly larger packet delays for this scheme as compared to the fixed network. In some sense, the Gupta-Kumar result in theorem 9.9 has a smaller throughput, but also has a smaller packet delay, since the delays depend on successful packet transmissions over the route and not the change in node topology. Hence a natural question to ask is whether there exists a fundamental trade-off between delay and throughput in ad hoc networks. This question was recently studied in El Gamal et al. (2004), where the authors quantified this trade-off.

In order to quantify the trade-off there needs to be a formal definition of delay. In El Gamal et al. (2004) delay $D(n)$ is defined as the sum of the times spent in every relay node. This definition does not include the queueing delay at the nodes, just the delay incurred in successful transmission of the packet on each single hop of the route. Given this definition of delay, El Gamal et al. (2004) established that for a fixed random network of n nodes, the delay-throughput trade-off for $\lambda(n) = O(1/\sqrt{n \log(n)})$ is $D(n) = \Theta(n\lambda(n))$. For a mobile ad hoc network, when $\lambda(n) = \Theta(1)$, El Gamal et al. (2004) showed that $D(n) = \Theta(\frac{\sqrt{n}}{v(n)})$, where $v(n)$ is the velocity of the mobile nodes. Therefore, this quantifies the cost of higher throughput in mobile networks.

The theoretical developments in sections 9.4.1 and 9.4.2 indicate the strong interactions between the physical layer coding schemes and channel conditions and the networking issues of resource allocation and application design. This is an important insight we can draw for the design of wireless networks. Therefore, several problems which are traditionally considered as networking issues and are typically designed independent of the transmission techniques need to be reexamined in the context of wireless networks. As illustrated, diversity needs to be taken into account while solving these problems. Such an integrated approach is a major lesson learned from the theoretical considerations, and we develop another aspect of this through the study of source coding using route diversity in section 9.5.

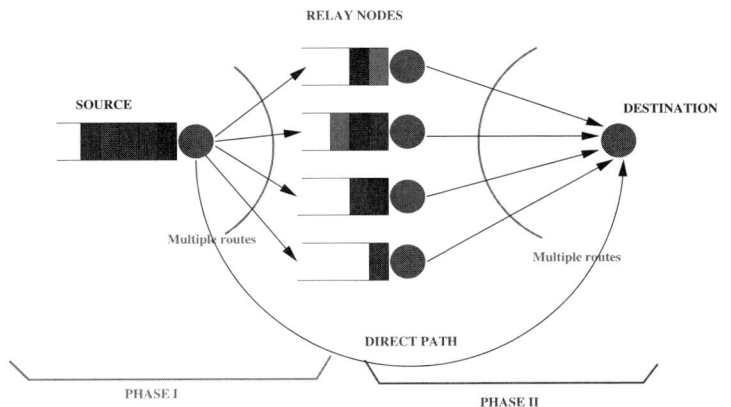

Figure 9.11   Multi-user diversity through relays.

9.5   Route Diversity

The interest in section 9.4.2 was the characterization of *long-term* throughput from source to destination. However, in applications such as sensor networks (see, for example, Pottie and Kaiser, 2000; Pradhan et al., 2002, and references therein), there could be node failures which lead to routes being disconnected through a transmission period. This might become particularly crucial when there are strong delay constraints, such as those in real-time data delivery. Such route failures can also occur in ad hoc networks (discussed in section 9.4.2) as well as in wired networks. In multihop relay strategies, we could utilize the existence of multiple routes from source to destination in order to increase the probability of successfully receiving the information at the destination within delay constraints despite route (path) failures. This is a form of *route diversity* (see fig. 9.12) and was first suggested by Maxemchuk (1975) in the context of wired networks. Note that in a broad sense, the multi-user diversity studied in mobile ad hoc networks in section 9.4.2 also utilizes the presence of multiple routes from source to destination. However, in that case the multiple routes were utilized to increase the long-term per S-D pair throughput. In the topic of this section we will utilize the multiple routes for low-delay applications.

We will examine this problem in the context of delivering a real-time source (like speech, images, video, etc.) with tight delay constraints. If the same information about the source is transmitted over both routes, then this is a form of repetition coding. However, when both routes are successful, there is no performance advantage. Perhaps a more sophisticated technique would be to send correlated descriptions of the source in the two routes such that each description is individually good, but they are different from one another so that if both routes are successful one gets a better approximation of the source. This is the basic idea behind *multiple description* (MD) source coding (El Gamal and Cover, 1982). This notion can be

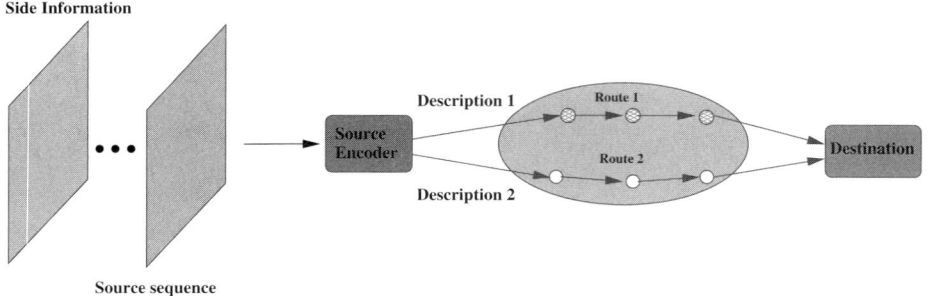

Figure 9.12   Route diversity.

extended to more than two descriptions as well, but in this section we will focus on the two-description case for simplicity. The idea is that the source is coded through several descriptions, where we require that performance (distortion) guarantees can be given to any subset of the descriptions and the descriptions mutually refine each other. This is the topic discussed in sections 9.5.1 and 9.5.2.

In a packet-based network such as the Internet, packet losses are inevitable due to congestion or transmission errors. If the data does not have stringent delay constraints, error recovery methods typically ensure reliability either through a repeat request protocol or through forward error correction (Keshav, 1997). Another technique is through scalable (or layered) coding techniques which send a lower-rate base layer or coarser description of the source and send refinement layers to enhance the description. Such a technique is again dependent on reliable delivery of the base layer, and if the base layer is lost, the enhancement layers are of no use to the receiver. Therefore, such layered techniques are again inherently susceptible to route failures. These arguments reemphasize the need to develop multiple description (MD) source coding schemes. Note that the layered coding schemes form a special case of such as MD coding scheme, where guarantees of performance are not given for individual layers, but the layers refine the coarser description of the source.

An important application for future wireless networks could be real-time video. There has been significant research into robust video coding in the presence of packet errors (Reibman and Sun, 2000). The main problem that arises in video is that the compression schemes typically have motion compensation, which introduces memory into the coded stream. Therefore, decoding the current video frame requires the availability of previous video frames. If previous frames are corrupted or lost, the decoder is required to develop methods to conceal such errors. This is an active research topic especially in the context of wireless channels (Girod and Farber, 2000). However, an appealing approach to this problem might be through route diversity and MD coding, and this is briefly discussed in section 9.5.1.

9.5.1   Multiple Description (MD) Source Coding

In order to formalize the requirement of the MD source coder, we study the setup shown in fig. 9.13. As mentioned earlier, we will illustrate the ideas using only the two-description MD problem. Given a source sequence $\{X(k)\}$, we want to design an encoder that sends two descriptions at rate R_1 and R_2 over the two routes such that we get *guaranteed approximations* of the source when either route fails, or when both succeed. In section 9.5.2 we develop techniques that achieve such an objective. In order to understand the fundamental bounds on the performance of such techniques, we need to examine the problem from an information-theoretic point of view. The main tool to do this is given in *rate-distortion* theory (Cover and Thomas, 1991). This theory describes fundamental limits of the trade-off between the rate of the representation of a source and the quality of the approximation. Not surprisingly, the origins of this theory are in Shannon (1948, 1958b). In order to give some of the basic ideas, we first make a short digression on the rudiments of this theory.

Given a source sequence $X^T = \{X(1), \dots, X(T)\}$ from a given alphabet $\mathcal{X}$, the *source encoder* needs to describe it using R bits per source sample (i.e., with a total of RT bits for the sequence). Equivalently we map the source to the index set $\mathcal{J} = \{1, \dots, 2^{RT}\}$. The goal is that given this description a decoder is able to *approximately* reconstruct the source sequence by the sequence $\hat{X}^T = \{\hat{X}(1), \dots, \hat{X}(T)\}$. This is accomplished by constructing a function $f : \mathcal{J} \to \hat{\mathcal{X}}^T$, and $\hat{\mathcal{X}}$ is the alphabet over which the reconstruction is done. Common examples for the alphabet are $\mathcal{X} = \mathbf{R} = \hat{\mathcal{X}}$, or the binary field. The *distortion measure* $\tilde{d}(X^T, \hat{X}^T)$ quantifies the quality of the approximation between the reconstructed and original source sequence. Typically, the distortion measure is a single-letter function constructed as

$$\tilde{d}(X^T, \hat{X}^T) = \frac{1}{T} \sum_{i=1}^{T} d(X(i), \hat{X}(i)), \tag{9.30}$$

where $d(X, \hat{X})$ denotes the quality of the approximation for each sample. Common examples are $d(X, \hat{X}) = |X - \hat{X}|^2$ and Hamming distance (Cover and Thomas, 1991).

The simplest framework to give performance bounds is to analyze the performance of a source encoder for an independent and identically distributed random source sequence. Typically, the interest is in the *average distortion* over the set of input sequences, for the given probability distribution associated with the source sequence. Therefore, the average distortion is $\mathrm{E}[\tilde{d}(X^T, \hat{X}^T)]$, and the problem becomes one of quantifying the smallest rate R that be used to describe the source with average fidelity D, asymptotically in the block length T. This is called the rate-distortion function $R(D)$ and can be given an operational meaning by proving that there exist source codes that can achieve this fundamental bound (Cover and

(margin note: rate-distortion function)

Thomas, 1991). The central result in single source rate-distortion theory is that $R(D)$ is characterized as

$$R(D) = \min_{p(\hat{x}|x):\mathrm{E}[d(x,\hat{x})] \le D} I(X; \hat{X}), \tag{9.31}$$

where, as before, $I(X; \hat{X})$ represents the mutual information between X and $\hat{X}$ (Cover and Thomas, 1991). A simple instantiation of this result is the special case where we want $D = 0$, i.e., the lossless case. In this case, one can see that $R(0) = H(X)$, where $H(X)$ is the entropy of the source. Another important special case is when the source sequence comes from a Gaussian distribution, $X \sim \mathcal{N}(0, \sigma_x^2)$, and we are interested in the squared error distortion metric, i.e., $d(X, \hat{X}) = |X - \hat{X}|^2$. In this case, equation 9.31 evaluates to $R(D) = \frac{1}{2} \log \frac{\sigma_x^2}{D}$ for $D \le \sigma_x^2$ and zero otherwise. Another way of writing this is in terms of the distortion-rate function $D(R)$, which characterizes the smallest distortion achievable for a given rate. In the Gaussian case we see that $D(R) = \sigma_x^2 2^{-2R}$. We will interchangeably consider these two quantities.

The result in equation 9.31 guarantees only that the average distortion does not exceed D. However, under some regularity conditions, the rate-distortion function remains the same even when we require that the probability of the distortion $\tilde{d}(X^T, \hat{X}^T)$ exceeding D to go to zero (Berger, 1977; Cover and Thomas, 1991). The characterization of the rate-distortion function given in equation 9.31 has also been extended in many other ways including sources with memory (Cover and Thomas, 1991).

Armed with this background, we can now formulate the question on the fundamental rate-distortion bounds on multiple description (MD) source coding. The *multiple description* source encoder needs to produce two descriptions of the source using R_1, R_2 bits per source sample respectively. We can formally describe the problem by requiring that the reconstructions $\{\hat{X}_1(k)\}, \{\hat{X}_2(k)\}, \{\hat{X}_{12}(k)\}$ use these descriptions to approximately reconstruct the source (see fig. 9.13). As in the "single description" case, we accomplish this by constructing functions

$$f_1 : \mathcal{J}_1 \to \hat{\mathcal{X}}^T, \ \ f_2 : \mathcal{J}_2 \to \hat{\mathcal{X}}^T, \ \ f_{12} : \mathcal{J}_1 \times \mathcal{J}_2 \longrightarrow \hat{\mathcal{X}}^T, \tag{9.32}$$

where $\mathcal{J}_i = \{1, \dots, 2^{R_i T}\}, i = 1, 2$, and $\hat{\mathcal{X}}$ is the alphabet over which the reconstruction is done. We want the approximations to give average fidelity guarantees of

$$\mathrm{E}[\tilde{d}(X^T, \hat{X}_1^T)] \le D_1, \ \ \mathrm{E}[\tilde{d}(X^T, \hat{X}_2^T)] \le D_2, \ \ \mathrm{E}[\tilde{d}(X^T, \hat{X}_{12}^T)] \le D_{12}. \tag{9.33}$$

The rate-distortion question in this context is to characterize the bounds on the tuple $(R_1, D_1, R_2, D_2, D_{12})$. Therefore, we are interested in characterizing the achievable *rate-distortion region* described by the tuple $(R_1, D_1, R_2, D_2, D_{12})$. As can be seen, this seems like a much more difficult question than the single-description problem for which there is a complete characterization. As a matter of fact, the complete characterization of the MD rate region is still an open question.

This problem was formalized in 1979, and in El Gamal and Cover (1982), a

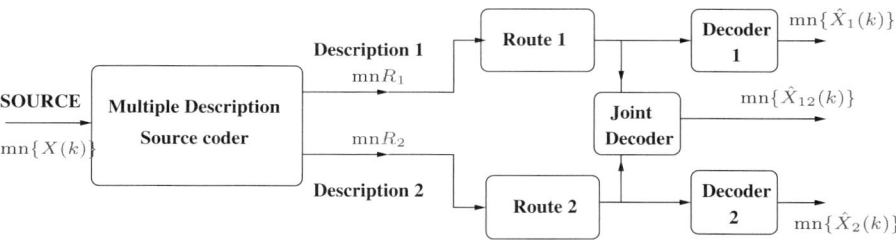

Figure 9.13 Multiple description (MD) source coding.

theorem was proved which demonstrated a region of the tuple $(R_1, D_1, R_2, D_2, D_{12})$ for which MD source codes exist.

Theorem 9.12

(El Gamal and Cover, 1982) Let $X(1), X(2), \ldots$ be a sequence of i.i.d. finite alphabet random variables drawn according to a probability mass function $p(x)$. If the distortion measures are $d_m(x, \hat{x}_m), m = 1, 2, 12$ then an achievable rate region for tuples $(R_1, R_2, D_1, D_2, D_{12})$ is given by the convex hull of the following.

$$R_1 \geq I(X; \hat{X}_1), \quad R_2 \geq I(X; \hat{X}_2), \quad R_1 + R_2 \geq I(X; \hat{X}_{12}, \hat{X}_1, \hat{X}_2) + I(\hat{X}_1; \hat{X}_2) \quad (9.34)$$

for some probability mass function $p(x, \hat{x}_1, \hat{x}_2, \hat{x}_{12})$ such that $\mathrm{E}[d_t(X, \hat{X}_t)] \leq D_t, \ t = 1, 2, 12$.

∎

This region was further improved in Zhang and Berger (1987) to a larger region for which MD source codes exist. However, what is unknown is whether these characterizations completely exhaust the set of tuples that can be achieved, i.e., a converse for the MD rate-distortion region. There are some special cases for which there are further results (Ahlswede, 1985; Fu and Yeung, 2002, and references therein). There has also been recent work on achievable rate-regions for more than two descriptions (Pradhan et al., 2004; Venkataramani et al., 2003). However, in these cases as well the complete characterization is unknown.

The *only* case for which the MD region is completely characterized is that for memoryless Gaussian sources with squared error distortion measures and specifically for two descriptions.[12] In Ozarow (1980), it was shown that the two-description MD region given in El Gamal and Cover (1982) was also applicable to the Gaussian case with squared error distortion where the alphabet is not finite. Moreover it was shown that the region in theorem 9.12 was in fact the complete characterization by proving a converse (outer bound) to the rate region. In this context the source was modeled as a sequence of i.i.d. Gaussian random variables $X \sim \mathcal{N}(0, \sigma_x^2)$ and the squared error distortion measure was chosen, i.e., $d_m(x, \hat{x}_m) = |x - \hat{x}_m|^2, m = 1, 2, 12$. Therefore, specializing the result in theorem 9.12 to the Gaussian case yields the following complete characterization of the set of all achievable tuples $(R_1, R_2, D_1, D_2, D_{12})$ (El Gamal and Cover, 1982; Ozarow, 1980):

$$D_1 \geq \sigma_x^2 e^{-2R_1}, \quad D_2 \geq \sigma_x^2 e^{-2R_2}, \tag{9.35}$$

$$D_{12} \geq \frac{\sigma_x^2 e^{-2(R_1+R_2)}}{1 - \left[\sqrt{\left(1 - \frac{D_1}{\sigma_x^2}\right)\left(1 - \frac{D_2}{\sigma_x^2}\right)} - \sqrt{\left(\frac{D_1}{\sigma_x^2}\right)\left(\frac{D_2}{\sigma_x^2}\right) - e^{-2(R_1+R_2)}}\right]^2}.$$

In order to interpret this result, consider the following. As seen before, for a single-description Gaussian problem, the minimum distortion for a given rate is $D(R) = \sigma_x^2 2^{-2R}$. Therefore, the distortions D_1, D_2 clearly need to be governed by the single-description bound, and this explains the first two inequalities in equation 9.35. However, in the MD problem we also need to bound the distortion D_{12} when both descriptions are available. From the single-description bound it is clear that we would have $D_{12} \geq D(R_1 + R_2) = \sigma_x^2 2^{-2(R_1+R_2)}$. Therefore, a natural question is whether this bound on D_{12} can be achieved with equality. However, the result in theorem 9.12 shows that this is not possible unless $D_1 = \sigma_x^2$ or $D_2 = \sigma_x^2$. Here is where the tension between the two descriptions manifests itself. We examine the tension in the symmetric case, when we have $D_1 = D_2 = D$, $R_1 = R_2 = R$ and specialize it for the unit variance source $\sigma_x^2 = 1$. If we want the individual descriptions to be as efficient as possible (i.e., $D = e^{-2R}$), then we see that $D_{12} \geq \frac{D}{2-D}$, which is far larger than $D(R_1 + R_2) = e^{-2(R_1+R_2)} = D^2$. For small D, we see that D_{12} is approximately $\frac{D}{2}$, which is much larger than D^2. Therefore, if we ask that the individual descriptions be close to optimal themselves, then they do not mutually refine each other very well. This reveals the tension between getting small the distortions D_1, D_2 of individual descriptions and a small D_{12}. We need to make the individual descriptions coarser in order to get more mutual refinement in D_{12}.

One important real-time application is that of video coding. This can be viewed as a sequence of individual frames which are correlated to each other. The traditional way of encoding video is by describing the "current" frame differentially with respect to the previous frame. This is done through a block-matching technique where the "closest" (in terms of squared distance) blocks from the previous frame are matched to blocks in the current frame, and then only the differences are transmitted. The rationale behind this idea is that blocks are only relatively displaced due to motion of objects in the video and hence this mechanism is called *motion compensation* in the literature (Reibman and Sun, 2000). Note that in this scheme, the encoder *explicitly* uses the knowledge of the previous frame. Clearly, when there are packet/route errors and the previous frame is not received at the destination, the reconstruction is difficult since the previous reference frame is not available. Therefore, several fixes to this problem have been developed over the past two decades (see Girod and Farber, 2000, and references therein).

In a more abstract framework, we can think of the video as a sequence of correlated random variables which we are trying to describe efficiently. In Witsenhausen and Wyner an alternate approach was taken by considering the video coding problem as a source coding problem with *side information*. In this setting, after encoding and transmitting the "previous" frame, the "current" frame develops an encoder which does *not* explicitly depend on the knowledge of the previous frame. The basic idea of this scheme arises from encoding schemes and decoding described in Slepian and Wolf (1973) and Wyner and Ziv (1976). Since the encoder does not explicitly use the side-information (previous frame) it can be designed such that the computational complexity is shifted from the encoder to the decoder. Such an architecture is attractive for applications where the encoder needs to be simple but the decoder can be more complex. This idea has been developed comprehensively in Puri and Ramchandran (2003), where practical coding techniques are developed with such applications in mind.

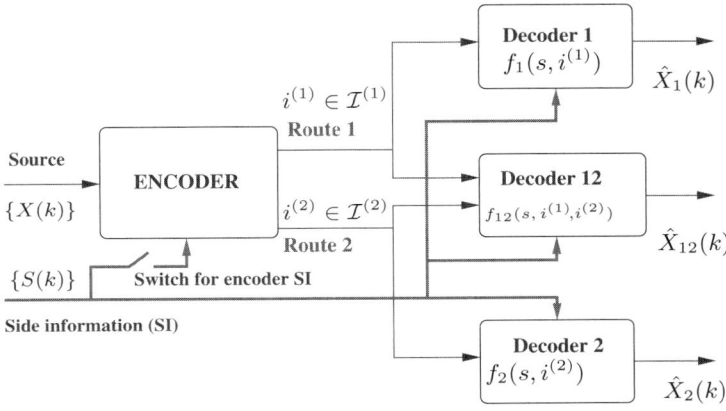

Figure 9.14   Multiple description source coding with side information.

However, even with this idea the robustness to route failures which is inherent to MD coding is not captured. Motivated by this, Diggavi and Vaishampayan (2004) considered the MD problem with side information (see fig. 9.14). In this abstract setting, we want to encode a source $\{X(k)\}$ when the decoder has knowledge of a correlated process $\{S(k)\}$ as side-information. For example, in the setting of Witsenhausen and Wyner and Puri and Ramchandran (2003), the side information could be the previous frame. In order to describe the source in the presence of route diversity, we can pose an MD problem, but now with side information as shown in fig. 9.14. Clearly this is a generalization of the MD problem and an achievable rate region was established for this problem in Diggavi and Vaishampayan (2004).

Theorem 9.13
Let $(X(1), S(1)), (X(2), S(2))\ldots$ be drawn i.i.d. $\sim Q(x, s)$. If *only* the decoder has access to the side information $\{S(k)\}$, then $(R_1, R_2, D_1, D_2, D_{12})$ is achievable if there exist random variables (W_1, W_2, W_{12}) with probability mass function $p(x, s, w_1, w_2, w_{12}) = Q(x, s)p(w_1, w_2, w_{12}|x)$, that is, $S \leftrightarrow X \leftrightarrow (W_1, W_2, W_{12})$

form a Markov chain, such that

$$R_1 > I(X; W_1|S), \quad R_2 > I(X; W_2|S) \tag{9.36}$$
$$R_1 + R_2 > I(X; W_{12}, W_1, W_2|S) + I(W_1; W_2|S)$$

and there exist reconstruction functions f_1, f_2, f_{12} which satisfy

$$D_1 \geq \mathrm{E}[d_1(X, f_1(S, W_1))], \quad D_2 \geq \mathrm{E}[d_2(X, f_2(S, W_2))] \tag{9.37}$$
$$D_{12} \geq \mathrm{E}[d_{12}(X, f_{12}(S, W_{12}, W_1, W_2))].$$

∎

This result gives an achievable rate region, but the complete characterization for this problem is open. A slightly improved region to theorem 9.13 is also found in Diggavi and Vaishampayan (2004). However, it is unknown whether this region exhausts the achievable rate region. But for the case when both the source and the side information are jointly Gaussian, and we are interested in the squared error distortion, a complete characterization of the rate-distortion region was obtained in Diggavi and Vaishampayan (2004).

In more detail, the result was the following. Let $(X(1), S(1)), (X(2), S(2)) \ldots$ be a sequence of i.i.d. jointly Gaussian random variables. With no loss of generality this can be represented by

$$S(k) = \alpha\left[X(k) + U(k)\right], \tag{9.38}$$

where $\alpha > 0$ and $\{X(k)\}, \{U(k)\}$ are independent Gaussian random variables with $\mathrm{E}[X] = 0 = \mathrm{E}[U]$, $\mathrm{E}[X^2] = \sigma_X^2$, $\mathrm{E}[U^2] = \sigma_U^2$. As considered in theorem 9.13, only the decoder has access to the side information $\{S(k)\}$. If the distortion measures are $d_m(x, \hat{x}_m) = ||x - \hat{x}_m||^2, m = 1, 2, 12$ then it is shown in Diggavi and Vaishampayan (2004) that the set of all achievable tuples $(R_1, R_2, D_1, D_2, D_{12})$ are given by

$$D_1 > \sigma_\mathcal{F}^2 e^{-2R_1}, \quad D_2 > \sigma_\mathcal{F}^2 e^{-2R_2}, \quad D_{12} > \frac{\sigma_\mathcal{F}^2 e^{-2(R_1+R_2)}}{1 - (\sqrt{\tilde{\Pi}} - \sqrt{\tilde{\Delta}})^2}, \tag{9.39}$$

where $\sigma_\mathcal{F}^2 = \frac{\sigma_X^2 \sigma_U^2}{\sigma_X^2 + \sigma_U^2}$ and $\tilde{\Pi}, \tilde{\Delta}$ are given by

$$\tilde{\Pi} = \left(1 - \frac{D_1}{\sigma_\mathcal{F}^2}\right)\left(1 - \frac{D_2}{\sigma_\mathcal{F}^2}\right), \quad \tilde{\Delta} = \left(\frac{D_1}{\sigma_\mathcal{F}^2}\right)\left(\frac{D_2}{\sigma_\mathcal{F}^2}\right) - e^{-2(R_1+R_2)}. \tag{9.40}$$

The result in equation 9.39 also shows that the rate-distortion region in this case is the same as that achieved when *both* encoder and decoder have access to the side information. That is, in the Gaussian case, the rates that can be achieved are the same whether the switch in fig. 9.14 is open or closed. In Wyner and Ziv (1976) it was shown that in the single-description Gaussian case, the decoder-only side information rate-distortion function coincided with that when both encoder and decoder were informed of the side information. The result (eq. 9.39) establishes that this is also true in the Gaussian two-description problem with decoder side information. However, the encoding and decoding techniques to achieve these rate tuples are very different when the encoder has access to the side information than when it does not. This shows that there might be efficient mechanisms to construct MD video coders which are robust to route failures. Some of the code constructions that bring this idea to fruition are discussed in section 9.5.2.

9.5.2   Quantizers for Route Diversity

The results given in section 9.5.1 show the existence of codes that can achieve the rate tuples given in theorems 9.12 and 9.13, but there are no *explicit* constructions. In this section we explore explicit coding schemes which utilize the presence of route diversity.

As seen in section 9.5.1, the single-description rate-distortion function quantifies the fundamental limits of the trade-off between the rate of the representation of a source and its average fidelity. The result in equation 9.30 showed the existence of such codes. Explicit constructions of these codes are called *quantizers* (Gersho and Gray, 1992; Gray and Neuhoff, 1998). More formally, quantizers map a sequence $\{X(1), \ldots, X(T)\}$ of source samples into a "representative" reconstruction $\{\hat{X}(1), \ldots, \hat{X}(T)\}$ through an *explicit* mapping which is typically computationally efficient. *Scalar quantizers* operate on a single source sample $X(k)$ at a time. Most current systems use scalar quantizers (Jayant and Noll, 1984). However, rate-distortion theory tells us that using sequences is important, and hence *vector quantizers* use sequences of source samples, i.e., $T > 1$ for quantization. Quantization techniques for single description have been quite well studied and understood (Gersho and Gray, 1992; Gray and Neuhoff, 1998; Jayant and Noll, 1984).

The rudiments of the MD coding ideas arose in the 1970s at Bell Laboratories. Jayant (1981) proposed and analyzed a very simple idea of channel splitting. The basic idea was to oversample a speech signal and send the odd samples through one channel and the even ones through another. However, this technique is not very efficient in terms of rate. Many of such simple coding techniques were being considered at Bell laboratories, but the ideas were not archived. These questions actually motivated the information-theoretic formulation of the MD problem described in section 9.5.1. The systematic study of coding for multiple descriptions was initiated in Vaishampayan (1993). Its publication resulted in a spurt of recent activity on the topic (see, for example, Diggavi et al., 2002c; Goyal and Kovacevic, 2001; Ingle and Vaishampayan, 1995; Vaishampayan et al., 2001, and references therein). More recently the utility of MD coding in conjunction with route diversity has also created interest in the networking community (see Apostolopoulos and Trott, 2004, and references therein).

The basic idea introduced in Vaishampayan (1993) constructed scalar quantizers for the MD problem. This was done specifically for the symmetric case, where $D_1 = D_2$ and $R_1 = R_2$. This symmetric construction was extended to structured (lattice) vector quantizers in Vaishampayan et al. (2001). The symmetric case has been further explored by several other researchers (Goyal and Kovacevic, 2001; Ingle and Vaishampayan, 1995). The importance of structured quantizers is in the computational complexity of the source encoder. For example, just as in channel

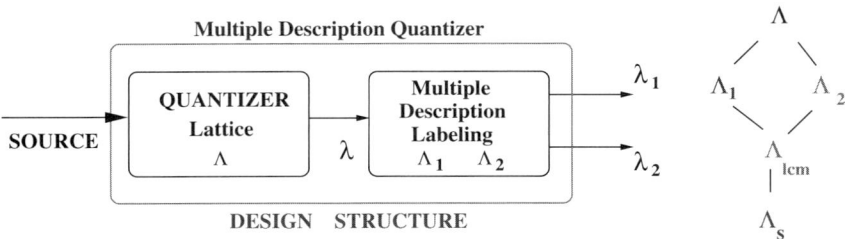

Figure 9.15   Structure of multiple description quantizer.

coding, trellis-based structures are also important in source coding. Such structures have also been proposed for the symmetric MD problem (Buzi, 1994; Jafarkhani and Tarokh, 1999). In general, unstructured quantizers based on training on some source samples can also be constructed, but the computational complexity of such techniques is much higher than structured (lattice) quantizers and therefore they are less attractive in practice. Such unstructured quantizers have been considered in the literature (Fleming et al., 2004). Our focus in this chapter will be on structured quantizers for which we have computationally efficient encoders as well as techniques to analyze their performance.

In general we would like to design MD quantizers that can attain an arbitrary rate-distortion tuple, and not just the symmetric case. This is motivated by applications where the multiple routes have disparate capacities (and therefore rate requirements) as well as different probabilities of route failures. In these cases, we need to design *asymmetric* MD quantizers which give graceful degradation in performance with route failures. Such a structure was studied in Diggavi et al. (2002c), and is depicted in fig. 9.15.

We illustrate the ideas of MD quantizer design from Diggavi et al. (2002c), using a scalar example. In fig. 9.16, the first line represents a uniform scalar quantizer. If we take a single source sample $X(k) \in \mathbf{R}$, then the uniform quantizer maps this sample to the closest "representative" point $\hat{X}$ on the one-dimensional (scaled integer) lattice Λ. Loosely, a T-dimensional lattice is a set of regularly spaced points in $\mathbf{R}^T$ for which any point can be chosen as the origin and the set of points would be the same. A more precise notion is based on the set of points forming an additive group (Conway and Sloane, 1999). Each of the representative points is given a unique label λ and this label is transmitted to the receiver. The transmission rate depends on the number of labels. Typically a finite set of points $2M$ is used to represent the labels. In a straightforward manner, this translates to a rate of $\log(2M)$ bits per source sample. If the source either has finite extent or a finite second-order moment, such a quantizer would have a bounded squared error distortion. If the representative points are separated by a distance of Δ, then the worst-case squared error distortion between a source sample and the representative is $\frac{\Delta^2}{4}$ for source samples $X(k) \in [-\frac{(M+1)\Delta}{2}, \frac{(M+1)\Delta}{2}]$. For a uniform distribution of the source in the region $X(k) \in [-\frac{(M+1)\Delta}{2}, \frac{(M+1)\Delta}{2}]$, the average distortion is $\frac{\Delta^2}{12}$

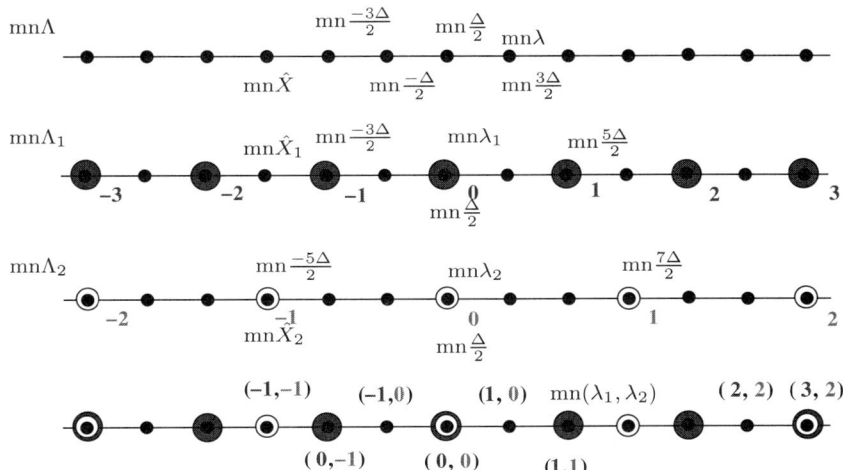

Figure 9.16 Scalar quantizer labeling example. The top line is a uniform scalar quantizer that maps source points $X(k)$ to a set of discrete representatives $\hat{X}$. The second and third lines show coarser uniform scalar quantizers. The last line puts together the combination of the coarser quantizers to give an ordered pair (λ_1, λ_2) as a label to every lattice point λ in the fine quantizer.

(Gersho and Gray, 1992).

The mapping described above is a single-description uniform scalar quantizer. The MD scalar quantizer needs to map every source sample to an *ordered pair* of representation points $(\hat{X}_1, \hat{X}_2)$. The labels (λ_1, λ_2) of this pair are used to send information over the two routes. For example, we could send the label λ_1 over the first route and label λ_2 over the second route. Now, in fig. 9.16 we have illustrated this by choosing coarser scalar quantizers in the second and third lines for the representations $\hat{X}_1$ and $\hat{X}_2$ respectively. These quantizers are also one-dimensional lattices Λ_1 and Λ_2 respectively. These representations $\hat{X}_1, \hat{X}_2$ in themselves give coarser information about the source sample, i.e., have a larger distortion than the "finer" quantizer Λ shown in the first line. Now, we need to represent the source sample $X(k)$ by a pair of representation points from Λ_1 and Λ_2. We want to choose this pair in such a way that if either of the labels is lost due to route failure, then we are still guaranteed a certain distortion. However, if both labels are received, i.e., both routes are successful, then we need to get a smaller distortion. This means that the label pair have to mutually refine each other's representations.

One such labeling technique is illustrated in fig. 9.16. Each point in the coarser lattices $\hat{X}_1$ in Λ_1 and $\hat{X}_2$ in Λ_1 is given a label λ_1 and λ_2 respectively. The idea is then to give a pair of labels (λ_1, λ_2) to each of the points on the fine lattice Λ. Every lattice point in Λ gets a *unique* label pair (λ_1, λ_2). Once this *labeling function* is constructed, then we can form the multiple description (MD) scalar quantizer by doing the following two steps. First, reduce the source sample $X(k) \in \mathbb{R}$ to its closest representative in Λ, $\hat{X}$ with label λ, i.e., apply a uniform scalar quantizer to $X(k)$. Given this $\hat{X}$, and the labeling function, we know the pair (λ_1, λ_2) that represents $\hat{X}$. The second step is to associate $\hat{X}_1$ with the reconstruction given by the label λ_1 in the first coarse quantizer Λ_1, and similarly for $\hat{X}_2$ in Λ_2. These operations are what the structure in fig. 9.15 represents. Therefore, in this design, the main task is to construct the labeling function for each point in Λ. Given the label pair (λ_1, λ_2), the encoder sends the index associated with λ_1 on route 1 and the index for λ_2 on route 2.

Before describing the labeling function, we examine the decoder structure in the MD scalar quantizer described above. First recall that the labeling function is designed so that any particular pair (λ_1, λ_2) is *uniquely* associated with a particular λ. Therefore, if *both* routes succeed, then the receiver is able to reconstruct λ and as a consequence $\hat{X}$. This means that the distortion in this case is that associated with the fine quantizer Λ, i.e., the average fidelity is $\frac{\Delta^2}{12}$. Now suppose route 1 succeeds and route 2 fails, then the receiver has only λ_1 and does not know λ_2. For example, suppose in fig. 9.16, the label pair $(-1, 0)$ was chosen at the source encoder, i.e., $\lambda_1 = -1, \lambda_2 = 0$. Now, the receiver knows that the encoder was trying to send one of the two points $(-1, 0)$ or $(-1, -1)$ and since route two failed it does not know which. More generally, in this situation, the receiver knows that λ belongs to the set of points in Λ which have the *same* first label λ_1 but have *different* second label λ_2. Now, assume that the decoder uses the reconstruction $\hat{X}_1$ associated with label $\lambda_1 = -1$ in Λ_1 (see second line in fig. 9.16). Therefore, for this particular example, the worst-case error due to this choice is $\frac{9}{4}\Delta^2$. This example also shows that the labeling function directly affects the decoder distortion. The design of the labeling function is the central part of the MD quantizer. The reconstruction $\hat{X}_1$ can use the mean of the set of all points in Λ associated with the *same* first label λ_1, which may improve the distortion. Note that in general this might not coincide with the reconstruction associated with λ_1. For design simplicity this reconstruction need not be taken into account in designing the labeling function, but rather can be used only at the decoder to improve the final distortion.

In general, we would need to construct a labeling function for all the points in Λ. However, we describe a particular design which solves a smaller problem and then expands its solution to Λ (Diggavi et al., 2002c; Vaishampayan et al., 2001). We will illustrate this idea using the example shown in fig. 9.16.

In the last line of fig. 9.16, we have depicted the overlay of the two coarse one-dimensional lattices Λ_1, Λ_2 along with Λ. We see that there is a repetitive pattern after every six points in Λ. This is not a coincidence, because Λ_1 was formed by taking every second point in Λ and Λ_2 by taking every third. The least common multiple is 6 and therefore we would expect the pattern to repeat. The basic idea is to just form a labeling function for these six points and then "shift" these labels to tile the entire lattice Λ. For example, in fig. 9.16, consider the point which we have labeled as $(2, 2)$ on the last line. This was done in the following manner. Notice that the repeating pattern of six points can be anchored by the points where both the Λ_1 and Λ_2 points coincide. In fig. 9.16, these are the points which have overlapped circles on the last line. We can think of all points in Λ with respect to these anchor points. For example, the point labeled $(2,2)$ is one point to the left of such an overlap point and is "equivalent" to the point labeled $(-1,0)$. More precisely, it is in the same *coset* as the other point with respect to the "intersection" lattice Λ_s, which is formed by the anchor points. Therefore, we get the label by shifting the label of $(-1,0)$ with respect to its cosets. In this case, note that $\lambda_1 = -1$ in Λ_1 is two points to the left of the anchor point $(0,0)$. Therefore, the corresponding point with respect to the anchor point $(3,2)$ is $\lambda_1 = 2$ and hence the first label for the point of interest is $\lambda_1 = 2$. Next, the corresponding point of the label $\lambda_2 = 0$ in Λ_2 with respect to the anchor point $(3,2)$ is $\lambda_2 = 2$. This gives us the label $(2,2)$ which is shown in the fig. 9.16. In a similar manner, given the labeling for the six points, we can construct the labeling for all points in Λ by the shifting technique described above. Actually, the six points correspond to the discrete Voronoi region of the point $(0,0)$ of the intersection lattice of the anchor points. Therefore, we can focus on constructing labels for the points in the Voronoi region of the intersection lattice. Note that in the example of fig. 9.16, the intersection lattice had an index of six which is exactly the least common multiple of the indices of lattices Λ_1, Λ_2 in Λ. This is also true when the indices of Λ_1, Λ_2 in Λ are not coprime (Diggavi et al., 2002c).

Let $V_{\Lambda_s:\Lambda}(0)$ be defined as the Voronoi region of the intersection lattice. Our problem is to develop the labeling function for the points in $V_{\Lambda_s:\Lambda}(0)$ in order to satisfy the individual distortion constraints D_1, D_2. This is accomplished by using a Lagrangian formulation in Diggavi et al. (2002c). This formulation reduces to finding the labeling scheme $\alpha(\lambda) = (\alpha_1(\lambda), \alpha_2(\lambda))$ so as to minimize,

$$\sum_{\lambda \in V_{\Lambda_s:\Lambda}(0)} \left[\gamma_1 \|\lambda - \alpha_1(\lambda)\|^2 + \gamma_2 \|\lambda - \alpha_2(\lambda)\|^2 \right]. \tag{9.41}$$

For this minimization problem we need to choose the appropriate labels $(\alpha_1(\lambda), \alpha_2(\lambda)) = (\lambda_1, \lambda_2)$. This is done by observing the following identity.

$$\gamma_1 \|\lambda - \lambda_1\|^2 + \gamma_2 \|\lambda - \lambda_2\|^2 = \frac{\gamma_1 \gamma_2}{\gamma_1 + \gamma_2} \|\lambda_2 - \lambda_1\|^2 + (\gamma_1 + \gamma_2) \|\lambda - \frac{\gamma_1 \lambda_1 + \gamma_2 \lambda_2}{\gamma_1 + \gamma_2}\|^2.$$

This results in the following design guideline. The labeling problem is split into two parts: (1) Choose $|V_{\Lambda_s:\Lambda}(0)|$ "shortest" pairs (λ_1, λ_2) (*not* all pairs of (λ_1, λ_2) are used). (2) Assign these pairs to lattice points $\lambda \in V_{\Lambda_s:\Lambda}(0)$. The second design can be solved very efficiently using linear programming methods. The solution of this labeling problem illustrates an important feature of the MD quantizer design that is quite distinct from the single-description case. It can happen that particular labels of each description can be noncontiguous, i.e., not all points λ which get the same label—say, λ_1—need to occur contiguously. This is quite different from the single-description case, where the labels are assigned to contiguous intervals. Also, the labels generated in this systematic manner are nontrivial and difficult to handcraft.

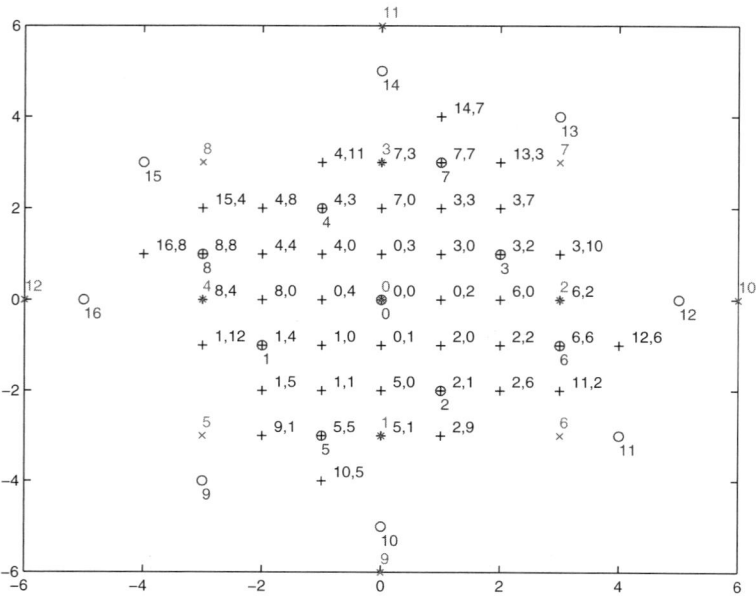

Figure 9.17 Labels for a two-dimensional integer lattice example.

The labeling scheme described for the scalar quantizer actually illustrates a more general principle which is applicable to MD vector quantizers (Diggavi et al., 2002c). We use a chain of lattices as illustrated in fig. 9.15, i.e., we use a fine lattice Λ and two coarser sublattices Λ_1, Λ_2. These lattices have an intersection lattice Λ_{lcm} one of whose Voronoi regions is what we label. The idea of using sublattice shifts as done above to generate the labels using only the labels of this Voronoi region can also be generalized (Diggavi et al., 2002c). One such example of the labels of the Voronoi region for a two-dimensional lattice is shown in fig. 9.17. Therefore, the vector quantizer proceeds as follows. We first reduce point $X^T \in \mathbb{R}^T$ using a fine lattice Λ, and then using the labeling function we find (λ_1, λ_2). Then as before λ_1 is sent over the first route and λ_2 is sent over the second route. The decoder also proceeds in a manner similar to the scalar quantizer described above.

As seen above, the crux of the MD quantizer design problem is to construct the appropriate labeling function. In Diggavi et al. (2002c), it is shown that an appropriate labeling function, along the lines described for the scalar quantizer, can be constructed very efficiently using a linear program. In fact, Diggavi et al. (2002c) shows that such a labeling scheme is very close to being optimal in terms of the rate distortion result given in theorem 9.12 in the high-rate regime.

9.5.3   Network Protocols for Route Diversity

In order to utilize route diversity in a network, one of the most important components is clearly the design of MD source coding techniques studied in section 9.5.2. However, an equally important question is the design of routing techniques that can enable the use of MD source coding. In this section we briefly examine these issues from a networking point of view.

In order to create route diversity, we need to have multiple routes which are disjoint, in that they do not share common links. This can be done through IP *source routing* (Keshav, 1997). Source routing is a technique whereby the sender of a packet can specify the route that a packet should take through the network. In the typical IP routing protocol, each router will choose the next hop to forward the packet by examining the destination IP address. However, in source routing, the "source" (i.e., the sender) makes some or all of these decisions. In strict source routing (which is virtually never used), the sender specifies the exact route the packet must take. The more common form is *loose source record route* (LSRR), in which the sender gives one or more hops that the packet must go through. Therefore, the sender can take an MD code and send each of the descriptions using different routes by explicitly specifying them the IP source routing protocol. An alternate technique might be to use an *overlay* network where there is an application that collects the different descriptions and sends them through different relay nodes in order to create route diversity. This discussion shows that creating route diversity is architecturally not difficult even using the provisions within the IP (Keshav, 1997).

This discussion from a networking point of view also exposes the inherent interactions required between the routing and application layers of the networking protocol stack. Such "interlayer" interactions become particularly important in wireless networks, where route failures could occur more frequently than in wired networks. Therefore, in this case diversity, albeit at a much higher layer in the IP stack, again becomes quite important.

9.6   Discussion

In this chapter we studied the emerging role of diversity with respect to three disparate topics. The idea of using multiple instantiations of randomness attempts to turn the presence of randomness to an advantage. For example, in multiple-antenna diversity, the degrees of freedom provided by the space diversity is utilized for increased rate or reliability. In mobile ad hoc networks, the random mobility is utilized to route information from source to destination.

To realize the benefits promised by the use of diversity, we need to have interactions across networking layers. For example, in opportunistic scheduling (studied in section 9.4.1) the transmission rates that can be supported by the physical layer interact with the resource allocation (scheduling), which is normally only a functionality of the data-link layer. In the multi-user diversity studied in mobile ad hoc networks (see section 9.4.2) the routing of the packets interacted with the physical layer transmission. Finally, the MD source coding studied in section 9.5 necessitated an interaction between source coding (application-layer functionality) and routing.

These examples of cross-layer protocols are increasingly becoming important in reliable network communication. Diversity is the common thread among several of these cross-layer protocols. The advantages of using diversity in these contexts are just beginning to be realized in practice. There might be many more areas where the ideas of using diversity could have an impact, and this is a topic of ongoing research.

Notes

[1] The term *sufficient statistics* refers to a function (perhaps many-to-one) which does not cause loss of information about the random quantity of interest.

[2] To be precise, we need to sample equation 9.1 at a rate larger than $2(W_I + W_s)$, where W_I is the input bandwidth and W_s is the bandwidth of the channel time variation (Kailath, 1961).

[3] In passband communication, a complex signal arises due to in-phase and quadrature phase modulation of the carrier signal, see Proakis (1995).

[4] This can be seen by noticing that for $M_t = 1$, a sufficient statistics is an equivalent scalar channel, $\tilde{y}^{(b)} = \mathbf{h}^{(b)*}\mathbf{y}^{(b)} = ||\mathbf{h}^{(b)}||^2\mathbf{x}^{(b)} + \mathbf{h}^{(b)*}\mathbf{z}^{(b)}$. In this chapter, $|h|^2 = \bar{h}h$, where $\bar{h}$ denotes complex conjugation, and for a vector $\mathbf{h}$ we denote its 2-norm by $||\mathbf{h}||^2 = \mathbf{h}^*\mathbf{h}$, where $\mathbf{h}^*$ denotes the Hermitian transpose and $\mathbf{h}^t$ denotes ordinary transpose.

[5] The assumption that $\{\mathbf{H}^{(b)}\}$ is i.i.d. is not crucial. This result is (asymptotically) correct even when the sequence $\{\mathbf{H}^{(b)}\}$ is a mean ergodic sequence (Ozarow et al., 1994). We use the notation $\mathbf{H}$ to denote the channel matrix $\mathbf{H}^{(b)}$ for a generic block b.

[6] For a matrix $\mathbf{A}$, we denote its determinant as $det(\mathbf{A})$ and $|\mathbf{A}|$, interchangeably.

[7] In Foschini (1996), a similar expression was derived without illustrating the converse to establish that the expression was indeed the capacity.

[8] Here the notation $o(1)$ indicates a term that goes to zero when $SNR \to \infty$.

[9] For an information rate of R bits per transmission and a block length of T, we define the codebook as the set of 2^{TR} codeword sequences of length T.

[10] A constellation size refers to the alphabet size of each transmitted symbol. For example, a QPSK modulated transmission has constellation size of 4.

[11] We use the notation $f(n) = \Theta(g(n))$ to denote $f(n) = O(g(n))$ as well as $g(n) = O(f(n))$. Here $f(n) = O(g(n))$ means $\lim\sup_{n\to\infty}|\frac{f(n)}{g(n)}| < \infty$.

[12] Interestingly, this result is specifically for two descriptions and does not immediately extend to the general case.

10   Designing Patterns for Easy Recognition: Information Transmission with Low-Density Parity-Check Codes

Frank R. Kschischang and *Masoud Ardakani*

10.1   Introduction

Coding for information transmission over a communication channel may be defined as the art of designing a (large) set of codewords such that (i) any codeword can be selected for transmission over the channel, and (ii) the corresponding channel output with very high probability identifies the transmitted codeword. Low-density parity-check codes represent the current state-of-the-art in channel coding. They are a family of codes with flexible code parameters, and a code structure that can be fine-tuned so that decoding can occur at transmission rates approaching the information-theoretical limits established by Claude Shannon, yet with "practical" decoding complexity. In this chapter—which is aimed at the non-expert—we show that these codes are easy to describe using probabilistic graphical models, and that their simplest decoding algorithms (the "sum-product" or "belief-propagation" algorithm, and variations thereof) can be understood as message-passing in the graphical model. We show that a simple Gaussian approximation of the messages passed in the decoder leads to a tractable code-optimization problem, and that solving this optimization problem results in codes whose performance appears to approach the Shannon limit, at least for some channels.

Communication channels are typically modeled (with no essential loss of generality) in discrete time. At each unit of time a channel accepts (from the transmitter) a "channel input symbol" and produces (for the receiver) a corresponding "channel output symbol" according to some probabilistic channel model. Information is usually transmitted by using the channel many times, i.e., by transmitting many channel input symbols. In so-called "block coding," messages are mapped to sequences $(x_1, \ldots, x_n)$ of channel inputs of a fixed block-length n. A code is a set of "valid codewords" agreed upon by the transmitter and receiver prior to communication. A code is typically a (carefully selected) subset of the set of all possible channel inputs of length n. Transmission of a codeword gives rise (at the receiver) to a "received word," an n-tuple $(y_1, \ldots, y_n)$ of channel output symbols. The task of the receiver is to infer from the received word, which codeword—and hence which message—was (ideally, most likely) transmitted. Alternatively, if the message consists of many symbols, the receiver may wish to infer the most likely value of each message symbol.

By attempting to determine which codeword was most likely transmitted, a decoding algorithm attempts to solve a noisy pattern recognition problem. There are, therefore, some similarities between the fields of coding theory and pattern recognition. However, a key difference that makes the two fields quite distinct is the fact that the set of valid codewords is under the control of the system designer in the former case, but not (usually) in the latter case. In other words, in coding theory the system designer is given the luxury of choosing the set of patterns to be recognized by the decoding algorithm. A major theme in coding theory research is, therefore, to optimize or fine-tune the structure of the code for effective recognition (decoding) by some particular class of decoding algorithms.

Another major difference between coding theory and typical pattern-recognition problems is the sheer number of patterns to be recognized by the decoding algorithm. In many pattern-recognition problems, the number of different patterns (or pattern classes) is relatively small. In coding theory, the numbers can be extraordinarily large. The transmission of k bits corresponds to the selection, by the transmitter, of one codeword from a code of 2^k possible codewords. Thus, transmission of single bit requires a code of just two codewords; transmission of two bits requires a code of four codewords, and so on. Typical values of k for codes used in practice range from less than a dozen bits to tens of thousands of bits, and hence the number of different codewords to be "recognized" can be as large as $2^{10,000}$ or more! Despite these huge numbers, decoding algorithms routinely make rapid decoding decisions, reliably producing decoded information at many megabits per second.

A key parameter of a code is its *rate*. A code of 2^k codewords, each having block-length n, is said to have a rate of $R = k/n$ bits/symbol (or bits per channel-use). Clearly the rate of a code is a measure of the "speed" at which information is transmitted, normalized per channel-use. To convert from bits per symbol to bits per second, one needs to know the number of symbols that may be transmitted per second, a value that typically scales linearly with the "channel bandwidth." Channel bandwidths can vary greatly, depending on the application; thus, from the point of view of code design, it is more appropriate to focus on code rate measured in bits/symbol (rather than bits/s).

Given a particular channel, one would clearly like to make the code rate as large as possible. On the other hand, one also desires to make reliable decoding decisions, i.e., decisions for which the probability of error approaches zero. At first glance, it may seem that there should be a trade-off between transmission rate and reliability, i.e., for a fixed k, intuition would suggest that for some sufficiently large n it should be possible to design a code so that the probability of decoding error can be made smaller than any chosen $\epsilon > 0$.

This would certainly be true for the transmission of $k = 1$ bit over a binary symmetric channel with "crossover probability" $p < 1/2$. Such a channel accepts $x_i \in \{0, 1\}$ at its input and produces a corresponding $y_i \in \{0, 1\}$ at its output. Each transmitted symbol is independently "flipped" with probability p, i.e., with probability p we have $y_i \neq x_i$. A single bit can be transmitted with a repetition code of two codewords $\{000 \cdots 0, 111 \cdots 1\}$, where both codewords have the same length n. This code can be decoded according to a "majority rule:" if the majority of the symbols in the received word are zero, then decode to the all-zero codeword; otherwise decode to the all-one codeword. It is easy to see that if n is made sufficiently large, then the probability of error under the majority rule can be made arbitrarily small for any $p < 1/2$. In this example, there is a smooth trade-off between rate and reliability; as the rate decreases, the reliability increases, and one might be inclined to believe that such is the trade-off in general.

In fact, although there *is* a fundamental trade-off between code rate and reliability, the trade-off is abrupt (like a step-function), not smooth. In his seminal 1948 paper (Shannon, 1948), Claude E. Shannon established the remarkable fact that typical communication channels are characterized by a so-called *channel capacity*, C, with the property that reliable communication (i.e., communication with probability of error approaching zero) is possible for *every* $R < C$. More precisely, Shannon showed that for every $R < C$ and every $\epsilon > 0$, by choosing a sufficiently large block-length n, there exists a block code of length n with at least 2^{nR} codewords and a decoding algorithm for this code that yields a probability of decoding error smaller than ϵ. Conversely, one may also show that if $R > C$, then, even using an algorithm that minimizes error probability, it is impossible to achieve arbitrarily small probability of error. Thus, to achieve arbitrarily good reliability, it is *not* necessary to have $R \to 0$, but rather only to have $R < C$.

Information theory allows us to refine our stated goal of coding for information transmission over a communication channel with capacity C. The goal is to design a code such that (i) any codeword can be selected for transmission over the channel, (ii) the corresponding channel output can be processed by an algorithm of "practical" complexity to identify (with very high probability) the transmitted codeword, and (iii) the rate of the code is "close" to C. In the remainder of this chapter, we will show how low-density parity-check (LDPC) codes achieve this goal.

10.2   A Brief Introduction to Coding Theory

Low-density parity-check codes are binary linear block codes (though it is possible to define non-binary versions as well). Accordingly, we begin by defining what this means.

Let F_2 denote the finite field of two elements $\{0, 1\}$, closed under modulo-two integer addition and multiplication. This field has the simplest possible arithmetic: for all $x \in F_2$, under addition we have $0 + x = x$ and $1 + x = \bar{x}$, where $\bar{x}$ denotes the complement of x (thus a two-input F_2-adder is an exclusive-OR gate), and under multiplication we have $0 \cdot x = 0$ and $1 \cdot x = x$ (thus a two-input F_2-multiplier is an AND gate). For every positive integer n, we let F_2^n denote the set of n-tuples with components from F_2, which forms a vector space over F_2 equipped with the usual component-wise vector addition and with multiplication by scalars from F_2. As is the convention in coding theory, we will always think of such vectors as *row vectors*.

By definition, a *binary linear block code* of block-length n and dimension k is a k-dimensional subspace of F_2^n. It follows that a binary linear block code is itself closed under vector addition and multiplication by scalars; in particular, the sum of two codewords is another codeword, and the code certainly always contains the all-zero vector $(0, 0, \ldots, 0)$. (Although it is certainly possible to define nonlinear codes, i.e., codes that are general sub*sets* of F_2^n, not necessarily sub*spaces*, most codes used in practice are linear codes.) A binary linear code of length n and dimension k will be denoted as an $[n, k]$ code. Such a code has 2^k codewords, and hence has rate $R = k/n$. We will only ever consider such codes with $k > 0$. From now on, when we write "code," we mean binary linear block code.

How can an $[n, k]$ code be specified? One way is to observe that such a code $\mathcal{C}$ is a k-dimensional vector space, and hence it has a basis (in general, many bases), i.e., a set $\{\mathbf{v}_1, \mathbf{v}_2, \ldots, \mathbf{v}_k\}$ of linearly independent vectors that span $\mathcal{C}$. These vectors can be collected together as the rows of a $k \times n$ matrix G, called a *generator matrix* for $\mathcal{C}$, with the evident property that $\mathcal{C}$ is the row space of G. For every distinct $\mathbf{u} \in F_2^k$ we obtain a distinct codeword $\mathbf{v} = \mathbf{u}G \in \mathcal{C}$. Hence, a generator matrix yields a way to implement an *encoder* for $\mathcal{C}$, by simply mapping a message $\mathbf{u} \in F_2^k$ under matrix multiplication by G to the codeword $\mathbf{v} = \mathbf{u}G \in \mathcal{C}$. Thus a code $\mathcal{C}$ may be specified by providing a generator matrix for $\mathcal{C}$.

Another way to specify an $[n, k]$ code $\mathcal{C}$—and the one we will use to define low-density parity-check codes—is to view $\mathcal{C}$ as the solution space of some homogeneous system of linear equations in n variables $X_1, X_2, \ldots, X_n$. Since, in F_2, there are only two possible scalars, the structure of any one such equation is exceedingly simple: it is always of the form

$$X_{i_1} + X_{i_2} + \cdots + X_{i_m} = 0, \tag{10.1}$$

where $\{i_1, i_2, \ldots, i_m\}$ is some subset of $\{1, 2, \ldots, n\}$. Such an equation is sometimes referred to as a *parity check*, since it specifies that the "parity" (the number of ones) in the subset of the variables indexed by $\{i_1, \ldots, i_m\}$ should be even, i.e., (10.1) is satisfied if and only if an even number of X_{i_1}, X_{i_2}, $\ldots$ X_{i_m} take value one. If we define the n-tuple h as the vector with value one in components $i_1, i_2, \ldots, i_m$, and value zero in all other components, then (10.1) may also be written as $(X_1, \ldots, X_n)h^T = 0$, where h^T denotes the "transpose" of h.

Given a system of $(n - k)$ parity-check equations, we may collect the corresponding h-vectors to form the rows of an $(n-k) \times n$ matrix H, called a *parity-check matrix*. The set

$$\mathcal{C} = \{\mathbf{x} : \mathbf{x}H^T = \mathbf{0}\}$$

of all possible solutions to this system of equations, i.e., the set of vectors that satisfy all parity checks, is then a code of length n and dimension *at least* k (and possibly more, if the rows of H are not linearly independent). Thus a code $\mathcal{C}$ may be specified by providing a parity-check matrix for $\mathcal{C}$.

Note that, whereas a generator matrix for $\mathcal{C}$ gives us a convenient encoder, a parity-check matrix H for $\mathcal{C}$ gives us a convenient means of testing a vector for membership in the code, since a given vector $\mathbf{r} \in F_2^n$ is a codeword if and only if $\mathbf{r}H^T = 0$. More generally, parity-check matrices are useful for decoding since, if $\mathbf{r}$ is a non-codeword, it is the structure of parity-check failures in the so-called *syndrome* $\mathbf{r}H^T$ that provides evidence about which bits of $\mathbf{r}$ need to be changed in order to recover a valid codeword.

Low-density parity-check (LDPC) codes have the special property that they are defined via a parity-check matrix that is *sparse*, i.e., by an H matrix that has only a small number of nonzero entries. If the H matrix has a fixed number of one in each row and a fixed number of ones in each column, then the corresponding code is called a *regular* LDPC code; otherwise, the code is an irregular code. As an example, one of the earliest families of LDPC codes—the so-called (3,6)-regular LDPC codes, defined by R. G. Gallager at MIT in the early 1960s (Gallager, 1963)—have an H matrix with exactly 6 ones in each row and 3 ones in each column. If we take n reasonably large, say $n = 2000$, the matrix contains just $3n = 6,000$ ones, whereas a binary matrix of the same size generated by flipping a fair coin for each matrix entry, would on average contain one million ones. Thus we see that the H matrix is very sparse indeed.

Why is sparseness important? The answer lies in the nature of the decoding algorithm, which we describe next.

10.3   Message-Passing Decoding of LDPC Codes

10.3.1   From Codes to Graphs

The relationship between variables and equations can be visualized using a graph, such as the one shown in Figure 10.1. This graph (called a Forney-style factor graph (Forney, 2001; Loeliger, 2004)) consists of various vertices and edges (as is conventional in graph theory), and (somewhat unconventionally) also includes a number of "half-edges," which are edges incident on a single vertex only. The half-edges are denoted as '$\perp$' in Figure 10.1.

Edges and half-edges represent binary variables. A "configuration" is an assignment of a binary value to each edge and half-edge. Certain configurations will be regarded as "valid configurations," and all others as invalid. Vertices in the graph represent "local constraints" that the variables must satisfy in order to form a valid configuration. So-called "equality constraints" (or "equality vertices"), denoted with '=' in Figure 10.1, constrain all neighboring variables (i.e., all incident edges) to take on the same value in every valid configuration, whereas so-called

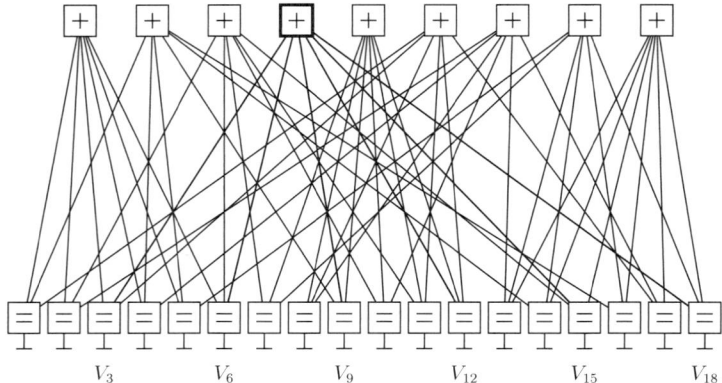

Figure 10.1 A factor graph for a (very small) (3,6)-regular LDPC code. Each edge (and half-edge) represents a binary variable. The boxes labeled = are equality constraints that enforce the rule that all incident edges are to have the same value in every valid configuration. The boxes labeled + are parity-check constraints that enforce the rule that all incident edges are to have even parity (zero-sum, modulo two).

"parity-check constraints" (or "check vertices"), denoted with '+' in Figure 10.1, constrain the neighboring variables to form a configuration having an even number of ones, i.e., a modulo-two sum of zero. The valid configurations are precisely those that satisfy *all* local constraints.

The half-edges represent the codeword symbols v_1, v_2, ..., v_{18}. Half-edges can be viewed as the "interface" (or "read-out") between the configuration space induced by the internal structure of the graph, and the desired "external" behavior. Equivalently, the full edges in the graph may be regarded as hidden (or "auxiliary" or "state") variables, and the half-edges as observed or primary variables. The equality constraints essentially serve to "copy" the value of each codeword symbol in a valid configuration to the neighboring (full) edges. Each of the parity-check constraints implements a single parity-check equation; for example, the highlighted parity-check constraint in Figure 10.1 essentially implements the equation

$$v_3 + v_6 + v_9 + v_{12} + v_{15} + v_{18} = 0;$$

however, instead of involving the variables v_3, v_6, etc., directly, the highlighted parity-check constraint vertex involves copies of these variables.

It should now be clear that the set of valid configurations projected on the half-edges in Figure 10.1 form a binary linear code satisfying the 9 different parity-check equations implemented by the check vertices. Indeed, as the reader may verify by tracing edges, this code is defined by the parity-check matrix

$$H = \begin{bmatrix} 1 & 1 & 1 & 1 & 1 & 1 & 0 & 0 & 0 & 0 & 0 & 0 & 0 & 0 & 0 & 0 & 0 & 0 \\ 1 & 0 & 0 & 1 & 1 & 0 & 0 & 0 & 1 & 0 & 0 & 0 & 0 & 1 & 0 & 1 & 0 & 0 \\ 0 & 1 & 0 & 0 & 0 & 1 & 1 & 0 & 0 & 1 & 1 & 0 & 0 & 0 & 1 & 0 & 0 & 0 \\ 0 & 0 & 1 & 0 & 0 & 1 & 0 & 0 & 1 & 0 & 0 & 1 & 0 & 0 & 1 & 0 & 0 & 1 \\ 0 & 0 & 0 & 0 & 0 & 0 & 1 & 1 & 1 & 1 & 1 & 1 & 0 & 0 & 0 & 0 & 0 & 0 \\ 1 & 0 & 1 & 0 & 0 & 0 & 0 & 1 & 0 & 0 & 1 & 1 & 0 & 0 & 0 & 0 & 1 & 0 \\ 0 & 1 & 0 & 1 & 0 & 0 & 0 & 1 & 0 & 1 & 0 & 0 & 1 & 0 & 0 & 0 & 1 & 0 \\ 0 & 0 & 0 & 0 & 1 & 0 & 1 & 0 & 0 & 0 & 0 & 0 & 1 & 1 & 0 & 1 & 0 & 1 \\ 0 & 0 & 0 & 0 & 0 & 0 & 0 & 0 & 0 & 0 & 0 & 0 & 1 & 1 & 1 & 1 & 1 & 1 \end{bmatrix}.$$

It is clear that H encodes the incidence structure of the graph, with an edge corresponding to each nonzero entry of H. If a nonzero entry occurs in row i and column j, then the edge connects the ith check vertex with the jth equality vertex. Because of the correspondence between H and the factor graph, sparseness of H implies sparseness of the graph and *vice versa*.

10.3.2 Channel Models

As noted above, communication channels are typically modeled in discrete time: a channel accepts a channel input x_i and produces a corresponding channel output y_i, according to some probabilistic model.

For example, the binary symmetric channel described earlier accepts, at time i, a binary digit $x_i \in \{0,1\}$ at its input, and produces a binary digit $y_i \in \{0,1\}$ at its output, with the property that $y_i = x_i$ with probability $1 - p$ (and therefore $y_i \neq x_i$ with probability p). The parameter p is called the *cross-over* probability of the channel. The binary symmetric channel is assumed to be *memoryless*, which means that, given the ith channel input x_i, the channel output y_i is independent of all other channel inputs and outputs, i.e., (assuming n channel inputs and outputs in total)

$$p(y_i|x_1,\dots,x_n,y_1,\dots,y_{i-1},y_{i+1},\dots,y_n) = p(y_i|x_i).$$

The capacity $C(p)$ of the binary symmetric channel with crossover probability p is given by (Shannon, 1948)

$$C(p) = 1 - \mathcal{H}(p) \text{ bits/symbol}$$

where $\mathcal{H}(p)$ denotes the binary entropy function

$$\mathcal{H}(p) = -p\log_2 p - (1-p)\log_2(1-p).$$

The capacity is plotted in Figure 10.2(a).

We will also consider the binary-input additive white Gaussian noise (AWGN) channel, which at time i accepts a channel input $x_i \in \{-1, 1\}$, and produces the (real-valued) output $y_i = x_i + n_i$, where n_i is a zero-mean Gaussian random variable with variance σ^2. This channel is also assumed to be memoryless, and serves as a model of the widely implemented continuous-time transmission schemes known as binary phase-shift keying (BPSK) and quadrature phase-shift keying (QPSK).

The capacity $C(\sigma)$ of the binary-input additive white Gaussian noise channel with noise variance σ^2 is given by

$$C(\sigma) = 1 - \frac{1}{\sqrt{2\pi}} \int_{-\infty}^{\infty} \exp(-u^2/2) \log_2(1 + \exp(-2(\sigma u + 1)/\sigma^2)) \mathrm{d}u \text{ bits/symbol.}$$

This function is plotted in Figure 10.2(b). Also plotted is the function

$$\frac{1}{2} \log_2(1 + 1/\sigma^2),$$

which represents the capacity of an additive white Gaussian noise channel in which the channel input is constrained to have unit second moment, but is unconstrained in value.

Instead of using the noise variance σ as a parameter of an AWGN channel, one often encounters the so-called "bit-energy to noise-density ratio," denoted E_b/N_0. This terminology arises in the context of continuous-time AWGN channels, in which

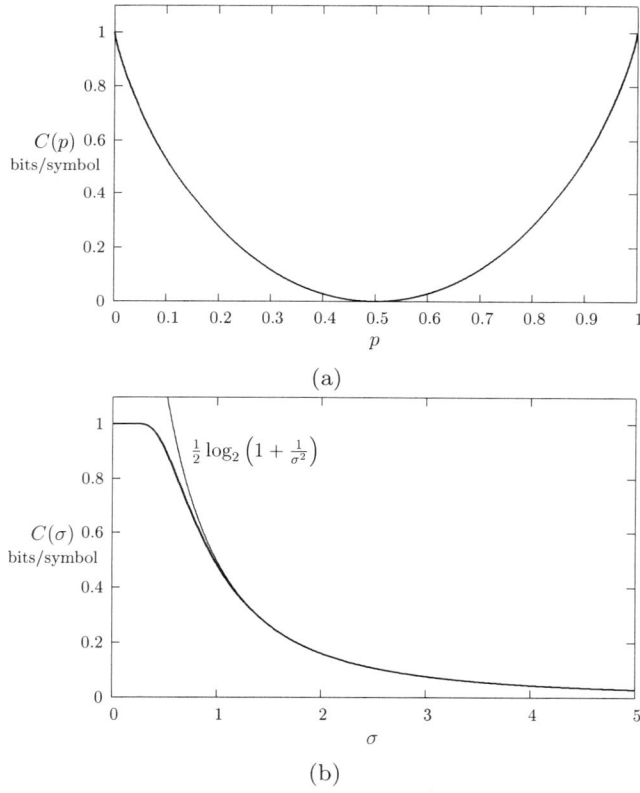

Figure 10.2 Capacity as a function of channel parameter: (a) the binary symmetric channel with crossover probability p; (b) the binary-input additive white Gaussian noise channel with noise variance σ^2.

the one-sided noise power spectral density is typically parameterized by the value N_0. For a code of rate R and a noise variance of σ^2, we have

$$\frac{E_b}{N_0} = \frac{1}{2R\sigma^2}.$$

Often the value of E_b/N_0 is quoted as a value in decibels (dB), i.e., the value quoted is $10\log_{10}(E_b/N_0)$. Thus, for example, a code of rate R operating in an AWGN channel with an E_b/N_0 of x dB is operating in a channel of noise variance

$$\sigma^2 = \frac{10^{-x/10}}{2R}.$$

In the coding literature one also often encounters the term "Shannon limit." The Shannon limit for a code of rate R is the value of the channel parameter corresponding to the worst channel that (in principle) could be used with a code of that rate, i.e., the channel parameter for which the capacity of the channel is R. In the case of AWGN channel, the performance of a coding scheme is often quoted in terms of a distance (in dB) from the Shannon limit. Thus, if the code achieves acceptable performance at a noise variance σ^2, but the corresponding Shannon limit is $\sigma_0^2 > \sigma^2$, then the distance to the Shannon limit is given as $10\log_{10}(\sigma_0^2/\sigma^2)$ dB.

10.3.3   From Graphs to Decoding Algorithms

Most decoding algorithms for low-density parity-check codes operate by passing "messages" along the edges of the graph describing the code. We will begin by providing an intuitive description of this process.

Initially, messages are derived from the channel outputs. The channel outputs are translated into a a "belief" about the value of the corresponding codeword symbol, where a "belief" is a guess about the value (zero or one) along with a measure of confidence in that guess. Unfortunately, due to channel noise, some of these beliefs are, in fact, erroneous. Beliefs about each codeword symbol are communicated to the check vertices. By enforcing the rule that in a valid configuration the modulo-two sum of the bit values is zero, the checks can update the beliefs. For example, if the beliefs received at a check form a configuration that does not satisfy the zero-sum rule, then the bit with the least confidence could be informed that it should probably alter its belief. The process of sending messages from equality vertices to check vertices and back again is called an "iteration," and after several iterations (depending on the code and the noise in the channel), the beliefs about the symbols, with high probability, reflect the transmitted configuration.

Now it becomes clear why sparseness in the graph is important. Firstly, the total amount of computation required per iteration is proportional to the number of edges in the decoding graph, and this number is exactly equal to the number of nonzero entries in the H matrix. Thus, the more sparse the graph, the smaller the decoding complexity. Secondly, sparseness helps to make it difficult for short cycles in the graph (which can sometimes cause reinforcement of erroneous beliefs) to influence the decoder unduly.

We also notice that this decoding algorithm naturally supports parallelism, as the message transfer between checks and variables can, in principle, all occur simultaneously. This observation is the foundation for a number of hardware implementations for LDPC decoders, in which processing nodes correspond directly to factor-graph vertices, and wires connecting these nodes correspond directly to factor-graph edges.

We will now give a more precise description of message-passing decoding, starting with the so-called *sum-product* algorithm. See (Kschischang et al., 2001) for more details.

Messages passed on an edge during decoding are probability mass functions for the corresponding binary variable. A probability mass function $p(x)$ for a binary variable X can be encoded with just a single parameter (e.g., $p(0)$, $p(1)$, $p(0) - p(1)$, $p(0)/p(1)$, etc.), and hence messages are real-valued scalars. A very commonly used parametrization, is the log-likelihood ratio (LLR), defined as $\ln(p(0)/p(1))$. Note that the sign of an LLR value indicates which symbol-value (0 or 1) is more likely, and so can be used to make a decision on the value of that symbol. The magnitude of the LLR can be interpreted as a measure of confidence in the decision; a large magnitude indicates a large disparity between $p(0)$ and $p(1)$, and hence a greater confidence in the truth of the decision. The "neutral message" corresponding to the uniform distribution will be denoted as μ_0. If an LLR representation is used, then $\mu_0 = 0$.

Messages are always directed (along an edge or half-edge) in the graph. A message on a half edge directed to a vertex v will be denoted as $\mu_{\to v}$, and a message on a full edge directed from a vertex v_1 to a vertex v_2 will be denoted as $\mu_{v_1 \to v_2}$. We will denote the set of neighbors of a vertex v_1 as $N(v_1)$. If $v_2 \in N(v_1)$, then $N(v_1) \setminus \{v_2\}$ is the set of neighbors of v_1 excluding v_2.

Initialization: The decoding algorithm is initialized by sending neutral messages on all edges, i.e., $\mu_{v_1 \to v_2} = \mu_{v_2 \to v_1} = \mu_0$ for every pair of vertices v_1, v_2 connected by an edge. The half-edges are initialized with so-called "intrinsic" or "channel" messages, corresponding to the received channel output. In particular, for a binary symmetric channel with crossover probability p, the LLR associated with channel output $y \in \{0, 1\}$ is given by

$$\lambda(y) = (-1)^y \ln \left(\frac{1}{p} - 1 \right),$$

and this is the initial message sent toward the corresponding equality vertex in the graph. Similarly, for the binary-input AWGN channel with noise variance σ^2, the LLR associated with channel output $y \in \mathrm{R}$ is given by

$$\lambda(y) = 2y/\sigma^2,$$

assuming that transmission of a zero corresponds to the $+1$ channel input, and transmission of a one corresponds to the -1 channel input.

Local Updates: Messages are updated at the vertices according to the principle that the message $\mu_{v_1 \to v_2}$ sent from vertex v_1 to its neighbor v_2 is a function of the messages directed toward v_1 on all edges *other* than the edge $\{v_1, v_2\}$. This principle is one of the pillars that leads to an analysis of the decoder; furthermore, this principle leads to optimum decoding in a cycle-free graph (see Kschischang et al., 2001).

Assuming that messages are represented as LLR values, then the message sent by the sum-product algorithm from an equality vertex v_1 to a neighboring check vertex v_2 is given as

$$\mu_{v_1 \to v_2} = \mu_{\to v_1} + \sum_{v' \in N(v_1) \setminus \{v_2\}} \mu_{v' \to v_1}, \tag{10.2}$$

where $\mu_{\to v_1}$ denotes the channel message received along the half-edge connected to v_1. Similarly, the message sent from a check vertex v_2 to a neighboring equality vertex v_1 is given as

$$\mu_{v_2 \to v_1} = 2 \tanh^{-1} \left(\prod_{v' \in N(v_2) \setminus \{v_1\}} \tanh(\mu_{v' \to v_2}/2) \right). \tag{10.3}$$

The hyperbolic tangent functions involved in this update rule are actually performing a change of message representation: if λ is the LLR value $\ln(p(0)/p(1))$, then $\tanh(\lambda/2) = p(0) - p(1)$, the probability difference. The product of $\tanh(\cdot)$ functions in (10.2) actually implements a product of probability differences, which can itself be seen as the difference in probabilities of local configurations having even parity with those having odd parity.

Messages received on full edges at an equality vertex are referred to as "extrinsic" messages. Extrinsic messages, unlike the "intrinsic" channel message reflect the structure of the code and change from iteration to iteration; when decoding is successful, the quality (magnitude in an LLR implementation) of the extrinsic messages improves from iteration to iteration. In this way, the extrinsic messages can "overwhelm" erroneous channel messages, leading to successful decoding.

Update Schedule: The order in which messages are updated is referred to as the "update schedule." A commonly used schedule is to send messages from each equality vertex toward the neighboring check vertices (in any order), and then to send messages in the opposite direction (in any order). One complete such update is referred to as an "iteration," and usually many iterations are performed before a decoding decision is reached. Other update schedules can lead to faster convergence (Sharon et al., 2004; Xiao and Banihashemi, 2004), but we will not consider these here.

Termination: Decoding decisions are based on *all* of the messages (both intrinsic and extrinsic) directed toward each equality vertex v. Decisions are made on a symbol-by-symbol basis. With LLR messages, the decision statistic is given as

$$\mu = \mu_{\to v} + \sum_{v' \in N(v)} \mu_{v' \to v}.$$

If $\mu > 0$, then the corresponding codeword symbol value is chosen to be 0; otherwise it is chosen to be 1. Usually iterations are performed until these decisions yield a valid codeword, or until the number of iterations reaches some allowed maximum.

Note that the total computational complexity (assuming a fixed maximum number of iterations, and a fixed distribution of vertex degrees) scales linearly with the block length of the code.

In addition to the sum-product algorithm, a number of other (often simpler) message-passing algorithms have been studied. These include the "min-sum" algorithm and Gallager's "decoding algorithm B," which are described next.

Min-sum algorithm: In the min-sum algorithm, the update rule at an equality vertex is the same as the sum-product algorithm (10.2), but the update rule at a check vertex v_2 is simplified to

$$\mu_{v_2 \to v_1} = \min_{v' \in N(v_2) \backslash \{v_1\}} |\mu_{v' \to v_2}| \cdot \prod_{v' \in N(v_2) \backslash \{v_1\}} \text{sign}(\mu_{v' \to v_2}). \qquad (10.4)$$

Notice that the $\tanh^{-1}$ of the product of tanh's is approximated as the minimum of the absolute values times the product of the signs. This approximation becomes more accurate as the magnitude of the messages is increased.

Gallager's decoding algorithm B: In this algorithm, introduced by Gallager (1963), the message alphabet is $\{0, 1\}$. In other words, the messages communicate "decisions" only, without an associated reliability.

The update rule at a check vertex v_2 is

$$\mu_{v_2 \to v_1} = \bigoplus_{v' \in N(v_2) \backslash \{v_1\}} \mu_{v' \to v_2}, \qquad (10.5)$$

where $\oplus$ represents the modulo-two sum of binary messages.

At an equality vertex v_1 of degree $d_v + 1$, the outgoing message $\mu_{v_1 \to v_2}$ is

$$\mu_{v_1 \to v_2} = \begin{cases} \overline{\mu_{\to v_1}} \text{ if } \exists v_1', v_2', \dots, v_b' \in N(v_1) \backslash \{v_2\} : \mu_{v_1' \to v_1} = \cdots = \mu_{v_b' \to v_1} = \overline{\mu_{\to v_1}}, \\ \mu_{\to v_1} \text{ otherwise} \end{cases}$$

$$(10.6)$$

where b is an integer in the range $\lfloor \frac{d_v - 1}{2} \rfloor < b < d_v$. Here, the outgoing message of an equality vertex is the same as the intrinsic message, unless at least b of the extrinsic messages disagree. The value of b may change from one iteration to another. The optimum value of b for a (d_v, d_c)-regular LDPC code (i.e., a code with an H-matrix having d_c ones in every row and d_v ones in every column) was computed by Gallager

(1963) and is the smallest integer b for which

$$\frac{1-p}{p} \leq \left[\frac{1 + (1 - 2p_e)^{d_c-1}}{1 - (1 - 2p_e)^{d_c-1}}\right]^{2b-d_v+1}, \tag{10.7}$$

where p and p_e are channel crossover probability (intrinsic message error rate) and extrinsic message error rate, respectively. It can be proved that Algorithm B is the best possible binary message-passing algorithm for regular LDPC codes.

10.4   LDPC Decoder Analysis

10.4.1   Decoding Threshold

For a binary symmetric channel with parameter $p < 1/2$ and an AWGN channel with parameter σ, the performance of an iterative decoder degrades with increasing channel parameter. Richardson and Urbanke (2001) studied the performance of families of low-density parity-check codes with a fixed proportion of check and equality vertices of certain degree. In the limit as the block length goes to infinity (so that the neighbors, next-neighbors, next-next-neighbors, etc., of each vertex, taken to a particular depth, can be assumed to form a tree), they show that the family exhibits a threshold phenomenon: there is a "worst-channel" for which (almost all) members of the family have a vanishing error probability as the block length and number of iterations go to infinity. This channel condition is called the *threshold* of the code family.

For example, the threshold of the family of (3,6)-regular codes on the AWGN channel under sum-product decoding is 1.1015 dB, which means that if an infinitely long (3,6)-regular code were used on an AWGN channel, convergence to zero error rate is almost surely guaranteed whenever E_b/N_0 is greater than 1.1015 dB. If the channel condition is worse than the threshold, a non-zero error rate is assured.

In practice, when finite-length codes are used, there is a gap between the E_b/N_0 required to achieve a certain (small) target error probability and the threshold associated with the given family, but this gap shrinks as the code length increases. The main aim in the asymptotic analysis of families of LDPC codes is to determine the threshold associated with the family, and one of the aims of code design is to choose the parameters of the family so that the threshold can be made to approach channel capacity.

10.4.2   Extrinsic Information Transfer (EXIT) Charts

An iterative decoder can be thought of as a "black box" that at each iteration takes two sources of knowledge about the transmitted codeword—the intrinsic information and the extrinsic information—and attempts to obtain an "improved" knowledge about the transmitted codeword. The "improved" knowledge is then used as the extrinsic information for the next iteration. When decoding is successful, the extrinsic information gets better and better as the decoder iterates. Therefore, in all methods of analysis of iterative decoders, statistics of the extrinsic messages at each iteration are studied.

For example, one might study the evolution of the entire probability density function (pdf) of the extrinsic messages from iteration to iteration. This is the most complete (and probably most complex) analysis, and is known as density evolution (Richardson and Urbanke, 2001). However, as an approximate analysis, one may study the evolution of a representative or an approximate parametrization of the true density.

An example of this approach is to use so-called "extrinsic information transfer" (EXIT) charts (Divsalar et al., 2000; El Gamal and Hammons, 2001; ten Brink, 2000, 2001). In EXIT-chart analysis, instead of tracking the density of messages, one tracks the evolution of a single parameter—a measure of the decoder's success—iteration by iteration. For example one might track the "signal-to-noise ratio" of the extrinsic messages (Divsalar et al., 2000; El Gamal and Hammons, 2001), their error probability (Ardakani and Kschischang, 2004) or the mutual information between messages and decoded bits (ten Brink, 2000). Initially, the term "EXIT chart" was used when tracking mutual information; however, the use of this term was generalized in (Ardakani and Kschischang, 2004) to the tracking of other parameters as well.

Let s denote the message-parameter being tracked, and let s_0 denote the parameter associated with the channel messages. If s_i denotes the parameter associated with the extrinsic messages at the input of the ith iteration, then an EXIT chart is the function $f(s_i, s_0)$ that gives the value of the message parameter s_{i+1} at the output of the ith iteration, i.e., we have

$$s_{i+1} = f(s_i, s_0).$$

In the remainder of this chapter, we use EXIT charts based on tracking the message error rate, as we find them most useful for our applications. Thus we track the proportion of messages that give the "wrong" value for the corresponding symbol. If p_{in} denotes this proportion at the input of an iteration, and p_{out} denotes this proportion at the output of an iteration, then we have

$$p_{out} = f(p_{in}, p_0),$$

where p_0 denotes the proportion of "wrong" channel messages.

For a fixed p_0 this function can be plotted using p_{in}-p_{out} coordinates. Usually EXIT charts are presented by plotting both f and its inverse f^{-1}, as this makes the visualization of the decoder easier. Figure 10.3 shows the concept. As can be seen from the figure, decoder progress can be visualized as a series of steps, shuttling between f and f^{-1} (as p_{out} of one iteration becomes p_{in} of the next). It can be seen that using EXIT charts, one can study how many iterations are required to achieve a target message error rate.

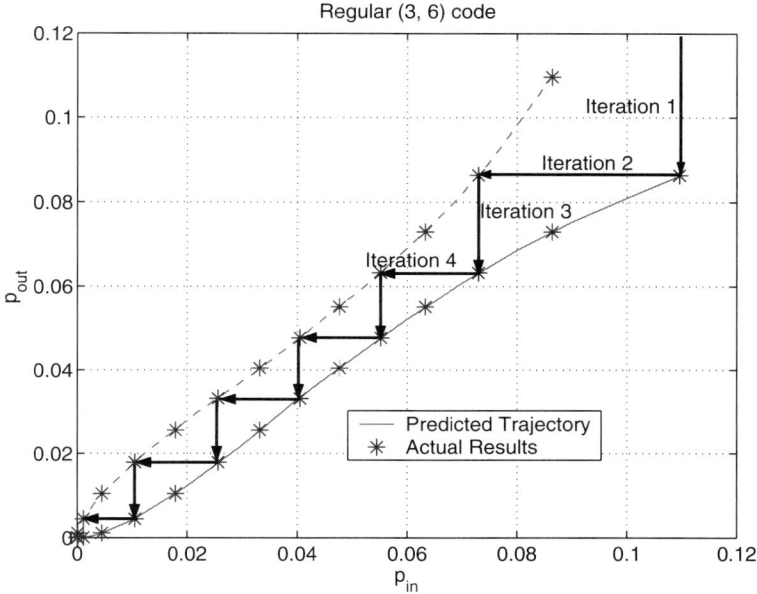

Figure 10.3   An EXIT chart based on message error rate. Simulation results are for a randomly generated (3,6)-regular code of block length 200,000 on an AWGN with $E_b/N_0 = 1.75$ dB.

The region of the graph between f and f^{-1} is referred to as the "decoding tunnel." If the "decoding tunnel" of an EXIT chart is closed, i.e., if f and f^{-1} cross at some large p_{in}, so that for some p_{in} we have $p_{out} > p_{in}$, successful decoding to a small error probability does not occur. In such cases we say that the EXIT chart is closed (otherwise, it is open). An open EXIT chart always lies below the 45-degree line $p_{out} = p_{in}$. As p_0 gets worse (i.e., as the channel degrades), the decoding tunnel becomes tighter and tighter, and hence the decoder requires more and more iterations to converge to a target error probability. Eventually, when p_0 is bad enough, the tunnel closes completely. This condition gives an estimate of the code threshold. i.e., we may estimate the threshold p_0^* as the worst channel condition for which the tunnel is open by defining p_0^* as

$$p_0^* = \arg\sup_{p_0}\{f(p_{in}, p_0) < p_{in}, \text{ for all } 0 < p_{in} \le p_0\}.$$

EXIT chart analysis is not as accurate as density evolution, because it tracks just a single parameter as the representative of a pdf. For many applications, however, EXIT charts are very accurate. For instance in (ten Brink, 2000, 2001), EXIT charts are used to approximate the behavior of iterative turbo decoders on a Gaussian channel very accurately. In (Ardakani and Kschischang, 2004) it is shown that, using EXIT charts, the threshold of convergence for LDPC codes on AWGN channel can be approximated within a few thousandths of a dB of the actual value.

In next section we show that EXIT charts can be used to design irregular LDPC codes which perform not more than a few hundredths of a dB worse than those designed by density evolution. One should also notice that when the pdf of messages can truly be described by a single parameter, e.g., in the so-called "binary erasure channel," EXIT chart analysis is equivalent to density evolution.

10.4.3 Gaussian Approximations

There have been a number of approaches to one-dimensional analysis of sum-product decoding of LDPC codes on the AWGN channel (Ardakani and Kschischang, 2004; Chung et al., 2001; Divsalar et al., 2000; Lehmann and Maggio, 2002; ten Brink and Kramer, 2003; ten Brink et al., 2004), all of them based on the observation that the pdf of the decoder's LLR messages is approximately Gaussian. This approximation is quite accurate for messages sent from equality vertices, but less so for messages sent from check vertices. In this subsection we describe an accurate one-dimensional analysis for LDPC codes based on a Gaussian assumption only for the messages sent from the equality vertices.

Because AWGN channels and binary symmetric channels treat 0's and 1's symmetrically, and because LDPC codes are binary linear codes, the behavior of a sum-product decoder is independent of which codeword was transmitted (Richardson and Urbanke, 2001). In the analysis of a decoder, we may therefore assume that the all-zero codeword (equivalent to the all-$\{+1\}$ channel word) is transmitted.

A probability density function $f(x)$ is called *symmetric* if $f(x) = e^x f(-x)$. In (Richardson and Urbanke, 2001) it has been shown that if the LLR of the channel messages is symmetric, then all messages sent in sum-product decoding are symmetric. A Gaussian pdf with mean m and variance σ^2 is symmetric if and only if $\sigma^2 = 2m$. As a result, a symmetric Gaussian density can be expressed by a single parameter.

Under the assumption that the all-zero codeword was transmitted over an AWGN channel, it turns out that the intrinsic LLR messages have a symmetric Gaussian density with a mean of $2/\sigma^2$ and a variance of $4/\sigma^2$, where σ^2 is the variance of the Gaussian channel noise. It follows that under sum-product decoding, all messages remain symmetric. In addition, since the update rule at the equality vertices is the summation of incoming messages, according to the *central limit theorem*, the density of the messages at the output of equality vertices tends to be Gaussian, so it seems sensible to approximate them with a symmetric Gaussian.

To avoid a Gaussian assumption on the output of check vertices, we consider one whole iteration at once. That is to say, we study the input-output behavior of the decoder from the input of the iteration (messages from equality vertices to check vertices) to the output of that iteration (messages from equality vertices to check vertices). Figure 10.4 illustrates the idea. In every iteration we assume that the input and the output messages shown in Figure 10.4, which are outputs of equality vertices, are symmetric Gaussian. We start with Gaussian distributed messages at

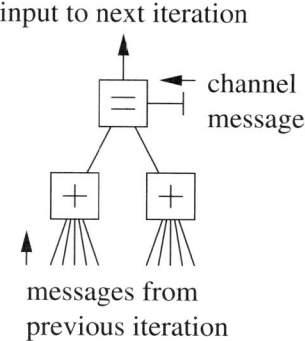

input to next iteration

channel
message

messages from
previous iteration

Figure 10.4   A depth-one tree for a (3,6)-regular LDPC code.

the input of the iteration and compute the pdf of messages at the output. This can be done by "one-step" density evolution. Then we approximate the actual output pdf with a symmetric Gaussian. Since we assume that the all-zero codeword is transmitted, the negative tail of this density reflects the message error rate. As a result, we can track the evolution of message error rate and represent it in an EXIT chart. This technique led to the results shown in Figure 10.3, showing the close agreement between simulation results and the decoding behavior predicted by the EXIT-chart analysis.

We refer to the method of approximating only the output of equality vertices with a Gaussian (but not the output of the check vertices) as the "semi-Gaussian" approximation.

10.4.4   Analysis of Irregular LDPC Codes

A single depth-one tree cannot be defined for irregular codes, since not all vertices (even of the same type) have the same degree. If the check degree distribution is fixed, each equality vertex of a fixed degree gives rise to its own depth-one tree. Irregularity in the check vertices is taken into account in these depth-one trees. For any fixed check degree distribution, we refer to the depth-one tree associated with a degree i equality vertex as the "degree i depth-one tree."

For reasons similar to the case of regular codes, we assume that at the output of any depth-one tree, the pdf of LLR messages is well-approximated by a symmetric Gaussian. As a result, the pdf of LLR messages at the input of check vertices can be approximated as a mixture of symmetric Gaussian densities. The weights of this mixture are determined by the proportion of equality vertices of each degree. Nevertheless, at the output of a given equality vertex, the distribution is still close to Gaussian and so the semi-Gaussian method can be used to find the EXIT charts corresponding to the equality vertices of different degrees. In other words, for any i, using the "degree i depth-one tree," an EXIT chart associated with equality vertices of degree i can be found. We call such EXIT charts *elementary EXIT charts*. We have

$$p_{out,i} = f_i(p_{in}, p_0),$$

where $p_{out,i}$ is the message error rate at the output of degree i equality vertices.

Now, using Bayes' rule, p_{out} for the mixture of all equality vertices can be computed as

$$p_{out} = \sum_{i \geq 2} Pr(\text{degree} = i) Pr(\text{error}|\text{degree} = i)$$

$$= \sum_{i \geq 2} \lambda_i f_i(p_{in}, p_0), \tag{10.8}$$

where λ_i denotes the proportion of edges incident on equality vertices of degree i.

Thus we obtain the important result that the overall EXIT chart can be obtained as a weighted linear combination of elementary EXIT charts. A similar formulation can be used when the mean of the messages is the parameter tracked by the EXIT chart (Chung et al., 2001). It has been shown in (Tuechler and Hagenauer, 2002) that when the messages have a symmetric pdf, mutual-information also combines linearly to form the overall mutual-information.

10.5   Design of Irregular LDPC Codes

We now describe how this EXIT-chart framework may be used to design irregular LDPC codes. In this framework, the design problem can be simplified to a linear program. Let λ_i denote the proportion of factor graph edges incident on an equality vertex of degree i, and let ρ_j denote the proportion of edges incident on a check vertex of degree j.

We formulate the design problem for an irregular LDPC code as that of shaping an EXIT chart from a group of elementary EXIT charts (according to (10.8)) so that the rate of the code is maximized, but subject to the constraint that the resulting EXIT chart remains open, i.e., so that $f(x) < x$ for all $x \in (0, p_0]$, where p_0 is the initial message error rate at the decoder.

It can be shown that the rate of an LDPC code is at least $1 - \frac{\sum \rho_j/j}{\sum \lambda_i/i}$ and hence, for a fixed check degree distribution, the design problem can be formulated as the following linear program:

$$\text{maximize:} \quad \sum_{i \geq 2} \lambda_i/i$$
$$\text{subject to:} \quad \lambda_i \geq 0,$$
$$\sum_{i \geq 2} \lambda_i = 1 \text{ and}$$
$$\forall p_{in} \in (0, p_0] \left(\sum_{i \geq 2} \lambda_i f_i(p_{in}, p_0) < p_{in} \right).$$

In the above formulation, we have assumed that the elementary EXIT charts are given. In practice, to find these curves we need to know the degree distribution of the code. We need the degree distribution to associate every input p_{in} to its equivalent input pdf, which is in general assumed to be a Gaussian mixture. In

Table 10.1 A List of Irregular Codes Designed for the AWGN Channel by the Semi-Gaussian Method

Degree sequence	Code 1	Code 2	Code 3	Code 4	Code 5	Code 6
$d_{v_1}, \lambda_{d_{v_1}}$	2, .1786	2, .1530	2, .1439	2, .1890	2, .2444	2, .3000
$d_{v_2}, \lambda_{d_{v_2}}$	3, .3046	3, .2438	3, .1602	3, .1158	3, .1687	3, .1937
$d_{v_3}, \lambda_{d_{v_3}}$	5, .0414	7, .1063	5, .1277	4, .1153	4, .0130	4, .0192
$d_{v_4}, \lambda_{d_{v_4}}$	6, .0531	10, .2262	6, .0219	6, .0519	5, .1088	7, .2378
$d_{v_5}, \lambda_{d_{v_5}}$	7, .0007	14, .0305	7, .0279	7, .0875	7, .1120	14, .0158
$d_{v_6}, \lambda_{d_{v_6}}$	10, .4216	19, .0001	8, .0103	14, .0823	14, .1130	15, .0114
$d_{v_7}, \lambda_{d_{v_7}}$	—	23, .1293	12, .1551	15, .0007	15, .0577	20, .0910
$d_{v_8}, \lambda_{d_{v_8}}$	—	32, .0736	30, .0004	16, .0001	25, .0063	25, .0002
$d_{v_9}, \lambda_{d_{v_9}}$	—	38, .0372	37, .3525	39, .3573	25, .0109	30, .0232
$d_{v_{10}}, \lambda_{d_{v_{10}}}$	—	—	40, .0001	40, .0001	40, .1652	40, .1077
d_c	40	24	22	10	7	5
Threshold σ	0.5072	0.6208	0.6719	0.9700	1.1422	1.5476
Rate	0.9001	0.7984	0.7506	0.4954	0.3949	0.2403
Gap to Shannon limit (dB)	0.1308	0.0630	0.0666	0.1331	0.1255	0.2160

other words, prior to the design, the degree distribution is not known and as a result, we cannot find the elementary EXIT charts to solve the linear program above.

To solve this problem, we suggest a recursive solution. At first we assume that the input message to the iteration has a single symmetric Gaussian density instead of a Gaussian mixture. Using this assumption, we can map every message error rate at the input of the iteration to a unique input pdf and so find f_i curves for different i. (It is interesting that even with this assumption the error in approximating the threshold of convergence, based on our observations, is less than 0.3 dB and the codes which are designed have a convergence threshold of at most 0.4 dB worse than those designed by density evolution. One reason for this is that when the input of a check vertex is mixture of symmetric Gaussians, due to the computation at the check vertex, its output is dominated by the Gaussian in the mixture having smallest mean.)

After finding the appropriate degree distribution based on the single Gaussian assumption we use this degree distribution to find the correct elementary EXIT charts based on a Gaussian mixture. Now we use the corrected curves to design an irregular code. In this level of design, the designed degree distribution is close to the degree distribution used in finding the elementary EXIT charts. Therefore, analyzing this code with its actual degree distribution shows minor error. One can continue these recursions for higher accuracy. However, in our examples after one iteration of design the designed threshold and the exact threshold differed less than 0.01 dB.

We have designed a number of irregular codes with a variety of rates using this Gaussian approximation. The results are presented in Table 10.1. In the design of the presented codes, we have avoided any equality vertices or check vertices with degrees higher than 40. Table 10.1 suggests that for code rates more than 0.25, the method is quite successful. For rates greater than 0.85, getting close to capacity requires high-degree check vertices. To show that our method can actually handle high-rate codes, we designed a rate 0.9497 code, which uses check vertices of degree 120 but no equality vertices of degree greater than 40. The degree sequence for this code is $\lambda = \{\lambda_2 = 0.1029, \lambda_3 = 0.1823, \lambda_6 = 0.1697, \lambda_7 = 0.0008, \lambda_9 = 0.1094, \lambda_{15} = 0.0240, \lambda_{35} = 0.2576, \lambda_{40} = 0.1533\}$. The threshold of this code is at $\sigma = 0.4462$, which means it has a gap of only 0.0340 dB from the Shannon limit.

10.6   Conclusions and Future Prospects

Low-density parity-check (LDPC) codes are a flexible family of codes with a simple decoding algorithm. As we have shown in this chapter, the structure of the codes can be fine-tuned to allow for decoding even at channel parameters that approach the Shannon limit. Because of their excellent properties, LDPC codes have attracted an enormous research interest, and there is now a large body of literature which we have not attempted to survey in this chapter.

Low-density parity-check codes are now beginning to emerge in a variety of communication standards, including, e.g., the DVB-S2 standard for digital video broadcasting by satellite (Eroa et al., 2004). An important research direction is that of finding LDPC codes with relatively short block lengths (a few thousand bits, say), but which still have excellent iterative decoding performance. For further reading in the area of LDPC codes, we recommend the books of Lin and Costello (2004) and MacKay (2003) and the survey article of (Richardson and Urbanke, 2003) as excellent starting points.

11   Turbo Processing

Claude Berrou, Charlotte Langlais, and Fabrice Seguin

Turbo processing is the way to process data in communication receivers so that no information stemming from the channel is wasted. The first application of the turbo principle was in error correction coding, which is an essential function in modern telecommunications systems. A novel structure of concatenated codes, nicknamed turbo codes, was devised in the early 1990s in order to benefit from the turbo principle. Turbo codes, which have near-optimal performance according to the theoretical limits calculated by Shannon, have since been adopted in several telecommunications standards.

The turbo principle, also called the message-passing principle or belief propagation, is exploitable in signal processing other than error correction, such as detection and equalization. More generally, every time separate processors work on data sets that have some link together, the turbo principle may improve the result of the global processing. In digital circuits, the turbo technique is based on an iterative procedure, with multiple repeated operations in all the processors considered. Another more natural possibility is the use of analog circuits, in which the exchange of information between the different processors is continuous.

11.1   Introduction

Error correction coding, also known as channel coding, is a fundamental function in modern telecommunications systems. Its purpose is to make these systems work even in tough physical conditions, due for instance to a low received signal level, interference, or fading. Another important field of application for error correction coding is mass storage (computer hard disk, CD and DVD-ROM, etc.), where the ever-continuing miniaturization of the elementary storage pattern makes reading the information more and more tricky.

Error correction is a digital technique, that is, the information message to protect is composed of a certain number of digits drawn from a finite alphabet. Most often, this alphabet is binary, with logical elements or bits 0 or 1. Then, error correction coding, in the so-called *systematic* way, involves adding some number of redundant logical elements to the original message, the whole being called a codeword. The mathematical law that is used to calculate the redundant part of the codeword is specific to a given code. Besides this mathematical law, the main parameters of a code are as follows.

- The code rate: the ratio between the number of bits in the original message and in the codeword. Depending on the application, the code rate may be as low as 1/6 or as high as 9/10.

- The minimum Hamming distance (MHD): the minimum number of bits that differ from one codeword to any other. The higher the MHD, the more robust the associated decoder confronted with multiple errors.

- The ability of the decoder to exploit soft (analog) values from the demodulator, instead of hard (binary) values. A soft value (that is, the sign and the magnitude) carries more information than a hard value (only the sign).

- The complexity and the latency of the decoder.

Since the seminal work by Shannon on the potential of channel coding (Shannon, 1948), many codes have been devised and used in practical systems. The state of the art, in the early 1990s, was the coding construction depicted in fig. 11.1. This is called "standard concatenation" and is made up of a serial combination of a Reed-Solomon (RS) encoder, a symbol interleaver, and a convolutional encoder. The corresponding decoder (fig. 11.1) is composed of a Viterbi decoder, a symbol de-interleaver, and an RS decoder. This concatenated scheme works nicely because the Viterbi decoder can easily benefit from soft samples coming from the demodulator, while the RS decoder can withstand residual bursty errors that may come from the Viterbi decoder. Nevertheless, although the MHD of the concatenated code is very large, the decoder does not provide optimal error correction. Roughly, the performance is 3 or 4 dB from the theoretical limit. Where does this loss come from?

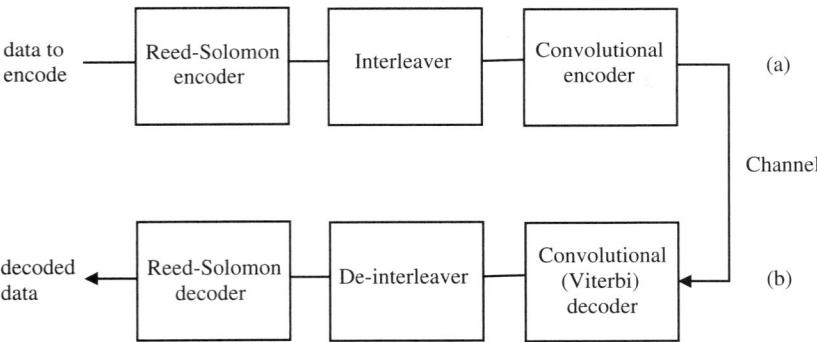

Figure 11.1 Standard concatenation of a Reed-Solomon encoder and a convolutional encoder, and the associated decoder.

The inner Viterbi decoder, processing analog-like input samples, is locally optimum, that is, it derives the maximum benefit from the redundancy added by the convolutional encoder. The outer RS decoder, which is also locally optimum, benefits from the work of the inner decoder and from the redundancy added by the RS encoder. Both decoders are optimum, each of them separately, but their association is not optimal: the Viterbi decoder does not exploit the redundancy offered by the RS codeword. A global decoder, which would use the whole redundancy in one processing step, would be extremely complex and is not realistic.

The way to contemplate near-optimal decoding of the standard concatenated scheme is to enable the inner decoder to benefit from the work done by the outer decoder, using a kind of feedback. This observation is at the root of turbo processing, "turbo" being used to refer to the way the power of a turbo engine is increased by the reuse of its exhaust gases. This being said, the concatenated decoding scheme of figure 11.1 does not easily lend itself to such a feedback principle. The code has to be devised in order to enable bidirectional exchanges between the two component decoders.

11.2 Random Coding and Recursive Systematic Convolutional (RSC) Codes

The theoretical limits were calculated by Shannon on the basis of random coding, which has since remained the reference in the matter of error correction. The systematic random encoding of a message having k information bits and producing a codeword with n bits may be achieved in the following way.

As a first step, once and for all, k binary words with $n - k$ bits are drawn at random and memorized. These k words will constitute the basis of a vector space, the ith random word ($1 \leq i \leq k$) being associated with the information message containing only zeros (the "all-zero" message) except in the ith place. The redundant part of any codeword is obtained by calculating the sum modulo two of random words whose address i is such that the ith bit of the original message is one. The coding rate is $R = k/n$.

This very simple construction leads to a very large MHD. Because two codewords differ at least by one information bit, and thanks to the random feature of the redundant part, the mean distance is $1 + \frac{n-k}{2}$. Nevertheless, the MHD of the code being a random value, its different realizations may be less than this mean value. A realistic approximation of the actual MHD is $\frac{n-k}{4}$. Such large values, for instance 100 for $n = 2k = 800$, are unreachable when using practical codes. Fortunately, these large MHDs are not necessary for common communications systems (Berrou et al., 2003).

The device depicted in fig. 11.2 is called a recursive systematic convolutional (RSC) code, whose length, the number of memory elements, is denoted ν. This encoder is based on the principle and the random features of the linear feedback register (LFR), also called a pseudo-random generator. When choosing an appropriate set for the feedback taps, the period P of the LFR is maximum and equal to $P = 2^\nu - 1$. For sufficiently large values of ν, P can be much higher than any length of messages to process, by several orders of magnitude. Therefore, the RSC code could then be assimilated to a quasi-perfect random code. Values of ν larger than 30 or 40 would be sufficient to make the RSC code equivalent to a random code.

The message $d = \{d_0, \ldots d_i, \ldots, d_{k-1}\}$ to be encoded feeds the LFR input and is transmitted as symbols X, as the systematic part of the codeword. The redundant or parity part is provided by the summation modulo two of certain binary values from the register. Using the D (delay) formalism, the redundant symbols Y are expressed as

$$Y(D) = \frac{G_2(D)}{G_1(D)} d(D), \tag{11.1}$$

where

$$\begin{aligned}
G_1(D) &= 1 + \sum_{j=1}^{\nu-1} G_1^{(j)} D^j + D^\nu \quad \text{and} \\
G_2(D) &= 1 + \sum_{j=1}^{\nu-1} G_2^{(j)} D^j + D^\nu
\end{aligned} \tag{11.2}$$

are the polynomials defining the taps for recursivity and parity construction. $G_1^{(j)}$ (resp. $G_2^{(j)}$) is equal to 1 if the register tap at level j ($1 \leq j \leq \nu - 1$) is used in the construction of recursivity (resp. parity), and 0 otherwise. $G_1(D)$ and $G_2(D)$ are generally defined in octal forms. For instance, $1 + D^3 + D^4$ is referred to as polynomial 23.

Convolutional encoding exhibits a side effect at the end of the coding process, which may be detrimental to decoding performance regarding the last bits of the message. In order to take its decision, the decoder uses information carried by current, past, and subsequent symbols, and the subsequent symbols do not exist at the end of the block. This point is known as the termination problem. Among several solutions to cope with this problem, the classical one consists in adding

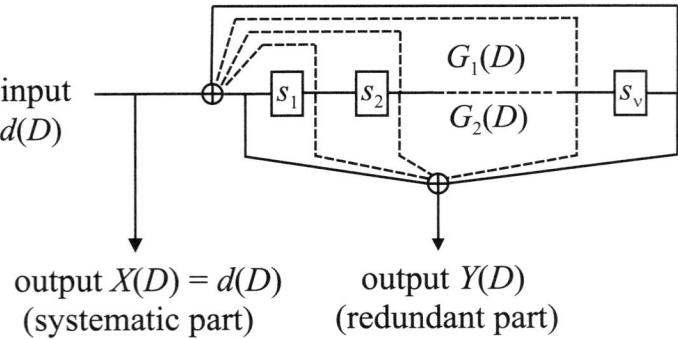

Figure 11.2   Recursive systematic convolutional (RSC) encoder, with length ν.

dummy information bits, called tail bits, to make the encoder return to the "all-zero" state. Another more elegant technique is tail-biting, also called circular, termination (Weiss et al., 2001). This involves allowing any state as the initial state and encoding the sequence, containing k information bits, so that the final state of the encoder register will be equal to the initial state. The trellis of the code (the temporal representation of the possible states of the encoder, from time $i = 0$ to $i = k - 1$) can then be regarded as a circle. In what follows, we will refer to *circular recursive systematic convolutional* (CRSC) codes, the circular version of RSC codes. Thus, without having to pay for any additional information, and therefore without impairing spectral efficiency, the convolutional code has become a real block code, in which, for each time i, the past is also the future, and vice versa.

RSC or CRSC codes, like classical nonrecursive convolutional codes, are linear codes. Thanks to the linearity property, the code characteristics are expressed with respect to the all-zero sequence. In this case, any nonzero sequence $d(D)$, accompanied by redundancy $Y(D)$, will represent a possible error pattern for the coding/decoding system, one meaning a binary error. Equation 11.1 indicates that only a fraction of sequences $d(D)$, which are multiples of $G_1(D)$, lead to short length redundancy. We call these particular sequences return-to-zero (RTZ) sequences (Podemski et al., 1995), because they force the encoder, if initialized in state 0, to retrieve this state after the encoding of $d(D)$. In what follows, we will be interested only in RTZ patterns, assuming that the decoder will never decide in favor of a sequence whose distance from the all-zero sequence is very large. The fraction of sequences $d(D)$ that are RTZ is exactly

$$p(\mathrm{RTZ}) = 2^{-\nu}, \tag{11.3}$$

because the encoder has 2^{ν} possible states and an RTZ sequence finishes systematically at state 0.

The shortest RTZ sequence is $G_1(D)$ or its shifted version. Any RTZ sequence, in the block of k bits with circular termination, may be expressed as

$$\mathrm{RTZ}(D) = G_1(D) \sum_{i=0}^{k-1} a_i D^i \quad \mathrm{mod}\ (1 + D^k), \tag{11.4}$$

where a_i takes value 0 or 1. Operation modulo $(1 + D^k)$ transforms all D^x monomials in the resulting product into $D^{x \mod k}$, for any integer x, so that all exponents are between 0 and $k - 1$.

The minimum number of 1's belonging to an RTZ sequence is two. This is because $G_1(D)$ is a polynomial with at least two nonzero terms, and equation 11.4 then guarantees that $\mathrm{RTZ}(D)$ also has at least two nonzero terms. The number of 1's in a particular RTZ sequence is called the input weight and is denoted w. We then have $w_{\min} = 2$ for RSC codes, and the RTZ sequences with weight 2 are of the general form

$$\mathrm{RTZ}_{w=2}(D) = D^\tau(1 + D^{p^P}) \mod (1 + D^k), \tag{11.5}$$

where τ is the starting time, p any positive integer, and P the period of the encoder, as previously introduced.

RTZ sequences with odd weight may either exist or not, depending on the expression of $G_1(D)$. RTZ sequences with even weight always exist, especially of the form

$$\mathrm{RTZ}_{w=2l}(D) = \sum_{j=1}^{l-1} D^{\tau_j}(1 + D^{p_j^P}) \mod (1 + D^k), \tag{11.6}$$

that is, as a combination of l any weight-2 RTZ sequences, with τ_j and p_j as any positive integers. This sort of composite RTZ sequence has to be considered closely when trying to design good permutations for turbo codes, as explained in section 11.5.

What we are searching for is a very long RSC code, having a large period P, in order to take advantage of quasi-perfect random properties. But such codes cannot be decoded, due to the too-large number of states to consider and to process. That is why other forms of random-like codes, a little more sophisticated, have to be devised.

11.3   Turbo Codes

In the previous section, we saw that the probability that any given sequence is an RTZ sequence for a CRSC encoder is $1/2^\nu$. Now, if we encode this sequence N times (fig. 11.3 with $\nu = 3$), each time in a different order and drawn at random by permutation Π_j ($1 \leq j \leq N$) (the first order may be the natural order), the probability that the sequence remains RTZ for all encoders is lowered to $1/2^{N\nu}$. For example, with $\nu = 3$ and $N = 7$, this probability is less than 10^{-6}. This technique is known as a multiple parallel concatenation of CRSC codes (Berrou et al., 1999). Of course, to deal with realistic coding rates (around $1/2$), some puncturing has to be performed, that is, not all the redundant symbols Y_j are used to form the codeword. For instance, if $R = 1/2$, each component encoder provides only k/N parity bits. If the message is not RTZ, after permutation Π_j, the average weight of sequence $\{Y_j\}$ is $\frac{k}{2N}$ (assuming that every other bit in $\{Y_j\}$ is 1, statistically). This guarantees a large distance when one permuted sequence, at least, is not RTZ.

Fortunately, it is possible to obtain quasi-optimum performance with only two encodings (fig. 11.3), and this is a classical turbo code (Berrou et al., 1993). For bit error rates (BERs) higher than around 10^{-5}, the permutation may still be drawn at random but, for lower rates, a particular effort has to be made in its design. The way the permutation is devised fixes the MHD $d_{\min}$ of the turbo code, and therefore the achievable asymptotic gain G_a offered by the coding scheme, according to the

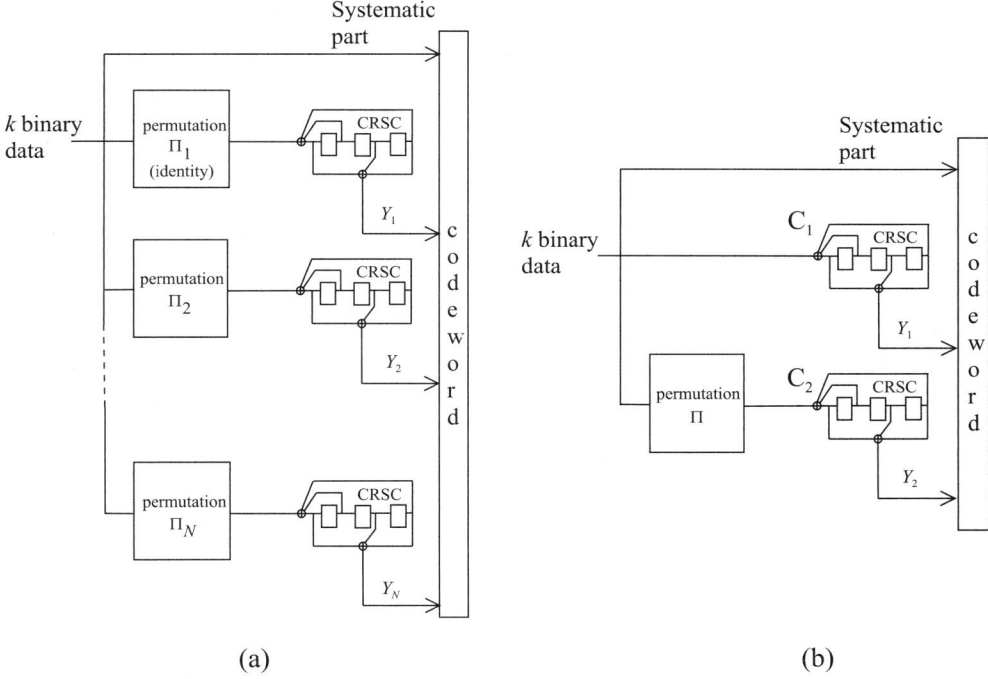

(a)   (b)

Figure 11.3 (a) In this multiple parallel concatenation of circular recursive systematic convolutional (CRSC) codes, the block containing k information bits is encoded N times. The probability that the sequence remains of the return-to-zero (RTZ) type after the N permutations, drawn at random (except the first one), is very low. The properties of this multiconcatenated code are very close to those of random codes. (b) The number of encodings can be limited to two, provided that permutation Π is judiciously devised. This is a classical turbo code.

well-known approximation

$$G_a \approx 10\log(Rd_{\min}) \tag{11.7}$$

The natural coding rate of a turbo code is $R = 1/3$. In order to obtain higher rates, certain redundant symbols are punctured. For instance, Y_1 and Y_2 symbols are transmitted alternately to achieve $R = 1/2$.

A particular turbo code is defined by the following parameters.

■ m, the number of bits in the input words. Applications known so far consider binary ($m = 1$) and double-binary ($m = 2$) input words (see section 11.6).

■ The component codes C_1 and C_2 (code memory ν, recursivity and redundancy polynomials). The values of ν are 3 or 4 in practice and the polynomials are generally those that are recognized as the best for simple unidimensional convolutional coding, that is, (15,13) for $\nu = 3$ and (23,35) for $\nu = 4$, or their symmetric forms.

- The permutation function, which plays a decisive role when the target BER is lower than about 10^{-5}. Above this value, the permutation may follow any law, provided of course that it respects at least the scattering property (the permutation may be the regular one, for instance).

- The puncturing pattern. This has to be as regular as possible, like for simple convolutional codes. In addition to this rule, the puncturing pattern is defined in close relationship with the permutation function when very low errors rates are sought for.

11.4   Turbo Decoding

Decoding a composite code by a global single process is not possible in practice, because of the tremendous number of states to consider. A joint probabilistic process by the decoders of C_1 and C_2 has to be elaborated, following a kind of divide-and-conquer strategy. Because of local latency constraints, this joint process is worked out in an iterative manner in a digital circuit. Analog versions of the turbo decoder are also considered, offering several advantages, as explained in section 11.7.

Turbo decoding relies on the following fundamental criterion, which is applicable to all so-called message-passing or belief-propagation algorithms (McEliece et al., 1998):

When having several probabilistic machines work together on the estimation of a common set of symbols, all the machines have to give the same decision, with the same probability, about each symbol, as a single (global) decoder would.

To make the composite decoder satisfy this criterion, the structure of fig. 11.4 is adopted. The double loop enables both component decoders to benefit from the whole redundancy.

The components are soft-in-soft-out (SISO) decoders, permutation Π and inverse permutation Π^{-1} memories. The node variables of the decoder are logarithms of likelihood ratios (LLRs), also simply called log-likelihood ratios. An LLR related to a particular binary datum d_i ($0 \leq i \leq k-1$) is defined, apart from a multiplying factor, as

$$\mathrm{LLR}(d_i) = \ln\left(\frac{\Pr(d_i = 1)}{\Pr(d_i = 0)}\right). \tag{11.8}$$

The role of a SISO decoder is to process an input LLR and, thanks to local redundancy (i.e., y_1 for DEC1, y_2 for DEC2), to try to improve it. The output LLR of a SISO decoder, for a binary datum, may be simply written as

$$\mathrm{LLR}_{\mathrm{out}}(d_i) = \mathrm{LLR}_{\mathrm{in}}(d_i) + z(d_i), \tag{11.9}$$

where $z(d_i)$ is the *extrinsic* information about d_i, provided by the decoder. If this

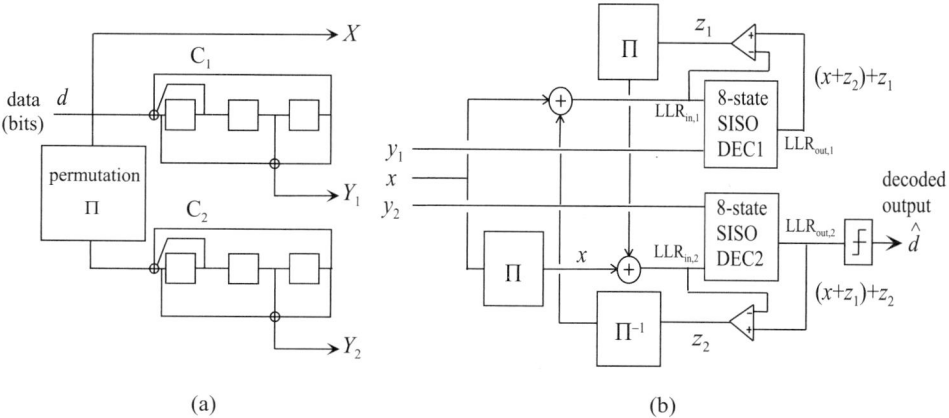

Figure 11.4   An 8-state turbo code (a) and its associated decoder (b), with a basic structure assuming no delay processing.

works properly, $z(d_i)$ is most of the time negative if $d_i = 0$, and positive if $d_i = 1$.

The composite decoder is constructed in such a way that only extrinsic terms are passed by one component decoder to the other. The input LLR to a particular decoder is composed of the sum of two terms: the information symbols (x) stemming from the channel, also called the *intrinsic* values, and the extrinsic terms (z) provided by the other decoder, which serve as a priori pieces of information. The intrinsic symbols are inputs common to both decoders, which is why extrinsic information does not contain them. In addition, the outgoing extrinsic information does not include the incoming extrinsic information, in order to minimize correlation effects in the loop. The subtractors in fig. 11.4 are used to remove intrinsic and extrinsic information from the feedback loops. Nevertheless, because the blocks have finite length, correlation effects between extrinsic and intrinsic values may exist and degrade the decoding performance.

The practical course of operation is:

Step 1: process the data peculiar to one code, say C_2 (x and y_2), by decoder DEC2, and store the extrinsic pieces of information (z_2) resulting from the decoding in a memory. If data are missing because of puncturing, the corresponding values are set to analog 0 (neutral value).

Step 2: process the data specific to C_1 (x, deinterleaved z_2 and y_1) by decoder DEC1, and store the extrinsic pieces of information (z_1) in a memory. By properly organizing the read/write instructions, the same memory can be used for storing both z_1 and z_2.

Steps 1 and 2 make up the first iteration.

Step 3: process C_2 again, now taking interleaved z_1 into account, and store the updated values of z_2.

And so on.

The process ends after a preestablished number of iterations, or after the decoded block has been estimated as correct, according to some stop criterion (see Matache et al., 2000, for possible stopping rules). The typical number of iterations for the decoding of convolutional turbo codes is four to ten, depending on the constraints relating to complexity, power consumption, and latency.

According to the structure of the decoder, after p iterations, the output of DEC1 is

$$\text{LLR}_{\text{out}1,p}(d_i) = (x + z_{2,p-1}(d_i)) + z_{1,p}(d_i),$$

where $z_{u,p}(d_i)$ is the extrinsic piece of information about d_i, yielded by decoder u after iteration p, and the output of DEC2 is

$$\text{LLR}_{\text{out}2,p}(d_i) = (x + z_{1,p-1}(d_i)) + z_{2,p}(d_i).$$

If the iterative process converges toward fixed points, $z_{1,p}(d_i) - z_{1,p-1}(d_i)$ and $z_{2,p}(d_i) - z_{2,p-1}(d_i)$ both tend to zero when p goes to infinity. Therefore, from the equations above, both LLRs become equal, which fulfills the fundamental condition of equal probabilities provided by the component decoders for each datum d_i. As for the proof of convergence itself, one can refer to various papers dealing with the theoretical aspects of the subject, such as Weiss and Freeman (2001) and Duan and Rimoldi (2001). An important tool for the analysis of convergence is the EXIT chart (ten Brink, 2001). EXIT, which stands for extrinsic information transfer, considers both SISOs decoders in the turbo decoder as nonlinear transfer functions of extrinsic information, in a statistical way.

Turbo decoding is not optimal. This is because, during the first half-iteration, an iterative process has obviously to begin with only a part of the redundant information available (either y_1 or y_2). Furthermore, correlation effects between noises affecting intrinsic and extrinsic terms may be detrimental. Fortunately, loss due to suboptimality is small, about some tenths of one dB.

There are two families of SISO algorithms, those based on the Viterbi algorithm (Battail, 1987; Hagenauer and Hoeher, 1989), which can be used for high-throughput continuous-stream applications; the others based on the APP (a posteriori probability, also called MAP or BCJR) algorithm (Bahl et al., 1974) or its simplified derived versions (Robertson et al., 1997) for block decoding. If the full APP algorithm is chosen, it is better for extrinsic information to be expressed by probabilities instead of LLRs, which avoids calculating a useless variance for extrinsic terms.

In practice, depending on the kind of SISO algorithm chosen, some tuning operations (multiplying, limiting) on extrinsic information are added to the basic structure to ensure stability and convergence within a small number of iterations.

11.5   Permutation

In a turbo code, permutation plays a double role:

1. It must ensure maximal scattering or spreading of adjacent bits, in order to minimize the correlation effects in the message passing between the two component decoders.

2. It contributes greatly to the value of the MHD. Between a badly designed and a well-designed permutation, the MHD may differ by a factor of 2 or 3.

Let us consider the binary turbo code represented in fig. 11.4, with permutation falling on k bits. The worst permutation we can imagine is permutation identity, which minimizes the coding diversity (i.e., $Y_1 = Y_2$). On the other hand, the best permutation that could be used, but which probably does not exist (Svirid, 1995), could allow the concatenated code to be equivalent to a sequential machine whose irreducible number of states would be 2^{k+6}. There are actually $k + 6$ binary storage elements in the structure, k in the permutation memory and 6 in the encoders. Assimilating this machine to a convolutional code would give a very long code and very large minimum distances, for usual values of k. From the worst to the best of permutations, there is great choice between the $k!$ possible combinations, and we still lack a sound theory about this. Nevertheless, good permutations have already been designed to elaborate normalized turbo codes using pragmatic approaches.

11.5.1   Regular Permutation

Maximum spreading (criterion 1 above) is achieved by regular permutation. For a long time, regular permutation was almost exclusively seen as rectangular (linewise writing and columnwise reading in an ad hoc memory, fig. 11.5). When using CRSC codes as the component codes of a turbo code, circular permutation, based on congruence properties, is more appropriate. Circular permutation, for blocks having k information bits (fig. 11.5), is devised as follows. After writing the data in a linear memory, with address i ($0 \leq i \leq k - 1$), the block is likened to a circle, both extremities of the block ($i = 0$ and $i = k - 1$) then being contiguous. The data are read out such that the jth datum read was written at the position i given by

$$i = \Pi(j) = Pj \quad \mathrm{mod}\ k,$$

where the skip value P is an integer, relatively prime with k.

We define the total spatial distance (or span) $S(j_1, j_2)$ as the sum of the two spatial distances, before and after permutation, for a given pair of positions j_1 and j_2:

$$S(j_1, j_2) = f(j_1, j_2) + f(\Pi(j_1), \Pi(j_2)), \tag{11.10}$$

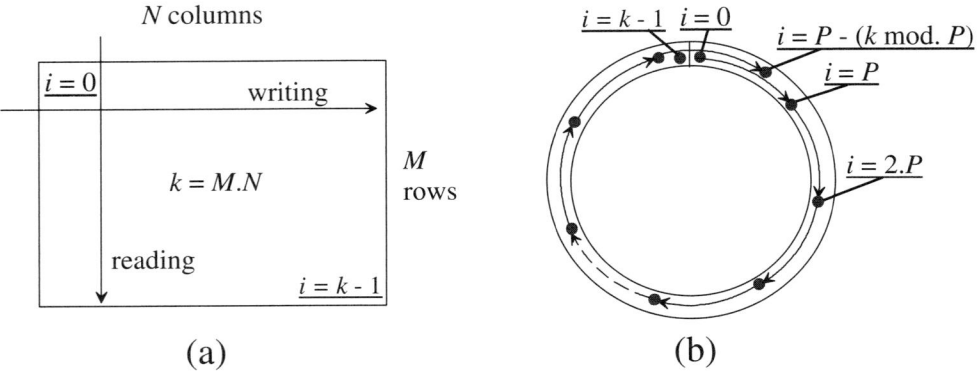

Figure 11.5 Rectangular (a) and circular (b) permutation.

where

$$f(u, v) = \min\{|u - v|, k - |u - v|\}. \tag{11.11}$$

Finally, we denote $S_{\min}$ the minimum value of $S(j_1, j_2)$ for all possible pairs j_1 and j_2:

$$S_{\min} = \min_{j_1, j_2}\{S(j_1, j_2)\}. \tag{11.12}$$

With regular permutation, the value of P that maximizes $S_{\min}$ (Boutillon and Gnaedig, 2005) is

$$P_0 = \sqrt{2k}, \tag{11.13}$$

with the condition:

$$k = \frac{P_0}{2} \mod P_0, \tag{11.14}$$

which gives:

$$S_{\min} = P_0 = \sqrt{2k}. \tag{11.15}$$

In practice, to comply as far as possible with the criterion of maximum total spatial distance, P is chosen as an integer close to P_0, and prime with k.

11.5.2   Real Permutations

Let us recall that a decoder of an RSC code is only sensitive to error sequences of the RTZ type, as introduced in section 11.2. Then, real permutations have to best satisfy the following ideal rule:
If a sequence is RTZ before permutation, then it is not RTZ after permutation, and vice versa.

In this case, at least one of the component decoders in fig. 11.4 is able to recover from the errors. But the previous rule is impossible to comply with, and a more realistic target is this one:

If a sequence is short RTZ before permutation, then either it is not RTZ or it is long RTZ after permutation, and vice-versa.

The dilemma in the design of a good permutation lies in the need to satisfy this practical rule for two distinct classes of codewords, which require conflicting treatment. The first class contains all nonzero codewords (again with reference to the "all zero" codeword) that are not combinations of simple RTZ sequences, and a good permutation for this class is as regular as possible, which ensures maximum spreading. This type of sequence has low input weight ($w \leq 3$).

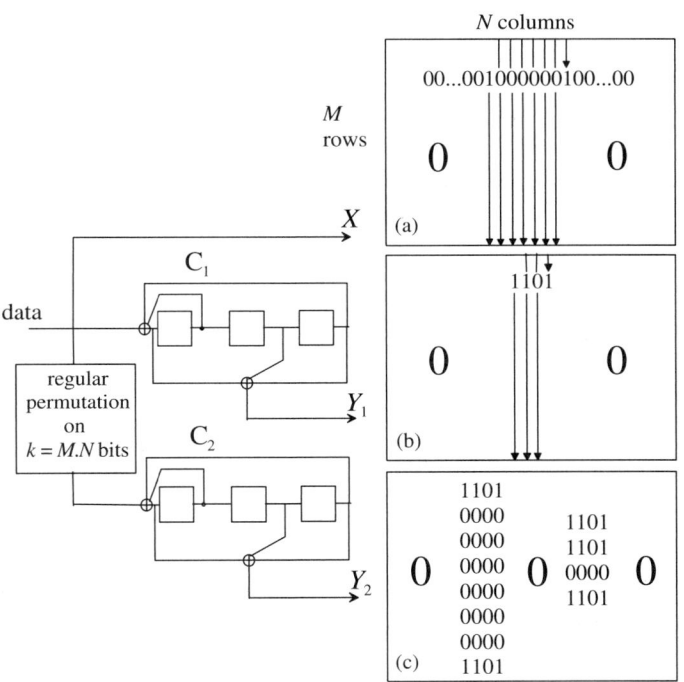

Figure 11.6 Some possible RTZ (return to zero) sequences for both encoders C_1 and C_2, with $G_1(D) = 1 + D + D^3$ (period $L = 7$). (a) With input weight $w = 2$; (b) with $w = 3$, (c) with $w = 6$ or 9.

The second class encompasses all codewords that are combinations of simple RTZ sequences, and nonuniformity (controlled disorder) has to be introduced into the permutation function to obtain a large MHD. Figure 11.6 illustrates the situation, showing the example of a 1/3 rate turbo code, using component binary encoders with code memory $\nu = 3$ and periodicity $L = 2^\nu - 1 = 7$. For the sake of simplicity, the block of k bits is organized as a rectangle with M rows and N columns ($M \approx N \approx \sqrt{k}$). Regular permutation is used, that is, data are written linewise and read columnwise.

Figure 11.6a depicts a situation where encoder C_1 (the horizontal one) is fed by an RTZ sequence with input weight $w = 2$. Redundancy Y_1 delivered by this encoder is poor, but redundancy Y_2 produced by encoder C_2 (the vertical one) is very informative for this pattern, which is also an RTZ sequence but whose span is $7M$ instead of 7. The associated MHD would be around $\frac{7M}{2}$, which is a large value for typical sizes k. With respect to this $w = 2$ case, the code is said to be "good" because $d_{\min}$ tends to infinity when k tends to infinity.

Figure 11.6b deals with a weight-3 RTZ sequence. Again, whereas the contribution of redundancy Y_1 is not high for this pattern, redundancy Y_2 gives relevant information over a large span, of length $3M$. The conclusions are the same as for the above case.

Figure 11.6c shows two examples of sequences with weights $w = 6$ and $w = 9$, which are RTZ sequences for encoder C_1 as well as for encoder C_2. They are obtained by a combination of two or three minimal-length RTZ sequences. The weight of redundant bits is limited and depends neither on M nor on N. These patterns are typical of codewords that limit the MHD of a turbo code when using a regular permutation.

In order to "break" rectangular patterns, some disorder has to be introduced into the permutation rule while ensuring that the good properties of regular permutation, with respect to low weights, are not lost. This is the crucial problem in the search for good permutation, which has not yet found a definitive answer. Nevertheless, some good permutations have already been devised for recent applications, e.g., Technical Specification Groups, IMT-2000 (3GPP, 1999; TIA/EIA/IS, 1999), and DVB (, DVB,D)). Further details about non-regular permutations can be found in Crozier and Guinand (2003) and Berrou et al. (2004).

11.6 Applications of Turbo Codes

Depending on the constraints imposed by the application (performance, throughput, latency, complexity, etc.), error correction codes can be divided into many families. We will consider here three domains, related to error rates:

Medium error rates (corresponding roughly to $10^{-2} > \text{BER} > 10^{-6}$ or $1 > \text{FER} > 10^{-4}$):
This is typically the domain of automatic repetition request (ARQ) systems and is also the more favorable level of error rates for turbo codes. To achieve near-optimum performance, eight-state component codes are sufficient. Figure 11.7 depicts the practical binary turbo code used for these applications and coding rates equal to or lower than 1/2. For higher rates, the double-binary turbo code of figure

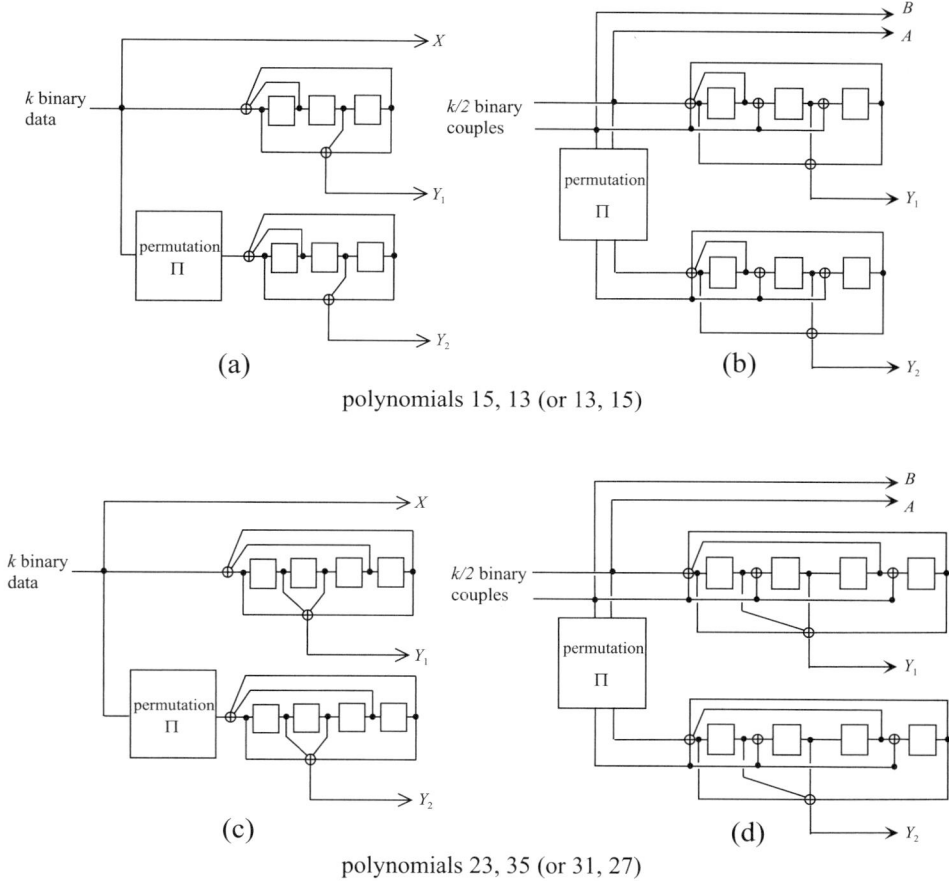

polynomials 15, 13 (or 13, 15)

polynomials 23, 35 (or 31, 27)

Figure 11.7 The four turbo codes used in practice. (a) 8-state binary, (b) 8-state double-binary, both with polynomials 15, 13 (or their symmetric form 13, 15), (c) 16-state binary, (d) 16-state double-binary, both with polynomials 23, 35 (or their symmetric form 31, 27). Binary codes are suitable for rates lower than 1/2, double-binary codes for rates higher than 1/2.

11.7 is preferable (Berrou and Jézéquel, 1999). For each of them, one example of performance, in frame error rate (FER) as a function of signal-to-noise ratio E_b/N_0, is given in fig. 11.8 (UMTS: $R = 1/3$, $k = 640$ and DVB-RCS: $R = 2/3$, $k = 1504$).

Low error rates ($10^{-6} >$ BER $> 10^{-11}$ or $10^{-4} >$ FER $> 10^{-9}$): sixteen-state turbo codes perform better than eight-state ones, by about 1 dB, for an FER of 10^{-7} (see fig. 11.8). Depending on the sought-for compromise between performance and decoding complexity, one can choose either one or the other. Figures 11.7d and 11.7d depict the 16-state turbo codes that can be used, the binary one for low rates, the double-binary one for high rates. In order to obtain

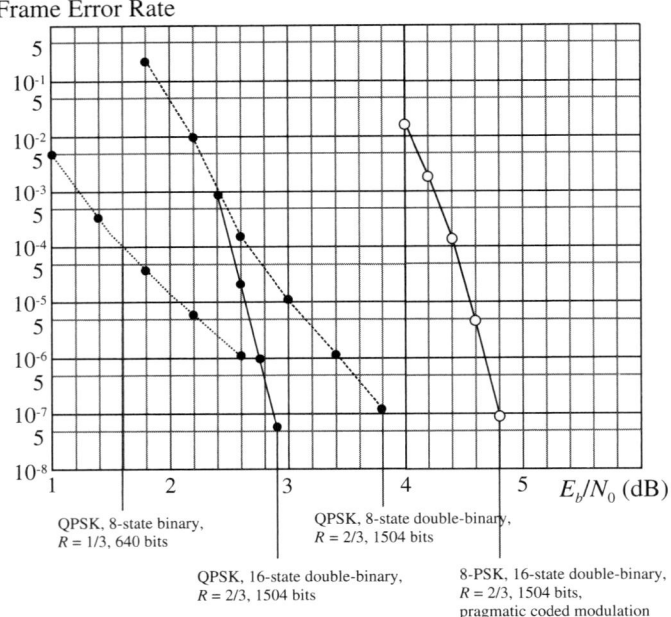

Frame Error Rate

QPSK, 8-state binary,
$R = 1/3$, 640 bits

QPSK, 8-state double-binary,
$R = 2/3$, 1504 bits

QPSK, 16-state double-binary,
$R = 2/3$, 1504 bits

8-PSK, 16-state double-binary,
$R = 2/3$, 1504 bits,
pragmatic coded modulation

Figure 11.8 Some examples of performance, expressed in FER, achievable with turbo codes on Gaussian channels. In all cases: decoding using the Max-Log-APP algorithm with 8 iterations and 4-bit input quantization.

good results at low error rates, the permutation function must be very carefully devised.

An example of performance, provided by the association of 8-PSK (phase-shift-keying) modulation and the turbo code of fig. 11.7d, is also plotted in fig. 11.8, for $k = 1,054$ and a spectral efficiency of 2 bit/s/Hz. This association is made according to the pragmatic approach, that is, the codec is the same as the one used for binary modulation. It just requires binary-to-octary conversion, at the transmitter side, and the converse at the receiver side.

Very low error rates $(10^{-11} > \text{BER}$ or $10^{-9} > \text{FER})$:
The largest minimum distances that can be obtained from turbo codes, for the time being, are not sufficient to prevent a slope change in the $\text{BER}(E_b/N_0)$ or $\text{FER}(E_b/N_0)$ curves, at very low error rates. Compared to what is possible today, an increase of MHDs by roughly 25% would be necessary to make turbo codes attractive for this type of application, such as optical transmission or mass storage error protection.

Table 11.1 summarizes the normalized applications of convolutional turbo codes, known to date. The first three codes of fig. 11.8 have been chosen for these various systems.

Table 11.1 Current Known Applications of (Convolutional) Turbo Codes

Application	Turbo Code	Termination	Polynomials	Rates
CCSDS (deep space)	binary, 16-state	tail bits	23, 33, 25, 37	1/6, 1/4, 1/3, 1/2
UMTS, CDMA2000 (3G mobile)	binary, 8-state	tail bits	13, 15, 17	1/4, 1/3, 1/2
DVB-RCS (Return channel over satellite)	double-binary, 8-state	circular	15, 13	1/3 up to 6/7
DVB-RCT (Return channel over terrestrial)	double-binary, 8-state	circular	15, 13	1/2, 3/4
Inmarsat (M4)	binary, 16-state	no	23, 35	1/2
Eutelsat (Skyplex)	double-binary, 8-state	circular	15, 13	4/5, 6/7
IEEE 802.16 (WiMAX)	double-binary, 8-state	circular	15, 13	1/2 up to 7/8

11.7 Analog Turbo Processing

The ever-improving performance of A/D and D/A data converters coupled with the achievements of the Moore's law have resulted today in fully digital processing. Thus, digital signal processors and other digital programmable devices have superseded the analog circuits traditionally used in telecommunications transceivers. The only blocks that have not yet been totally replaced by digital counterparts are the front-end blocks such as amplifiers and oscillators. One may ask if this tendency is the most adapted to the design of some receiver functions, in particular error correction having to cope with spurious analog signal. From fig. 11.9, which represents a generic digital transceiver, the following comments about the nature of the signal through the chain can be made. First the data to send are processed by the channel encoder, which adds redundant bits in order to make these data more resilient to channel noise. The resulting digital signal is next modulated and up-converted using a high-frequency carrier. This results in an analog signal that is transmitted over the channel. As the signal propagates through the channel, it is altered by various noise sources which are themselves analog (for example weather or electromagnetic conditions, interference). On the receiver side the corrupted data are amplified and down-converted in baseband. This is again analog processing. A crucial choice has now to be made. Should the signal be digitized or should it remain analog? Should it take a hard decision or a soft one? The loss of information due to the quantization does not plead in favor of a digital solution. This explains the motivation of some laboratories that propose the implementation of channel decoders in analog form

(Hagenauer, 1997a; Lustenberger et al., 1999; Moerz et al., 2000). These studies have not only proved the validity of the concepts but have also shown significant gains in terms of speed, power consumption, and silicon area over digital solutions (Gaudet and Gulak, 2003; Moerz et al., 2000).

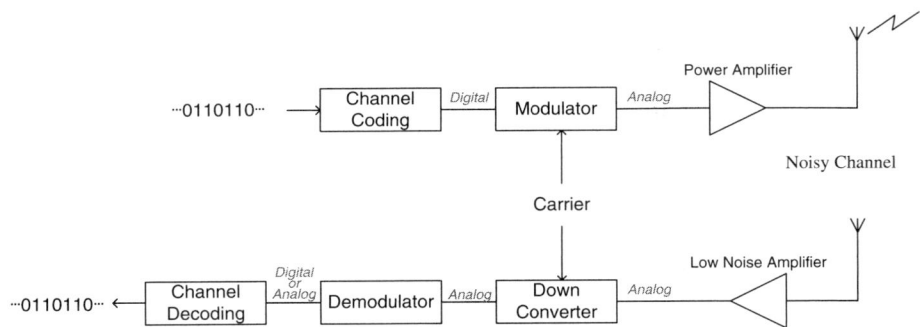

Figure 11.9   Generic digital transceiver.

Historically, the analog implementation of error correction algorithms began with some research on the soft-output Viterbi algorithm (SOVA) (Battail, 1987; Hagenauer and Hoeher, 1989). Nevertheless the Viterbi algorithm does not easily lend itself to analog implementation, for which the APP algorithm (Anderson and Hladick, 1998; Bahl et al., 1974) is preferred. It uses only sum/product operators and logarithm/exponential functions. The latter is necessary to convert probabilities into log-likelihood ratios (LLR), as introduced in section 11.4, and vice versa. LLRs are available at the output of a soft demodulator such as that described in Seguin et al. (2004). If X is a binary random variable and x its observation, the LLR is defined by

$$\text{LLR}(X) = \ln \left(\frac{\Pr(X = 1|x)}{\Pr(X = 0|x)} \right). \tag{11.16}$$

The exponential and natural logarithm functions are readily available from a bipolar junction transistor (BJT) biased in the forward active region. The collector current I_C depends on base-emitter voltage V_{BE} according to the well-known relation:

$$I_C \approx I_S \exp \left(\frac{V_{BE}}{U_T} \right), \tag{11.17}$$

where I_S is the saturation current and U_T the thermal voltage. When connected as a diode (fig. 11.10a), the transistor produces a voltage V between collector and emitter that depends on current I as

$$V \approx U_T \ln \left(\frac{I}{I_s} \right). \tag{11.18}$$

Associating a current with a probability, and a voltage with an LLR, it is thus

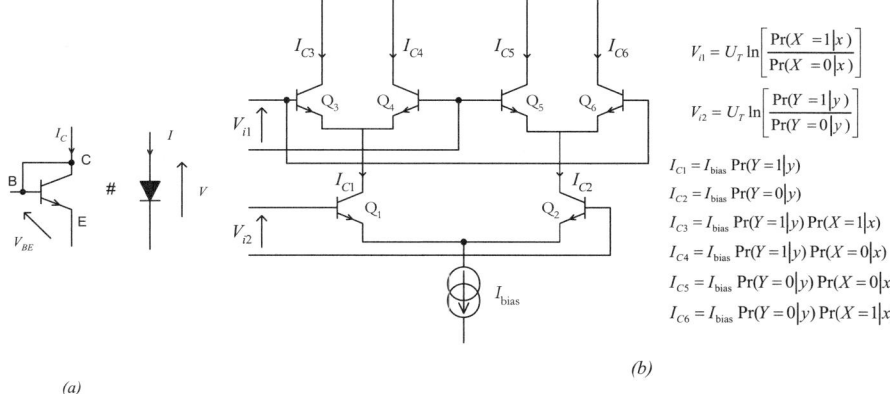

Figure 11.10 Basic structures for analog decoders. (a) Diode connected bipolar transistor; (b) Gilbert cell.

possible to convert LLRs into probabilities and probabilities into LLRs, by using differential structures with transistors and diodes. Currents can also be easily added or multiplied in order to satisfy the APP operations. The basic computing block of the decoder is the well-known analog multiplier (fig. 11.10b) called the *Gilbert cell* (Gilbert, 1968). This cell can convert LLRs into probabilities and multiply them by each other at the same time. The collector currents of transistors Q_3, Q_4, Q_5, Q_6 which represent information can also be summed at a certain node of the circuit. When adding a diode to the collector of each transistor, information can be reconverted into LLR form. Thus the APP algorithm can be implemented using a BJT-based network that directly maps the code trellis.

11.7.1 Implementation of the APP Algorithm

The encoder, at time i ($0 \leq i \leq k - 1$), and the trellis section of an $R = 1/2$ four-state recursive systematic convolutional (RSC) code are depicted in fig. 11.11. This trellis is assumed to be circular, that is, the states at the beginning and at the end of the encoding are equal.

After receiving the LLRs stemming from the channel, the frame containing $n = 2k$ symbols is decoded by feeding the decoder inputs in parallel and by letting the analog network converge toward a stable state. The topology of the circular decoder is given in fig. 11.12. The on-chip network is the direct translation of the APP algorithm (Anderson and Hladick, 1998). It is divided into as many sections as the k information bits to decode. Each section is built from several modules:

- a Γ module to compute the branch metrics,

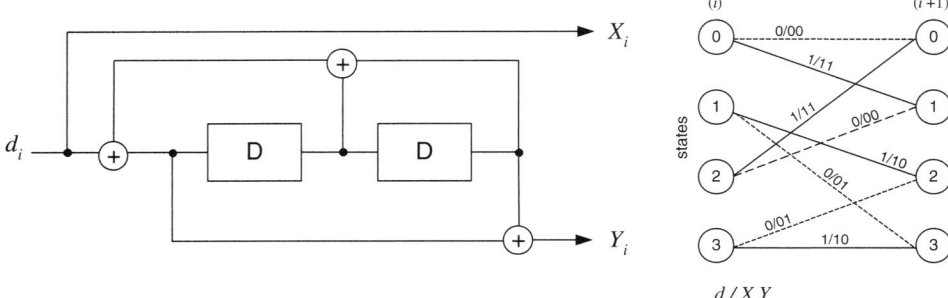

Figure 11.11   A 4-state RSC encoder with code rate $R = 1/2$. The input information symbol X is transmitted together with the redundant symbol (parity bit) Y. A trellis section is also shown, whose branches are labeled with the encoded symbols.

- an A module to compute the forward metrics,

- a B module to compute the backward metrics,

- and a **Dec** module that takes a final hard decision on the value of the information bit.

The input samples $\mathrm{LLR}(X_i)$ and $\mathrm{LLR}(Y_i)$ are associated with the ith couple of transmitted symbols X_i and Y_i. The forward and backward metrics α_i and β_{i+1} are yielded, for each branch of the trellis, by the adjacent trellis sections. The outputs of the section are the metrics α_{i+1} and β_i, which are used as inputs to the adjacent sections, as well as the hard decision $\hat{d}_i$. Moreover, in order to implement a turbo decoder, an additional module—**Extr**—module is required to compute extrinsic information $\mathrm{LLR}_{\mathrm{ext}}(X_i)$. These values are then used as inputs for the Γ module of another APP decoder.

As examples, the Γ and A modules of the four-state decoder are illustrated in fig. 11.13. The branch metrics γ are directly obtained by using the outputs of the Gilbert cell fed with the LLRs. Let $\alpha_i(s)$ be the forward metric associated with the state s of the ith section. Let $\gamma_i(s', s)$ be the branch metric between any state s' of the ith section and one linked state s of the $(i + 1)$st section. Then the four forward metrics of each section i between 0 and $k - 1$ are recursively computed as follows:

$$\alpha_{i+1}(s) = \sum_{s'=0}^{3} \alpha_i(s')\gamma(s', s). \tag{11.19}$$

Therefore, the Gilbert cell has to be extended to perform the four multiplications and additions, as shown in fig. 11.13.

Note that the structures presented require relatively few transistors. This leads to low silicon area and low power consumption, and opens up the way to fully parallel processing. This later property is essential for the design of high-speed decoders.

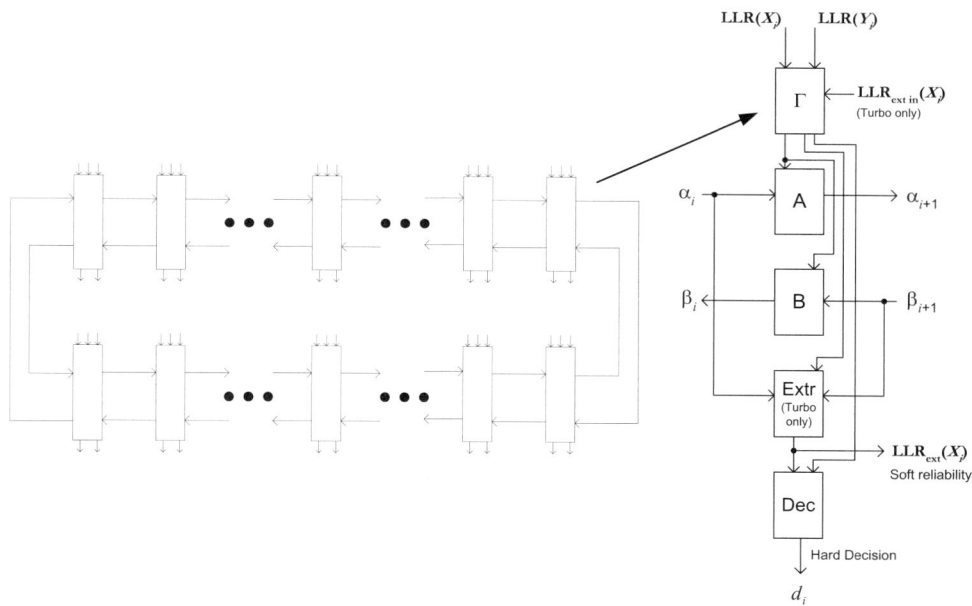

Figure 11.12   Analog circular APP decoder. There are as many sections as information bits to decode. Each section is made up of modules related to the APP algorithm operations.

11.7.2   The Next Step: The Analog Turbo Decoder

As a stand-alone decoder of a simple convolutional code, the APP algorithm does not offer outstanding performance. However, once used in a turbo architecture (Berrou et al., 1993), the APP algorithm reaches its full potential. As shown in fig. 11.14, the two APP decoders exchange information (the so-called extrinsic information), through interleavers, on the reliability of the received data. In a digital version of the turbo decoder, these exchanges are clocked, and the decoding process is repeated as many times as necessary to reach a solution. The complexity and the latency of the decoding process are proportional to the number of iterations, which could be drawbacks for some applications. As can be easily seen, the turbo architecture is well suited for analog implementation since it is a simple feedback system that does not require any internal clocking. In the analog version, the exchanges of extrinsic information are continuous in time. This property, associated with the high degree of parallelism mentioned previously, leads to potential throughput of several Gbit/s, together with low complexity and latency.

To give an example, the architecture of a complete DVB-RCS turbo decoder (, DVB) was simulated using behavioral models. The decoding of the smallest frame of this standard (48 double-binary symbols and rate 1/2) confirms the high capacity of the analog turbo decoder (Arzel et al., 2004). Besides the gains in throughput,

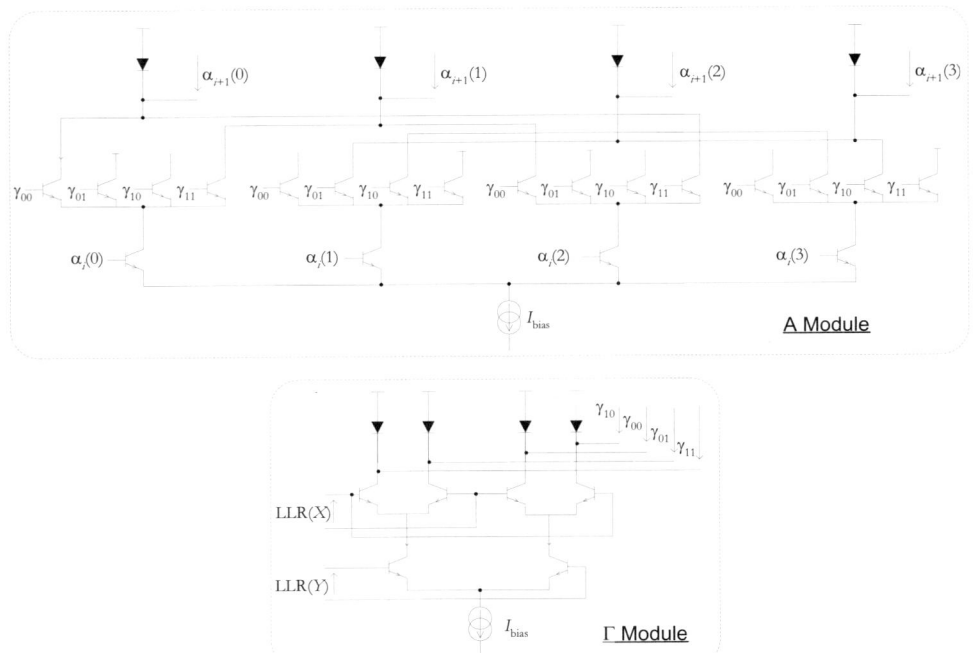

Figure 11.13   Transistor-level design of Γ and A modules.

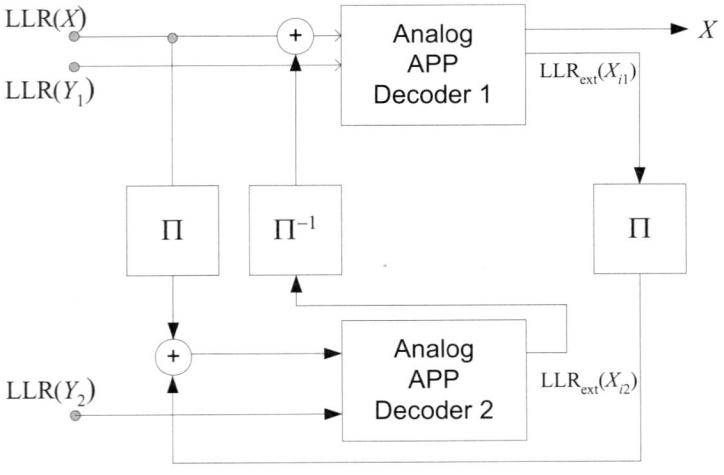

Figure 11.14   The analog turbo decoder.

complexity, and latency, fig. 11.15 illustrates that analog decoding also yields a gain in performance (about 0.1 dB for a BER of 10^{-4}) compared to the digital version. The equivalent digital circuit uses floating-point number representation and runs for 15 iterations, which provides maximum iterative performance. The analog decoder performs better because it benefits from continuous time: there is no iteration but continuous sharing of extrinsic information.

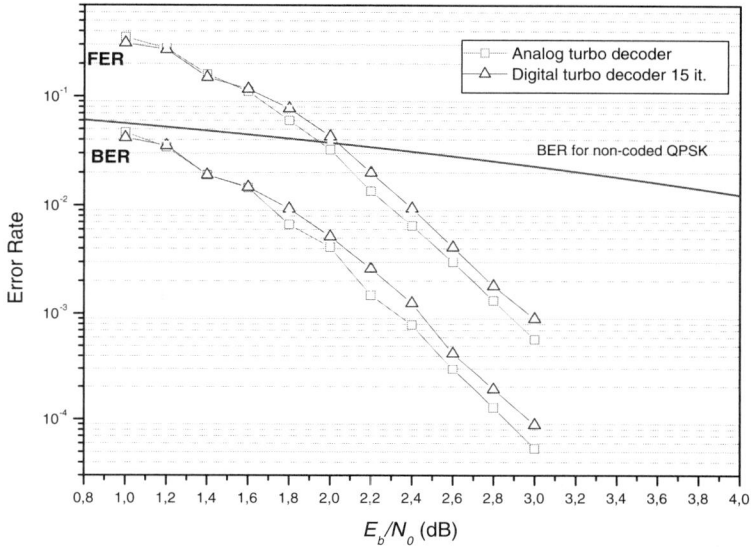

Figure 11.15   Bit and frame error rate curves for the DVB-RCS double-binary turbo code. Frames with $k = 96$ information bits and $R = 1/2$. Analog and digital decoding simulation results are compared.

In conclusion, the ability of the analog decoder to use soft time, in addition to soft input samples, enhances the error correction of turbo codes while increasing data rates and reducing complexity and latency.

11.8   Other Applications of the Turbo Principle

For the sake of clarity we assume a point-to-point transmission with only one transmitter and one receiver. However, this discussion can be extended to more complex situations such as wireless broadcast transmission with a point-to-multipoint transmission. This communication system involves different tasks. The aim of each task can be different and sometimes contradictory but the system must globally address the following problem: Transmit digital information over the propagation channel with maximum rate and minimum error probability. Actually, the heart of this problem is the propagation channel because its physical nature always leads to a limited available frequency bandwidth and a non-ideal frequency response. Consequently, the channel corrupts the transmitted signal by introducing distortion, noise disturbances, and other interference.

The general block diagram shown in fig. 11.16 depicts the basic elements of a digital communications system. The source encoder converts the original message into a sequence with as few binary bits as possible in order to save bandwidth and to optimize the transmission rate. Unlike the source encoder, the channel encoder adds redundancy to the compressed message to form codewords. As detailed in the previous sections, the purpose of this function is to increase the reliability of the transmitted data that will be distorted and impaired by the channel. Finally the digital modulator converts the codewords into waveforms that are compatible with the channel. At the receiver part the demodulator performs a conversion of the waveforms into an analog-like sequence. This sequence is then fed to the decoder, which attempts to reconstruct the compressed message from the constraints of the code. Finally, the source decoder reconstructs the original message from the knowledge of the source encoding algorithm.

The above description is in fact a simplified version of a digital communication system, and several other functions have been omitted. The functions in a receiver can be classified into three categories: estimation, detection, and decoding (fig. 11.17).

By estimation, we mean the estimation of certain channel parameters. Carrier

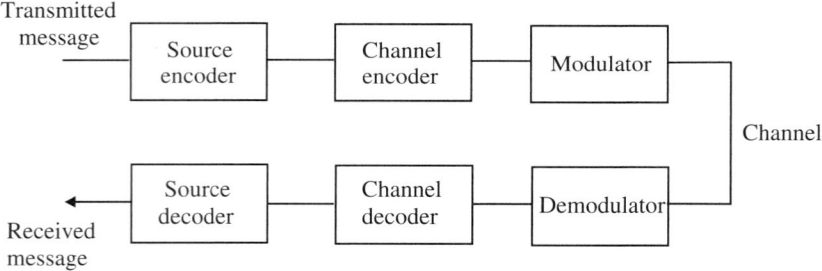

Figure 11.16   Basic representation of a digital communications system.

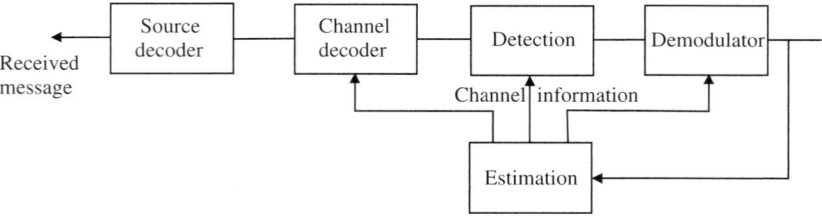

Figure 11.17   A more detailed receiver.

phase recovery, frequency offset estimation, timing recovery, and channel impulse response estimation belong to this category. Detection involves all processing of interference introduced by the channel such as intersymbol interference, multiple access interference, cochannel interference, and multi-user interference. Whatever the type of interference, the same mathematical models can be exploited and algorithms based on identical structures and criteria can be derived. The data are finally recovered by channel decoding.

In this conventional representation, each processing step is performed separately. Detection and estimation modules ignore channel and source coding and work as if the received symbols were mutually independent.

Information theory informs us that the maximum likelihood (ML) criterion should be globally applied over the whole receiver in order to minimize the error probability, which is the final target. Unfortunately, with so many factors to take into account, the complexity becomes prohibitive and the problem totally intractable. Consequently, the optimal ML approach is never implemented in practice and system engineers prefer the conventional receiver of fig. 11.17 despite its suboptimality. Nevertheless, the loss of performance due to this suboptimality is kept as low as possible thanks to the use of soft information. As already mentioned at the beginning of this chapter, the process of making hard decisions discards a part (i.e., the magnitude, as an image of the reliability) of the inherent information present in the received data. For instance, instead of making hard decisions, it is better for the detector to deliver soft information to the decoder.

11.8.1   Feedback Process: The Answer to Suboptimality

Using soft information in the conventional receiver is not sufficient to reach the same performance as that of the optimal ML receiver. The loss of performance essentially comes from the fact that the tasks at the front of the receiver, typically detection and estimation, do not benefit from the work of the channel decoder.

In a turbo decoder, this problem has been solved by introducing a feedback loop between the component decoders. This loop enables a bidirectional exchange of soft information between these decoders. Each decoder benefits from the work done by the other. The turbo principle can also be applied to solve the suboptimality issue of the conventional receiver. This approach was originally proposed by Douillard et al. (1995) to jointly process equalization and decoding.

In a turbo receiver, an iterative process replaces the sequential process of the conventional receiver of fig. 11.17. The received data are now processed several times by the detector, the estimator, and the decoder. Each pass through the estimation/detection/decoding scheme is called an iteration as in a turbo decoder. The iterative receiver is illustrated in fig. 11.18. At the first iteration, the detection and the estimation functions do not benefit from any a priori information about the transmitted data. The soft information provided to the decoder is then the same as in the conventional sequential receiver. In turn, the decoder provides its own soft information for estimation/detection, creating a feedback loop. From the second iteration, this feedback information acts as a priori information for estimation/detection that improves their results. Figure 11.18 also indicates a turbo process between the channel decoder and the source decoder. In this situation, constraints inherent to the source encoder may help the channel decoder work.

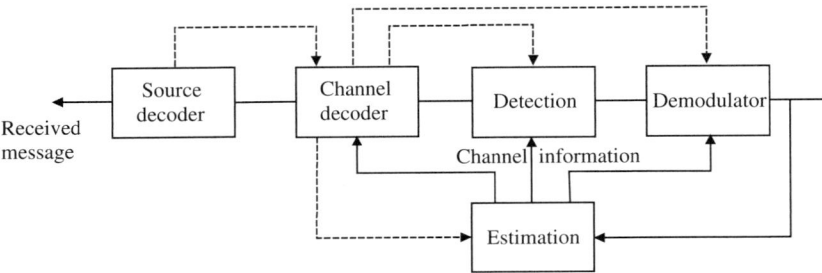

Figure 11.18   Illustration of a simplified turbo receiver. The permutation functions are not represented.

Thus a turbo receiver involves the repetition of information exchanges a number of times, the channel decoder playing a central role thanks to the amount of redundancy it benefits from. If the SNR is not too bad, the turbo receiver produces new and better estimates at each iteration. Nevertheless, this performance improvement is bounded by the ML performance, which is impassable.

In section 11.5, the role and the importance of the permutation function were emphasized. In the turbo receiver, permutations (which are not represented in fig. 11.18) also play a crucial role by minimizing the correlation effects in consecutive pieces of information exchanged by the different processors. In particular, an appropriate permutation located after the decoder avoids the consequences of possible residual error bursts.

The implementation of the turbo receiver assumes that all functions are able to accept and deliver soft extrinsic information. In particular, the APP algorithm is particularly well suited to the decoder, especially if it is a turbo decoder.

11.8.2   Probabilistic Detection and Estimation Algorithms

Actually, before the invention of turbo coding and decoding, probabilistic algorithms with *a priori* probabilities for conventional detection or estimation tasks were not needed and so, they were not known! Consequently, before implementing an iterative process between estimation/detection and decoding tasks, probabilistic algorithms for estimation and detection have to be elaborated. In fact, the APP algorithm suits the problem of detection well, whereas the expectation maximization (EM) algorithm (Moon, 1996), also a probabilistic algorithm, is more adapted to the problem of estimation. Lots of iterative digital receivers are based on these two algorithms, whose major advantage is optimality. Furthermore, iterative exchanges between detection, or estimation, and decoding can be implemented directly thanks to a priori probabilities. However, these algorithms are computationally demanding, and the resulting complexity of the receiver may become prohibitive. In this situation, suboptimal alternatives are needed.

11.8.3   Suboptimal Algorithms

The principles of these alternatives were already known for conventional detection and estimation. They are suboptimal because they are based on criteria other than minimizing the error probability, such as minimizing the mean square error (minimum mean square error, MMSE) or maximizing the SNR. But these suboptimal algorithms also lead to less complex implementations, often in the form of filter-based structures. As these structures were traditionally used in the conventional receiver of fig. 11.17, no a priori input was required. In order to exchange information with the channel decoder, detection and estimation algorithms have to take into account a priori probabilities and also deliver probabilities.

Two solutions have been considered in the literature to implement soft-in/soft-out detectors and estimators. The first one involves modifying or adapting preexisting techniques. The second one is to derive totally new architectures.

Let us consider the first approach. The problem can be formulated as: How can the a priori input sequence be utilized in a conventional estimation/detection algorithm, whereas so far they only involved the processing of the received data? In fact, several estimation or detection algorithms already use feedback of estimated data, for example, the DFE (decision feedback equalizer) for combating intersymbol interference (Proakis, 2000) and the SIC (successive interference cancellation) receiver for multi-user transmission (see Dai and Poor, 2002, for instance). An illustration of these locked-up schemes is given in fig. 11.19. The loop generally involves hard decisions on symbols, provided by a hard slicer, that are utilized to improve the estimation/detection result. The derivation of the algorithms is generally based on the assumption that the hard decisions fed back to the detection/estimation algorithm correspond exactly to the transmitted symbols. In a turbo process, the decoder

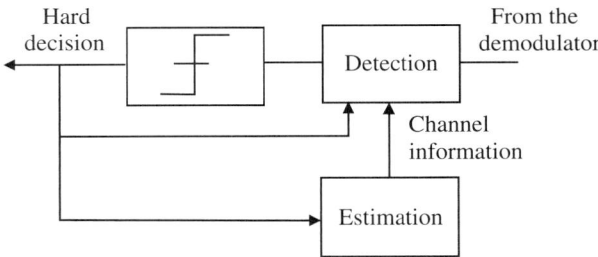

Figure 11.19 Illustration of a conventional locked-up estimation/detection scheme.

can replace the hard slicer and now provide improved feedback information. The detection or estimation algorithms are not modified and are still based on the assumption of perfect feedback. As the channel decoder provides probabilities and the detection/estimation algorithms symbol estimates, a mapper is needed between the channel decoder and the estimation/detection algorithms. This mapper generates an estimate of transmitted symbols from the a posteriori or *extrinsic* probabilities provided by the decoder. This estimation is then directly used as feedback into the suboptimal detection and estimation algorithms, as explained previously.

However, more potential gain is available if the assumption of perfect feedback is dropped in the derivation of the algorithms. Indeed, the assumption of perfect feedback in a turbo process is not valid since the data estimates are improved iteration after iteration. Perfect feedback is achieved only when the convergence of the turbo process is attained. Consequently, retaining the assumption of perfect feedback in the derivation of the detection/estimation algorithms, as in the first approach, generally leads to nonoptimal solution. A new derivation of the algorithm becomes necessary. The derivation is still based on the initial suboptimal criteria, MMSE for example. But, in addition, feedback decoder probabilities, replacing the perfect feedback assumption, are introduced into the derivation. Very powerful algorithms are thus derived, generally different from conventional algorithms and often based on an interference canceller structure. Furthermore, they offer very significant complexity advantages over probabilistic algorithms.

Iterative schemes based on well-designed suboptimal algorithms achieve very good performance and can potentially lead to exactly the same performance bound as iterative schemes based on probabilistic algorithms. The major difference is the convergence speed. More iterations are sometimes necessary to achieve the performance bound.

In conclusion, the turbo principle has found numerous applications in communication receivers. It has proved its capacity to achieve performance very close to that of the ML receiver, while requiring significantly reduced complexity.

11.8.4   Further References

Because the literature in the domain of turbo receivers is huge, we will restrict references to pioneering papers or overview papers.

As already mentioned, the turbo principle was extended for the first time to the problem of joint equalization and decoding in 1995 (Douillard et al., 1995). This new turbo receiver, called a turbo detector, was based on an APP equalizer and an APP channel decoder, and demonstrated quasi-optimal performance. The second successful attempt concerned coded modulation. Robertson and Wörz (1996) introduced an efficient coding scheme for high-order modulations based on the parallel concatenation of Ungerboeck codes (Ungerboeck, 1982). This technique, named turbo trellis coded modulation (TTCM), provides good performance on AWGN channels.

Later, Glavieux et al. (1997) proposed a low complexity version of a turbo equalizer based on adaptive MMSE filters . For the first time, a nonprobabilistic algorithm was introduced into a turbo receiver. Here, a priori information is taken into account in the filter coefficient computation thanks to an adaptive algorithm such as the least mean square (LMS) algorithm. At the same time, Hagenauer introduced the "turbo" principle, extending this concept to several tasks in a communication system: joint source and channel decoding, coded modulation, and multi-user detection (Hagenauer, 1997b).

Following these pioneering works, a huge number of papers was then devoted to the turbo principle. For instance, a bit-interleaved coded modulation (BICM) turbo decoder for ergodic channels has been proposed by ten Brink et al. (1998) and Chindapol et al. (1999) independently. This technique, called iterative demapping or BICM-ID (iterative decoder), is based on a feedback loop between the demapper, which computes LLRs, and the channel decoder. The a priori information provided by the decoder is used in the soft demapper to remove the assumption of independent and identically distributed coded bits. This technique needs non-Gray mapping in order to provide performance gains.

In the domain of synchronization, Langlais and Hélard (2000) have addressed the problem of carrier phase recovery for turbo decoding at low SNRs, by using tentative decisions from the first component decoder in the carrier phase recovery loop. As this system does not exploit the iterative structure of the turbo decoder, Lottici and Luise (2002) then developed a carrier phase recovery embedded in turbo decoding. Recently, Barry et al. (2004) have proposed an overview of methods for implementing timing recovery with turbo decoding.

In the past few years many significant developments have arisen in the field of iterative multi-user detection. A number of references can be found in the paper by Poor (2004). Finally, the turbo principle can also be applied to multiple-antenna detection as in MIMO (multi-input multi-output) systems. The famous BLAST system from Bell Labs has thus been transposed in a turbo receiver, in which the multi-antenna detector exchanges information with the channel decoder (Haykin et al., 2004). Optimal and suboptimal multiple-antenna detectors are also presented, leading to various implementation complexities.

12   Blind Signal Processing Based on Data Geometric Properties

Konstantinos Diamantaras

12.1   Introduction

Blind signal processing deals with the outputs of unknown systems excited by unknown inputs. At first sight the problem seems intractable, but a closer look reveals that certain signal properties allow us to extract the inputs or to identify the system up to some, usually not important, ambiguities. Linear systems are mathematically most tractable and, naturally, they have attracted most of the attention. Depending on the type of the linear system, blind problems arise in a wide variety of applications, for example, in digital communications (Diamantaras and Papadimitriou, 2004a,b; Diamantaras et al., 2000; Godard; Papadias and Paulraj, 1997; Paulraj and Papadias, 1997; Shalvi and Weinstein, 1990; Talwar et al., 1994; Tong et al., 1994; Torlak and Xu, 1997; Treichler and Agee, 1983; Tsatsanis and Giannakis, 1997; van der Veen and Paulraj, 1996; van der Veen et al., 1995; Yellin and Weinstein, 1996), in biomedical signal processing (Choi et al., 2000; Cichocki et al., 1999; Jung et al., 1998; Makeig et al., 1995, 1997; McKeown et al., 1998; Vigário et al., 2000), in acoustics and speech processing (Douglas and Sun, 2003; Parra and Spence, 2000; Parra and Alvino, 2002; Shamsunder and Giannakis, 1997), etc. Many recent books on the subject (Cichocki and Amari, 2002; Haykin, 2001a,b; Hyvärinen et al., 2001) provide extensive discussion on related problems and methods.

The most general finite, linear, time invariant (LTI) system is expressed by a multichannel convolution of length L, operating on a discrete vector signal $\mathbf{s}(k) = [s_1(k), \cdots, s_n(k)]^T$,

$$\mathbf{x}(k) = \sum_{i=0}^{L-1} \mathbf{H}_i \, \mathbf{s}(k - i). \tag{12.1}$$

The FIR *filter taps* $\mathbf{H}_i$ are complex matrices, in general, of size $m \times n$, $m \geq 1$. Thus the output is an m-dimensional complex vector $\mathbf{x}(k)$. For $n, m > 1$, equation 12.1 describes a linear, discrete, multi-input multi-output (MIMO) system.

12.1.1　Types of Mixing Systems

We shall study two special cases of system 12.1 sharing many similarities but also having some special characteristics as described below:

Instantaneous Mixtures: In this case we have more than one source and more than one observation, i.e., $m, n > 1$, but there is no convolution involved, so $L = 1$. The output vector is produced by a linear, instantaneous transformation:

$$\mathbf{x}(k) = \mathbf{H}\mathbf{s}(k). \tag{12.2}$$

This type of system is also called *memoryless*.

Single-Input Single-Output (SISO) Convolution: In this case we have exactly one source and one observation, so $m, n = 1$, but the convolution is nontrivial, i.e., $L > 1$.

$$x(k) = \sum_{i=0}^{L-1} h_i\, s(k - i). \tag{12.3}$$

equation 12.3 describes a linear, SISO FIR filter.

12.1.2　Types of Blind Problems

Regardless of the specific system type, there are two kinds of blind problems which are of interest here, depending on whether we desire to extract the input signals or the system parameters.

Blind Source Extraction: In this type of problem our goal is to recover the source(s) given the observation signal $\mathbf{x}(k)$ or $x(k)$. If there are more than one source the problem is called *blind source separation* (*BSS*). In the case of BSS the linear system may be either instantaneous or convolutive (general MIMO). In the case of *blind deconvolution* (*BD*) we want to invert a linear filter which, of course, operates on its input via the convolution operator, hence the name *deconvolution* attributed to this problem. The problem is very important, for example, in wireless communications, where n transmitted signals corrupted by intersymbol interference (ISI), multi-user interference (MUI), and noise are received at m antennas.

The source separation/extraction problem has an inherent ambiguity in the order and the scale of the sources: the original signals can not be retrieved in their original order or scale unless some further information is available. For example, if the source samples (symbols) are drawn from a known finite alphabet then there is no ambiguity in the scale. If however, the alphabet is symmetric with respect to zero, then there exists a sign ambiguity since both signals $s(k)$ and $-s(k)$ are plausible. Furthermore, the ordering ambiguity is always present if the problem involves more than one source.

Blind System Identification: In this type of problem our goal is to obtain the system parameters rather than recovering the source signals. If the system is memoryless then our goal is to recover the mixing matrix $\mathbf{H}$. If the system involves nontrivial convolution then the goal is to extract the filter taps $h_0, \ldots, h_{L-1}$, or $\mathbf{H}_0, \ldots, \mathbf{H}_{L-1}$.

12.1.3 Approaches to Blind Signal Processing

Typically, blind problems are approached either using statistical properties of the signals involved, or exploiting the geometric structure of the data constellation, as described next.

Higher-Order Methods: According to the central limit theorem, the system output—which is the sum of many input samples—will approach the Gaussian distribution, irrespective of the input distribution. A characteristic property of the Gaussian distribution is that all higher-order cumulants (for instance, the kurtosis) are zero. If the inputs are not normally distributed, their higher-order cumulants will be nonzero, for example positive, and so equation 12.1 will work as a "cumulant reducer." Clearly, the blind system inversion—the linear transform that will recover the sources from the output—should function as a "cumulant increaser," i.e., it should maximize the absolute cumulant value for a given signal power. In fact, this is the basic idea behind all higher-order methods.

1. *Second-Order Methods*: Alternatively, second-order methods can be applied when the sources have colored spectra, regardless of their distribution. If the source colors are not identical then the time-delayed covariance matrices have a certain eigenvalue structure which reveals the mixing operator, in the memoryless case. This information can be used for recovering the sources as well. In the dynamic case, things are more complicated, although, again, second-order methods have been proposed based on the statistics of either the frequency or the time domain.

2. *A Third Approach: Exploiting The Signal Geometry*: Neither higher-order nor second-order methods exploit the cluster structure or shape of the input data when such a structure or shape exists. Consider for example a source signal $s(k)$ whose samples are drawn from a finite alphabet $\mathcal{A}_M = \{\pm 1, \cdots \pm (M/2)\}$ ($M=$ even). Let the SISO FIR filter described in equation 12.3 be excited by $s(k)$. Writing N equations ($N \geq L$) of the form (in equation 12.3) for N consecutive values of k, we obtain the following matrix equation:

$$
\begin{bmatrix} x(k) \\ x(k+1) \\ \vdots \\ x(k+N-1) \end{bmatrix} = \begin{bmatrix} s(k) & s(k-1) & \cdots & s(k+1-L) \\ s(k+1) & s(k) & & s(k+2-L) \\ \vdots & \vdots & & \vdots \\ s(k+N-1) & s(k+N-2) & \cdots & s(k+N-L) \end{bmatrix} \begin{bmatrix} h_0 \\ h_1 \\ \vdots \\ h_{L-1} \end{bmatrix}
$$

(12.4)

$$
\mathbf{x} = \mathbf{Sh}
$$

(12.5)

where the $N \times L$ Toeplitz matrix $\mathbf{S}$ involves $(N + L - 1)$ unknown input symbols. It is possible, in principle, to identify $\mathbf{h}$ in a deterministic way by an exhaustive search over all M^{N+L-1} possible $\mathbf{S}$'s such that $\min_{\mathbf{h}} \|\mathbf{x} - \mathbf{Sh}\|^2 = 0$. Although it is highly impractical, this observation tells us that there is more to blind signal processing than statistical processing. If the sources, for example, have a certain structure which produces clusters in the data cloud, or the input distribution is bounded (e.g., uniform), then one can exploit the geometric properties of the output constellation and derive fast and efficient deterministic algorithms for blind signal processing. These methods are treated in this chapter. In particular, section 12.2 discusses blind methods for systems with finite alphabet sources. The discussion covers both the instantaneous and the convolutive mixtures and it is based on the geometric properties of the data cloud. Section 12.3 discusses the case of continuous-valued sources that are either sparse or have a specific input distribution, for example, uniform. Our discussion on continuous sources covers only the case of instantaneous systems. Certainly there is a lot of room for innovation along this line of research since many issues, today, remain open.

12.2   Finite Alphabet Sources

Blind problems involving sources with finite alphabets (FA) have drawn a lot of attention, because such types of signals are common in digital communications. Popular modulation schemes, for instance, quadrature amplitude modulation QAM), pulse amplitude modulation (PAM) and binary phase-shift keying (BPSK), produce signals with limited numbers of symbols. A large body of literature exists on the instantaneous mixture problem, not only because it is the simplest one but also because most methods dealing with the more realistic convolutive mixture problem lead to the solution of an instantaneous problem. In Anand et al. (1995), the blind separation of binary sources from instantaneous mixtures is approached using separate clustering and bit-assignment algorithms. An extension of this method is presented in Kannan and Reddy (1997), where a maximum likelihood (ML) estimate of the cluster centers is provided. Talwar et al. (1996) presented two iterative least-squares methods: ILSP (iterative least squares with projection) and ILSE (iterative least squares with enumeration) for the BSS of binary sources. The same problem is treated in van der Veen (1997), where the real analytical constant modulus algorithm (RACMA) is introduced based on the singular value decomposition (SVD) of the observation matrix. In Pajunen (1997), an iterative algorithm is proposed for the blind separation of more binary sources than sensors. Finite alphabet sources and instantaneous mixtures are discussed in Belouchrani and Cardoso (1994) where a ML approach is proposed using the EM algorithm. Grellier and Comon (1998) introduce a polynomial criterion and a related minimization algorithm to separate FA sources. In all the above methods the geometric properties of the data cloud are not explicitly used. Geometrical concepts, such as the relative distances between the cluster centers, were introduced in Diamantaras (2000) and Diamantaras and Chassioti (2000). It turns out that just one observation signal is sufficient for blindly separating n binary sources, in the noise-free case, under mild assumptions. A similar algorithm based on geometric concepts was later proposed in Li et al. (2003).

In this section we shall study the geometric structure of data constellations generated from linear systems operating on signals with finite alphabets. We'll find that the geometry of the obtained data cloud contains information pertaining to the generating linear operator. This information can be exploited either for the blind extraction of the system parameters or for the blind retrieval of the original sources.

12.2.1   Instantaneous Mixtures Of Binary Sources

The simplest alphabet is the two-element set, or *binary alphabet* $\mathcal{A}_a = \{-1, 1\}$. We shall assume that the samples of some source signals are drawn from $\mathcal{A}_a$, and the signals will be called *binary antipodal* or, simply, *binary*. In digital communications the carrier modulation scheme using symbols from $\mathcal{A}_a$ is called binary phase-shift keying (BPSK). The reader is encouraged to verify that our results can be easily generalized to any type of binary alphabet, for example, the nonsymmetric set $\mathcal{A}_b = \{0, 1\}$.

In this subsection we shall concentrate on problem type 1, i.e., on linear memoryless mixtures of many sources, $n > 1$. Depending on the number of output signals (observations) m, we treat three distinct cases: $m = 1$; $m = 2$; and $m > 2$.

12.2.1.1   A Single Mixture

The instantaneous mixture of n sources linearly combined into a single observation is described by the following equation:

$$x(k) = \sum_{i=1}^{n} h_i \, s_i(k) = \mathbf{h}^T \mathbf{s}(k), \qquad (12.6)$$

$$\mathbf{h} = [h_1 \cdots h_n]^T, \qquad \mathbf{s}(k) = [s_1(k) \cdots s_n(k)]^T.$$

We assume that the mixing coefficients h_i are real and that $s_i(k) \in \mathcal{A}_a$. If the coefficients are complex, then the problem corresponds to the case $m = 2$, which is treated later. We start by studying the noise-free system since our primary interest is to investigate the structural properties of the signals and not to develop methods to combat the noise. Of course, eventually, the development of a viable algorithm will have to deal with the noise issue.

Equation 12.6 can be seen as the projection $\tilde{x}(k)$ of $\mathbf{s}(k)$ along the direction of the normal vector $\tilde{\mathbf{h}}$, scaled by $\|\mathbf{h}\|$:

$$x(k) = \|\mathbf{h}\| \, \tilde{x}(k), \tag{12.7}$$
$$\tilde{x}(k) = \tilde{\mathbf{h}}^T \mathbf{s}(k), \tag{12.8}$$
$$\tilde{\mathbf{h}} = \mathbf{h}/\|\mathbf{h}\|. \tag{12.9}$$

The set of values of $x(k)$ will be called the *constellation* of $x(k)$ and it will be denoted by $\mathcal{X}$. It is a set of (at most) 2^n points in 1-D space, R.

two sources   In order to facilitate our understanding of the geometric structure of $\mathcal{X}$, let us start by assuming that there are only $n = 2$ sources. Thus, there exist four possible realizations of the vector $\mathbf{s}(k)$, which form the source constellation $\mathcal{S} = \{\mathbf{s}^{--}, \mathbf{s}^{-+}, \mathbf{s}^{+-}, \mathbf{s}^{++}\}$, where

$$\mathbf{s}^{--} = [-1, -1]^T, \ \mathbf{s}^{-+} = [-1, 1]^T, \ \mathbf{s}^{+-} = [1, -1]^T, \ \text{and} \ \mathbf{s}^{++} = [1, 1]^T.$$

Consequently, the output constellation $\mathcal{X}$ also consists of four distinct values:

$$x^{--} = \|\mathbf{h}\| \, \tilde{x}^{--} = \mathbf{h}^T \mathbf{s}^{--},$$
$$x^{-+} = \|\mathbf{h}\| \, \tilde{x}^{-+} = \mathbf{h}^T \mathbf{s}^{-+},$$
$$x^{+-} = \|\mathbf{h}\| \, \tilde{x}^{+-} = \mathbf{h}^T \mathbf{s}^{+-},$$
$$x^{++} = \|\mathbf{h}\| \, \tilde{x}^{++} = \mathbf{h}^T \mathbf{s}^{++}.$$

Figure 12.1 shows the projections $\tilde{x}^{--}$, $\tilde{x}^{-+}$, $\tilde{x}^{+-}$, $\tilde{x}^{++}$, of the source constellation $\mathcal{S}$ for four different normal mixing vectors $\tilde{\mathbf{h}}$. It is obvious that the relative distance between the points on the projection line is a function of the angle θ between the projection line and the horizontal axis. Apparently, the problem involves a lot of symmetry. In particular, it is straightforward to verify that we obtain the same output constellation $\mathcal{X}$ for the angles $\pm\theta$, $\pm(\pi/2 - \theta)$, $\pm(\pi - \theta)$, and $\pm(3\pi/2 - \theta)$, (any θ). This multiple symmetry is the result of the interchangeability of the two sources, s_1 and s_2, as well as the invariance of the source constellation to sign changes. These ambiguities are, however, acceptable, since it is not possible to recover the original source order or the original source signs. Both the source order and the sign are unobservable as it is eminent from the following relations:

$$x(k) = [\pm h_1, \pm h_2] \, [\pm s_1(k), \pm s_2(k)]^T,$$
$$= [\pm h_2, \pm h_1] \, [\pm s_2(k), \pm s_1(k)]^T.$$

Therefore, let us assume, without loss of generality, that the mixing vector $\mathbf{h}$ satisfies the following constraint

$$h_1 > h_2 > 0. \tag{12.10}$$

Under this assumption, the elements of $\mathcal{X}$ are ordered:

$$x^{--} = -h_1 - h_2 < x^{-+} = -h_1 + h_2 < x^{+-} = +h_1 - h_2 < x^{++} = +h_1 + h_2. \tag{12.11}$$

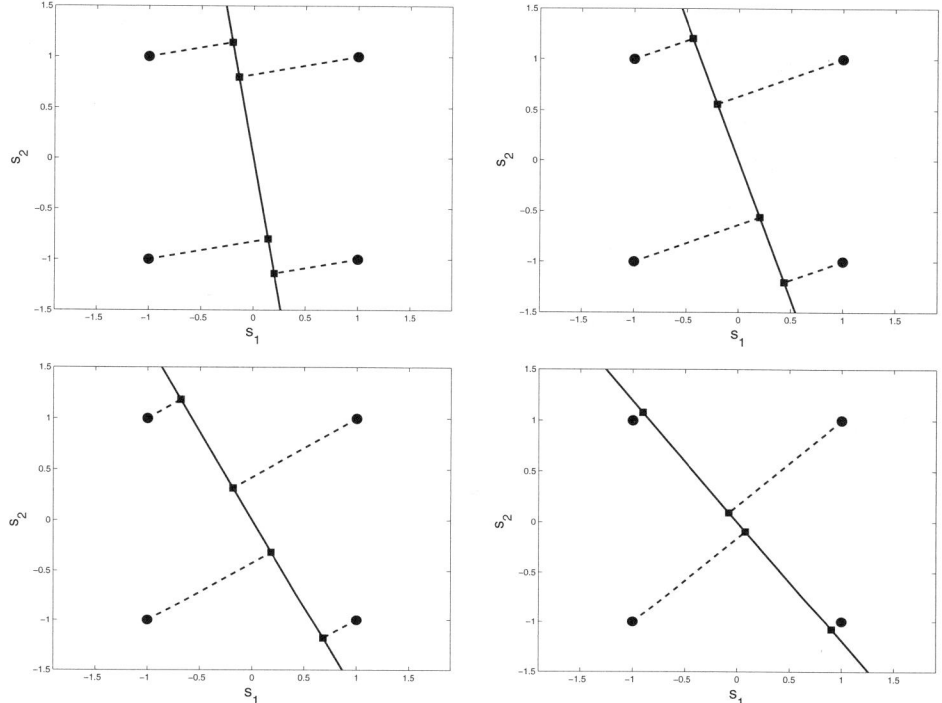

Figure 12.1   The source constellation (circles) of two independent binary sources is projected on four different directions. The relative distances of the projection points (marked by squares) is clearly a function of the slope of the projection line.

Indeed, the first and third inequalities in 12.11 are obvious since the mixing coefficients are positive. The second inequality is also true since $x^{+-} - x^{-+} = 2(h_1 - h_2) > 0$.

Thus, by clustering the (observable) output sequence $\{x(1), x(2), x(3), \cdots\}$ we obtain four cluster points c_1, c_2, c_3, c_4, which can be arranged in increasing order and set into one-to-one correspondence with the elements of $\mathcal{X}$.

$$c_1 = x^{--} < c_2 = x^{-+} < c_3 = x^{+-} < c_4 = x^{++} \tag{12.12}$$

$$c_1 = -c_4; \qquad c_2 = -c_3 \ .$$

Then using equation 12.11 we can recover the mixing parameters:

$$h_1 = (c_3 - c_1)/2, \tag{12.13}$$

$$h_2 = (c_2 - c_1)/2. \tag{12.14}$$

Example 12.1

Figure 12.2 shows the position of the cluster points $c_1, \ldots, c_4$, for the random mixing vector $\mathbf{h} = [0.9659, 0.2588]^T$. According to equations 12.11 and 12.12, these cluster points are

$$
\begin{aligned}
c_1 &= x^{--} &=& -1.2247 \\
c_2 &= x^{-+} &=& -0.7071 \\
c_3 &= x^{+-} &=& 0.7071 \\
c_4 &= x^{++} &=& 1.2247
\end{aligned}
$$

By computing the distances between the pairs (c_3, c_1) and (c_2, c_1), we obtain directly the unknown mixing parameters:

$$
\begin{aligned}
(c_3 - c_1)/2 &= 0.9659 = h_1 \\
(c_2 - c_1)/2 &= 0.2588 = h_2
\end{aligned}
$$

■

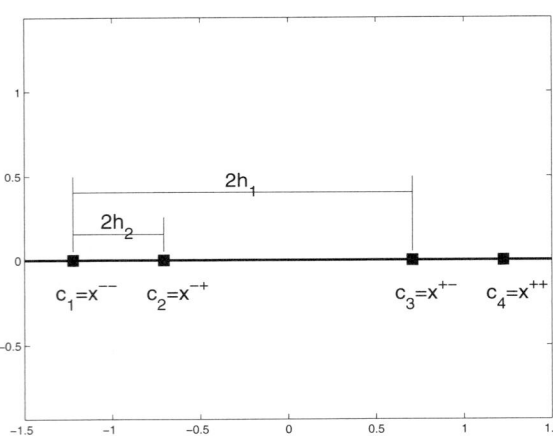

Figure 12.2 The distances c_3–c_1 and c_2–c_1 between the cluster points are equal to twice the size of the unknown mixing parameters.

If our aim is to identify the mixing parameters h_1, h_2, then equations 12.13 and 12.14 have achieved our goal. If, in addition, we want to extract the hidden sources then we may estimate each input sample $\mathbf{s}(k)$, separately, by finding the binary vector $\mathbf{b} = [b_1, b_2]^T \in \mathcal{A}_a^2$, so that $\mathbf{h}^T \mathbf{b}$ best approximates $x(k)$. This corresponds to the following binary optimization problem,

$$
\hat{\mathbf{s}}(k) = \arg \min_{\mathbf{b} \in \mathcal{A}_a^2} |x(k) - \mathbf{h}^T \mathbf{b}|, \quad \text{for all } k. \tag{12.15}
$$

Luckily the above optimization problems are decoupled, for different k, and therefore the solution is trivial.

more than 2 sources

The whole idea can be extended to more than two sources using recursive *system deflation*. This process iteratively identifies and removes the two smallest mixing parameters, thus eventually reducing the problem to either the two-input case, which is solved as above; or the single-input case, which is trivial. Our linear mixture model is again the one described in 12.6 with $n > 2$ and some real mixing vector $\mathbf{h} = [h_1, \cdots, h_n]^T$.

As before, without loss of generality, we shall assume that the mixing parameters are positive and arranged in decreasing order:

$$h_1 > h_2 > \cdots > h_n > 0. \tag{12.16}$$

We have already shown that for $n = 2$ the centers c_i are arranged in increasing order. For $n > 2$ things are a bit more complicated. Let us define $\mathbf{B}^{(n)}$ to be the $2^n \times n$ matrix whose ith row $\mathbf{b}_i^{(n)T}$, is the binary representation of the number $(i - 1) \in \{0, \cdots, 2^n - 1\}$:

$$\mathbf{B}^{(n)} \triangleq \begin{bmatrix} -1 & -1 & \cdots & -1 & -1 \\ -1 & -1 & \cdots & -1 & 1 \\ -1 & -1 & \cdots & 1 & -1 \\ \vdots & \vdots & & \vdots & \vdots \\ 1 & 1 & \cdots & -1 & 1 \\ 1 & 1 & \cdots & 1 & -1 \\ 1 & 1 & \cdots & 1 & 1 \end{bmatrix}. \tag{12.17}$$

Although the sequence $\{c_1, \cdots, c_{2^n}\}$, of the centers

$$c_i = \mathbf{b}_i^{(n)T} \mathbf{h} = \sum_{j=1}^{n} b_{ij}^{(n)} h_j, \quad i = 1, \cdots, 2^n \tag{12.18}$$

is not exactly arranged in increasing (or decreasing) order, there is a lot of structure in the sequence as summarized by the following facts (Diamantaras and Chassioti, 2000):

■ The first three centers $c_1 < c_2 < c_3$ are the three smallest values in the sequence c_i. Similarly, the last three centers $c_{2^n-2} < c_{2^n-1} < c_{2^n}$ are the three largest values in the sequence $\{c_i\}$.

■ The sequence $c_1, \ldots, c_{2^n}$, defined by 12.18 consists of consecutive quadruples, each arranged in increasing order:

$$c_{4i+1} < c_{4i+2} < c_{4i+3} < c_{4i+4}, \quad i = 0, \cdots, 2^{n-2} - 1$$

The smallest element of the ith quadruple is

$$c_{4i+1} = [\sum_{j=1}^{n-2} b_{4i+1,j}^{(n)} h_j] - h_{n-1} - h_n. \tag{12.19}$$

- The differences

$$\delta_1 = c_{4i+2} - c_{4i+1} = 2h_n \tag{12.20}$$

$$\delta_2 = c_{4i+3} - c_{4i+1} = 2h_{n-1} \tag{12.21}$$

$$\delta_3 = c_{4i+4} - c_{4i+1} = 2(h_{n-1} + h_n) \tag{12.22}$$

between the members of the ith quadruple are independent of i.

- Since

$$c_2 = c_1 + 2h_n, \quad \text{and}$$

$$c_3 = c_1 + 2h_{n-1}$$

the two smallest mixing parameters h_{n-1}, h_n can be retrieved using the values of the three smallest centers c_1, c_2, and c_3:

$$h_n = (c_2 - c_1)/2 \ , \tag{12.23}$$

$$h_{n-1} = (c_3 - c_1)/2 \ . \tag{12.24}$$

Once we have obtained h_{n-1} and h_n we can define a new sequence $\{c_i'\}$ by picking the first elements of each quadruple shifted by the sum $(h_{n-1} + h_n)$, thus obtaining

$$c_i' = c_{4(i-1)+1} + h_{n-1} + h_n = \sum_{j=1}^{n-2} b_{4(i-1)+1,j}^{(n)} h_j, \quad i = 1, \cdots, 2^{n-2}. \tag{12.25}$$

Notice however, that the first $n - 2$ bits of the $[4(i-1)+1]$-th row of $\mathbf{B}^{(n)}$ are all the bits of the ith row of $\mathbf{B}^{(n-2)}$. In other words,

$$b_{4(i-1)+1,j}^{(n)} = b_{ij}^{(n-2)}, \quad j = 1, \cdots, n-2$$

therefore,

$$c_i' = \mathbf{b}_i^{(n-2)^T} \mathbf{h} = \sum_{j=1}^{n-2} b_{ij}^{(n-2)} h_j, \quad i = 1, \cdots, 2^{n-2}. \tag{12.26}$$

Using these facts the following recursive algorithm is constructed:

Algorithm 12.1 : n binary sources, one observation

Step 1: Compute the centers c_i and sort them in increasing order.

Step 2: Compute h_n, h_{n-1}, from equations 12.23 and 12.24.

Step 3: Compute the differences δ_i, using equations 12.20, 12.21, and 12.22.

Step 4: Remove the set $\{c_1, c_2, c_3, c_1 + \delta_3\}$ from the sequence $\{c_i\}$. Set $c_1' = c_1 + h_n + h_{n-1}$ as the first element of a new sequence $\{c_i'\}$.

Step 5: Repeat until all elements have been removed:

Find the smallest element c_j of the remaining sequence $\{c_i\}$;

Remove the set $\{c_j, c_j + \delta_1, c_j + \delta_2, c_j + \delta_3\}$ from $\{c_i\}$;

Keep $c_j + h_n + h_{n-1}$ as the next element of the sequence $\{c'_i\}$.

At the end, the new sequence $\{c'_i\}$ will be four times shorter than the original $\{c_i\}$.
Step 6: Recursively repeat the algorithm for the new sequence $\{c'_i\}$ and for a new $n' = n - 2$ to obtain $h_{n'} = h_{n-2}$, $h_{n'-1} = h_{n-3}$. Eventually, $n' = 2$ or $n' = 1$. ■

Steps 4 and 5 are the basic recursion which reduces the problem size from n to $n-2$ by replacing the sequence c_i by c'_i. At step 6, we will iteratively obtain the pairs (h_n, h_{n-1}), (h_{n-2}, h_{n-3}), ..., until we reach the case where $n' = 2$ or $n' = 1$. The case for $n' = 2$ sources was treated in the previous subsection. The case for $n' = 1$ is trivial since it involves only one source. In this case, the observation is simply a scaled version of the input, $x(k) = h_1 s_1(k)$, thus, the estimation of h_1 and $s(k)$ is easy: we have $h_1 = |x(k)|$ (since $|s(k)| = 1$ and $h_1 > 0$) and so $s(k) = x(k)/h_1$.

Example 12.2

Consider the following system with four sources and one observation:

$$x(k) = -0.4326 s_1(k) + 1.2656 s_2(k) + 0.1553 s_3(k) - 0.2877 s_4(k).$$

The mixing vector $\mathbf{h} = [-0.4326, 1.2656, 0.1553, -0.2877]$ does not satisfy equation 12.16. The algorithm will recover the vector $\hat{\mathbf{h}} = [1.2656, 0.4326, 0.2877, 0.1553]$, which does satisfy equation 12.16, and it is identical to $\mathbf{h}$ except for the permutation and sign changes of its elements.
Step 1: The sorted sequence of centers is

$$c = \{ \quad -2.1412, -1.8306, -1.5658, -1.2760, -1.2552, -0.9654, -0.7006, -0.3900,$$
$$0.3900, 0.7006, 0.9654, 1.2552, 1.2760, 1.5658, 1.8306, 2.1412 \}$$

Step 2: Using equations 12.23 and 12.24 we compute $\hat{h}_3 = 0.2877$, $\hat{h}_4 = 0.1553$.
Step 3: Using equations 12.20, 12.21, and 12.22, we obtain $\delta_1 = 0.3106$, $\delta_2 = 0.5754$, $\delta_3 = 0.8860$.
Step 4: Remove $\{c_1, c_2, c_3, c_1 + \delta_3\} = \{-2.1412, -1.8306, -1.5658, -1.2552\}$ from c. Set $c'_1 = -1.6982$. New sorted sequence:

$$c = \{ \quad -1.2760, -0.9654, -0.7006, -0.3900,$$
$$0.3900, 0.7006, 0.9654, 1.2552, 1.2760, 1.5658, 1.8306, 2.1412 \}.$$

Step 5: Remove $\{c_1, c_1 + \delta_1, c_1 + \delta_2, c_1 + \delta_3\} = \{-1.2760, -0.9654, -0.7006, -0.3900\}$ from c. Set $c'_2 = -0.8330$. New sorted sequence:

$$c = \{0.3900, 0.7006, 0.9654, 1.2552, 1.2760, 1.5658, 1.8306, 2.1412\}.$$

Step 6: Remove $\{c_1, c_1 + \delta_1, c_1 + \delta_2, c_1 + \delta_3\} = \{0.3900, 0.7006, 0.9654, 1.2760\}$. Set $c'_3 = 0.8330$. New sorted sequence:

$$c = \{1.2552, 1.5658, 1.8306, 2.1412\}.$$

Step 6: Remove $\{c_1, c_1 + \delta_1, c_1 + \delta_2, c_1 + \delta_3\} = \{1.2552, 1.5658, 1.8306, 2.1412\}$. Set $c_4' = 1.6982$. New sorted sequence:

$$c = \emptyset.$$

The new sequence $c' = \{-1.6982, -0.8330, 0.8330, 1.6982\}$ yields the estimates of the remaining mixing parameters $\hat{h}_1 = 1.2656$, $\hat{h}_2 = 0.4326$. ■

12.2.1.2   Two Mixtures

In the case of $m = 2$ mixtures the observed data $\mathbf{x}(k)$ lie in the two-dimensional space R^2. Although it is possible to see each mixture separately as a single-mixture–multiple-sources problem, as the one treated in the previous subsection, this is not the most efficient approach to the problem. It turns out that the 2D structure of the output constellation reveals the mixing operator $\mathbf{H}$ in a very elegant and straightforward way. To see that, let us start by considering the data constellation of a binary antipodal signal $s_1(k)$ (fig. 12.3a). The constellation actually consists of two points on the real axis: $s^- = -1$ and $s^+ = 1$. Next, consider a linear transformation from R^1 to R^2 which maps $s_1(k)$ to a vector signal $\mathbf{x}^{(1)}(k) = [x_1^{(1)}(k), x_2^{(1)}(k)]^T$:

$$\mathbf{x}^{(1)}(k) = \mathbf{h}_1 s_1(k). \tag{12.27}$$

The linear operator $\mathbf{h}_1 = [h_{11}, h_{12}]^T$ is a two-dimensional vector shown in fig. 12.3b. The constellation of $\mathbf{x}(k)$ is shown in fig. 12.3c, and it also consists of two points $\mathbf{x}^- = -\mathbf{h}_1 = s^-\mathbf{h}_1$ and $\mathbf{x}^+ = \mathbf{h}_1 = s^+\mathbf{h}_1$.

Now let us look at shape of the data cloud corresponding to the linear combination of several binary antipodal sources $s_1(k), \ldots, s_n(k)$. It is instructive to study the shape of this cloud as n increases gradually from $n = 2$ and upward. The linear mixture of $n = 2$ sources

$$\mathbf{x}^{(2)}(k) = \mathbf{h}_1 s_1(k) + \mathbf{h}_2 s_2(k) \tag{12.28}$$

has the geometric structure shown in fig. 12.4b, for the mixing vectors $\mathbf{h}_1$, $\mathbf{h}_2$, shown in fig. 12.4a. The data cluster contains four points: $\mathbf{x}^{++} = s^+\mathbf{h}_1 + s^+\mathbf{h}_2$, $\mathbf{x}^{+-} = s^+\mathbf{h}_1 + s^-\mathbf{h}_2$, $\mathbf{x}^{-+} = s^-\mathbf{h}_1 + s^+\mathbf{h}_2$, and $\mathbf{x}^{--} = s^-\mathbf{h}_1 + s^-\mathbf{h}_2$.

Adding a third source $s_3(k)$ with the mixing vector $\mathbf{h}_3$, the data mixture

$$\mathbf{x}^{(3)}(k) = \mathbf{h}_1 s_1(k) + \mathbf{h}_2 s_2(k) + \mathbf{h}_3 s_3(k) \tag{12.29}$$

has the constellation shown in fig. 12.5. Now the data cluster contains eight points: $\mathbf{x}^{+++} = s^+\mathbf{h}_1 + s^+\mathbf{h}_2 + s^+\mathbf{h}_3$, $\mathbf{x}^{++-} = s^+\mathbf{h}_1 + s^+\mathbf{h}_2 + s^-\mathbf{h}_3$, $\mathbf{x}^{+-+} = s^+\mathbf{h}_1 + s^-\mathbf{h}_2 + s^+\mathbf{h}_3$, $\mathbf{x}^{+--} = s^+\mathbf{h}_1 + s^-\mathbf{h}_2 + s^-\mathbf{h}_3$, $\mathbf{x}^{-++} = s^-\mathbf{h}_1 + s^+\mathbf{h}_2 + s^+\mathbf{h}_3$, $\mathbf{x}^{-+-} = s^-\mathbf{h}_1 + s^+\mathbf{h}_2 + s^-\mathbf{h}_3$, $\mathbf{x}^{--+} = s^-\mathbf{h}_1 + s^-\mathbf{h}_2 + s^+\mathbf{h}_3$, and $\mathbf{x}^{---} = s^-\mathbf{h}_1 + s^-\mathbf{h}_2 + s^-\mathbf{h}_3$.

By simple inspection of figures 12.3, 12.4, and 12.5, one can make the following useful observations:

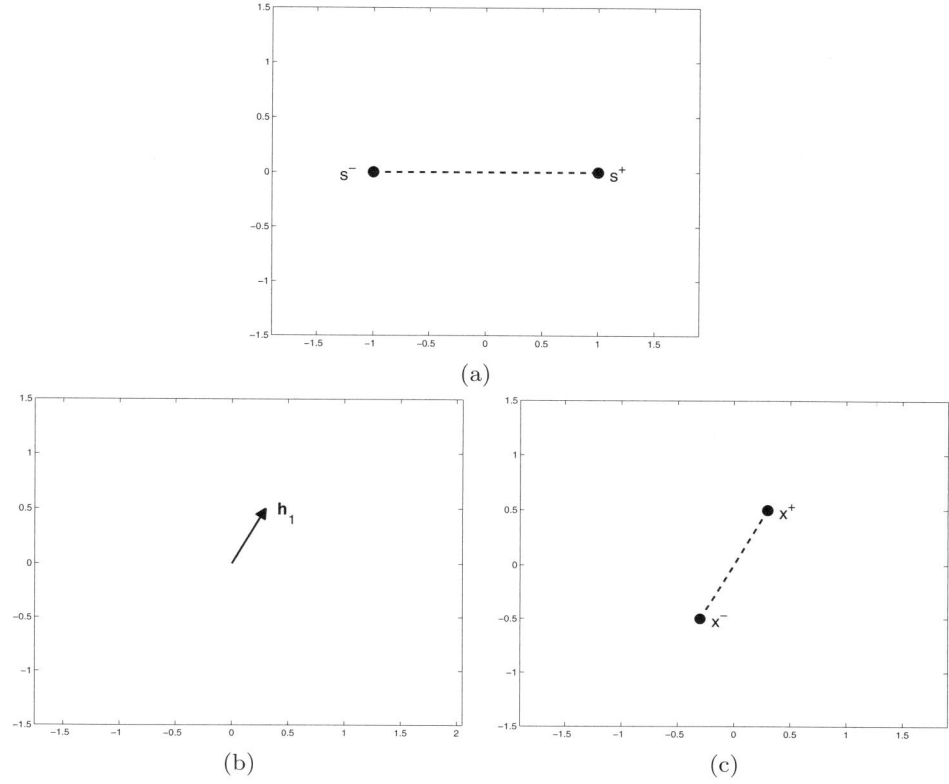

Figure 12.3   (a) Data constellation of a binary antipodal signal $s_1(k)$. (b) Linear transformation vector $\mathbf{h}_1$. (c) Data constellation of the transformed signal $\mathbf{x}^{(1)}(k) = \mathbf{h}_1 s_1(k)$.

1. The number of cluster points is 2^n, where n is the number of binary sources.

2. The data constellation is a symmetric, self-repetitive figure. While the symmetry is obvious, the self-repetitive structure can be seen by comparing, for example, fig. 12.5b against fig. 12.4b. The first consists of two copies of the latter shifted by the vectors $-\mathbf{h}_3$ and $\mathbf{h}_3$. The same is true for figs. 12.4b and 12.3c except that the shift is by the vectors $-\mathbf{h}_2$ and $\mathbf{h}_2$.

3. For every cluster point there exist n copies at the directions $\mathbf{h}_1$ or $-\mathbf{h}_1$, and $\mathbf{h}_2$ or $-\mathbf{h}_2, \ldots$, and $\mathbf{h}_n$ or $-\mathbf{h}_n$.

It is even more interesting and, in fact, very useful to study the properties of the *convex hull* of the data constellation set. By definition, the convex hull of a set of points in 2D space is the smallest polygon that contains them or, in other words, the *bounding polygon* for these points. Figures 12.6a–c show the convex hulls H_1, H_2, and H_3 for the data constellations corresponding to the mixtures $\mathbf{x}^{(1)}$, $\mathbf{x}^{(2)}$, and $\mathbf{x}^{(3)}$, respectively. Let d be the distance between the two alphabet symbols.

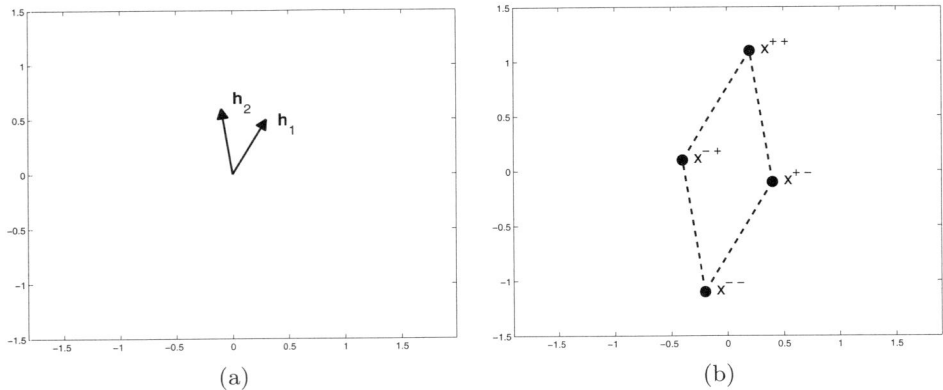

Figure 12.4 (a) Mixing vectors $\mathbf{h}_1$, $\mathbf{h}_2$. (b) Data cluster for the mixture $\mathbf{x}^{(2)}(k) = \mathbf{h}_1 s_1(k) + \mathbf{h}_2 s_2(k)$ of two binary antipodal sources.

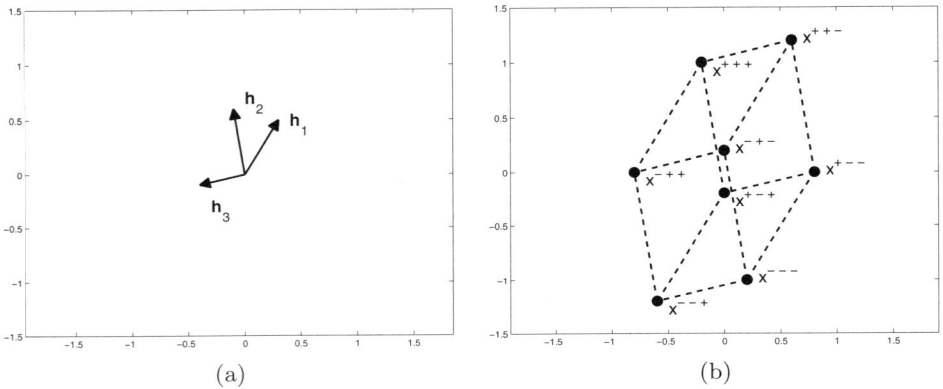

Figure 12.5 (a) Mixing vectors $\mathbf{h}_1$, $\mathbf{h}_2$, $\mathbf{h}_3$. (b) Data cluster for the mixture $\mathbf{x}^{(3)}(k) = \mathbf{h}_1 s_1(k) + \mathbf{h}_2 s_2(k) + \mathbf{h}_3 s_3(k)$ of three binary antipodal sources.

It can be shown that any convex hull H satisfies the following properties (for the proof, see Diamantaras, 2002):

1. Every edge $\mathbf{e}$ of H is parallel to some mixing vector $\mathbf{h}_i$, $i \in \{0, 1, \cdots, n\}$. Also, $\mathbf{e}$ has length $d\|\mathbf{h}_i\|$. For the binary antipodal alphabet $\mathcal{A}_a$, we have $d = 2$.

2. Every vector $\mathbf{h}_i$ corresponds to a pair of edges, i.e., it is parallel to two edges $\mathbf{e}_i$ and $\mathbf{e}'_i$ of equal length $d\|\mathbf{h}_i\|$. It follows that H has $2n$ edges.

3. H is symmetric. If the alphabet is symmetric around 0 (e.g., $\mathcal{A}_a$) then the center of symmetry is the point $\mathbf{x}_O = 0$. Otherwise, the center of symmetry is a nonzero point $\mathbf{x}'_O \in \mathrm{R}^m$.

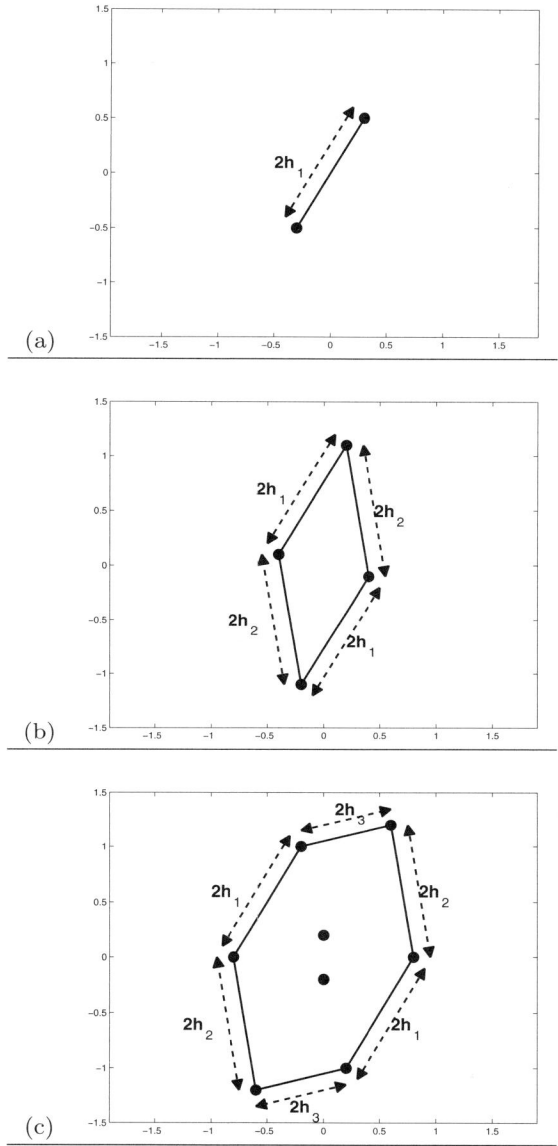

Figure 12.6   Convex hulls for data constellations of mixtures of n binary sources.
(a) $n = 1$, (b) $n = 2$, (c) $n = 3$.

These important results show that there is a two-to-one correspondence between the edges of the convex hull and the unknown mixing vectors: there is a pair of edges parallel to each mixing vector, and furthermore, the edges have length equal to d times the length of their corresponding mixing vectors. Thus we easily come to the following procedure for the identifying the $\mathbf{h}_i$'s:

Algorithm 12.2 : n **binary sources, two observations**

Step 1: Find the constellation set $\mathcal{X}$ of the 2D mixture $\mathbf{x}(k)$.
Step 2: Compute the convex hull H of $\mathcal{X}$.
Step 3: H consists of $2n$ edge pairs $\{\mathbf{e}_i, \mathbf{e}'_i\}$, $\mathbf{e}_i \| \mathbf{e}'_i$, $i = 1, \cdots, n$. The number of sources is n.
Step 4: The mixing vectors are: $\mathbf{h}_i = \mathbf{e}_i/d$, up to an unknown ordering and sign.

Of course, the original order and sign of the vectors are irretrievable. As we have seen, this is a general, problem-inherent limitation and it is not specific to this (or any other) particular method. In fact, the limitation cannot be overcome without additional information regarding the sources or the mixing operators.

12.2.1.3   More Than Two Mixtures

It is not difficult to see that the whole convex hull idea can be extended to the case where $m \geq 3$. Again, the edges of the convex hull will be parallel to the mixing vectors $\mathbf{h}_i$ except that, now, the convex hull lies in R^m. The algorithms for computing the convex hull in m-dimensional spaces are not as simple as the ones for the 2D case. For a comprehensive discussion of this topic see Preparata and Shamos (1985).

12.2.2   Instantaneous Mixtures of M-ary Alphabet Sources

The results of section 12.2.1 can be easily extended to M-ary signals, i.e., signals whose alphabet contains M discrete and equally distributed values. For example, the alphabet $\mathcal{A}_5 = \{-1, -1/2, 0, 1/2, 1\}$ contains $M = 5$ symbols symmetrically distributed around 0. Similar results, as in the binary case, hold here as well. Again, the convex hull directly connects the constellation geometry with the unknown mixing vectors. Let d be the distance between the maximum and minimum symbols in the M-ary alphabet $\mathcal{A}_M$

$$d = \max\{\mathcal{A}_M\} - \min\{\mathcal{A}_M\}.$$

Also let H be the convex hull of the constellation $\mathcal{X}$ of the mixture $\mathbf{x}(k) = \mathbf{h}_1 s_1(k) + \cdots + \mathbf{h}_n s_n(k)$. The the following statements are true (see fig. 12.7):

1. The number of cluster points is M^n, where n is the number of M-ary sources.

2. The data constellation is a symmetric, self-repetitive figure.

3. Every edge $\mathbf{e}$ of H is parallel to some mixing vector $\mathbf{h}_i$, $i \in \{0, 1, \cdots, n\}$, and $\mathbf{e}$ has length $d\|\mathbf{h}_i\|$.

4. Every vector $\mathbf{h}_i$ corresponds to a pair of edges, i.e., it is parallel to two edges $\mathbf{e}_i$ and $\mathbf{e}'_i$ of equal length $d\|\mathbf{h}_i\|$. It follows that H has $2n$ edges.

5. H is symmetric. For alphabets symmetric around zero the center of symmetry is $\mathbf{x}_O = 0$.

We may use algorithm 12.2 without modifications for the solution of the M-ary case as well.

12.2.3   Noisy Data

The analysis of the previous subsections pertains to systems with noiseless outputs. In most applications however, the observation is burdened with noise, either because the system itself is noisy or the receiving device introduces errors in the measurements. The additive noise model is commonly used for describing the observation error:

$$\mathbf{x}(k) = \mathbf{Hs}(k) + \mathbf{v}(k). \tag{12.30}$$

Without loss of generality, and for the sake of visualization, we shall focus on the two-output case. An entirely similar discussion holds for the cases $n = 1$ or $n > 2$. The vector signal $\mathbf{v}(k) = [v_1(k), \cdots, v_n(k)]^T$, contains the noise components $v_i(k)$ for each observed output signal $i = 1, \cdots, n$. The constellation of $\mathbf{x}$ is now less crisp since the true centers are surrounded by a cloud of points (fig. 12.8a). The methods presented in sections 12.2.1 and 12.2.2 can still be applied preceded by a clustering process that will estimate the actual centers from the noisy data cloud. Such clustering methods include the ISODATA or K-means algorithm (Duda et al., 2001; Lloyd, 1982; MacQueen, 1967), the EM algorithm (Dempster et al., 1977), the neural gas algorithm (Martinetz et al., 1993), Kohonen's self-organizing feature maps (SOM) (Kohonen, 1989), RBF neural networks (Moody and Darken, 1989), and many others. For a detailed treatment of clustering methods refer to Theodoridis and Koutroubas (1998). Figure 12.8b shows the estimation of the true centers using the K-means algorithm in a system with three binary inputs and two linear output mixtures with noise power at 15dB. Notice that the estimation errors inside the convex hull do not affect the results. It is only the errors at the boundary that are significant. We apply the blind identification method discussed earlier in this section using the estimated centers provided by K-means, obtaining the results shown in table 12.1.

12.2.4   Convolutive Mixtures of Binary Sources

The convolutive mixtures of binary sources are described by the output of the MIMO FIR system (eq. 12.1). The blind problems related to such systems are con-

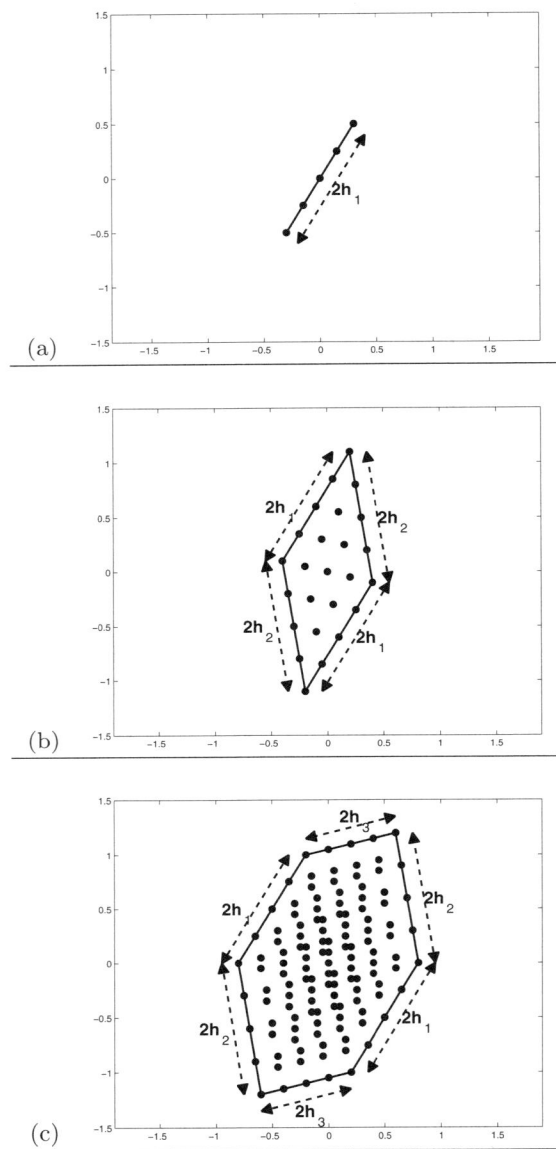

Figure 12.7 Convex hulls of mixture constellations from n M-ary sources $(M = 5)$. The source symbols are drawn from the alphabet $\{-1, -0.5, 0, 0.5, 1\}$, with maximum distance $d = 2$. (a) $n = 1$, (b) $n = 2$, (c) $n = 3$.

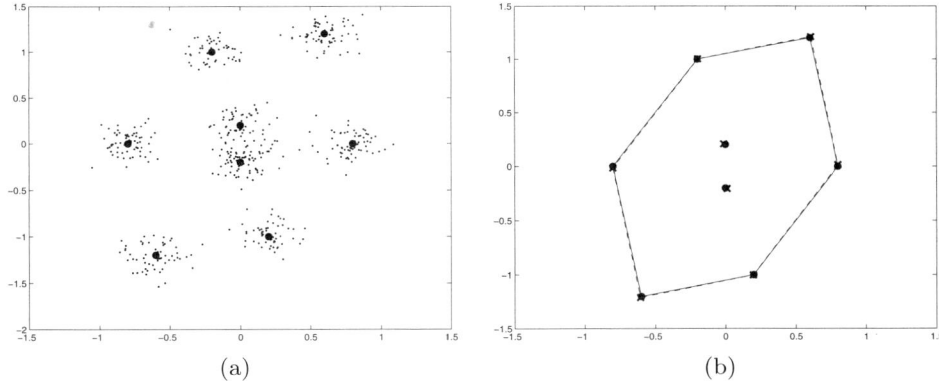

Figure 12.8   (a) Data constellation for a noisy memoryless linear system with 3 binary inputs and 2 outputs (mixtures). The noise level is 15 dB. Superimposed are the true cluster centers marked with "o". (b) True cluster centers (o) and estimated cluster centers (x) using the K-means algorithm. Also shown is the true convex hull (solid line) and the estimated convex hull (dashed line).

Table 12.1   True and Estimated Mixing Vectors

$\mathbf{h}_1$	$\hat{\mathbf{h}}_1$	$\mathbf{h}_2$	$\hat{\mathbf{h}}_2$	$\mathbf{h}_3$	$\hat{\mathbf{h}}_3$
0.3000	0.3017	-0.1000	-0.0960	-0.4000	0.3917
0.5000	0.4902	0.6000	0.6089	-0.1000	0.1009

siderably more difficult than the corresponding instantaneous mixture problems, but at the same time, they are much more important. Convolutive mixing models, for example, can describe multipath and crosstalk phenomena in wireless communications, being in that sense much more realistic than instantaneous models. In this section we shall approach the blind source separation and blind system identification problems of MIMO FIR models using the geometric properties of the data constellation. We shall treat, first, the simpler single-input single-output (SISO) problem and then continue on to the multi-input single-output (MISO) case. The proper MIMO problem is not explicitly discussed since it can be seen as a multitude of m decoupled MISO problems.

12.2.4.1   Blind SISO Deconvolution as Instantaneous Blind Source Separation

In this subsection we shall use the results of the previous sections to solve the blind SISO identification and deconvolution problems. Our approach is to relate any given SISO system with an overdetermined instantaneous mixtures model, hence the same methods can be applied as in sections 12.2.1. Let us consider a linear, FIR, single-input single-output (SISO) system with a binary antipodal input $s(k)$,

$$x(k) = \sum_{i=0}^{L-1} h_i s(k-i) \tag{12.31}$$

We shall assume that the impulse response h_i, $i = 0, \cdots, L-1$, is real. Let us create a vector sequence $\mathbf{x}(k)$ using time-windowing of length m on the output sequence $x(k)$

$$\mathbf{x}(k) = [x(k), \cdots, x(k-m+1)]^T. \tag{12.32}$$

Then using the system 12.31 we have

$$\mathbf{x}(k) = \mathbf{H}\mathbf{s}(k), \tag{12.33}$$

where $\mathbf{H}$ is the *Toeplitz system matrix*

$$\mathbf{H} = \begin{bmatrix} h_0 & h_1 & \cdots & h_{L-1} & 0 & \cdots & 0 \\ 0 & h_0 & \cdots & h_{L-2} & h_{L-1} & 0 & 0 \\ & & \ddots & & & \ddots & \\ 0 & \cdots & 0 & h_0 & \cdots & h_{L-2} & h_{L-1} \end{bmatrix} \tag{12.34}$$

and

$$\mathbf{s}(k) = [s(k), \ s(k-1), \cdots, \ s(k-m-L+2)]^T. \tag{12.35}$$

Now, equation 12.33 describes m linear instantaneous mixtures $x_i'(k) = x(k-i)$, $i = 0, \cdots, m-1$, of n sources $s_j'(k)$ defined as follows:

$$s_j'(k) = s(k-j+1), \quad j = 1, 2, \cdots, n = m+L-1.$$

Thus, we have successfully transformed the problem into the same form treated in section 12.2.1:

$$x_i'(k) = \sum_{j=1}^{n} h_{ij} s_j'(k), \tag{12.36}$$

where h_{ij} is the (i, j)th element of $\mathbf{H}$. Equivalently, we can write

$$\mathbf{x}(k) = \sum_{j=1}^{n} \mathbf{h}_j s_j'(k), \tag{12.37}$$

where the mixing vectors $\mathbf{h}_1, \ldots, \mathbf{h}_n$ are the columns of $\mathbf{H}$. Given the above formulation, the results of Section 12.2.1 apply directly to this problem. There are, however, some special points to be noted:

1. For any nontrivial FIR filter of length $L > 1$, the number of observations x_1', ..., x_m' is necessarily less than the number of sources $s_1', \ldots, s_n'$, since $n = m+L-1 > m$.

2. The mixing vectors have no arbitrary form. For example, $\mathbf{h}_1$ has the form $[\times, 0, \cdots, 0]^T$ and $\mathbf{h}_n$ has the form $[0, \cdots, 0, \times]^T$.

3. The sources are not independent. In fact, any one is a shifted version of any other.

Next, we shall give examples for two cases: $m = 1$; and $m = 2$.

Example 12.3 *Time Window of Length* $m = 1$

Suppose that we observe the output $x(k)$ of a SISO filter $\mathbf{h} = [-0.4937, -1.1330, 0.7632, 0.1604]^T$ excited by the binary input $s(k)$. Using algorithm 12.1 we shall identify the filter with the necessary permutation and sign changes so that the estimated taps will be positive and arranged in decreasing order. Thus we shall obtain $\hat{\mathbf{h}} = [1.1330, 0.7632, 0.4937, 0.1604]^T$ and so

$$x(k) = \hat{h}_1 \hat{s}'_1(k) + \hat{h}_2 \hat{s}'_2(k) + \hat{h}_3 \hat{s}'_3(k) + \hat{h}_4 \hat{s}'_4(k)$$
$$= (-h_2)(-s'_2(k)) + h_3 s'_3(k) + (-h_1)(-s'_1(k)) + h_4 s'_4(k).$$

Obviously, the estimated sources $\hat{s}'_i$ correspond to the true "sources" s'_i as follows:

$$\hat{s}'_1(k) = -s'_2(k) = -s(k-1),$$
$$\hat{s}'_2(k) = s'_3(k) = s(k-2),$$
$$\hat{s}'_3(k) = -s'_1(k) = -s(k),$$
$$\hat{s}'_4(k) = s'_4(k) = s(k-3).$$

Since the signals $\hat{s}'_i$ are shifted versions of the original source, $s(k)$, it is easy to recover their correct order and relative sign changes by computing for each signal, the time shift with maximum correlation to an arbitrary reference, for example, $\hat{s}'_1$. Applying the same ordering and sign changes to $\hat{\mathbf{h}}$, we obtain $\pm \mathbf{h}$. ∎

Example 12.4 *Time Window of Length* $m = 2$

Consider the same SISO filter as before and let us use time-windowing of length $m = 2$ to obtain the vector sequence $\mathbf{x}(k)$:

$$\mathbf{x}(k) = \begin{bmatrix} x(k) \\ x(k-1) \end{bmatrix}$$

$$= \begin{bmatrix} -0.4937 & -1.1330 & 0.7632 & 0.1604 & 0 \\ 0 & -0.4937 & -1.1330 & 0.7632 & 0.1604 \end{bmatrix} \begin{bmatrix} s(k) \\ s(k-1) \\ s(k-2) \\ s(k-3) \\ s(k-4) \end{bmatrix}.$$

Using algorithm 12.2 we estimate the original mixing vectors $\mathbf{h}_1 = [-0.4937, 0]^T$, $\mathbf{h}_2 = [-1.1330, -0.4937]^T$, $\mathbf{h}_3 = [0.7632, -1.1330]^T$, $\mathbf{h}_4 = [0.1604, 0.7632]^T$, $\mathbf{h}_5 = [0, 0.1604]^T$, but with an arbitrary order and sign change. The estimated mixing vectors can be put in the correct order by observing that the true system parameters satisfy the following:

$$h_{1,1} = h_{2,2} = -0.4937,$$

$$h_{2,1} = h_{3,2} = -1.1330,$$

$$h_{3,1} = h_{4,2} = 0.7632,$$

$$h_{4,1} = h_{5,2} = 0.1604,$$

$$h_{5,1} = h_{1,2} = 0 .$$

Since the sign of each estimated vector is arbitrary, we compare the absolute values, $|\hat{h}_{i,1}|$ against $|\hat{h}_{j,2}|$, and we change the signs of either $\hat{\mathbf{h}}_i$ or $\hat{\mathbf{h}}_j$, as necessary, so that $\hat{h}_{i,1} = \hat{h}_{j,2}$. Once the correct order of the mixing vectors is retrieved we automatically obtain the correct filter impulse response (up to a sign). Subsequently, the system input, $s(k)$, is retrieved using standard (nonblind) deconvolution methods. ■

12.2.4.2   Blind SISO Identification

An alternative approach for identifying the impulse response $\mathbf{h} = [h_0, \cdots, h_{L-1}]^T$ of a general SISO system (eq. 12.31) has been proposed by Yellin and Porat (1993). The method is not based on constellation geometry but rather on the properties of the successor values of "equivalent" observations. The source symbols $s(k)$ may be drawn from an M-ary alphabet $\mathcal{A}_M = \{\pm 1, \cdots, \pm(M/2)\}$ (M is even). Before we proceed we need to introduce the concept of *equivalence* between two observations:

Definition 12.1   Observation Equivalence

Two observations $x(k)$ and $x(l)$ are said to be equivalent if the input values that produce them according to 12.31 are identical: $s(k - i) = s(l - i)$, for all $i = 0, \cdots, L - 1$.

Note that two equivalent observations are necessarily equal, but the converse may not be true. Indeed, it is possible that two equal observations $x(k) = x(l)$, are produced by two different strings of input symbols $[s(k), \cdots, s(k - L + 1)] \neq [s(l), \cdots, s(l - L + 1)]$.

   Consider four sets of $(N + 1)$ consecutive observations from equation 12.31: $\mathcal{X}_j = \{x(j), x(j + 1), \cdots, x(j + N)\}$, $\mathcal{X}_k = \{x(k), x(k + 1), \cdots, x(k + N)\}$, $\mathcal{X}_l = \{x(l), x(l+1), \cdots, x(l+N)\}$, $\mathcal{X}_m = \{x(m), x(m+1), \cdots, x(m+N)\}$. Further assume that the pairs $\{x(j), x(k)\}$ and $\{x(l), x(m)\}$ are equivalent. Define

$$\begin{aligned} \sigma_{jki} &= [s(j + i) - s(k + i)]/2, \\ \sigma_{lmi} &= [s(l + i) - s(m + i)]/2; \quad i = 1, \cdots, N \end{aligned} \tag{12.38}$$

and note that $\sigma_{jki}, \sigma_{lmi} \in \mathcal{A}_M^0 = \mathcal{A}_M \cup 0$. Let the following conditions be true:

1. σ_{jk1}, σ_{lm1} are nonzero and coprime, i.e., their greatest common divisor is 1;

2. for all $\alpha, \beta \in \mathcal{A}_M$

$$\left| \frac{\sigma_{jk1}}{\sigma_{lm1}} \right| = \left| \frac{\alpha}{\beta} \right| \Rightarrow |\alpha| = |\sigma_{jk1}|, |\beta| = |\sigma_{lm1}|;$$

3. for all $\alpha, \beta \in \mathcal{A}_M^0$,

$$\frac{\sigma_{jk1}}{\sigma_{lm1}} \neq \frac{\sigma_{jki} - \alpha}{\sigma_{lmi} - \beta}, \qquad \text{for all } i = 2, \cdots, N.$$

The method starts by identifying the first filter tap h_0 up to a sign, and continues by recursively identifying the remaining taps given the previous ones. Begin with the remark that $x(j)$ and $x(k)$ are equivalent, so $[s(j), \cdots, s(j-L+1)] = [s(k), \cdots, s(k-L+1)]$. Then, the *successor values* of $x(j)$, $x(k)$, can be written as

$$x(j + 1) = h_0 s(j + 1) + \sum_{i=1}^{L-1} h_i s(j + 1 - i),$$

$$x(k + 1) = h_0 s(k + 1) + \sum_{i=1}^{L-1} h_i s(k + 1 - i),$$

so,

$$\frac{x(j + 1) - x(k + 1)}{2} = \sigma_{jk1} h_0. \qquad (12.39)$$

Similarly, for $x(l + 1)$, $x(m + 1)$:

$$\frac{x(l + 1) - x(m + 1)}{2} = \sigma_{lm1} h_0, \qquad (12.40)$$

and so,

$$\frac{x(j + 1) - x(k + 1)}{x(l + 1) - x(m + 1)} = \frac{\sigma_{jk1}}{\sigma_{lm1}}. \qquad (12.41)$$

By condition 2, the ratio $|\sigma_{jk1}/\sigma_{lm1}|$ is produced by a unique enumerator-denominator pair in $\mathcal{A}_M$. Thus both values σ_{jk1} and σ_{lm1} can be uniquely identified, up to a sign, leading to the magnitude estimation of h_0 by:

$$|h_0| = \frac{|x(j + 1) - x(k + 1)|}{2|\sigma_{jk1}|} = \frac{|x(l + 1) - x(m + 1)|}{2|\sigma_{lm1}|}. \qquad (12.42)$$

Without loss of generality, we may assume that $h_0 > 0$, and proceed to the estimation of h_1 as follows: Write the second successors of $x(l)$, $x(k)$, as

$$x(j + 2) = h_0 s(j + 2) + h_1 s(j + 1) + \sum_{i=2}^{L-1} h_i s(j + 1 - i),$$

$$x(k+2) = h_0 s(k+2) + h_1 s(k+1) + \sum_{i=2}^{L-1} h_i s(k+1-i),$$

hence,

$$\frac{x(j+2) - x(k+2)}{2} = \sigma_{jk2} h_0 + \sigma_{jk1} h_1. \tag{12.43}$$

Similarly,

$$\frac{x(l+2) - x(m+2)}{2} = \sigma_{lm2} h_0 + \sigma_{lm1} h_1. \tag{12.44}$$

The pair of equations 12.43 and 12.44 involve three unknowns: σ_{jk2}, σ_{lm2}, h_1. However, it turns out that since the first two unknowns come from the discrete set $\mathcal{A}_M^0$ and condition 3 is true, the solution is unique. Indeed, assume there existed two different solutions $\{\sigma_{jk2}^{(1)}, \sigma_{lm2}^{(1)}, h_1^{(1)}\}$, $\{\sigma_{jk2}^{(2)}, \sigma_{lm2}^{(2)}, h_1^{(2)}\}$. Then by equations 12.43 and 12.44 we have

$$(\sigma_{jk2}^{(2)} - \sigma_{jk2}^{(1)}) h_0 = (h_1^{(2)} - h_1^{(1)}) \sigma_{jk1}, \tag{12.45}$$

$$(\sigma_{lm2}^{(2)} - \sigma_{lm2}^{(1)}) h_0 = (h_1^{(2)} - h_1^{(1)}) \sigma_{lm1}. \tag{12.46}$$

Thus,

$$\frac{\sigma_{jk1}}{\sigma_{lm1}} = \frac{\sigma_{jk2}^{(2)} - \sigma_{jk2}^{(1)}}{\sigma_{lm2}^{(2)} - \sigma_{lm2}^{(1)}},$$

which is impossible, according to condition 3. Therefore, there exists a unique solution to equations 12.43 and 12.44. From these equations it follows that

$$h_1 = \left(\frac{x(j+2) - x(k+2)}{2} - \sigma_{jk2} h_0 \right) / \sigma_{jk1} = \left(\frac{x(l+2) - x(m+2)}{2} - \sigma_{lm2} h_0 \right) / \sigma_{lm1},$$

so the unique h_1 can be obtained by finding the intersection between the sets

$$\begin{aligned} \mathcal{F}_1 &= \left\{ \frac{x(j+2) - x(k+2)}{2\sigma_{jk1}} + \frac{\alpha h_0}{\sigma_{jk1}}; \ \alpha \in \mathcal{A}_M^0 \right\} \\ \mathcal{F}_2 &= \left\{ \frac{x(l+2) - x(m+2)}{2\sigma_{lm1}} + \frac{\beta h_0}{\sigma_{lm1}}; \ \beta \in \mathcal{A}_M^0 \right\} \end{aligned}.$$

This is computationally trivial since the two sets are finite with few elements. Inductively, for h_i, $i > 2$, and given the values for $h_0, \ldots, h_{i-1}$, we form the "deflated" successors

$$\bar{x}(j+i+1) = x(j+i+1) - \sum_{p=1}^{i-1}\sigma_{jk(p+1)}h_{i+p} \tag{12.47}$$

$$\bar{x}(k+i+1) = x(k+i+1) - \sum_{p=1}^{i-1}\sigma_{jk(p+1)}h_{i+p} \tag{12.48}$$

$$\bar{x}(l+i+1) = x(l+i+1) - \sum_{p=1}^{i-1}\sigma_{lm(p+1)}h_{i+p} \tag{12.49}$$

$$\bar{x}(m+i+1) = x(m+i+1) - \sum_{p=1}^{i-1}\sigma_{lm(p+1)}h_{i+p} \tag{12.50}$$

and we obtain a set of two equations similar to 12.43 and 12.44:

$$\frac{\bar{x}(j+i+1) - \bar{x}(k+i+1)}{2} = \sigma_{jk(i+1)}h_0 + \sigma_{jk1}h_i, \tag{12.51}$$

$$\frac{\bar{x}(l+i+1) - \bar{x}(m+i+1)}{2} = \sigma_{lm(i+1)}h_0 + \sigma_{lm1}h_i. \tag{12.52}$$

which are solved in a similar fashion, producing the unknown tap h_i. Thus, the whole approach is summarized in the following algorithm

Algorithm 12.3 *Yellin and Porat*

Step 1: Collect T observation measurements.
Step 2: Find pairs of equivalent measurements. Estimate h_0 according to equation 12.42.
Step 3: Estimate h_1 using h_0 and the pairs of equivalent observations.
Step 4: Continue with the estimation of $h_2, \ldots, h_n$ given the previous estimates.
Step 5: Use the estimated impulse response to deconvolve the observation sequence and obtain the system input.

Remark

■ The choice of pairs of equivalent observations (step 2 in algorithm 12.3) is far from trivial. The indices j, k, l, m, must satisfy various constraints so that the assumptions of the method are met. First, according to condition A, we must have $\sigma_{jk1}, \sigma_{lm1} \neq 0$, implying that $x(j+1) \neq x(k+1)$, $x(l+1) \neq x(m+1)$. Second, according to condition C, for all $i = 2, \cdots, N$ the ratios $\sigma_{jki}/\sigma_{lmi}$ should not be equal to $\sigma_{jk1}/\sigma_{lm1}$. A thorough discussion on the implementation details is in the original paper (Yellin and Porat, 1993).

■ The method can be easily extended to handle complex input constellations (such as QAM) and/or complex filter taps.

■ For the special case of i.i.d. input signals it is estimated that a sufficient batch size that guarantees $E > 2$ equivalent pairs of measurements is $T = 2.44E^{0.61}NM^{N/2}$.

■ It is difficult to satisfy condition C if the source alphabet is binary ($\mathcal{A}_M = \mathcal{A}_a$), because there is a limited choice for the values of σ_{jki}, σ_{lmi}, which belong to the set $\mathcal{A}_a^0 = \{-1, 0, 1\}$.

12.2.4.3   MISO Systems: Direct Source Extraction

The blind source extraction directly from the output of a multi-input single-output (MISO) system is treated in Diamantaras and Papadimitriou (2005). This work is an extension of earlier work on SISO systems (Diamantaras and Papadimitriou, 2004a). The key to the approach in both cases is the structure of the successor values of equivalent observations induced by the fact that the sources are binary. Subsequently, we shall present the results for the more general MISO case. Let us consider a multi-input single-output (MISO) model described by the following equation:

$$x(k) = \sum_{i=0}^{L-1} \mathbf{h}_i^T \mathbf{s}(k - i), \tag{12.53}$$

where $\mathbf{h}_i$ for $i = 0, \ldots, L-1$, are a set of unknown real n-dimensional mixing vectors or *filter taps*. The source vector signal $\mathbf{s}(k) = [s_1(k), \ldots, s_n(k)]^T$ is composed of n independent binary antipodal signals: $s_i(k) \in \mathcal{A}_a$. The observations of the mixtures are real-valued scalars. For each k, the vector $\mathbf{s}(k)$ can take one of 2^n values denoted by $\mathbf{b}_i^{(n)}$, $i = 1, \ldots, 2^n$. The vector $\mathbf{b}_i^{(n)T}$ is the ith row of the matrix $\mathbf{B}^{(n)}$ defined in equation 12.17.

Let us extend the concept of observation equivalence, defined before for SISO systems, to MISO systems by simply replacing the scalar inputs with vector inputs. Each observation $x(k)$ is generated by the linear combination of L n-dimensional source vectors, therefore, the observation space $\mathcal{X} \ni x(k)$ is a discrete set consisting of, at most, 2^M elements, $M = nL$. The cardinality $|\mathcal{X}|$ will be less than 2^M if and only if there exist two different L-tuples $\{\mathbf{b}_{j_0}^{(n)}, \cdots, \mathbf{b}_{j_{L-1}}^{(n)}\}$ and $\{\mathbf{b}_{l_0}^{(n)}, \cdots, \mathbf{b}_{l_{L-1}}^{(n)}\}$, of binary vectors such that $\sum_{i=0}^{L-1} \mathbf{h}_i^T \mathbf{b}_{j_i}^{(n)} = \sum_{i=0}^{L-1} \mathbf{h}_i^T \mathbf{b}_{l_i}^{(n)}$. The following avoids this situation:

Assumption 12.1 Two observations $x(k)$, $x(l)$, are equivalent if and only if they are equal.   ■

Hence, $|\mathcal{X}| = 2^M$. In other words, to every observation value $r \in \mathcal{X}$ corresponds a unique L-tuple $\{\bar{\mathbf{b}}_0(r), \cdots, \bar{\mathbf{b}}_{L-1}(r)\}$ of consecutive source vectors that generates this observation. No other observation value $r' \in \mathcal{X}$ corresponds to the same L-tuple of binary vectors. For any $x(k) = r$, we have

$$x(k) = \sum_{i=0}^{L-1} \mathbf{h}_i^T \bar{\mathbf{b}}_i(r), \tag{12.54}$$

since, by definition,

$$\bar{\mathbf{b}}_i(r) = \mathbf{s}(k - i), \qquad \text{for } i = 0, \cdots, L - 1.$$

Now the *successor observation*, $x(k+1)$, can be written as

$$x(k+1) = \mathbf{h}_0^T \mathbf{s}(k+1) + \sum_{i=1}^{L-1} \mathbf{h}_i^T \mathbf{s}(k-(i-1))$$

$$= \mathbf{h}_0^T \mathbf{s}(k+1) + \sum_{i=1}^{L-1} \mathbf{h}_i^T \bar{\mathbf{b}}_{i-1}(r). \tag{12.55}$$

Since $\mathbf{s}(k+1)$ is an n-dimensional binary antipodal vector, $x(k+1)$ can take one of the following 2^n possible values:

$$y_p(r) = \mathbf{h}_0^T \mathbf{b}_p^{(n)} + \sum_{i=1}^{L-1} \mathbf{h}_i^T \bar{\mathbf{b}}_{i-1}(r), \quad p = 1, \cdots, 2^n. \tag{12.56}$$

Note that the successor values $y_p(r)$ do not depend on the specific time index k but only on the observation value r. Therefore, each observation value r creates a class of successors $\mathcal{Y}(r)$ with cardinality $|\mathcal{Y}(r)| = 2^n$. Furthermore, we have $\sum_{p=1}^{2^n} \mathbf{b}_p^{(n)} = 0$, so the mean $\bar{y}(r)$ of the members of $\mathcal{Y}(r)$ is

$$\bar{y}(r) = \frac{1}{2^n} \sum_{p=1}^{2^n} y_p(r)$$

$$= \frac{1}{2^n} \left(\mathbf{h}_0^T \sum_{p=1}^{2^n} \mathbf{b}_p^{(n)} + 2^n \sum_{i=1}^{L-1} \mathbf{h}_i^T \bar{\mathbf{b}}_{i-1}(r) \right)$$

$$= \sum_{i=1}^{L-1} \mathbf{h}_i^T \bar{\mathbf{b}}_{i-1}(r). \tag{12.57}$$

Now, let us replace every $x(k) = r$ with the mean $\bar{y}(r)$ to obtain a new sequence

$$x^{(2)}(k) = \sum_{i=1}^{L-1} \mathbf{h}_i^T \bar{\mathbf{b}}_{i-1}(r)$$

$$x^{(2)}(k) = \sum_{i=1}^{L-1} \mathbf{h}_i^T \mathbf{s}(k-i+1). \tag{12.58}$$

The new MISO system 12.58 has the same taps as the original system 12.53 except that it is shorter since $\mathbf{h}_0$ is missing. We will say that the system has been deflated. An additional but trivial difference is that the source sequence is time-shifted. Based on the discussion above, the whole *filter-* or *system-deflation* method, is summarized as follows:

Algorithm 12.4 System Deflation

Step 1: For every $r \in \mathcal{X}$ locate the set of time instances $\mathcal{K}(r) = \{k : x(k) = r\}$.
Step 2: Find the successor set $\mathcal{Y}(r) = \{x(k+1) : k \in \mathcal{K}(r)\}$. This set must contain 2^n distinct values $y_1(r), \ldots, y_{2^n}(r)$.
Step 3: Compute the mean $\bar{y}(r) = 1/2^n \sum_{i=1}^{2^n} y_i(r)$.
Step 4: Replace $x(k)$ by $\bar{y}(r)$, for all $k \in \mathcal{K}(r)$. ∎

Clearly, for this method it is essential that all observation-successor pairs $[r, y_i(r)]$, $i = 1, \cdots, 2^n$ will appear, at least once, in the output sequence x. Applying the deflation method $L-1$ times, the system will be eventually reduced to an multi-input single-output instantaneous problem:

$$x^{(L)}(k) = \mathbf{h}_{L-1}^T \mathbf{s}(k - L + 1). \tag{12.59}$$

The BSS problem of the type in 12.59 has been treated in section 12.2.1.

The main disadvantage of this method stems from the assumption that the data set must contain every possible observation-successor pair. As the size of the MISO system increases this assumption requires exponentially larger observation data sets. An alternative approach starts by observing that for any $r \in \mathcal{X}$ the *centered successors*,

$$c_i = y_i(r) - \bar{y}(r) = \mathbf{h}_0^T \mathbf{b}_i^{(n)} \quad i = 1, \cdots, 2^n, \tag{12.60}$$

are independent of r. Thus every observation has the same set of centered successors. We shall refer to the set $C = \{c_i; \ i = 1, \cdots, 2^n\}$ as the *centered successor constellation set* of system 12.53. C can be easily computed by first obtaining $\mathcal{Y}(r)$, for any r, and then subtracting the mean $\bar{y}(r)$ from each element $y_i(r) \in \mathcal{Y}(r)$. Note that C is symmetric in the sense that $c \in C \Leftrightarrow -c \in C$.

Now, for every observation value $r = x(k) \in \mathcal{X}$ we have

$$x(k) = \mathbf{h}_0^T \mathbf{s}(k) + \sum_{l=1}^{L-1} \mathbf{h}_l^T \mathbf{s}(k - l) \tag{12.61}$$

$$r = \mathbf{h}_0^T \mathbf{b}_i^{(n)} + \sum_{l=1}^{L-1} \mathbf{h}_l^T \bar{\mathbf{b}}_l(r), \quad \text{some } i \tag{12.62}$$

$$= c_i + \sum_{l=1}^{L-1} \mathbf{h}_l^T \bar{\mathbf{b}}_l(r), \quad \text{some } i \tag{12.63}$$

Furthermore, due to the symmetry of the constellation set, there exists a "dual" observation value $r^d \in \mathcal{X}$ such that

$$r^d = -c_i + \sum_{l=1}^{L-1} \mathbf{h}_l^T \bar{\mathbf{b}}_l(r) \tag{12.64}$$

$$r^d = r - 2c_i. \tag{12.65}$$

Assume that for every observation $r \in \mathcal{X}$, there exists a unique index $j \in \{1, \cdots, 2^n\}$ such that $r - 2c_j \in \mathcal{X}$. Then the dual value r^d can be identified by testing all $r - 2c_j$, $j = 1, \cdots, 2^n$, for membership in the observation space $\mathcal{X}$. Let us now replace $x(k)$ by the average of r, r^d, to obtain

$$\tilde{x}^{(2)}(k) = (r + r^d)/2 = r - c_j = \sum_{l=1}^{L-1} \mathbf{h}_l^T \bar{\mathbf{b}}_l(r). \tag{12.66}$$

Note that $\mathbf{b}_l(r) = \mathbf{s}(k - l)$, so

$$\tilde{x}^{(2)}(k) = \sum_{l=1}^{L-1} \mathbf{h}_l^T \mathbf{s}(k - l). \tag{12.67}$$

Equation 12.67 describes a new, shortened MISO system,

Assumption 12.2 For *only one* $r_0 \in \mathcal{X}$, there exist at least 2^n k_i, $i = 1, \cdots, 2^n$ $\in \{1, 2, \cdots, K\}$ such that $x(k_i) = r_0$, $x(k_i + 1) = \sigma_i(r_0)$, $i = 1, \cdots, 2^n$. In addition to that, every possible value of $\mathcal{X}$ exists at least once in the data set. ■

Summarizing the above results, our second method for obtaining the deflated system 12.67 is described below:

Algorithm 12.5 System Deflation 2

Step 1: Locate an observation value r_0 for which 2^n distinct successors $\sigma_i(r_0)$, $i = 1, \cdots, 2^n$, exist in the data set.
Step 2: Compute the successor constellation set C according to equation 12.60.
Step 3: For every observation $r = x(k)$ find the (unique) value j for which $r - 2c_j \in \mathcal{X}$.
Step 4: Replace $x(k)$ by $r - c_j$. ■

Again, the $L - 1$ times repetition of this algorithm will reduce the system into a memoryless one,

$$\tilde{x}^{(L)}(k) = \mathbf{h}_{L-1}^T \mathbf{s}(k), \tag{12.68}$$

which can be treated as described in the previous section on MIMO systems.

Example 12.5 MISO System Identification and Source Separation

We shall demonstrate the application of the second method via a specific example. Assume that we observe the output $x(k)$ of a two-input one-output system (fig. 12.9a). The system has two binary inputs s_1, s_2, convolution length $L = 3$, and filter taps $\mathbf{h}_1 = [-0.9024, 1.5464]^T$, $\mathbf{h}_2 = [-0.6131, 0.7166]^T$, $\mathbf{h}_3 = [-0.4131, -0.1621]^T$. The output constellation contains $2^{nL} = 64$ clusters: $\mathcal{X} = \{\pm 4.3537, \pm 4.0295, \pm 3.5275, \cdots\}$. Already, the first value $r = -4.3537$ has $2^n = 4$ distinct successors in the output sequence $x(k)$. From those successor values the centered successor constellation set is easily computed to be

$$C = \{-2.4488, -0.6440, 0.6440, 2.4488\}.$$

After the deflation steps 2 and 3 we obtain a new sequence $x^{(2)}(k)$ (fig. 12.9b). Now the output constellation contains $2^{n(L-1)} = 16$ clusters: $\mathcal{X}^{(2)} = \{\pm 1.9049, \pm 1.5807, \pm 1.0787, \cdots\}$ and the centered successor constellation set is

$$C^{(2)} = \{-1.3297, -0.1035, 0.1035, 1.3297\}.$$

We use this set to obtain a second deflated signal $x^{(3)}(k)$ (fig. 12.9c). This signal actually corresponds to an instantaneous mixture of the two sources. The output

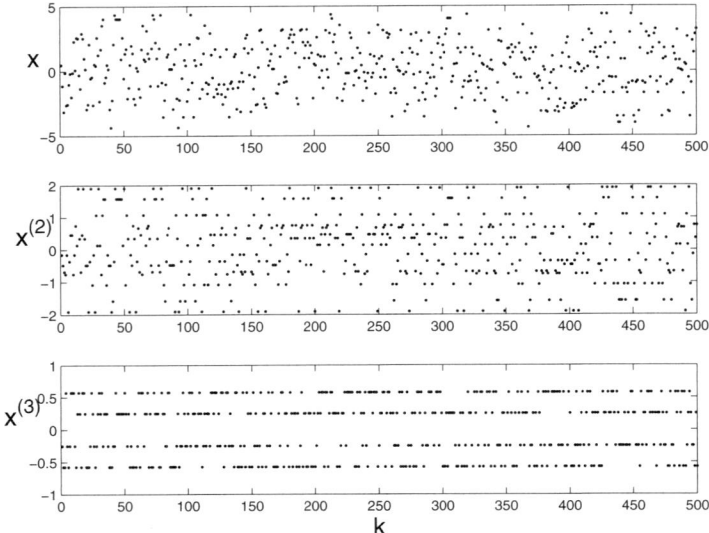

Figure 12.9 (Top) Output signal from a two-input-one-output FIR system of length $L = 3$. The output constellation contains $2^{nL} = 64$ distinct clusters. (Middle) First deflated signal with $2^{n(L-1)} = 16$ clusters. (Bottom) Second deflated signal with $2^{n(L-2)} = 4$ clusters. The last signal corresponds to an instantaneous mixture of the two sources.

constellation has only four clusters:

$$\mathcal{X}^{(3)} = \{-0.5752, -0.2510, 0.2510, 0.5752\}.$$

We may apply algorithm 12.1 to obtain an estimate of the mixing parameters and of the input signals as well. We obtain $\hat{h}_{3,1} = 0.4131 = -h_{3,1}$, $\hat{h}_{3,2} = 0.1621 = -h_{3,2}$. Subsequently performing the optimization (eq. 12.15) for the estimation of the sources we get perfect reconstruction (except for the sign):

$$\hat{s}_1(k) = -s_1(k),$$
$$\hat{s}_2(k) = -s_2(k). \quad \blacksquare$$

12.3   Continuous Sources

In section 12.2 we exploited the constellation structure of signals generated by linear systems with finite alphabet inputs. In many applications, however, the range of values of the source data is continuous. In this case the geometrical properties of the signals can still be exploited to derive efficient deterministic blind separation methods provided that the sources are sparse, or the input distribution is bounded, or the number of observations is $m = 2$.

12.3.1   Early Approaches: Two Mixtures, Two Sources

It is possible to generalize the geometric properties of binary signals described in section 12.2.1, when the sources symbols are bounded. We start with the simplest case of two instantaneous mixtures x_1, x_2, and two sources s_1, s_2, ($m = n = 2$):

$$\mathbf{x}(k) = [x_1(k),\, x_2(k)]^T = \mathbf{h}_1 s_1(k) + \mathbf{h}_2 s_2(k) \qquad (12.69)$$

We shall describe two of the earliest and most characteristic methods by Puntonet et al. (1995) and Mansour et al. (2001).

The Method of Puntonet et al. (1995)   The geometry of mixtures of binary signals bears similarity to the geometry of mixtures of bounded sources. Consider the mixing model 12.69 and let $s_1(k), s_2(k) \in [-B, B]$. The linear operation of equation 12.69 transforms the original square source constellation (fig. 12.10a) into a parallelogram-shaped constellation with edges parallel to the vectors $\mathbf{h}_1$ and $\mathbf{h}_2$ (fig. 12.10b).

The blind identification task is then equivalent to finding the edges of the convex hull of the output constellation. Puntonet et al. (1995) proposed a simple procedure for doing that. This procedure is composed of two steps:

Step 1: Locate the outmost corner $\mathbf{x}_O$ of the parallelogram by finding the observation with the maximum norm: $\mathbf{x}_O = \mathbf{x}(k_0)$, $k_0 = \arg\max_k \{\|\mathbf{x}(k)\|^2\}$.

Step 2: Translate the observations $\mathbf{x}'(k) = \mathbf{x}(k) - \mathbf{x}(k_0)$ such that $\mathbf{x}_O$ becomes the origin, and compute the slopes of the parallelogram by computing the minimum and maximum ratios: $r_{min} = \min_k (x_2'(k)/x_1'(k))$, $r_{max} = \max_k (x_2'(k)/x_1'(k))$. These are the ratios h_{12}/h_{11}, h_{22}/h_{21}, not necessarily in that order.

Once the slopes of the edges are determined, the mixing matrix is estimated by

$$\hat{\mathbf{H}} = \begin{bmatrix} 1 & 1/r_{min} \\ r_{max} & 1 \end{bmatrix}. \qquad (12.70)$$

Since $(r_{min}, r_{max}) = (h_{12}/h_{11})$ or (h_{22}/h_{21}) we have,

$$\hat{\mathbf{H}} = \begin{bmatrix} 1 & h_{21}/h_{22} \\ h_{12}/h_{11} & 1 \end{bmatrix} \text{ or } \begin{bmatrix} 1 & h_{11}/h_{12} \\ h_{22}/h_{21} & 1 \end{bmatrix}. \qquad (12.71)$$

Remember now that

$$\mathbf{H} = [\mathbf{h}_1,\, \mathbf{h}_2] = \begin{bmatrix} h_{11} & h_{21} \\ h_{12} & h_{22} \end{bmatrix},$$

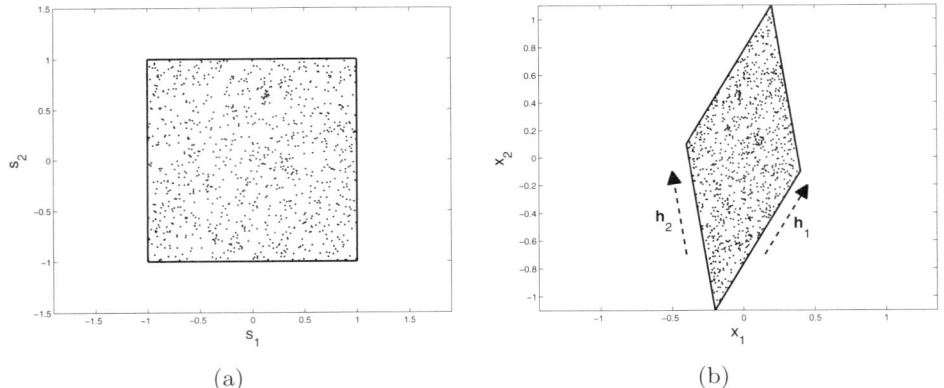

Figure 12.10   (a) Source constellation for two independent sources uniformly distributed between -1 and 1. (b) Output constellation after a 2×2 linear, memoryless transformation of the sources in panel a.

so,

$$
\hat{\mathbf{H}} = \mathbf{H} \left[\begin{array}{cc} 1/h_{11} & 0 \\ 0 & 1/h_{22} \end{array} \right] \text{ or } \hat{\mathbf{H}} = \mathbf{H} \left[\begin{array}{cc} 0 & 1/h_{12} \\ 1/h_{21} & 0 \end{array} \right].
$$

In either case, the source estimate $\hat{\mathbf{s}}(k) = \hat{\mathbf{H}}^{-1}\mathbf{x}(k)$ will be

$$
\hat{\mathbf{s}}(k) = [h_{11}s_1(k), \, h_{22}s_2(k)]^T \text{ or } \hat{\mathbf{s}}(k) = [h_{12}s_2(k), \, h_{21}s_1(k)]^T. \tag{12.72}
$$

Thus, the estimated sources will be equal to the true ones except for the usual unspecified scale and order.

Note that the method works even if the source pdf is semibounded, for example, bounded only from below. In that case the parallelogram is open-ended but the visible corner is sufficient for identifying the two slopes. The main drawbacks of this approach are two: it cannot generalize to more sources or observations, and it will not work if the source pdf is not bounded (for example, Gaussian, Laplace, etc).

The Method of Mansour et al. (2001)   Another simple procedure for the solution of the 2×2 instantaneous BSS problem has been proposed by Mansour et al. (2001). The transformation $\mathbf{s} \mapsto \mathbf{x}$ described by equation 12.69 represents a skew, rotation, and scaling of the original axes in 2 dimensions. The first step of the procedure is to remove the skew by *prewhitening* $\mathbf{x}$ using the covariance matrix $\mathbf{R}_x = E\{\mathbf{x}(k)\mathbf{x}(k)^T\}$. If $\mathbf{R}_x = \mathbf{L}_x \mathbf{L}_x^T$ is the Cholesky factorization of $\mathbf{R}_x$, let

$$
\mathbf{z}(k) = \mathbf{L}_x^{-1}\mathbf{x}(k). \tag{12.73}
$$

The mapping $\mathbf{x} \mapsto \mathbf{z}$ is called *prewhitening transformation* because the output vector $\mathbf{z}(k)$ is white: $\mathbf{R}_z = \{\mathbf{z}(k)\mathbf{z}(k)^T\} = \mathbf{L}_x^{-1}\mathbf{R}_x\mathbf{L}_x^{-T} = \mathbf{I}$. The prewhitening transformation (eq. 12.73) makes the axes become orthogonal again, but the rotation and the scaling remains. The next step is to compensate for the rotation by computing the angle θ of the furthermost point of the constellation of $\mathbf{z}$ from the origin. We consider two cases:

- The sources are uniformly distributed, say, between -1 and 1 (fig. 12.11a). The source constellation is a square and the angle θ corresponds to a corner of the square. Therefore, in order to compensate for θ, the corner should return to its original position at $\frac{\pi}{4}$. This is achieved by the following orthogonal transformation:

$$\mathbf{y}(k) = \left[\begin{array}{cc} \cos(\frac{\pi}{4} - \theta) & -\sin(\frac{\pi}{4} - \theta) \\ \sin(\frac{\pi}{4} - \theta) & \cos(\frac{\pi}{4} - \theta) \end{array} \right] \mathbf{z}(k). \tag{12.74}$$

- The sources are super-Gaussian, i.e., $\mathrm{kurt}(s_i) = E[s_i^4] - 3(E[s_i^2])^2 > 0$, $i = 1, 2$ (fig. 12.11b). The constellation of $\mathbf{s}$ in this case is "pointy" along the directions $[\pm 1, 0]$ and $[0, \pm 1]$. The angle θ corresponds to one of the "hands" of the X-shaped constellation for $\mathbf{x}$. Clearly, θ should be reduced to 0. This is done by the following rotation transformation:

$$\mathbf{y}(k) = \left[\begin{array}{cc} \cos(-\theta) & -\sin(-\theta) \\ \sin(-\theta) & \cos(-\theta) \end{array} \right] \mathbf{z}(k). \tag{12.75}$$

In both cases there remains an unknown scaling of the sources which cannot be removed since it is unobservable in all BSS problems.

12.3.2   Sparse Sources, Two Mixtures

Another special case of continuous sources that can be successfully treated using geometric methods is the case of sparse sources. A signal $s_i(k)$ is *sparse* if it is equal to zero most of the time. The sparseness of the s_i is measured by the *sparseness probability*

$$p_S(s_i) = Pr\{s_i(k) = 0\}.$$

Values of p_S closer to 1 correspond to more sparse data, whereas values closer to 0 represent dense data. Consider now the typical instantaneous mixing model:

$$\mathbf{x}(k) = \mathbf{H}\mathbf{s}(k), \tag{12.76}$$

assuming that all the sources are sparse. Then it is highly likely that there exist some time instances such that only one source is active at that instance. If, for example, only s_i is nonzero at time k, then $\mathbf{x}(k)$ is proportional to $\mathbf{h}_i$, the ith column of $\mathbf{H}$. The number of outputs m is not important, as long as $m \geq 2$. In fact, the number of outputs may even be less than the number of sources ($m < n$). In the subsequent discussion we shall use the convenient value $m = 2$ because it will

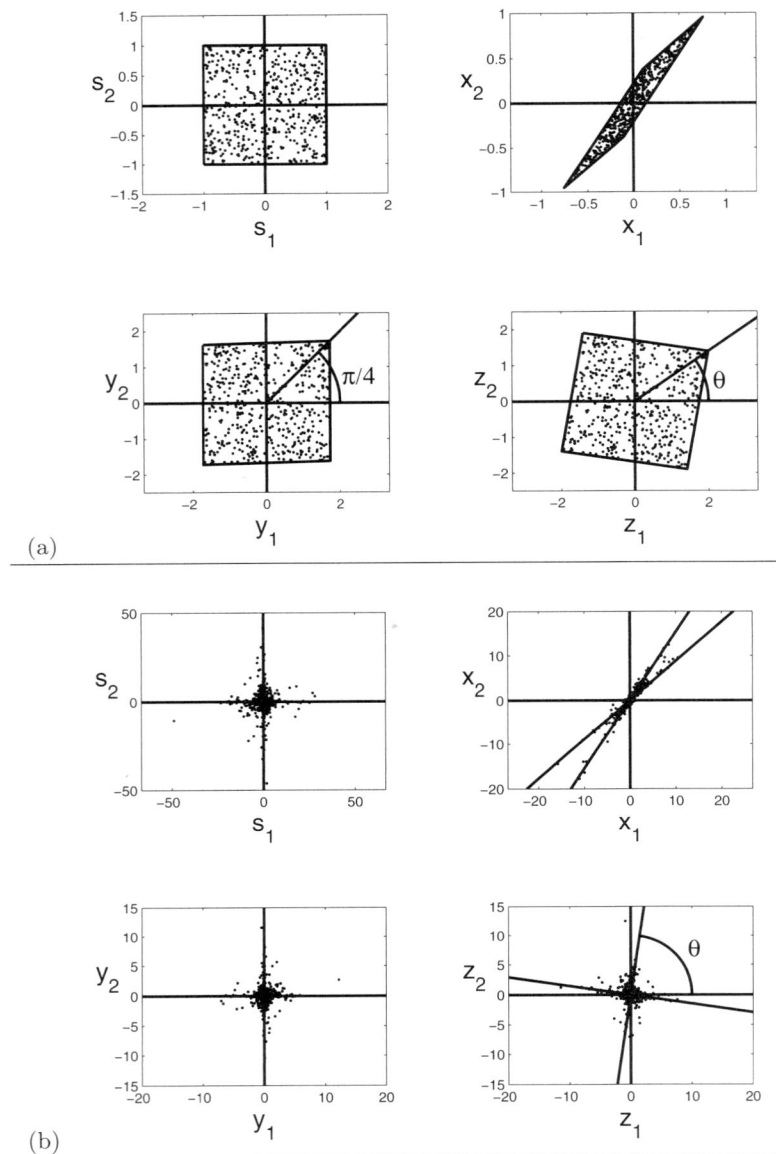

Figure 12.11 The linear, instantaneous transformation $\mathbf{s} \mapsto \mathbf{x}$ introduces skew, rotation, and scaling on the original axes. The whitening transform $\mathbf{x} \mapsto \mathbf{z}$ removes the skew, making the axes orthogonal again. Then the rotation can be removed by an orthogonal transformation $\mathbf{z} \mapsto \mathbf{y}$. (a) If the source distribution is uniform we must rotate so that θ becomes $\pi/4$. (b) If the source distribution is super-Gaussian then we must rotate so that θ becomes 0.

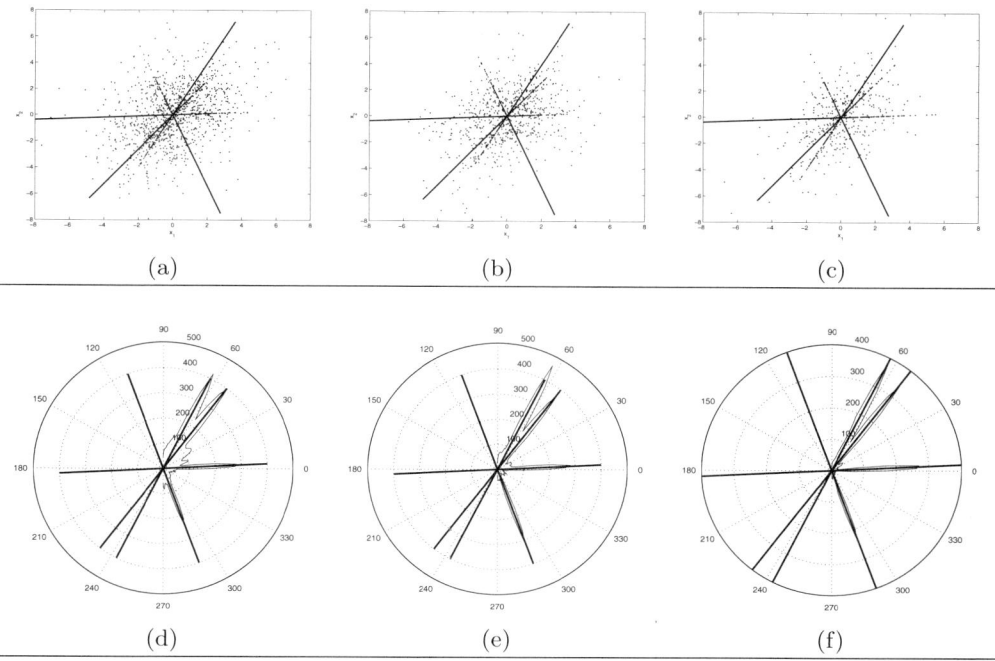

Figure 12.12 Output constellation for $m = 2$ outputs and $n = 4$ sparse sources. The top three plots correspond to different sparseness probabilities (a) $p_S = 0.6$, $p_S = 0.7$, (c) $p_S = 0.8$. The solid lines are the directions of the four vector-columns of H. The three bottom figures d, e, and f are polar plots of the data density (potential) function with spreading parameter $\sigma = 8$ corresponding to the constellations a, b, and c, respectively.

help us visualize the results. Bofill and Zibulevsky (2001) observed that the data are clustered along the directions of the mixing vectors $\mathbf{h}_i$, i.e., the columns of $\mathbf{H}$. Figure 12.12 shows the output constellation for the memoryless system (eq. 12.76) with $m = 2$ outputs, $n = 4$ sparse inputs, and different sparseness levels. As the sparseness of the inputs increases, the four clustering directions become more easily identifiable (see figures 12.12a,b,c).

Thus blind system identification is achieved by identifying the directions of maximum data density. Assuming that the sources are zero mean, so they can take both positive and negative values, the clustering will extend to the negative directions $-\mathbf{h}_i$ as well. Since, for each i, both opposing directions $\mathbf{h}_i$ and $-\mathbf{h}_i$ are equally probable, it is not possible to identify the "true" vector. This is a manifestation of the sign ambiguity which is inherent to the BSS problem. Not surprisingly, the ordering ambiguity is also present in the sense that there is no predefined order on the directions of maximum data density.

For $m = 2$, a practical algorithm has been proposed by Bofill and Zibulevsky (2001). For any two-dimensional vector $\mathbf{x} = [x_1,\, x_2]^T \neq 0$ let us define the angle of $\mathbf{x}$,

$$\theta(\mathbf{x}) = \arctan(x_2/x_1). \tag{12.77}$$

The directions θ where the random variable $\theta_k = \theta(\mathbf{x}(k))$ has the highest density are the directions of the mixing vectors. The density is estimated by the use of a potential function $U(\theta)$:

$$U(\theta) = \sum_k w(k)\, t(\theta - \theta_k; \sigma) \tag{12.78}$$

$$t(\alpha; \sigma) = \begin{cases} 1 - \frac{\alpha}{\pi/(4\sigma)}, & \text{for } |\alpha| < \frac{\pi}{4\sigma}; \\ 0, & \text{otherwise.} \end{cases} \tag{12.79}$$

where $w(k) = \|\mathbf{x}(k)\|$ is a weight putting more emphasis on more reliable data, t is a triangular function, and σ adjusts the angular width, i.e., the spread of each local contribution to the potential function (see figures 12.12d,e,f). The directions of the mixing vectors are identified as the peaks of the potential function. The number of sources need not be known in advance since it can be identified by the number of peaks.

Following the mixing matrix identification step, the sources can be estimated in a second step, using the N observation samples $\mathbf{x}_1, \ldots, \mathbf{x}(N)$. In the presence of noise the samples $\mathbf{s}(1), \ldots, \mathbf{s}(N)$, can be estimated by solving N small minimization problems:

$$\min_{\mathbf{s}(k)} \frac{1}{\sigma^2} \|\mathbf{H}\mathbf{s}(k) - \mathbf{x}(k)\|^2 + \lambda \sum_{j=1}^{n} |s_j(k)|, \quad \text{for } k = 1, \cdots, N. \tag{12.80}$$

The first term minimizes the square error (σ is the noise variance), while the second term is a penalty for nonsparsity. In the absence of noise, the optimization problem is formulated in a slightly different fashion:

$$\min_{\mathbf{s}(k)} \sum_{j=1}^{n} |s_j(k)|, \quad \text{subject to } \mathbf{x}(k) = \mathbf{H}\mathbf{s}(k). \tag{12.81}$$

Example 12.6 Separation of Full Ensemble Music

Bofill and Zibulevsky report a number of experiments with real data including mixtures of speech (four voices), music (five songs), single musical tones (six flutes), and simple melodies from a single musical instrument (six flute melodies). The results of these experiments are published online by Bofill. Here we shall present the separation experiment of five songs from two mixtures. The source data were 5-seconds-long excerpts from five full-ensemble music pieces extracted from standard CDs. The data were downsampled to 11,025 Hz monophonic and were preprocessed as follows:

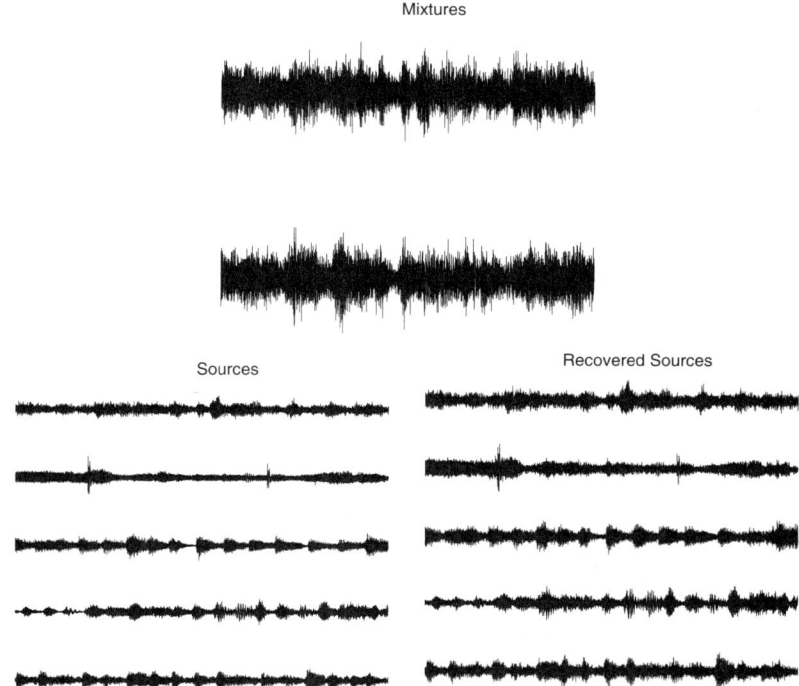

Figure 12.13 Blind separation of five full-ensemble music pieces from two linear mixtures.

- All sources were normalized to unit energy.

- Two mixtures were generated using a 2×5 mixing matrix $\mathbf{H}$. The five mixing vectors (the columns of $\mathbf{H}$) were formed with equally spaced angles.

- The mixtures were rescaled between -1 and 1 and processed in frames of length T with a hop distance d between starting points of successive frames.

- Each frame was transformed with FFT of length T and only the coefficients of the positive half spectrum were kept. All FFT segments were concatenated in a single vector which was the input to the separation algorithm.

In this particular experiment the frame parameters were $T = 4,096$ and $d = 1,228$ samples. The five sources, the two mixtures, and the five reconstructed signals are shown in fig. 12.13. The signal to reconstruction-error ratio, for a wide range of values of the smoothness parameter σ, was around 15 dB. ∎

12.3.3 Dense Sources, Two Mixtures: Geometric ICA

The data-density concepts for sparse or bounded sources cannot be directly extended to non-sparse sources. However, Theis et al. (2003a,b) and Jung et al. (2001)

have developed a theory relating the data densities in the polar coordinates with the mixing vectors $\mathbf{h}_i$, $i = 1, \cdots, n$, when the sources are nonsparse, provided that their pdf is symmetric, non-Gaussian, and unimodal (i.e., has only one peak). This theory applies to memoryless systems of the type in equation 12.76 with $m = 2$ outputs and $n \geq 2$ inputs. Extensions for $m > 3$ are possible but impractical due to the high computational cost and the extremely large required data sets.

In an analogous way to the sparse case, the method is based on the properties of the density (pdf) $\rho_{\bar{\Theta}}$ of the random variable

$$\bar{\theta} = \theta(\mathbf{x}) \quad \mathrm{mod} \ \pi,$$

where $\theta(\mathbf{x})$ is the angle of $\mathbf{x}$ defined in equation 12.77. Here, however, the peaks of the density may not have a one-to-one correspondence with the mixing vectors, especially when the number of sources is greater than the number of observations $(n > m)$. The basic result is that the angles

$$\theta_i = \theta(\mathbf{h}_i), \quad i = 1, \cdots, n \tag{12.82}$$

of the mixing vectors $\mathbf{h}_i$ satisfy the *geometric convergence condition (GCC)* defined below:

Definition 12.2 Geometric Convergence Condition
The set of angles $\{\theta_1, \ldots, \theta_n\}$, $\theta_i \in [0, \pi)$, satisfies the GCC if, for each i, θ_i is the median of ρ_Y restricted to the receptive field $\Phi(\theta_i)$. ■

Definition 12.3 Receptive Field
For a set of angles $\{\theta_1, \ldots, \theta_n\}$, $\theta_i \in [0, \pi)$, the receptive field $\Phi(\theta_i)$ is the set consisting of the angles θ closest to θ_i:

$$\Phi(\theta_i) = \{\theta \in [0, \pi) : |\theta - \theta_i| \leq |\theta - \theta_j| \text{ for all } j \neq i\}. \quad ■$$

Since the angles of the true mixing vectors satisfy the GCC we hope that we can find them by devising an algorithm which converges when the GCC is satisfied. This is exactly the aim of the geometric ICA algorithm (Theis et al., 2003a,b). This iterative algorithm works with a set of n unit-length vectors (and their opposites) and terminates only when the angles of these vectors are the medians of their corresponding receptive fields. It is conjectured that the only stable points of this algorithm are the true mixing vectors.

The algorithm starts by picking n random pairs of opposing vectors: $\{\mathbf{w}_i(0), \mathbf{w}'_i(0) = -\mathbf{w}_i(0)\}$, $i = 1, \cdots, n$. For each iteration k, a new observation vector $\mathbf{x}(k)$ is projected onto the unit circle:

$$\mathbf{z}(k) = \frac{\mathbf{x}(k)}{\|\mathbf{x}(k)\|}.$$

Then we locate the vector $\mathbf{w}_j(k)$ closest to $\mathbf{z}(k)$ and we update the pair $\mathbf{w}_j(k)$,

$\mathbf{w}'_j(k)$, as follows:

$$
\begin{aligned}
\mathbf{w}_j^{temp} &= \mathbf{w}_j(k) + \eta(k)\frac{\mathbf{z}(k)-\mathbf{w}_j(k)}{\|\mathbf{z}(k)-\mathbf{w}_j(k)\|}, \\
\mathbf{w}_j(k+1) &= \mathbf{w}_j^{temp}/\|\mathbf{w}_j^{temp}\|, \\
\mathbf{w}'_j(k+1) &= -\mathbf{w}_j(k+1).
\end{aligned}
\tag{12.83}
$$

The other $\mathbf{w}$'s are not updated in this iteration. It can be shown that the set $\mathcal{W} = \{\mathbf{w}_1(\infty), \ldots, \mathbf{w}_n(\infty)\}$ is a fixed point of this algorithm if and only if the angles $\theta(\mathbf{w}_1(\infty)), \ldots, \theta(\mathbf{w}_n(\infty))$ satisfy the GCC. We already know that the set $\mathcal{A} = \{\theta(\mathbf{h}_1), \ldots, \theta(\mathbf{h}_n)\}$ satisfies the GCC, therefore, we hope that, at convergence, $\{\theta(\mathbf{w}_1(\infty)), \ldots, \theta(\mathbf{w}_n(\infty))\} = \mathcal{A}$. If this is true then the vectors $\mathbf{w}_1(\infty), \ldots, \mathbf{w}_n(\infty)$ are parallel to the mixing vectors $\mathbf{h}_1, \ldots, \mathbf{h}_n$, although not necessarily in that order. Since the order and scale are insignificant, this is not a problem. If $m = n$, then the estimated matrix $\hat{\mathbf{H}}^{-1} = [\mathbf{w}_1(\infty), \ldots, \mathbf{w}_n(\infty)]^{-1}$ solves the BSS problem. In the overdetermined case $(m > n)$ the general algorithm for the source recovery is the maximization of $P(\mathbf{s})$ under the constraint $\mathbf{x} = \mathbf{Hs}$. This linear optimization problem can be approached using various methods, such as, for example, the one described in section 12.3.2.

The FastGEO Algorithm   An alternative way to find the mixing vectors is to design a function which is zero exactly when its arguments satisfy the GCC. Then we simply have to compute the zeros of this function, for example, by exhaustive search. This approach describes the so-called FastGEO algorithm (Jung et al., 2001; Theis et al., 2003a). Let us separate the interval $[0, \pi)$ into n subintervals with separating boundaries $\phi_1, \ldots, \phi_n$, and let θ_i be the median of $\bar{\theta}$ in the subinterval $[\phi_i, \phi_{i+1}]$,

$$
\theta_i = F_{\bar{\Theta}}^{-1}\left(\frac{F_{\bar{\Theta}}(\phi_i) + F_{\bar{\Theta}}(\phi_{i+1})}{2}\right), \quad i = 1, \cdots, n,
\tag{12.84}
$$

where $F_{\bar{\Theta}}$ is the cumulative distribution function of $\bar{\theta}$, $F_{\bar{\Theta}}^{-1}$ is the inverse function of $F_{\bar{\Theta}}$ (we assume it exists), and $\phi_{n+1} = \phi_1 + \pi$ (see fig. 12.14). Then the function

$$
\mu^{(n)}(\phi_1, \cdots, \phi_{n-1}) = \left[\frac{\theta_1 + \theta_2}{2} - \phi_2, \cdots, \frac{\theta_{n-1} + \theta_n}{2} - \phi_n\right]^T
\tag{12.85}
$$

is zero if and only if

$$
\frac{\theta_i + \theta_{i+1}}{2} = \phi_{i+1}, \qquad i = 1, \cdots, n-1
$$

for all i, and so by definition the receptive field $\Phi(\theta_i)$ is exactly the subinterval $[\phi_i, \phi_{i+1}]$ and θ_i is the median of its receptive field; in other words, the set $\{\theta_i, \ldots, \theta_n\}$ satisfies the GCC. For each set of separating boundaries $\{\phi_1, \ldots, \phi_{n-1}\}$ we compute the medians $\theta_1, \ldots, \theta_n$ by equation 12.84 and then the function $\mu^{(n)}(\phi_1, \cdots, \phi_{n-1})$ by equation 12.85. The FastGEO algorithm is the exhaustive search for the zeros of $\mu^{(n)}$.

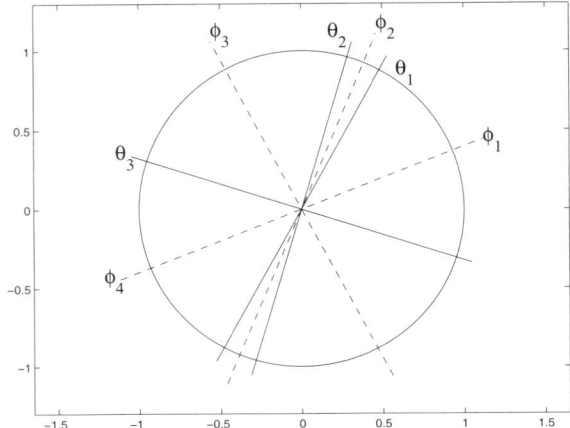

Figure 12.14 The angles θ_i of the mixing vectors $\mathbf{h}_i$ satisfy the geometric convergence condition if they are the median of the random variable $\theta(\mathbf{x})$ within the interval $\Phi(\theta_i) = [\phi_i, \phi_{i+1}]$. $\Phi(\theta_i)$ is called the receptive field of θ_i and it is the set consisting of the angles θ closest to θ_i.

Especially for $n = 2$ we let $\phi_1 = \phi$ and we have $\phi_2 = \phi + \pi/2$, so

$$\mu^{(2)}(\phi) = \frac{\theta_1 + \theta_2}{2} - (\phi + \pi/2).$$

Example 12.7

Let x_1, x_2 be two instantaneous mixtures of two uniform sources s_1, s_2. The mixtures were generated by the following mixing operator

$$\mathbf{H} = \begin{bmatrix} 0.0735 & 0.2913 \\ -0.3391 & 0.3725 \end{bmatrix}.$$

The distribution of the angle $y = \phi(\mathbf{x})$ is shown in fig. 12.15. The same figure shows the receptive field boundaries $\{\phi_1, \phi_2, \phi_3\} = \{77.0998, 167.0998, 257.0998\}$ (in degrees), corresponding to the angles $\{\theta_1, \theta_2\} = \{51.9759, 102.2237\}$ of the mixing vectors $\mathbf{h}_1 = [0.0735, -0.3391]^T$, $\mathbf{h}_2 = [0.2913, 0.3725]^T$. The angles θ_2 and $\theta_1 + 180$ are the medians of the angle distribution in the corresponding receptive fields. ∎

12.4 Conclusions

Blind signal processing (BSP) refers to a wide variety of problems where the output of a system is observable but neither the system nor the input is known. The large family of BSP problems includes blind signal separation (BSS), blind system or

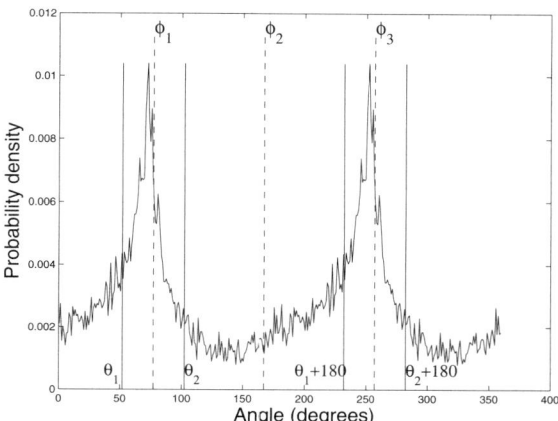

Figure 12.15   The distribution of the angle $\theta(\mathbf{x})$ for two instantaneous mixtures of two uniformly distributed sources. The receptive field boundaries are defined by the angles ϕ_i. The angles θ_i of the mixing vectors are the medians of each receptive field.

channel identification (BSI or BCI), and blind deconvolution (BD). Traditional approaches exploit statistical properties of second or higher order. Recently a third approach has emerged using the geometric properties of the data cloud. This approach exploits the finite alphabet property of the input data or the shape of the constellation depending on the probability density of the sources. In such an approach our basic tools are methods for data clustering and shape description such as the convex hull. The advantage of the geometric approach is the finite nature of the methodology following the clustering step. Typically, this methodology is fast for small problem sizes, i.e., for few sources or short channels. The main disadvantage is the combinatorial explosion which is incurred when the problem size grows large. To combat this drawback, channel-shortening methods may come to our assistance. The problem, however, is far from solved and many issues remain open. In this chapter we presented the main geometric principles used in blind signal processing. We presented a comprehensive literature survey of geometric methods and we outlined the basic methods for blind source separation, blind deconvolution, and blind channel identification.

13   Game-Theoretic Learning

Geoffrey J. Gordon

> *Whatever games are played with us, we must play no games with ourselves.*
> —Ralph Waldo Emerson

A game is a description of an environment where multiple decision makers can interact with one another. Each of the decision makers, called a *player*, may have its own goals; these goals may align with the goals of other players, conflict with them, or some combination of the two.

Some traditional examples of games are bridge, blackjack, chess, roulette, and poker. Less traditional examples include auctions, marketing campaigns, decisions about where to build a new factory, and various types of social interactions such as applying for a job. Finally, many popular games mix components of perception and physical skill with the problem of making good decisions; examples include football, freeze tag, paintball, and driving in traffic.

This chapter is about how to learn to play a game. We will discuss how a player can, by repeated interaction with its environment and with the other players, discover how to make decisions which achieve its goals as reliably as possible. Playing a real-life game such as football is far beyond the capability of any current artificial learning system, but we will at least begin to address the issues of exploration and generalization which arise in such a problem.

The difficulty of making good decisions in a game can range from trivial to nearly impossible. We can classify games according to several dimensions; each of these classifications affects the type of solution we can seek, the algorithms we can use, and the difficulty of finding a good plan of action.

one-step vs. sequential

In *one-step* games each player decides on its strategy all at once. In *sequential* games a player commits to its actions in several steps, and after each step it may find out something about the other players' choices. Of course, to learn about any game the players will usually need to play it several times; so, we can speak of a *repeated* game, either one-step or sequential.

(im)perfect information

In *perfect information* games, each player knows all the choices that the other players have already made. In games with *imperfect information*, the players know only some of the past choices of other players. For example, if an auction house is selling several copies of the same item one after another via sealed bids, the bidders will know the sale prices of the previous items but will not know the details of the previous bids.

(in)complete
information

In *complete information* games the players know all the details of the game they are playing: they know the structure of the game, the outcomes of all past external events that are relevant to their future payoffs, and what the payoffs are in any situation for themselves and for the other players. In *incomplete information* games some of the players are missing some of this information; for example, in bridge or poker the players don't see each other's cards.

In this chapter we will discuss all of these different types of games in turn. Each one presents different difficulties, so we will discuss various algorithms for learning to play them. Each of the algorithms provides different performance guarantees, so we will describe and compare the types of guarantees that are available.

We will start with one-step games. The standard representation of one-step games is the normal form, described in section 13.1. Given the normal form, classical game theory looks for distributions of play from which no player can alter its actions to improve its payoffs; such distributions are called equilibria, and we will discuss them in section 13.2. Equilibria are possible outcomes of learning, since learning players will not be satisfied as long as they think they can improve their reward. So, in section 13.3, we will review learning algorithms for one-step games and use the different types of equilibria to describe what happens when various learning algorithms play against one another.

From one-step games we will move to sequential games. In sequential games we can define additional types of equilibria, and we need to move to more complicated learning algorithms. Sections 13.4 and 13.5 cover these new equilibria and learning algorithms. Finally, we will conclude in section 13.6 with some examples of how game-theoretic learning algorithms have been applied to solve real-world problems. These problems range from poker to robotic soccer.

13.1 Normal-Form Games

Any game can in principle be described with the following information:

Normal form
representation

- A list of the players. We will assume that there are only finitely many of them.

- For each player, a list of the *actions* (also called *plays* or *pure strategies*) that it may choose. We will assume that there are finitely many pure strategies. Any probability distribution over pure strategies is called a *mixed strategy*.

- Given a *strategy profile* (that is, a pure strategy for every player), the *utility* or *payoff* which each player assigns to the resulting outcome. If there are external random events which affect utility, we only need to know the expected utility of each strategy profile.

This representation is called the *normal form* of the game. As with any general representation, the normal form may be nowhere near the most concise way to describe a game. Still, it does allow us to discuss many different games and algorithms in a general way.

We can represent a normal-form game with a table: there is one entry in the table for each strategy profile, and the entry is a vector which lists the utility of the resulting outcome for each player. For example, the following table represents the children's game rock-paper-scissors:

	R	P	S
R	$0,0$	$-1,1$	$1,-1$
P	$1,-1$	$0,0$	$-1,1$
S	$-1,1$	$1,-1$	$0,0$

The first entry on the second row of this table says that, if the row player chooses P (for "paper") while the column player chooses R (for "rock"), then the row player gets a payoff of 1 while the column player gets a payoff of -1. The payoffs can in general be random variables, but we are only interested in their expected values, so we will not bother to write out any other properties of their distributions.

common knowledge
The payoff table is assumed to be *common knowledge*. That is, every player knows it, every player knows that every player knows it, and so forth. If the payoff table is not common knowledge (that is, if some players have information about it that others don't), the game is called a Bayesian game; section 13.4.1 covers Bayesian games in more detail.

The simplest type of game to reason about is a two-player constant-sum game. *Constant sum* means that, for each strategy profile (i.e., for each entry in the table), the sum of the payoffs to the two players is constant. For example, rock-paper-scissors is a constant-sum game, since each utility vector sums to zero. Constant-sum normal-form games are one of the few types of game with a universally accepted and easy to compute solution concept (the minimax equilibrium; see section 13.2.1).

If the payoffs do not sum to a constant, or if there are more than two players, the game is called *general sum*. By convention a game with three or more players is always called general sum, even if the payoffs do sum to a constant, since minimax equilibrium doesn't make sense for multiplayer games.

Environments in which all players have the same payoffs are called *cooperative* or *team* games. Team games may appear easy, but they can be difficult to solve because of imperfect or incomplete information: while a player may believe that a particular strategy profile is best, it may not be able to trust the other players to agree. So, it may have to choose an action which appears suboptimal in order to try to reach a different but safer outcome.

13.2 Equilibrium

An equilibrium of a game is a self-reinforcing distribution over strategy profiles. That is, if it is common knowledge that all players are acting according to a given equilibrium, no one player wants to change how it plays. There are many different types of equilibrium, which differ in how they formalize the above definition.

Classical game theory takes the view that the best way to analyze a game is to determine what its equilibria are. We can justify this view if we assume that the game is common knowledge among the players, and that the players have common knowledge of each other's rationality. However, equilibria don't tell us everything if the players have limited computation (often called *bounded rationality*) or if some players disagree about the rules of the game. Also, a single game may have many equilibria, and it may not be clear how the players should (or can) select one.

In this chapter we take a slightly different view: we are more concerned with how a player may adapt its actions based on information about what the other players are doing. So, we will write down learning algorithms and analyze what happens when the players use these algorithms in different types of games. Still, the various ideas of equilibrium are important: for example, under appropriate circumstances some of the learning algorithms we describe below will converge toward various types of equilibrium play.

We have already mentioned one type of equilibrium, the *minimax* or *von Neumann* equilibrium for constant-sum matrix games. For general-sum games, there are at least two important types of equilibrium: the *Nash equilibrium* and the *correlated equilibrium*. And we will see even more types of equilibrium when we discuss sequential decision making in section 13.4 below.

In addition to the Nash and correlated equilibria, it is sometimes helpful to know the *safety value* of each player in the game. A player's safety value is the best payoff that it can guarantee itself no matter what the other players do. That is, even if the other players irrationally ignore their own payoffs, they cannot force the first player to accept less than its safety value. In any equilibrium, each player must have a payoff at least as high as its safety value: if it did not, it would switch to its safety strategy.

13.2.1 Minimax Equilibrium

In a minimax equilibrium the players are required to choose independent probability distributions over their strategies, say x for the row player and y for the column player. If the payoff to the row player is $r(x, y)$, then the minimax value of the game for the row player is

$$\min_y \max_x r(x, y) \tag{13.1}$$

and a minimax strategy for the row player is any value of x which achieves the maximum in 13.1. Neither player will wish to deviate from its set of minimax strategies, since any deviation will give the other player a strategy which gets strictly better than the minimax payoff. Any constant-sum matrix game has at least one minimax equilibrium. The set of these minimax equilibria is convex, all minimax equilibria have the same value, and $\min_y \max_x r(x, y) = \max_x \min_y r(x, y)$.

In the game of rock-paper-scissors described above, there is exactly one minimax equilibrium: both players play R, P, and S with equal probability. By taking the expectation over the different possible outcomes (each of the nine profiles RR, RP, RS, ... has probability 1/9) we can see that the average payoff is zero for both players. On the other hand, if one of the players deviates from the equilibrium, the other player has a response that nets better payoff: for example, if the row player picks R with probability 1/2 and P and S with probability 1/4 each, then the column player can choose P all the time. The column player will then get an expected payoff of

$$(1/2)1 + (1/4)0 + (1/4)(-1) = 1/4 > 0.$$

13.2.2 Nash equilibrium

In general-sum games the idea of minimax equilibrium no longer makes sense, so we must seek alternative types of equilibrium. Perhaps the best-known type of equilibrium for general-sum games is the Nash equilibrium.

In a Nash equilibrium we require the players to choose independent distributions over strategies. So, a Nash equilibrium is a profile of strategy distributions such that, if we hold the distributions fixed for every player except one, the remaining player can get no benefit by changing its play. There is always at least one Nash equilibrium for every game, and there may be many.

In a constant-sum game, Nash equilibria are the same as minimax equilibria. But in a general-sum game, a player's payoff may differ greatly from one Nash equilibrium to another, and the set of Nash equilibria may be nonconvex and difficult to compute.

To illustrate Nash equilibria, consider the game of "Battle of the Sexes". In this game, a husband and wife want to decide whether to go to the opera (O) or the football game (F). One of them (the row player) prefers opera, while the other (the column player) prefers football. But, they also prefer to be together; so, they have the following payoffs:

	O	F
O	4, 3	0, 0
F	0, 0	3, 4

.

This game has three Nash equilibria. Two of them are deterministic: both players go to the opera, or both go to football. The last one is mixed: the row player picks opera 3/7 of the time, while the column player picks opera 4/7 of the time.

(In this mixed strategy, each player's distribution makes the other player perfectly indifferent about whether to pick opera or football.)

13.2.3   Correlated Equilibrium

In real life many people would solve the Battle of the Sexes by flipping a coin to decide which event to go to. This strategy is not a Nash equilibrium: in a Nash equilibrium the players are not allowed to communicate before the game, so they cannot both see the same coin flip. Instead it is a correlated equilibrium, which is like a Nash equilibrium except that we drop the requirement of independence between the players' distributions over strategies.

More formally, consider a distribution P over the set of strategy profiles; P may contain arbitrary correlations between the strategies of the different players. Some external mechanism, which we will call the *moderator*, selects a strategy profile x according to P and reports to player i the action x_i that it is supposed to follow. P is a correlated equilibrium if player i has no incentive to play anything other than x_i (even after finding out that x_i was recommended, which may tell it something about the other players' strategies).

The coin-flip strategy is a correlated equilibrium in which the distribution P places weight $1/2$ on each of the profiles OO and FF. Another everyday example of a correlated equilibrium is a traffic light: we can model the light as being randomly red or green as we approach it.[1] Red is the moderator's recommendation to stop, while green means to go through the intersection without stopping. Given a red light it is not worth going through the intersection and risking a crash, while with a green light we can assume that the traffic on the cross street will stop and our best strategy is to maintain speed.

13.2.4   Equilibria in Battle of the Sexes

To gain more intuition for Nash and correlated equilibria we will illustrate how to compute them for the Battle of the Sexes. We will start by computing the correlated equilibria, which satisfy a set of linear equality and inequality constraints; we will then obtain the Nash equilibria by adding in some nonlinear constraints.

We can describe a correlated equilibrium in Battle of the Sexes with numbers a, b, c, and d representing the probability of the four strategy profiles OO, OF, FO, and FF:

	O	F
O	a	b
F	c	d

Suppose that the row player receives the recommendation O. Then it knows that the column player will play O and F with probabilities $\frac{a}{a+b}$ and $\frac{b}{a+b}$. (The denominator is nonzero since the row player has received the recommendation O.) The definition

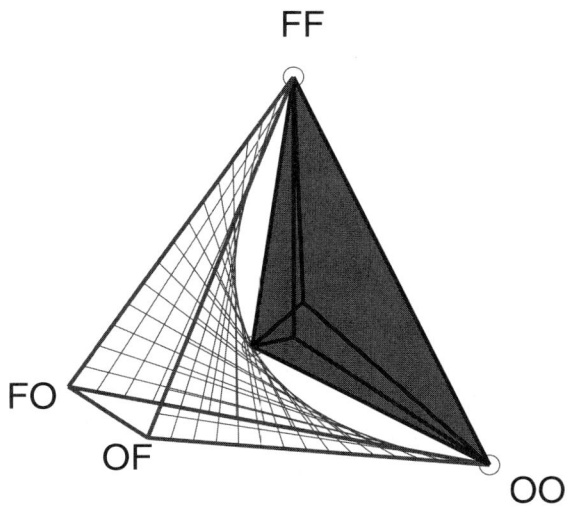

Figure 13.1 Equilibria in the Battle of the Sexes. The corners of the outlined simplex correspond to the four pure strategy profiles OO, OF, FO, and FF; the curved surface is the set of distributions where the row and column players pick independently; the convex shaded polyhedron is the set of correlated equilibria. The Nash equilibria are the points where the curved surface intersects the shaded polyhedron.

of correlated equilibrium states that in this situation the row player's payoff for playing O must be at least as large as its payoff for playing F.

In other words, in a correlated equilibrium we must have

$$4\frac{a}{a+b} + 0\frac{b}{a+b} \geq 0\frac{a}{a+b} + 3\frac{b}{a+b} \qquad \text{if } a+b>0.$$

Multiplying through by $a+b$ yields the linear inequality

$$4a + 0b \geq 0a + 3b \tag{13.2}$$

(We have discarded the qualification $a+b>0$ since inequality 13.2 is always true in this case.) On the other hand, by examining the case where the row player receives the recommendation F, we can show that

$$0c + 3d \geq 4c + 0d. \tag{13.3}$$

Similarly, the column player's two possible recommendations tell us that

$$3a + 0c \geq 0a + 4c \tag{13.4}$$

and

$$0b + 4d \geq 3b + 0d. \tag{13.5}$$

Intersecting the four constraints (13.2–13.5), together with the simplex constraints

$$a + b + c + d = 1.$$

and

$$a, b, c, d \geq 0$$

yields the set of correlated equilibria. The set of correlated equilibria is shown as the six-sided shaded polyhedron in fig. 13.1. (Figure 13.1 is adapted from Nau et al. (2004).)

For a game with multiple players and multiple strategies we will have more variables and constraints: one nonnegative variable per strategy profile, one equality constraint which ensures that the variables represent a probability distribution, and one inequality constraint for each ordered pair of distinct strategies of each player. (A typical example of the last type of constraint is "given that the moderator tells player i to play strategy j, player i doesn't want to play k instead.") All of these constraints together describe a convex polyhedron. The number of faces of this polyhedron is no larger than the number of inequality and nonnegativity constraints given above, but the number of vertices can be much larger.

The Nash equilibria for Battle of the Sexes are a subset of the correlated equilibria. The large tetrahedron in fig. 13.1 represents the set of probability distributions over strategy profiles. In most of these probability distributions the players' action choices are correlated. If we constrain the players to pick their actions independently, we are restricting the allowable distributions. The set of distributions which factor into independent row and column strategy choices is shown as a hyperbola in fig. 13.1. The constraints which define an equilibrium remain the same, so the Nash equilibria are the places where the hyperbola intersects the six-sided polyhedron.

13.3 Learning in One-Step Games

In normal-form games we have assumed that the description of the game is common knowledge: everyone knows all of the rules of the game and the motivations of the other players. In these types of games, learning may have several roles:

▪ To select a single equilibrium from the set of possible equilibria. Without prior communication the players have no way to decide which equilibrium to play, and the results of miscoordination can be disastrous.

▪ To accelerate computation. An agent with limited brainpower may decide to run a learning algorithm because the agent gets good payoffs faster by learning than it would by directly computing an equilibrium.

▪ To model limited players. While the other players may have known motivations, they may not be completely rational in pursuing these motivations; so, a learning algorithm may be able to take advantage of their bounded rationality.

In addition, if the description of the game is not common knowledge (see section 13.3.4), we can add two more reasons for learning:

- To discover the other players' motivations.

- To learn about our own payoffs or the distributions of random events.

We will start with the problem of identifying sets of equilibria, then move to various types of learning algorithms that can be used to find equilibria.

13.3.1 Computing Equilibria

The simplest equilibria to find are in two-player constant-sum games. In these games, minimax equilibria are the same as Nash equilibria. (Correlated equilibria are slightly more complicated but still easy to find; see Forges (1990) for more detail.)

The problem of finding the minimax value of a game is a linear program: write y for the strategy distribution of the column player and z for the value of the game to the row player. Write M for the matrix of payoffs to the row player, with one entry for each pair of strategies. Then the entries of the vector My are the payoffs to the row player for each of its strategies. To keep the row player from wanting to change strategies, these payoffs must all be less than or equal to z. Subject to these constraints, the column player wants to select its mixed strategy y to make z as small as possible:

$$
\begin{aligned}
\text{minimize} \quad & z \\
\text{subject to} \quad & My \leq \mathbf{1}z \\
& \mathbf{1}^\top y = 1 \\
& y \geq 0.
\end{aligned}
\tag{13.6}
$$

Here $\mathbf{1}$ denotes the vector $(1, 1, \ldots, 1)^\top$, so the last two lines ensure that y is a probability distribution.

A solution to the linear program 13.6 tells us a strategy for the column player that achieves the minimax value. (In fact, the optimal solutions to program 13.6 are the same as the optimal strategies for the column player; this fact constitutes a proof that the set of minimax strategies is convex.) If we want a strategy for the row player as well we can swap the roles of row and column players in program 13.6; or, we can simply look at the optimal dual variables for the constraint $My \leq \mathbf{1}z$, which many linear program solvers will supply at almost no additional cost.

The next simplest equilibria are the correlated equilibria in general-sum games. As shown in section 13.2, the set of correlated equilibria of a given game is convex; so, we can solve a linear program to find (say) a correlated equilibrium that maximizes the payoff to the first player, or one that maximizes the sum of payoffs to all players. (The latter criterion is called *social welfare*.) The linear program for finding correlated equilibria is bigger than the one for finding minimax equilibria: the former has a number of variables proportional to a^p for a game with a actions and p players, while the latter needs only about a variables. But both linear programs are polynomial in the size of the game, since we need to list pa^p numbers to specify the payoffs.

The complexity of finding Nash equilibria in general-sum games is an important open problem in the theory of algorithms. We know exponential-time algorithms to solve the problem, and we do not know any polynomial-time algorithms, but no one has proved that the problem is NP-complete. Simple questions related to Nash equilibria are known to be hard, however; for example, Conitzer and Sandholm (2003) show that counting the number of Nash equilibria is #P-complete, and that it is NP-complete to test whether there exists a Nash equilibrium with social welfare at least k.

Finding the Nash equilibria of a two-player general-sum game can be cast as a linear complementarity problem, that is, as a linear program with the addition of complementarity constraints of the form $pq = 0, p \geq 0, q \geq 0$. To get an intuition for why, consider any Nash equilibrium distributions x and y for the row and column players, and write z for the value of the game to the row player. If M is the matrix of row player payoffs, then

$$x^\top (z\mathbf{1} - My) = 0. \tag{13.7}$$

Both factors in equation 13.7 are always positive: x is a probability distribution, and each entry of $(z\mathbf{1} - My)$ is the difference between the value of the game and the value of a single action. So, equation 13.7 means that the row player will not put positive probability on any strategy which achieves less than the value of the game. Equation 13.7 is related to the linear program 13.6, in which the optimal primal and dual variables satisfy a similar complementarity constraint.

The classical algorithm for solving linear complementarity problems is the Lemke-Howson algorithm (Lemke, 1965). More recently, researchers have developed interior point methods for LCPs (Kojima et al., 1991), but no one has found an algorithm which has been proved to run in polynomial time.

For multiplayer general-sum games the complementarity problem becomes nonlinear and therefore even harder to solve, so people often turn to algorithms that are not based on complementarity problems. One such algorithm is *simplicial subdivision*. Suppose that we are given a continuous update rule which takes in a profile of mixed strategies and produces as output another profile of mixed strategies. By Brouwer's fixed point theorem, such an update rule must have at least one fixed point.

It is relatively easy to design update rules whose fixed points are exactly the Nash equilibria; see equation 13.9 and fig. 13.2 below for an example. Given such an update rule, one way to find Nash equilibria is to divide the set of profiles of mixed strategies into a mesh and search for a mesh cell whose center is near a fixed point. By searching a sequence of finer and finer meshes we can locate all of the

Nash equilibria to any desired level of accuracy; see McKelvey and McLennan for more details.

13.3.2 Hill-Climbing Algorithms

Let $r_i(a_1, a_2, \dots)$ be the payoff to player i when the players choose the actions $a_1, a_2, \dots$ in a normal-form game. For convenience of notation we will write $r_i(a_1, a_2, \dots) = r_i(a_i, a_{\neg i})$, where $a_{\neg i}$ stands for the action profile of all players except for i. Also for convenience we will write $r_i(x_i, a_{\neg i})$ for the expected reward when player i follows the mixed strategy x_i.

It is easy to see that r_i is a linear function of x_i when we hold $a_{\neg i}$ fixed, and that its gradient is

$$g_i = \frac{d}{dx_i} r_i(x_i, a_{\neg i}) = \begin{pmatrix} r_i(1, a_{\neg i}) \\ r_i(2, a_{\neg i}) \\ \vdots \end{pmatrix}, \qquad (13.8)$$

where $r_i(j, \dots)$ represents the payoff for the jth pure strategy. Many researchers have designed learning algorithms which observe $a_{\neg i}$ and alter x_i in the direction of g_i to attempt to improve performance. These algorithms have an informational advantage when compared to algorithms like linear programming: instead of needing to know all of the payoffs of the game, they just need to know g_i.

The simplest gradient-based algorithm is *best response*. Suppose we are playing the same game repeatedly, and that on the tth trial the ith player observes the gradient vector $g_i^{(t)}$. Then on the $(t+1)$st trial, player i will play a strategy

$$x_i^{(t+1)} \in \arg\max_{x \in \Delta} x \cdot g_i^{(t)},$$

where Δ is the simplex of probability distributions.

Unfortunately the best-response algorithm is not a very good one: imagine playing the zero-sum game "Matching Pennies," in which two players each choose heads (H) or tails (T) and the row player gets a point when the choices match. The payoff matrix for Matching Pennies is

	H	T
H	$1, -1$	$-1, 1$
T	$-1, 1$	$1, -1$

If the row player runs best response, starting with a prediction of T on the first trial, and if the column player plays the sequence HTHTHTHTHT ... , then the row player will lose on every trial.

A slightly smarter algorithm is *fictitious play* (Brown, 1951; Fudenberg and Levine, 1998). On the $(t+1)$st trial, fictitious play chooses

$$x_i^{(t+1)} \in \arg\max_{x \in \Delta} x \cdot \sum_{k=1}^{t} g_i^{(k)},$$

where Δ is the simplex of probability distributions. That is, it plays a best response to the average of all previous strategy profiles rather than to the most recent strategy profile.

Fictitious play has a very weak convergence property: in a constant-sum game, if two fictitious players play against each other, the average action $\frac{1}{t} \sum_{k=1}^{t} x_i^{(k)}$ of either player converges to a minimax equilibrium. Unfortunately a player who runs fictitious play is not guaranteed to achieve the minimax payoff even when the other players are also running fictitious play: it may play the right mixture of actions but always at the wrong time. For example, two fictitious players learning the game of Matching Pennies can play HTHTHT ... and THTHTH ... ; these sequences have the proper 50-50 mix of H and T, but give a payoff of -1 per trial to the row player instead of the minimax value of 0.

The flaw with both best response and fictitious play is that their strategy choice is deterministic (except possibly when there are exact ties between actions). Playing a deterministic strategy is often a bad idea: anyone who can predict your actions can take advantage of you. Even if the environment is not intentionally hostile, a deterministic strategy can result in poor performance, as in the example of two fictitious players playing Matching Pennies.

To get around this limitation, researchers have proposed a number of algorithms which are capable of learning to play mixed strategies. One of the simplest such algorithms is *gradient ascent*.

Suppose player i plays the mixed strategy $x_i^{(t)}$ at time t. Recall that player i's gradient at time t is $g_i^{(t)}$. At time $t+1$, gradient ascent computes the unnormalized strategy

$$\bar{x}_i^{(t+1)} = (1 - \alpha^{(t)})x_i^{(t)} + \alpha^{(t)} g_i^{(t)}$$

and then plays

$$x_i^{(t+1)} = P_\Delta(\bar{x}_i^{(t+1)}), \tag{13.9}$$

where $\alpha^{(t)}$ is a learning rate. The projection operator $P_\Delta(\bar{x})$ ensures that the recommended play is a legal probability distribution: it projects $\bar{x}$ onto the probability simplex Δ by minimum Euclidean distance.

Gradient ascent has much stronger performance guarantees than fictitious play and the other gradient-based algorithms described above. If we decrease the learning rate according to a schedule like $\alpha^{(t)} = 1/\sqrt{t}$, then a player which runs gradient ascent is guaranteed in the long run to achieve an average payoff at least as high as its safety value (Zinkevich, 2003). (See section 13.3.4 for additional algorithms with similar guarantees.) In a two-player two-action game the guarantee is even

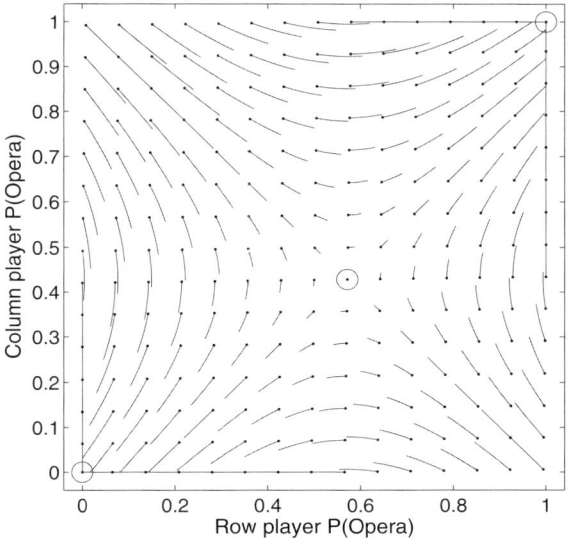

Figure 13.2   The gradient dynamics of the Battle of the Sexes. If we initialize the players' strategy profile at one of the small dots, the gradient ascent update (eq. 13.9) will move it along the corresponding line. The fixed points of the gradient dynamics are the Nash equilibria $(0,0)$, $(1,1)$, and $(4/7, 3/7)$ (marked with circles); the first two are stable fixed points while the last is an unstable fixed point.

stronger: Singh et al. (2000) proved that two gradient-ascent learners will achieve many properties of a Nash equilibrium in the limit, including the average payoffs and the average strategy.

The most current strategy $x_i^{(t)}$ may not converge when two gradient-ascent players learn simultaneously: Singh et al. (2000) showed that the joint strategies can enter a limit cycle, even in a two-player two-action game. If the strategies do converge, though, their limit must be a Nash equilibrium: the projected gradient $P_\Delta(x_i + g_i) - x_i$ is zero exactly when player i can get no benefit by changing its strategy. In other words, the Nash equilibria are exactly the fixed points of the update in equation 13.9; see fig. 13.2 for an example.

Since Nash equilibria can be difficult to find, it is interesting to look for modifications to the gradient ascent algorithm which make it converge more often. Bowling (2003); Bowling designed a family of algorithms called WoLF (for Win or Learn Fast) which adjust the learning rate $\alpha^{(t)}$ to encourage convergence. He proved convergence to Nash in self-play for two-player two-action games, and empirically observed convergence in larger games.

13.3.3   Types of Performance Guarantees

In section 13.3.2 we mentioned three types of performance guarantees for hill-climbing algorithms: bounds on their reward in arbitrary environments, convergence of their reward in self-play, and convergence of their strategy in self-play. When playing in an arbitrary environment the best guarantee we can hope for is to achieve our safety value, since the environment could be designed to thwart us at every turn. (Convergence in an arbitrary environment is not desirable, since the environment itself could be nonstationary.)

On the other hand, guarantees about what happen in self-play are not by themselves completely satisfying since we often don't have control over the learning algorithms that the other players are running. So, we might want to analyze the behavior of our algorithms when the other players use various classes of learning algorithms. For example, we might want to prove that we achieve our safety value in an arbitrary environment, but that we achieve the payoff of some equilibrium when the other players are in some sense rational.

Below we will describe two rationality properties called *no external regret* and *no internal regret*. No internal regret is the stronger property; that is, players which have the no-internal-regret property will also have no external regret.

We will demonstrate that we can achieve both of these no-regret properties by presenting learning algorithms which do so. If we achieve no regret, our payoffs will be at least as high as our safety value. If all players achieve no external regret (using whatever learning algorithms they choose), then (1) in a constant-sum game, their average strategies will converge to minimax strategies, and (2) in a general-sum game, if their average strategies converge, the result will be a Nash equilibrium. On the other hand, if all players achieve no internal regret (no matter what learning algorithms they use to do so) then their payoffs and their average strategy profile will converge to a correlated equilibrium.

While the performance guarantees for no-regret algorithms are strong, perhaps the strongest for any class of learning algorithms in games, there are still other performance criteria which we do not know how to achieve. For example, here are some open problems:

1. While the payoffs of the no-internal-regret algorithms will converge to correlated equilibrium payoffs, we do not know how to achieve convergence of the strategies themselves except in two-player two-action games.

2. Instead of a correlated equilibrium, we might like the stronger guarantee of convergence to the payoffs or strategies of a Nash equilibrium.

3. Instead of a Nash equilibrium, we might prefer some other subclass of correlated equilibria. For example, we might look for a correlated equilibrium which maximizes social welfare, or one which maximizes payoff to the first player.

The first of these problems is mostly cosmetic: with no-internal-regret algorithms our average strategy profile will be a correlated equilibrium even if none of the learning algorithms explicitly represents this average profile. Solving either of the latter two problems with an efficient algorithm, on the other hand, would require significant advances in our understanding: a learning algorithm which converges sufficiently rapidly to a Nash equilibrium would allow us to find a

Nash equilibrium in polynomial time, an important open question in complexity theory (see section 13.2). And, a learner which converges sufficiently rapidly to the correlated equilibrium that maximizes social welfare would prove $P = NP$, an even more important open question (Forges and von Stengel, 2002).

13.3.4   Regret

Let $P(a_i, a_{\neg i})$ be a distribution over strategy profiles. We will say that player i's *external regret* for not having played pure strategy j is

$$\rho_{ij}^{P} = E_{a \sim P}(r_i(j, a_{\neg i}) - r_i(a_i, a_{\neg i})).$$

That is, the external regret is the difference in the payoff that i would receive if it were to play action j instead of playing according to P. Also define the *overall external regret* to be the external regret versus the best strategy j:

$$\rho_i^P = \max_j \rho_{ij}^P.$$

External regret is an interesting quantity because it is related to Nash equilibrium. Consider a distribution P which factors into an independent strategy choice for each of the players. Then P is a Nash equilibrium if and only if the players all have $\rho_i^P \leq 0$: having $\rho_{ij}^P \leq 0$ is equivalent to saying that player i has no incentive to switch to strategy j.

Now define the *internal regret* for a pair of strategies j and k to be

$$\bar{\rho}_{ijk}^{P} = E_{a \sim (P|a_i = j)}(r_i(k, a_{\neg i}) - r_i(j, a_{\neg i})),$$

where $(P \mid a_i = j)$ is the distribution we get by conditioning $P(a)$ on the event $a_i = j$. That is, the internal regret is the benefit that player i would get by switching all of its plays of action j to action k instead. We can also define the *overall internal regret* to be the internal regret for the best strategy switch:

$$\bar{\rho}_i^P = \max_{ij} \bar{\rho}_{ijk}^P.$$

Internal regret is an interesting quantity because P is a correlated equilibrium if and only if all players have $\bar{\rho}_i^P \leq 0$: saying $\rho_{ijk}^P \leq 0$ is equivalent to saying that player i has no incentive to play k when recommended j.

Researchers have designed a variety of learning algorithms which attempt to achieve low internal or external regret. More precisely, if we look at the distribution P_t of joint plays after t time steps, these algorithms will attempt to make the average regret $\rho_i^{P_t}$ or $\bar{\rho}_i^{P_t}$ approach zero. Algorithms which achieve these properties are called (respectively) *no external regret* or *no internal regret* algorithms, or when it won't cause confusion, just *no regret* algorithms. In this section we will describe several no-external-regret algorithms; for an example of a no-internal-regret algorithm see Greenwald and Jafari (2003).

We have already seen our first no regret algorithm: the gradient ascent algorithm described in section 13.3.2 is no external regret. Our second no-regret algorithm is equally simple: it is called *follow the perturbed leader* or FPL. It was originally due to Hannan, and is described in detail and extended in Kalai and Vempala (2002).

FPL is a smarter version of the fictitious play algorithm described in section 13.3.2. Define the expected gradient

$$g_i(P) = E_{a \sim P} \begin{pmatrix} r_i(1, a_{\neg i}) \\ r_i(2, a_{\neg i}) \\ \vdots \end{pmatrix}. \tag{13.10}$$

Equation 13.10 is a natural generalization of equation 13.8.

With this notation, fictitious play chooses the best action based on the average gradient so far:

$$x_i^{(t)} = \arg\max_{x \in \Delta} x \cdot g_i(P_t).$$

Recall that the problem with fictitious play is that it is too predictable: for example, in Matching Pennies, a smart opponent can cause a fictitious player to guess wrong on every trial. To make fictitious play less predictable, we can add a little bit of noise. Pick a constant ϵ and let

$$\delta_j^{(t)} \sim \text{uniform}(0, \epsilon/\sqrt{t})$$

be a set of independent, uniformly distributed random variables for all actions j and trials t. Then FPL chooses

$$x_i^{(t)} = \arg\max_{x \in \Delta} x \cdot (g_i(P_t) + \delta^{(t)}).$$

That is, FPL finds the average payoff against P_t for each of its actions, adds some noise, and picks the action with the highest perturbed payoff.

If we add too little noise we will be predictable, but if we add too much noise we will play completely randomly. To strike a balance between these two extremes, as the number of trials t increases FPL adds less and less noise. FPL is a no-external-regret algorithm: if player i runs FPL, then $\rho_i^{P_t} = O(1/\sqrt{t})$, so $\rho_i^{P_t} \to 0$ as $t \to \infty$.

Our next no-regret algorithm is called *weighted majority*. It is due to Littlestone and Warmuth (1992), and was applied to games by Freund and Schapire (1996). Like FPL and fictitious play it is based on the average gradient so far: it chooses

$$x_i^{(t)} = (1/Z)e^{\beta t g_i(P_t)},$$

where Z is a normalizing constant, β is a learning rate, and e^g for a vector g means the vector whose ith component is the exponential of the ith component of g.[2]

With a fixed learning rate β, weighted majority's regret is $\rho_i^{P_t} = O(1)$, not low enough for the no-regret property. But if we know the number of trials t ahead of time, and if we set β in proportion to $1/\sqrt{t}$, then weighted majority achieves regret

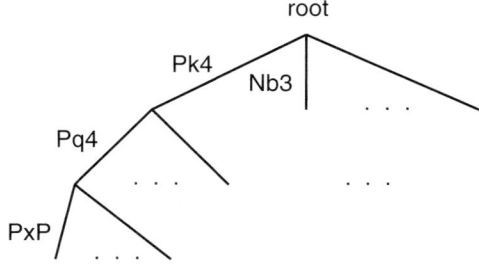

Figure 13.3 Part of the game tree for chess. Each node encodes a board position (including the identity of the player who is about to move) and each edge encodes an action. There can be a payoff at each edge for each player, although this payoff is not shown. A typical sequence of moves is Pk4, Pq4, PxP.

$O(1/\sqrt{t})$ and is therefore a no-external-regret algorithm.

Our final no-external-regret algorithm is called *regret matching* (Hart and Mas-Colell, 2001). Regret matching bases its plays directly on the current regrets. Define $[x]_+ = \max\{x, 0\}$; regret matching plays

$$x_i^{(t)} = (1/Z) \left[\rho_i^{P_t} \right]_+ ,$$

where Z is a normalizing constant. (If $\rho_i^{P_t} \leq 0$ then regret matching plays arbitrarily.) Regret matching has no parameters to tune, and achieves regret $\rho_i^{P_t} = O(1/\sqrt{t})$ after t trials.

13.4 Sequential Decisions

So far in this chapter we have considered only one-step games, that is, games where the players make a single simultaneous choice of action. In general, though, the players can have a sequence of interleaved choice points. We can analyze these sequential games by converting them to normal form, but as we will see below in section 13.4.2 there are some disadvantages to this approach: the normal-form representation can be very large, and the solutions we find will not take into account all the consequences of the sequential nature of the game.

The simplest type of sequential game is one with alternating moves and perfect information. Chess, checkers, and backgammon are examples of such games. We can solve perfect-information sequential games by *dynamic programming*. That is, we can build a *game tree* which encodes all possible sequences of play (see fig. 13.3), and work our way backward from the leaves of the tree to the root.

A leaf represents a position where there are no more moves available (for example, because one of the players has already won). It is easy to assign a value to a leaf, since the rules of the game tell us the payoffs to each player. Once we have assigned values to all the leaves there will be some positions where, no matter

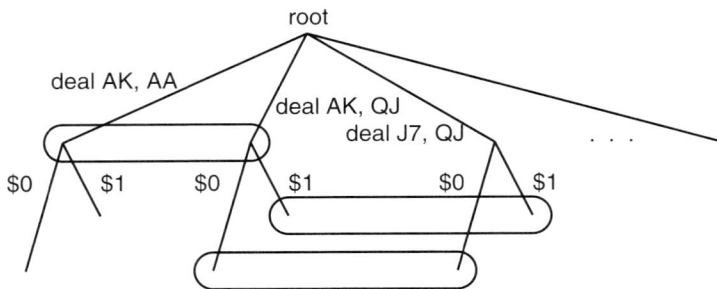

Figure 13.4 Part of the game tree for Texas Hold'em poker. Ovals indicate indistinguishable pairs of positions.

what move the player takes, the next position will be a leaf.[3] We can assign a value to such level-1 positions by picking the move which is most advantageous to the player whose turn it is. Once we have assigned values to all the level-1 positions, there will be some level-2 positions (i.e., positions where no matter what move the player takes, the next position will be either a leaf or a level-1 position). We can assign a value to the level-2 positions by considering all of their possible moves, and so on for level-3 and higher board positions.

If different players have different knowledge about the world, the result is an imperfect-information game like bridge or poker. We can write down game trees for imperfect-information games too; see fig. 13.4 for an example. The difference between an imperfect-information game tree and a tree for a game like chess is the **information sets** use of *information sets* to encode what each player knows and when. An information set is a collection of nodes where it is some player's turn to move, and which that player cannot distinguish from one another.

For example, in Texas Hold'em, the dealer gives each player two face-down cards before the first round of betting. The first player might see that its cards are AK; in this situation, the second player might hold a pair of aces, a QJ, or any other possible hand, and the first player cannot tell which. Similarly, once the first player has bet and it is the second player's turn to move, the second player can look at its QJ but will not know whether the first player holds AK or J7. (Of course, the second player can tell something about which hand the first player holds by reasoning that the first player will bet more aggressively on AK than on J7; this sort of reasoning is one of the challenges which makes learning in sequential games difficult.)

Dynamic programming no longer works for imperfect-information games: the best move at one position in the game tree can be influenced by action choices at other positions. One well-known example of this fact is bluffing in poker. A good poker strategy will bet strongly on some low-value hands, even though it knows it may get called and lose money; the reason it bluffs on low hands is so that the opponents won't immediately know to fold when it starts betting on a high-value hand. Changing the number and type of bluffs would require changing betting strategy on many other hands as well.

extensive form
games

We will consider two main types of imperfect-information games. In *extensive form games* the information sets satisfy *perfect recall*: consider two nodes A and B where player i is about to move, and call the descendants of A where i is about to move the i-descendants of A. If A and B are in different information sets, then all of the i-descendants of A are in different information sets from all of the i-descendants of B. And, if a and b are two different actions, then the i-descendants of A which we can reach by playing a are in different information sets than the i-descendants of A which we can reach by playing b.

stochastic games

Perfect recall means that a player never forgets any of its actions or observations. In *stochastic games*, on the other hand, the information sets encode a Markov property: the players forget their specific actions and observations and remember only a sufficient statistic of their history. Typically a stochastic game is represented with a set of S states; each player observes the state and all players simultaneously choose their actions. Based on the current state and all the actions, the world transitions to a new state and the process repeats.

In our previous notation, if there are k players, player i will play at steps $i, i + k, i + 2k, \cdots$. At all steps there are exactly S information sets. For each group of k steps the information set remains equivalent—any information available to one player is available to all players. And, given the information set at time $tk + 1$, the information set at step $(tk + 1) + k$ depends only on the actions of the players from times $tk + 1$ to $(t + 1)k$.

For both extensive-form games and stochastic games, we will assume that there is a fixed time horizon T before which the game is guaranteed to end. Without an assumption like this, the players could receive infinite expected total payoff, and they could delay bad outcomes indefinitely. It is possible to relax this *finite horizon* assumption, but we will not do so in this chapter.

13.4.1   Uncertainty

Bayesian games

An important special case of extensive-form game is the *Bayesian game*. Bayesian games are normal-form games with incomplete information. In a Bayesian game the entries of the payoff matrix are uncertain, and each player has its own private information about the payoffs.

For example, recall that in the Battle of the Sexes the row player prefers opera over football. But suppose that the opera might be either Verdi (V), which the row player likes, or Wagner (W), which the row player dislikes. The row and column players both have some information about which opera is playing: the column player knows for sure whether it's V or W, while the row player only knows with 80% certainty. We can represent this game, which we will call the "Return of the Battle of the Sexes", with three tables:

	Types				V, V or W, V				V, W or W, W	
	V	W			O	F			O	F
V	0.4	0.1		O	4, 3	0, 0		O	2, 2	0, 0
W	0.1	0.4		F	0, 0	3, 4		F	0, 0	3, 4

The first table describes the private information which might be available to each player. A player's private information is called its *type*, and the vector of all types is called the *type profile*. The table shows the prior probability of each type profile. For example, profile V, W has probability 0.1.

The actual payoffs of the game depend on the type profile. In this case we have assumed that the column player knows which opera is playing, so the payoffs are the same for V, V and W, V (and similarly for V, W and W, W). The payoffs when Verdi is playing are the same as for the original Battle of the Sexes; if Wagner is playing then the payoff for the play O, O is reduced to 2 for both players because of their frustration at going to the wrong opera.

All three of the tables are assumed to be common knowledge. The only private knowledge that each player has is its signal about which opera is playing.

A pure strategy in a Bayesian game is a function from a player's type to its action. For example, a strategy for the row player might be "if I think Verdi is playing then I will go to the opera, otherwise I will go to the football game." In the Return of the Battle of the Sexes, each player has $2^2 = 4$ pure strategies; in general, with t types and a actions there are a^t strategies.

Given a pure strategy profile we can compute the expected payoff to each player when they play according to that profile. So, we can convert a Bayesian game into its normal-form representation by evaluating all profiles of strategies. If there are p players then there will be a^{tp} entries in the normal-form matrix. The normal form for the Return of the Battle of the Sexes is

	OO	OF	FO	FF
OO	3, 2.5	1.8, 1.4	1.2, 1.1	0, 0
OF	2, 1.5	2.8, 2.8	0.7, 0.7	1.5, 2
FO	1, 1	0.5, 0.6	2, 2.4	1.5, 2
FF	0, 0	1.5, 2	1.5, 2	3, 4

The row and column labels are pure strategies; for example, the label OF means to play O on seeing V and F on seeing W. The top-left entry in the table tells us that, if both players always go to the opera no matter what their types are, the expected payoff for the row player is 3 and for the column player it is 2.5. These numbers are the averages of the payoffs for the play O, O in the two payoff tables, since both payoff tables occur with equal probability.

Bayes-Nash equilibrium   A Nash equilibrium in the normal-form game is called a *Bayes-Nash equilibrium* of the original Bayesian game (Harsanyi, 1967). In a Bayes-Nash equilibrium, no

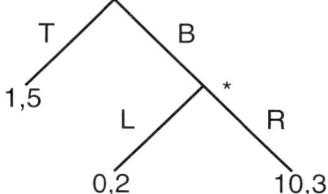

Figure 13.5 Subgame-perfect equilibrium. The Nash equilibrium T,L with payoffs 1, 5 is not subgame perfect, while the equilibrium B,R with payoffs 10, 3 is.

player has an expected incentive to deviate. That is, conditioned on its own type, a player can compute a posterior distribution over the other players' types. From this posterior along with the other players' distributions over their mappings from types to actions, the player can compute how often each other player will choose each of its actions. From these action distributions, in turn, it can compute an expected payoff for each of its own actions. In a Bayes-Nash equilibrium, the expected payoff for the recommended action is at least as high as the expected payoff for any other distribution over actions.

The translation from a correlated equilibrium in the normal-form game to an equilibrium in the Bayesian game is more subtle. The issues are the same as for general extensive-form games, so we will put off discussing them until section 13.4.2.

13.4.2 Equilibrium in Extensive Form Games

Just as for Bayesian games, we could try to compute equilibria in a general extensive-form game by converting it to normal form. Converting it to normal form means listing and evaluating all possible pure strategy profiles; a pure strategy for an extensive-form game is a deterministic function from information sets to actions.

There are two problems with this approach. The first is that the game might get very big: for example, the game of one-card poker mentioned in section 13.6.2 has only 26 information sets per player (if we use a deck with 13 different ranks) and 2 actions at each information set, but each player has approximately 67 million pure strategies. That means that the normal-form matrix has about (67 million)^2 entries in it.

The second problem is more subtle: neither Nash equilibria nor correlated equilibria translate exactly from normal-form to extensive-form games. The normal-form equilibria assume that there is only one choice point. In these equilibria we pick our strategy before the start of the game and prove that there is no incentive to deviate. But as we play out the game we gain more information. So, even if there was initially no incentive to deviate, an incentive may arise later. To handle the increased complexity of sequential decisions we will introduce two new types of equilibrium: the subgame perfect Nash equilibrium and the extensive-form correlated equilibrium.

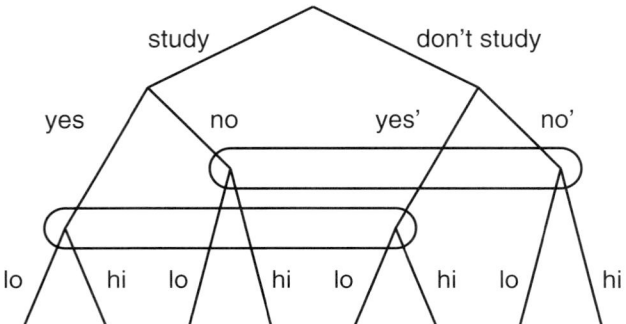

Figure 13.6   The Job Market game. The student wants to prove to the employer that s/he has studied in college by correctly answering a difficult question at the job interview.

To illustrate the need for subgame-perfect equilibria, consider the game of fig. 13.5. In this game the pure strategies for the first player are T and B, while those for the second player are L and R. The pair T,L is a Nash equilibrium in the normal form of this game: changing T to B lowers the first player's payoff from 1 to 0, while changing L to R doesn't affect payoffs since the second player's choice is irrelevant when the first player plays T.

There is something unsatisfying about the T,L equilibrium, though: if we have reached the node marked ∗, the second player will not want to play L since its payoff for R is higher (3 instead of 2). This situation is called an *incredible threat*: the second player threatens to play L to keep the first player from playing B, but the first player should have no reason to believe this threat.

subgame-perfect equilibrium
To remove the possibility of incredible threats, we define a Nash equilibrium to be *subgame perfect* if no player has an incentive to deviate in any subgame, whether or not that subgame is reachable under the equilibrium strategies. A *subgame* is a subtree of the game tree which doesn't break up information sets; that is, no information set has some members inside the subgame and other members outside of it. (Every game has at least the trivial subgame which contains the whole tree; some games have many more subgames as well, but other games don't have any proper subgames.)

One way to look for subgame-perfect equilibria is to modify the game so that all information sets are reachable under any strategy. For example, we could say that the game begins at a uniform random information set with probability ϵ; or we could say that a player's desired action is replaced with a uniform random action with probability ϵ at each step. (The latter modification is called *trembling hands*.) In either case for sufficiently small ϵ the payoff of any profile of mixed strategies will be arbitrarily close to its original payoff, but as long as $\epsilon > 0$ any equilibrium in the normal form of the modified game will be subgame perfect.

Just as for Nash equilibria, we can require subgame perfection for correlated equilibria. But when we generalize correlated equilibria to extensive-form games we need to consider the problem of hiding excess information as well. In a normal-form correlated equilibrium a moderator tells each player which pure strategy to follow. If we translate such an equilibrium to an extensive-form game, the moderator is required to tell each player its entire policy at the beginning of the game; this may be too much information too soon.

For example, in the "Job Market" game (due to (due to Forges and von Stengel, 2002, and pictured in fig. 13.6), a student first decides whether to study or goof off in college. Then, on a job interview, the employer asks a difficult question; the student must answer yes or no. Finally, the employer decides whether to offer a low or a high salary. The employer is willing to offer the high salary to a student who has studied, but not to a student who hasn't. The payoffs are such that the employer must be fairly certain that the student has studied before offering the high salary, while the student is only willing to study if it offers a good chance of getting the high salary.

This game has a single Nash equilibrium: since the employer can't observe whether the student has studied, it always offers a low salary, and so the student has no incentive to study. This Nash equilibrium is also the only correlated equilibrium in the normal form of this game. We might hope that we could design an equilibrium in which the student's yes or no answer (which the employer can perceive) is correlated with its decision of whether to study (which the employer cannot perceive). Unfortunately, in a correlated equilibrium which comes from the normal-form representation, the student gets both recommendations from the moderator at the same time: given a recommendation to study and answer yes, yes', a better play is don't study, yes, yes'.

To achieve an equilibrium in which the answer to the question is correlated with the study or don't study decision, the student must not find out whether to say yes until after it has chosen to study. In an *extensive-form correlated equilibrium* (EFCE), the moderator computes a recommended action for every information set, but only informs the player what action to take if the player reaches the appropriate information set.

There is an EFCE in the Job Market game in which the players follow the following four strategy profiles with equal probability:

study,yes,yes'	hi on yes, lo on no
study,yes,no'	hi on yes, lo on no
study,no,yes'	lo on yes, hi on no
study,no,no'	lo on yes, hi on no

In this equilibrium the correct answer is equally likely to be yes or no. The recommended answer if the student chooses study is perfectly correlated with the correct answer, while choosing don't study results in a random recommendation.

13.5   Learning for Sequential Decisions

Learning in sequential decision problems is much harder than learning in one-step games, so the available algorithms for sequential games are not as mature as their one-step counterparts are. We will start with algorithms for computing equilibria in extensive-form games. Then we will examine a generalization of no-external-regret learning to extensive-form games. Finally, we will turn to learning in stochastic games, where we will describe two classes of algorithm: *policy gradient* learning and *temporal difference* learning. (Policy gradient and temporal difference algorithms will work for extensive-form games as well, but for simplicity we will describe them as they apply to stochastic games.)

13.5.1   Sequence Weights

The most efficient algorithms for computing equilibria or for learning to play in extensive-form games are based on the *sequence form* representation of mixed strategies. The sequence form specifies distributions over a player's pure strategies via a set of *sequence weights*. Player i has one sequence weight for every pair (s, a), where s is one of i's information sets and a is an action available from s.

Sequence weights cannot represent every possible mixed strategy, but they can represent an important subset called *behavior strategies*. A behavior strategy is a function which maps an information set to a probability distribution over actions to take at that information set. The behavior strategies are rich enough that, if P is a distribution over i's pure strategies which achieves reward r (against some distribution over the strategy profiles of the other players), there will be a behavior strategy which achieves reward at least r.

Sequence weights are important for three reasons. First, they are a compact representation for strategies: if a player has n information sets and m actions, then the set of sequence weights and the set of behavior strategies will each have (about) mn dimensions, while the mixed strategies will be (about) m^n-dimensional. Second, sequence weights allow us to compute payoffs easily: if we hold the other players' strategies fixed, player i's payoff is a linear function of its sequence weights. Finally, the set of valid sequence weights for a given player is a convex polyhedron, so optimizing a linear function over the set of sequence weights is a linear program.

Given a distribution P over i's pure strategies, we will define the corresponding sequence weights as follows. Pick an information state s and an action a. Write $P(a \mid s)$ for the probability under P of selecting action a given that we have reached s. Let $\mathrm{anc}_i(s, a)$ be the set of ancestors of (s, a): that is, $(s', a') \in \mathrm{anc}_i(s, a)$ if and only if we must reach s' and have player i take action a' in order to reach (s, a). By convention, $(s, a) \in \mathrm{anc}_i(s, a)$. Given this notation, the sequence weights are

$$\phi^i_{s,a} = \prod_{(s',a') \in \mathrm{anc}_i(s,a)} P(a' \mid s').$$

With the above definition, we can verify that sequence weights satisfy the three properties that we have claimed for them. First, they are a compact representation of behavior strategies: we can easily recover the behavior strategy which corresponds to a vector of sequence weights, since

$$P(a \mid s) = \frac{\phi^i_{s,a}}{\phi^i_{s',a'}} = \frac{\phi^i_{s,a}}{\sum_b \phi^i_{s,b}}, \tag{13.11}$$

where (s', a') is the parent of s, that is, the last information set and action which player i visited before reaching s. (If the denominator in equation 13.11 is zero, we can set $P(a \mid s)$ arbitrarily, since there is no possibility of reaching information set s.)

Next, we can verify that the payoffs are a linear function of the sequence weights. Let l be a leaf of the game tree, and let $(s^i(l), a^i(l))$ be the last state and action which player i visited on the way to l. The probability of reaching leaf l is the product of the probabilities for all of the players' action choices along the path from the root to l. Since the product of i's action choice probabilities is $\phi^i_{s^i(l),a^i(l)}$, the overall probability of reaching l is

$$P(l) = \prod_{\text{players } j} \phi^j_{s^j(l),a^j(l)}. \tag{13.12}$$

(For convenience of notation, in equation 13.12 we are considering Nature to be an additional player who always plays the same behavior strategy.) Note that $P(l)$ in equation 13.12 is a linear function of ϕ^i if we hold ϕ^j fixed for $j \neq i$. Now, the total payoff to player i is

$$\sum_{\text{leaves } l} P(l) r_i(l), \tag{13.13}$$

where $r_i(l)$ is i's expected payoff conditional on reaching l. Since $r_i(l)$ is a constant, and since a sum of linear functions is linear, equation 13.13 means that i's expected payoff is a linear function of ϕ^i, as claimed.

Finally, we can check that the set of valid sequence weights is convex. This last fact is easy to verify: we can just list the linear constraints which bound the set. First, sequence weights are all positive:

$$\phi^i_{s,a} \geq 0 \qquad \forall s, a. \tag{13.14}$$

Second, if s is an initial information set for i (that is, if it is i's turn to play in s and i didn't have a turn prior to reaching s) then the weights of actions leaving s must sum to 1:

$$\sum_a \phi^i_{s,a} = 1 \qquad s \text{ is initial.} \tag{13.15}$$

Equation (13.15) holds because $\phi^i_{s,a} = P(a \mid s)$, and the probabilities of all the actions must sum to 1. Finally, if s has parent (s', a'), then the weights of actions

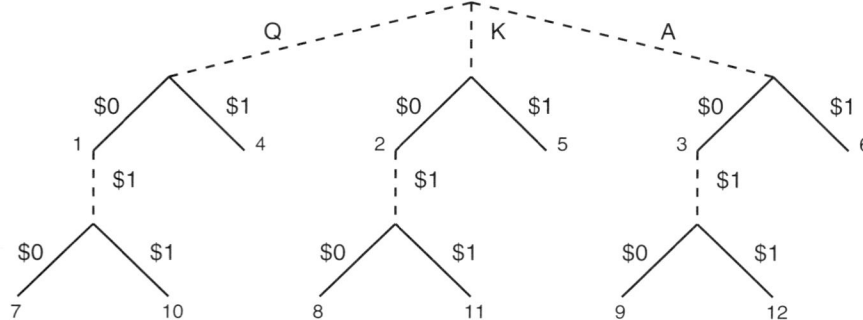

Figure 13.7 Sequences for the first player in the game of one-card poker with a three-card deck. Solid lines denote actions, while dashed lines denote observations. Small numbers refer to columns in fig. 13.8.

leaving s must sum to the weight of (s', a'):

$$\sum_a \phi^i_{s,a} = \phi^i_{s',a'} \qquad (s', a') \text{ parent of } s. \tag{13.16}$$

Equation 13.16 holds because $\phi^i_{s,a} = \phi^i_{s',a'} P(a \mid s)$, and the probabilities of all the actions must sum to 1. Any vector ϕ^i satisfying equations 13.14–13.16 can be translated into a behavior strategy via equation 13.11, so these constraints specify all and only the valid sequence weights.

13.5.2   Computing Equilibria

Using the sequence-form representation of strategies, we can generalize the algorithms for finding minimax, Nash, or correlated equilibria in one-step games. We can find a minimax equilibrium in an extensive-form game by solving a linear program. We can find a Nash equilibrium in a two-player game by solving a linear complementarity problem, or in a multiplayer game by solving a nonlinear complementarity problem. (The resulting Nash equilibria are not necessarily subgame perfect, but we can use the tricks described in section 13.4.2 to find subgame perfect equilibria.) Finally, we can find extensive-form correlated equilibria in a two-player game by solving a linear program.

In this section we will describe the algorithm for finding minimax equilibria in two-player constant-sum games. For Nash equilibria, see Koller et al. (1996). For correlated equilibria, see Forges and von Stengel (2002).

We have already shown that the payoff for each player is a linear function of its sequence weights, when we hold the other players' strategies fixed. In a two-player game, that means that we can write the payoff for either player as a bilinear function of the sequence weight vectors. If we write ϕ for the vector of sequence weights for the first player and ψ for the vector of sequence weights for the second player, then we can define matrices M and N such that $\psi^\top M \phi$ is the first player's payoff while $\phi^\top N \psi$ is the second player's payoff.

$$
\left(
\begin{array}{ccccccccc|c}
1 & & 1 & & & & & & & 1 \\
& 1 & & 1 & & & & & & 1 \\
& & 1 & & 1 & & & & & 1 \\
-1 & & & & & 1 & & 1 & & 0 \\
& -1 & & & & & 1 & & 1 & 0 \\
& & -1 & & & & 1 & & 1 & 0
\end{array}
\right)
$$

Figure 13.8 Sequence weight constraints $A\phi = a$ for the first player in one-card poker. The first row tells us that sequence weights ϕ_1 and ϕ_4 must add to 1; that is, after seeing Q, player 1 must either bet \$0 or bet \$1.

Now consider a zero-sum game such as one-card poker. (In a zero-sum game, $N = -M^\top$.) The first player's sequences for one-card poker are shown in fig. 13.7: the player first observes its card, either Q, K, or A. Then it decides whether to bet or pass. If it passes, the second player may raise, in which case the first player has the opportunity to fold (bet nothing) or call (match the bet).

The small numbers in fig. 13.7 mark the 12 different sequences for player 1. For example, the number 7 marks the sequence where player 1 was dealt a Q, bet \$0, saw that player 2 bet \$1, and folded. Figure 13.8 shows the consistency constraints $A\phi = a$ on player 1's sequence weights ϕ. (In addition to the consistency constraints, we must also have nonnegativity: $\phi \geq 0$.)

If we write $B\psi = b$ and $\psi \geq 0$ for the constraints on player 2's sequence weights, then the problem of finding a minimax equilibrium is

$$
\begin{aligned}
& \max_\phi \min_\psi \ \psi^\top M \phi \\
& A\phi = a \\
& B\psi = b \\
& \phi, \psi \geq 0
\end{aligned}
\qquad . \tag{13.17}
$$

We can turn program 13.17 into a linear program by introducing Lagrange multipliers for some of the constraints and solving for some of the variables. Start by adding Lagrange multipliers z and $\lambda \geq 0$ for the constraints $B\psi = b$ and $\psi \geq 0$, respectively:

$$
\begin{aligned}
& \max_{\phi, z, \lambda} \min_\psi \ \psi^\top M \phi + z^\top (B\psi - b) - \lambda^\top \psi \\
& A\phi = a \\
& \phi, \lambda \geq 0.
\end{aligned}
\tag{13.18}
$$

We can now optimize over ψ by setting the derivative of the objective to zero:

$$
0 = \frac{d}{d\psi} \left(\psi^\top M \phi + z^\top (B\psi - b) - \lambda^\top \psi \right) = M\phi + B^\top z - \lambda.
$$

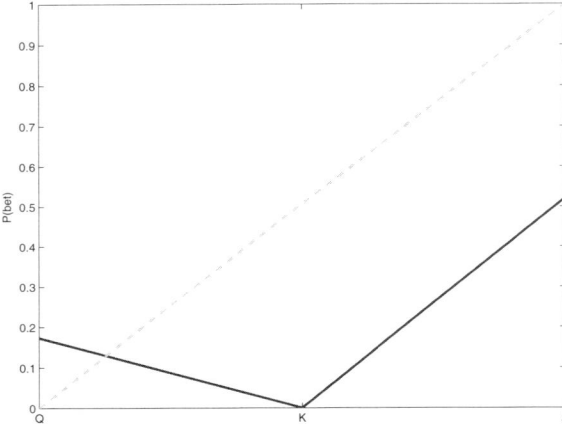

Figure 13.9 A minimax behavior strategy for the first player in one-card poker. Solid line: probability of betting after initial deal. Dashed line: probability of betting given a pass on the first round and a raise by the opponent.

Substituting in the above constraint, we have

$$
\begin{aligned}
&\max_{\phi,z,\lambda} \ -z^\top b \\
&A\phi = a \\
&M\phi + B^\top z - \lambda = 0 \\
&\phi, \lambda \geq 0
\end{aligned}
\tag{13.19}
$$

or equivalently

$$
\begin{aligned}
&\min_{\phi,z} \ z^\top b \\
&A\phi = a \\
&M\phi + B^\top z \geq 0 \\
&\phi \geq 0.
\end{aligned}
\tag{13.20}
$$

Any optimal vector ϕ for the linear program 13.20 tells us the sequence weights for a minimax strategy for player 1; see fig. 13.9. For ease of interpretation, fig. 13.9 shows a behavior strategy rather than the corresponding sequence weights; for example, the leftmost point on the solid line is ϕ_4, while the leftmost point on the dashed line is ϕ_{10}/ϕ_1. Notice that the strategy includes "bluffing" (the probability of betting on being dealt a Q is higher than the probability of betting on K) and "slow-playing" (the probability of betting when dealt an A is only a bit over 50%, even though player 1 is guaranteed to win in this situation).

13.5.3 Extensive-Form No Regret

We can generalize the definition of external regret to apply to extensive-form games: let $P(s_1, s_2, \dots)$ be a distribution over profiles of extensive-form pure strategies.

Then our regret for not having played the extensive-form pure strategy j is

$$\rho_{ij}^P = E_{s \sim P}(r_i(j, s_{\neg i}) - r_i(s_i, s_{\neg i})).$$

That is, ρ_{ij}^P is the difference in payoffs that player i would achieve by switching from its part of P to j.

The only problem with this extended definition of external regret is that there can be a very large number of extensive-form pure strategies, and we do not want to compare our performance explicitly against each one of them. Fortunately, our rewards for these pure strategies are not independent. If we hold the other players' strategies fixed, and if we represent each of our pure strategies by its sequence weights, then our reward for each strategy (and therefore our regret compared to it) will be a linear function of its weights.

To take advantage of this dependence among rewards, write ϕ for a vector of sequence weights for player i, satisfying $\phi \geq 0$ and $A\phi = a$ as in section 13.5.1. Write $r_i(\phi, s_{\neg i})$ for the expected reward of the behavior strategy which corresponds to ϕ when played against a fixed strategy $s_{\neg i}$ for the other players.

Now we can define the *external regret vector*, $\rho(P)$, to be any vector such that

$$\rho(P) \cdot \phi = E_{s \sim P}(r_i(\phi, s_{\neg i}) - r_i(s_i, s_{\neg i})). \tag{13.21}$$

That is, $\rho(P) \cdot \phi$ is our regret for not having played ϕ. (For details of how to construct a vector $\rho(P)$ satisfying equation 13.21, see Gordon (2005); we know such a vector exists because the regret is a linear function of ϕ.[4])

Just as before, our overall regret ρ^P is the maximum of our regrets against all pure strategies. Equivalently, since the set of valid sequence weights is convex and its corners represent the pure strategies, we can maximize over all valid sequence weights instead:

$$\rho^P = \max_{\phi \in \mathcal{H}} \rho(P) \cdot \phi,$$

where

$$\mathcal{H} = \{\phi \mid A\phi = a, \phi \geq 0\}.$$

And just as before, if we write P_t for the empirical distribution of strategy profiles after t repetitions of the game, we will say that we have *no external regret* if $\rho^{P_t} \to 0$ as $t \to \infty$.

The condition that the empirical average regret approaches zero is equivalent to a condition on the regret vector: if we define the *safe set* to be

$$\mathcal{S} = \{\rho \mid \rho \cdot \phi \leq 0 \; \forall \phi \in \mathcal{H}\},$$

then the condition $\rho^{P_t} \to 0$ is equivalent to $\rho(P_t) \to \mathcal{S}$.

Figure 13.10 shows two examples of sequence weight sets $\mathcal{H}$ and safe sets $\mathcal{S}$. On the left is a normal-form game, Battle of the Sexes. Normal-form games have one sequence for each action, and the sequence weights must form a probability

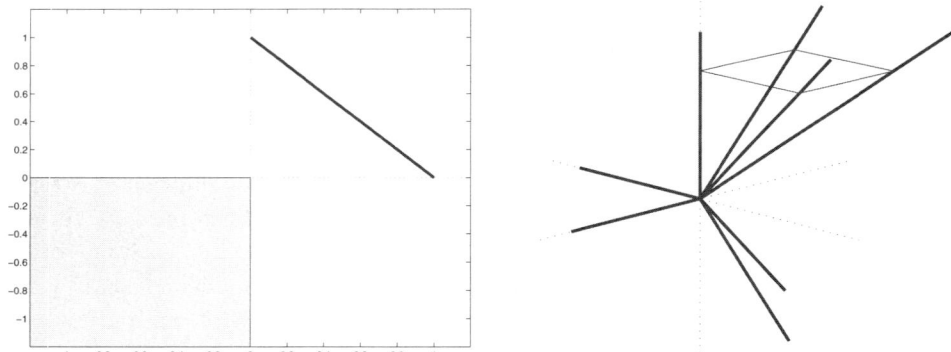

Figure 13.10 Sequence weight sets and safe sets. Left: Battle of the Sexes. Right: Return of the Battle of the Sexes.

distribution. So, the sequence weight set for Battle of the Sexes is the dark line segment $\phi_O + \phi_F = 1$, $\phi \geq 0$. The safe set is the negative orthant (shown shaded): if any component of ρ is positive, then there is some $\phi \in \mathcal{H}$ which has a positive dot product with ρ. The boundary rays of $\mathcal{S}$ are $(0, -1)$ and $(-1, 0)$; each of these rays is normal to an edge of the set $\bar{\mathcal{H}}$, which is defined as the union of $\lambda \mathcal{H}$ for all $\lambda \geq 0$. (For normal-form games, $\bar{\mathcal{H}}$ is the positive orthant.)

On the right is a Bayesian game, Return of the Battle of the Sexes. Bayesian games have one sequence for each pair of a type and an action, and the weights for each type must form a probability distribution. So, Return of the Battle of the Sexes has four sequences, VO, VF, WO, and WF; the weights for VO and VF must sum to 1, as must the weights for WO and WF. To avoid plotting in four dimensions, we will use the three-dimensional reduced representation of sequence weight vectors $(\phi_{VF}, \phi_{WF}, \phi_{VO} + \phi_{VF})$.[5] In the plot, the dotted lines are the coordinate axes. The feasible sequence weights are the square in the first octant. We have connected each corner of the square to the origin with a heavy line; these lines are the boundary rays of $\bar{\mathcal{H}}$. The remaining heavy lines are the boundary rays of the set $\mathcal{S}$: they are

$$\begin{pmatrix} -1 \\ 0 \\ 0 \end{pmatrix} \quad \begin{pmatrix} 0 \\ -1 \\ 0 \end{pmatrix} \quad \begin{pmatrix} \sqrt{2} \\ 0 \\ -\sqrt{2} \end{pmatrix} \quad \begin{pmatrix} 0 \\ \sqrt{2} \\ -\sqrt{2} \end{pmatrix}.$$

Each one of these boundary rays is orthogonal to one face of the set $\bar{\mathcal{H}}$.

As with normal-form games, there are several different algorithms which achieve no external regret in extensive-form games. An example of such an algorithm is extensive-form regret matching: take the vector $\rho(P_t)$, project it onto $\bar{\mathcal{H}}$ by minimum Euclidean distance, and normalize to get a valid sequence weight vector. (The projection requires solving a convex quadratic program, and so can be done in polynomial time. There is a possibility that the closest point to $\rho(P_t)$ in $\bar{\mathcal{H}}$ is the origin, which can't be normalized; in this case the algorithm selects its strategy arbitrarily.)

Extensive-form regret matching achieves average regret $O(1/\sqrt{t})$ after t trials (with constants that depend on the size of the largest sequence weight vector in $\mathcal{H}$). If we apply extensive-form regret matching to a normal-form game such as Battle of the Sexes, it is identical to the original regret matching algorithm: $\bar{\mathcal{H}}$ is the positive orthant, so the projection operation just zeros out the negative coordinates of $\rho(P_t)$.

Some additional no-regret algorithms are described in Gordon (2005). And, the FPL algorithm of Kalai and Vempala can be used in conjunction with the analysis of the current section to produce another no-regret algorithm for extensive-form games.

As is the case for normal-form games, playing two no-external-regret algorithms against each other in a constant-sum extensive-form game will lead to minimax equilibrium payoffs. The algorithms' strategies may not converge, but their average strategies over time will converge to their respective sets of minimax strategies.

13.5.4   Learning in Stochastic Games

If we are learning in a stochastic game, our output is a policy π which maps information sets (also called *states*) to distributions over actions. We can represent π directly as a parameterized function, $\pi(s;\theta)$, with adjustable parameters θ. Or, as described below, we can represent π indirectly through an evaluation function.

If we represent π directly then it turns out that we can compute the gradient

$$g_i(\theta) = \frac{d}{d\theta} E(r_i \mid \pi(\cdot;\theta))$$

by observing sequences of play under policy $\pi(\cdot;\theta)$. (For details see, for example, Sutton et al., 2000, .) Using this gradient we can easily implement a gradient ascent algorithm:

$$\theta^{(t+1)} = \theta^{(t)} + \alpha^{(t)} g_i(\theta^{(t)}).$$

This gradient ascent algorithm is called *policy gradient.*

Policy gradient generalizes gradient ascent in one-step games, since we can always take θ to be a probability distribution over pure strategies. As such, policy gradient inherits many of the properties of regular gradient ascent: for example, with a fixed learning rate two policy-gradient players can cycle forever around an equilibrium.

In addition, because of the more general policy representation, policy gradient has the problem of local optima: even if our current gradient is zero (or more generally normal to any active constraints on θ), there may be other representable policies which achieve better reward.

Despite the problem of local optima, policy gradient algorithms have been applied with some success to large-scale learning problems. One of the reasons for their success is their low requirements for information: they merely need to observe sequences of states and rewards generated from the current policy. In particular, they do not need to see the reward for any state-action pairs which the player did not experience.

Instead of representing its policy directly, a learner could represent an *evaluation function* instead. An evaluation function (also called a *value function*) maps information sets to estimates of the player's future payoffs.

If we were given a perfect value function (i.e., if we knew the player's expected payoff from any of its information sets) then the sequential learning problem would be decoupled into multiple one-step learning problems, one for each information set: by adding the player's immediate payoff to its expected future payoff (as reported by the value function), we can compute the total expected future payoff if the player chooses any particular action.

A simple algorithm for learning a value function is the method of temporal differences, also called TD(0). In TD(0) we are given a compact representation of the value function with adjustable parameters θ, say $V(s; \theta)$ for state s. If we observe a transition from state s to state s' with payoff r, we compute the temporal difference error

$$d = V(s; \theta) - [r + V(s'; \theta)]$$

and the gradient

$$g = \frac{d}{d\theta} V(s; \theta),$$

and then update

$$\theta \leftarrow \theta + \alpha g d.$$

The temporal difference error measures how far our prediction of total payoff at state s was from the (presumably more accurate) prediction which is available one step later at s'.

We can generalize TD(0) so that it uses information from multiple time steps to compute its update. The resulting algorithm, TD(λ), uses the parameter $\lambda \in [0, 1]$ to specify how to weight the different time steps: $\lambda = 0$ puts weight only on the current time step, while $\lambda = 1$ puts equal weights on all future time steps. TD(λ) for $0 < \lambda < 1$ can have lower variance in its updates than TD(0). (It is not obvious that this is true, because TD errors at adjacent time steps are not independent, but the lower variance has been observed experimentally.) For more information about TD(λ), see Sutton and Barto (1998).

Temporal difference learning has been proved to converge in some circumstances. For example, if we represent our value function using a linear combination of fixed basis functions, and if we train our TD learner on sequences of states and rewards sampled using a fixed vector of policies for all players, and if we decrease our learning rate over time, our value function will converge and have bounded error. In general, though, TD learning is often used in situations where its convergence guarantees don't apply. The most important determiner of success appears to be the use of sequences of states and rewards sampled using a fixed or slowly changing policy for each learner.

Basic TD learning doesn't attempt to model the other players directly. We can define a version of TD learning which assumes known motivations for the other players. For example, if we know that the other players are out to get us, we can construct a zero-sum game at each time step using our current value function and solve it to get an update for our parameters. The resulting algorithm is called minimax Q-learning; for details, see Littman (2001). That paper also describes a generalization of minimax Q learning called friend-or-foe Q-learning, in which each player is treated either as an enemy or an ally.

13.6 Examples

So far in this chapter we have discussed algorithms that learn in "small" games. (By "small" we mean games that can easily fit into the memory of our computer when represented in the format needed for our learning algorithm.) Most real-world games, of course, are not small.

In real games we need to deal with several new issues, all of which are poorly understood. First, we need to worry about approximation: we will not be able to represent every possible policy or every possible value function, so we need to design approximate representations. In addition to being able to represent good policies or value functions in our games, our representations must be compatible with whatever learning algorithm we use: if we naively mix a learning algorithm with an approximate representation, the combination can fail in unexpected ways.

Second, the problem of exploration becomes a pressing issue. In large games a clever opponent can hide whole regions of the state space from us while we are learning.

Finally, there is the question of negotiation. If we try to learn, there is always the risk that one of the other players will try to teach us. For example, in the Battle of the Sexes, if the opera-loving player always goes to the opera, then most learning algorithms will end up picking the opera as well. In more complicated games, groups of players may cooperate to guide other players toward desired strategies; it can be difficult for a player even to pick which coalition to join, much less to figure out how to convince the other players to pick the right strategies.

Clearly there are a lot of problems which game theorists don't yet know how to solve. But, game-theoretic learning algorithms can still be helpful as part of a larger intelligent system. In this section we will examine some case studies of game-theoretic learning algorithms in real-world learning problems.

13.6.1 Board Games

Two of the most famous examples of learning in games are Samuels's checker player Sutton and Barto (1998) and Tesauro's TD-Gammon (Tesauro, 1994). More recently, Baxter et al. (1997) described a system called KnightCap which learns to play chess.

All three of these systems learn by some variation of the method of temporal differences (section 13.5.4). That is, they maintain an evaluation function which estimates how likely they are to win the game starting from different board positions; based on experience from repeated play of the game, they adjust the evaluation function so that it more accurately predicts their actual winning percentage.

Both the checker and backgammon learners use straightforward $TD(\lambda)$ in self-play to update their evaluation functions. That is, suppose we write $V(x; \theta)$ for the evaluation function on board x with adjustable parameters θ. Then if we observe sequence of positions $x_1, x_2, x_3, \cdots$, the learners will update θ to try to bring $V(x_1; \theta)$ closer to

$$(1 - \lambda)(V(x_2) + \lambda V(x_3) + \lambda^2 V(x_4) + \dots). \qquad (13.22)$$

Samuels's program represented V with a linear combination of 31 human-designed features, and reportedly learned to play a decent amateur game of checkers. Tesauro's program represented V with a single-hidden-layer neural network, and learned to play expert-level backgammon. In fact, a version which added a few ply of look-ahead on top of the learned value function was able to compete at the level of the top human backgammon players in the world.

Encouraged by the success at backgammon, several people tried applying temporal difference learning to other games such as chess and go. Initial attempts were not successful; apparently backgammon has properties which make it more amenable to temporal difference learning than other games. It is not clear exactly what these properties are, but one possibility is backgammon's randomness: human players can't follow lines of play very many moves into the future because they can't consider all possible combinations of rolls of the dice. If this is true, then backgammon players have to rely more on pattern recognition than on deep search, making the learning problem easier.

In contrast to previous programs, the chess learning program KnightCap was able to improve its rating on Internet chess servers from 1650 to 2100 by learning from three days worth of experience. Further learning and optimizations such as an opening book brought its rating to 2500. The Internet ratings are not directly comparable to FIDE ratings, but they do correspond to very strong human-level play.

KnightCap's success can be attributed to three main differences from previous experiments. The first difference is its learning algorithm, called TD-Leaf(λ), which combines $TD(\lambda)$ with minimax look-ahead search. Suppose that our current position is x_1. Suppose we have searched forward some number of moves and found that the best sequence of play (the so-called *principal variation*) leads to the board position x_1^{leaf} at the fringe of the search tree. Similarly, suppose that the next position is x_2, and searching from x_2 leads to x_2^{leaf}. Then instead of updating θ to move $V(x_1; \theta)$ closer to an average of future values as in equation 13.22, TD-Leaf(λ) updates

$V(x_1^{\text{leaf}};\theta)$ to be closer to

$$(1-\lambda)(V(x_2^{\text{leaf}}) + \lambda V(x_3^{\text{leaf}}) + \lambda^2 V(x_4^{\text{leaf}}) + \dots). \qquad (13.23)$$

The second difference from previous experiments is that KnightCap was built on top of a fairly sophisticated chess-playing program. This program had a cleverly designed board representation from which it was able to calculate informative features; and its search algorithm, even before learning, was able to expand relevant portions of the game tree. The initial 1650 rating corresponded to a version of the program whose evaluation function depended only on material value, i.e., on how many pieces of each type were on the board.

The last difference is that, unlike previous programs, KnightCap did not learn by self-play. Instead it played human opponents who connected to the chess servers. This difference helped KnightCap explore better. Unlike backgammon, chess is deterministic. So, programs which learn by self-play can get stuck playing the same game over and over. Unpredictable human opponents help to jolt the learner out of local optima. Also, human players tend to select opponents near their own level of play, so as KnightCap improved it attracted more skilled opponents. This sequence of better and better opponents helped to *shape* KnightCap's learning: if it had played against the better opponents from the start, it might not have been able to understand why it was losing.

13.6.2 Poker

In contrast to board games like chess and backgammon, card games like poker and bridge require learners to reason about hidden information. It has been only recently that computers have been able to learn to play these sorts of games well; in this section we will look at a program which learns to play the game of poker.

Perhaps the best current artificial poker player is SparBot, developed by the University of Alberta games group (see Billings et al., 2003, which describes a family of prototypes for SparBot called PsOpti). SparBot plays Texas Hold'em, a popular and strategically interesting poker variant in which players combine hidden private cards with face-up common cards to form the best possible hand. SparBot plays the two-player version, which places a strong emphasis on deception and bluffing, but which is simpler than multiplayer poker because it is zero-sum. SparBot computes its strategy in advance and does not adapt its play online, but many of the techniques in SparBot are relevant to the question of learning while playing (and in fact the authors of SparBot are working on an adaptive cousin called VexBot).

Reduced versions of poker can be solved exactly. For example, the game of one-card poker (in which players are dealt a single card and bet on who holds the highest one) has $4k$ sequences per player with a k-card deck. No-regret learning algorithms such as the ones in section 13.5.3 are easily able to learn minimax strategies in self-play (Gordon, 2005). These strategies exhibit many of the same features as do strategies in full-size poker variants, such as bluffing and slow-playing; see fig. 13.9.

Slightly larger variants such as "Rhode Island Hold'em" (Shi and Littman, 2001) can be approximately solved with tricks like grouping together sets of similar hands and learning a single strategy for all of them.

To find an approximate minimax equilibrium for the full game of Texas Hold'em, SparBot uses several tricks. The first is grouping: at each stage of the game it evaluates the strength and the potential of its hand, divides all hands into a manageable number of groups based on these evaluations, and plays the same mixture of actions for all hands in the same group. (Hand strength measures the probability of winning with the current cards, while hand potential measures how much hand strength is likely to increase with future cards. For example, a full house is a strong hand, while a hand with four hearts and a spade may be weak but has good potential since another heart will produce a flush.)

The second trick is reducing the number of bets in a round of betting. The full game of Hold'em allows up to four bets per player per betting round, but restricting to three bets per player reduces the branching factor of the game tree without changing strategy too much.

The last trick is to split the game into temporally overlapping subgames. We can imagine a version of Hold'em in which the players give up control after the second-to-last round of betting: after this round is complete, we turn up the last card and have the players follow a predetermined strategy in the last betting round. The strategy for this reduced game will be similar in the early rounds to the strategy for the full game, but will start to diverge toward the end of the game. If we now fix the initial part of the strategy for this game, we arrive at another reduced version of Hold'em, in which the players gain control after the first round of betting and play all the way to the end.

By combining these tricks, SparBot reduces the size of the games it must solve from about 10^{18} to about 10^7 information sets. This size is small enough that the techniques of section 13.5 are able to find minimax equilibria in a reasonable amount of time. SparBot has played against a number of experienced and master human poker players, and is generally able to hold its own against them.

13.6.3   Robotics

Recently, robotic games such as RoboCup soccer have become popular. While the role of learning in most RoboCup teams is still small, researchers have examined various learning problems in robot soccer. For example, Stone et al. looked at the problem of deciding whether to try to shoot a goal or pass to a teammate (Stone and Veloso, 1996), and at the problem of playing keep-away (Stone et al., 2005). And, the University of Maryland developed a team whose entire high-level strategy was learned through self-play (Luke, 1998).

Bowling and Veloso (2003) looked at another interesting subproblem of RoboCup: maneuvering to keep an opponent away from a particular spot on the field. They set up two robots as shown in figures 13.11 and 13.12; robot A scores a point if it enters the dotted circle, while robot B scores a point if it keeps robot A

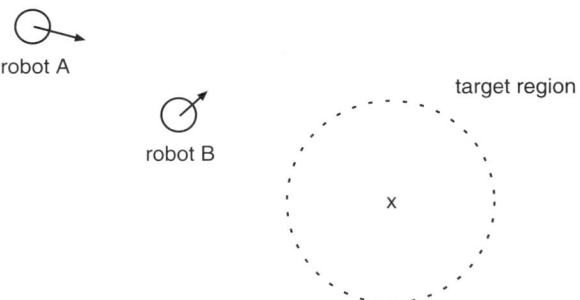

Figure 13.11   The keep-out game. Robot A tries to get to the target region, while robot B tries to block robot A. State is $x, y, \dot{x}, \dot{y}$ for each robot, minus one degree of freedom for rotational symmetry.

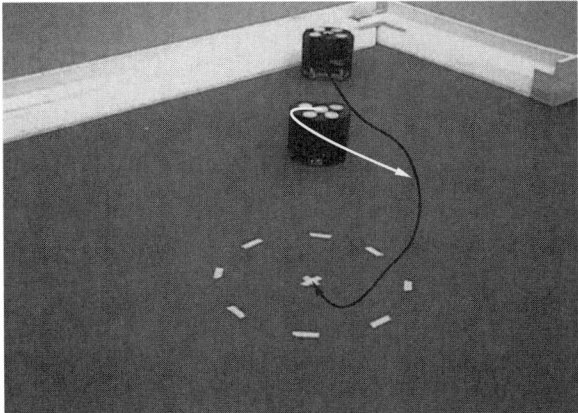

Figure 13.12   Robots playing the keep-out game. Reprinted with permission from Michael Bowling.

out of the circle for a preset length of time.

Representing an evaluation function for this game would be complex but not impossible. The state is described by seven continuous variables: the x and y positions and the x and y velocities of the two robots make eight degrees of freedom, but we lose one degree of freedom because the world is rotationally symmetric.

Instead of learning an evaluation function, though, the robots learned their policies directly. They picked from seven actions (defined by seven carefully selected target points which depended on the locations of the robots) ten times a second, according to a randomized policy with about 70,000 adjustable parameters. By observing the states and rewards encountered in practice games, they estimated the gradient of their payoffs with respect to the policy parameters, and adjusted the parameters according to a gradient ascent algorithm like the ones in section 13.3.2. While the convergence guarantees of section 13.3.2 do not hold in this more complicated situation, the robots were able to improve their performance both in simulation and in physical game-play.

13.7   Discussion

This chapter has examined the problem of learning in games. We divided games into one-step vs. sequential, and into constant-sum vs. general sum. Both of these divisions separate easier from more difficult learning problems: constant-sum one-step games are relatively well understood, while general-sum sequential games are an active research area.

For one-step games, we defined minimax equilibrium, Nash equilibrium, and correlated equilibrium. These different equilibria represent possible or desirable outcomes when several learning players play against one another. We described off-line algorithms which can compute these various equilibria from the description of a game. We also described learning algorithms, with varying information requirements, which can learn to beat fixed opponents, find various sorts of equilibria under various circumstances, or guarantee minimum performance levels against arbitrary opponents.

For sequential games, we defined subgame-perfect Nash equilibrium and extensive-form correlated equilibrium. We again described off-line algorithms which can compute equilibria, and learning algorithms with various types of guarantees, but in general the guarantees for learning sequential games are weaker than the corresponding guarantees for one-step games.

Finally, we examined several case studies of learning algorithms applied to real-world-sized games. By combining and extending the algorithms described in this chapter, these learners have been able to achieve human-level play in difficult learning environments.

In conclusion, for the simplest games, there are many successful learning algorithms with appealing performance guarantees. For more complicated games it is not even completely clear what representations we can use or what performance guarantees we would like to prove; still, there are successful practical learning algorithms for these games, as well as the beginnings of a theoretical analysis.

Notes

[1] Despite the apparently deterministic always-red-in-my-direction behavior of the traffic lights where I live.

[2] As originally defined, the weighted majority algorithm bases its plays on the rewards for the pure strategies rather than the regrets, but the two definitions are equivalent.

[3] Since we have assumed that the game tree is finite. Infinite game trees are also possible, but we will not discuss them here.

[4] We do not need to include a separate constant term since the constraints $A\phi = a$ are not trivial (they always include at least one constraint saying that the weights of the actions leaving some initial state sum to 1).

[5] We are also using the dot product which corresponds to this reduced representation, so lines which appear orthogonal in the figure are actually orthogonal.

14 Learning Observable Operator Models via the Efficient Sharpening Algorithm

Herbert Jaeger, Mingjie Zhao, Klaus Kretzschmar, Tobias Oberstein, Dan Popovici, and *Andreas Kolling*

Hidden Markov models (HMMs) today are the method of choice for black-box modeling of symbolic, stochastic time series with memory. HMMs are usually trained using the expectation-maximization (EM) algorithm. This learning algorithm is not entirely satisfactory due to slow convergence and the presence of many globally suboptimal solutions. Observable operator models (OOMs) present an alternative. At the surface OOMs appear almost like HMMs: both can be expressed in structurally identical matrix formalisms. However, the matrices and state vectors of OOMs may contain negative components, whereas the corresponding components in the world of HMMs are nonnegative probabilities. This freedom in sign gives OOMs algebraic properties that radically differ from those of HMMs, and leads to novel learning algorithms that are fast and yield asymptotically correct model estimates. Unfortunately, the basic versions of these algorithms are statistically inefficient, which has so far precluded a widespread use of OOMs. This chapter, first, gives a tutorial introduction to OOMs, and second, introduces a novel approach to OOM estimation called *efficiency sharpening* (ES). The ES method is iterative. In each iteration, the model estimated in the previous round is used to construct an estimator with a better statistical efficiency than the previous one. The computational load per iteration is comparable to one EM iteration, but only two to five iterations are typically needed. The chapter gives an analytical derivation of the ES principle and describes two learning algorithms that build on this principle, a simple "poor man's" version and a more complicated but superior version which is based on a suffix-tree representation of the training string. The quality of the latter algorithm is demonstrated on a task of learning a model of a long belletristic text, where OOM models markedly outperform HMM models in quality, requiring only a fraction of learning time.

The algorithms described in this chapter are available in a free Matlab implementation at http://www.faculty.iu-bremen.de/hjaeger/oom_research.html.

14.1 Introduction

Observable operator models (OOMs) are mathematical models of stochastic processes. In their basic version, they describe stationary, finite-valued, discrete-time processes—in other words, symbol sequences. We will restrict ourselves to this basic type of processes in this chapter.

A number of models for stochastic symbol sequences are widely used. Listed in order of increasing expressiveness, the most common are elementary Markov chains, higher-order Markov chains, and hidden Markov models (HMMs) (Bengio, 1999; Rabiner, 1990). Well-understood learning algorithms to estimate such models from data exist. Specifically, HMMs are usually trained by versions of the expectation-minimization (EM) algorithm (Dempster et al., 1977). HMMs currently mark the practical limit of analytical and algorithmic tractability, which has earned them a leading role in application areas such as speech recognition (Jelinek, 1998), biosequence analysis (Durbin et al., 1998) and control engineering (Elliott et al., 1995).

In this chapter we wish to establish OOMs as a viable alternative to HMMs—albeit as yet only for the case of modeling stationary symbol processes. We see three main advantages of OOMs over HMMs:

- The mathematical theory of OOMs is expressed purely in terms of linear algebra and admits a rigorous, transparent semantic interpretation.

- OOMs properly generalize HMMs, that is, the class of processes that have finite-dimensional OOM properly includes the processes characterized by finite-dimensional HMMs.

- New learning algorithms for OOMs, derived from a novel principle which we would like to call *efficiency sharpening* (ES), yields model estimates in a fraction of the computation time that EM-based algorithms require for HMM estimation. Furthermore, on most data sets that have been investigated so far, the OOM models obtained via ES are markedly more accurate than HMM models.

However, at the current early state of research there remain also painful shortcomings of OOMs. Firstly, the OOMs learned from data are prone to predict negative "probabilities" for some (rare) sequences, instead of small nonnegative values. Currently only heuristic methods to master this problem are available. Secondly, our OOM learning algorithms tend to become instable for large model dimensions. Again, heuristic coping strategies exist, which are detailed out in this chapter.

This chapter has two main parts. The first part (sections 14.2 through 14.9) contains a tutorial introduction to the basic theory of OOMs, including the basic version of the learning algorithm. This material has been published before (Jaeger, 2000) but has been almost completely rewritten with a more transparent notation and a new didactic approach. We hope that this tutorial part becomes the standard introductory text on OOMs. The second part (sections 14.10 through 14.15), as an original contribution, establishes the ES principle and two learning algorithms are derived from it. Two case studies round off the presentation.

14.2 The Basic Ideas behind Observable Operator Models

In this section we first describe the essence of OOMs in informal terms and then condense these intuitions into a mathematical formalism.

Envision a soccer-playing robot[1] engaged in a soccer game. In order to play well, the robot should make predictions about possible consequences of its actions. These consequences are highly uncertain, so in one way or the other they must be internally represented to the robot as a distribution of future trajectories (fig. 14.1a).

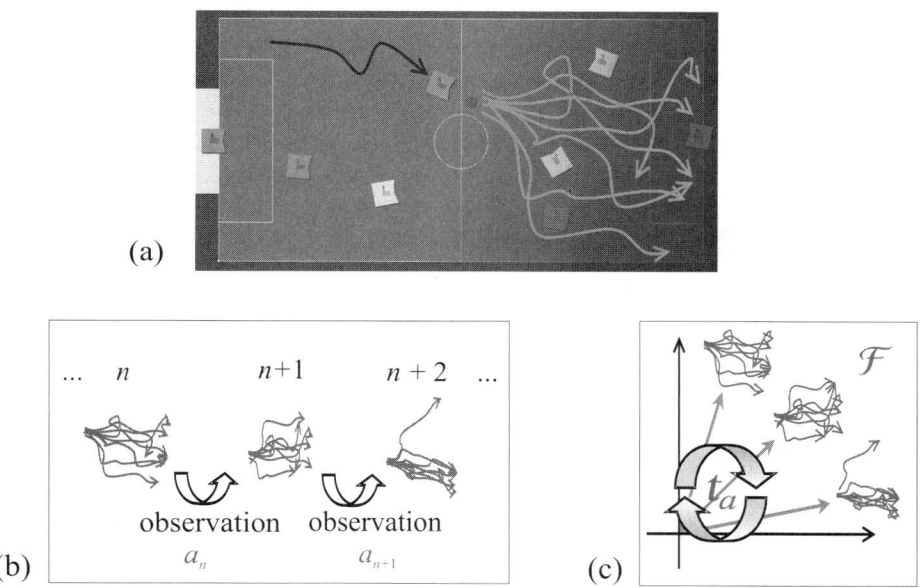

Figure 14.1 (a) A robot's future depicted as a "spaghetti bundle" of expected possible future trajectories. (b) The robot's expected futures change due to incoming observations a_n of information. (c) An operator τ_a associated with an observation a yields an update operation on the vector space of future distributions.

Soccer is a dynamic game, and the robot has to update its expectations about the future in an update cycle from time n to $n+1$, assuming a unit cycle time. OOMs are a mathematical model of this kind of update operation. Clearly the update is steered by the information that the robot collects during an update interval. This comprises incoming sensory information, communications from other robots, but also the robot's own issued motor commands or even results from some planning algorithms that run on it—in short, *everything* that is of some informational value for the expected future. We comprise all of these bits of information under the term of an *observation*. At the root of OOMs lies the assumption that *nothing but* the observation a_n between n and $n+1$ controls the update of future expectations, and that such update operations can be *identified with* observations (fig. 14.1b). Thus, in OOMs we have for every possible observation one operator that can be used to update expected futures. This identification of observations with update operators has given OOMs their name, *observable operator* models.

Mathematically, a future of a stochastic process is a probability distribution on the set of potential future trajectories after the current time n. Such distributions can be specified by a real-valued function f in various ways. For instance, f may be a probability density funciton, or one may use a function f which assigns probabilities to finite-length future sequences, that is, a function on words over the observation alphabet. At this point we do not care about the particular format of f, we only assume that some real-valued function can describe a future's distribution (for general abstract treatment see Jaeger, 1999)).

The real-valued functions f over some set can be added and multiplied with scalars and hence span a vector space F. Identifying observations with update operators on futures, and identifying futures with functions f which are vectors in F, we find that observations can be seen as operators on F. In the OOM perspective, each possible observation a is identified with an operator t_a on F (fig. 14.1c).

The key to OOMs is the observation that these *observable operators* are linear. We now give a formal treatment of the case where the stochastic process is of a particular simple kind, namely, discrete-time, finite-valued, and stationary. Let $(X_n)_{n\in\mathbb{N}}$, or for short, (X_n), be a stationary, discrete-time stochastic process with values in a finite alphabet $O = \{a^1, \dots, a^\alpha\}$ of possible observations.

We shall use the following shorthand. For $P(X_n = a_0, \dots, X_{n+r} = a_r)$ we write $P(a_0 \dots a_r)$ or more briefly $P(\bar{a})$. For conditional probabilities $P(X_n = b_0, \dots, X_{n+r} = b_r \mid X_{n-s} = a_0, \dots, X_{n-1} = a_{-1})$ we write $P(b_0 \dots b_r \mid a_0 \dots a_{s-1})$ or $P(\bar{b} \mid \bar{a})$. Unconditional probabilities $P(\bar{a})$ can be seen as conditional probabilities conditioned by the empty sequence ε, that is, $P(\bar{b}) = P(\bar{b} \mid \varepsilon)$.

The distribution of (X_n) is uniquely characterized by the probabilities of finite substrings, i.e., by all probabilities of the kind $P(\bar{b})$, where $\bar{b} \in O^*$ (O^* denotes the set of all finite strings over O including the empty string).

For every $\bar{a} \in O^*$, we define a real-valued function

$$f_{\bar{a}} : O^* \to \mathbb{R}, \tag{14.1}$$

$$\bar{b} \mapsto \begin{cases} P(\bar{b} \mid \bar{a}), & \text{if } P(\bar{a}) \neq 0, \\ 0, & \text{if } P(\bar{a}) = 0, \end{cases}$$

with the understanding that $f_{\bar{a}}(\varepsilon) = 1$ if $P(\bar{a}) > 0$, else it is 0.

A function $f_{\bar{a}}$ describes the future distribution of the process after an initial realization $\bar{a}$. In our robot illustration in fig. 14.1, $\bar{a}$ would correspond to the past that the robot has in its short-term memory (symbolized by the blue trajectory), and $f_{\bar{a}}$ would correspond to the "spaghetti bundle" of future trajectores, as anticipated at that moment. We call these $f_{\bar{a}}$ the *prediction functions* of the process.

Let F be the functional vector space spanned by the prediction functions.

Thus F can be seen as the (linear closure of the) space of future distributions of the process (X_t).

We now define the observable operators. In order to specify a linear operator on a vector space, it suffices to specify the values the operator takes on a basis of the vector space. Choose a set $(f_{\bar{a}_i})_{i \in I}$ of prediction functions that is a basis of F. Define, for every $a \in O$, a linear *observable operator* $t_a : F \to F$ by putting

$$t_a(f_{\bar{a}_i}) = P(a \,|\, \bar{a}) f_{\bar{a}_i a} \tag{14.2}$$

for all $i \in I$ ($\bar{a}a$ denotes the concatenation of the sequence $\bar{a}$ with a). It is easy to verify (Jaeger, 2000) that equation 14.2 carries over from basis elements $f_{\bar{a}_i}$ to all $\bar{a} \in O^*$:

Proposition 14.1
For all $\bar{a} \in O^*$, $a \in O$, the linear operator t_a satisfies the condition

$$t_a(f_{\bar{a}}) = P(a \,|\, \bar{a}) f_{\bar{a}a}. \tag{14.3}$$

Furthermore, the definition of observable operators does not depend on the choice of basis of F.

Intuitively, the observable operator t_a describes the change of knowledge about a process's future due to an incoming observation of a—which is just the idea of our update operators. A new ingredient that we find here is that the updated future distribution $f_{\bar{a}a}$ becomes weighted by $P(a \mid \bar{a})$. This circumstance can be used to express the probability of a sequence $P(a_0 \ldots a_r)$ in terms of the operators $t_{a_0}, \ldots, t_{a_r}$. Let $\sigma : F \to \mathrm{R}$ be the linear function that returns 1 on all basis vectors $f_{\bar{a}_i}$. Then the following proposition holds (proof in Jaeger, 2000).

Proposition 14.2
For all $a_0 \ldots a_r \in O^*$,

$$P(a_0 \ldots a_r) = \sigma \, t_{a_r} \cdots t_{a_0} \, f_\varepsilon. \tag{14.4}$$

Note that equations 14.3 and 14.4 are valid for any choice of basis vectors $f_{\bar{a}_i}$. Equation 14.4 is the fundamental equation of OOM theory. It reveals how the distribution of any stationary symbol process can be expressed purely by means of linear algebra. Furthermore, the observable operators and f_ε are uniquely determined by the the distribution of (X_t). This leads to the following definition:

Definition 14.1
Let $(X_n)_{n \in \mathrm{N}}$ be a stationary stochastic process with values in a finite set O. The structure $(F, (t_a)_{a \in O}, f_\varepsilon)$ is called the *observable operator model* of the process. The vectors $f_{\bar{a}}$ are called *states* of the process; the state f_ε is called the *initial state*. The vector space dimension of F is called the *dimension of the process*.

We will soon introduce matrix representations of OOMs. If we wish to distinguish the abstract OOMs introduced above from matrix representations, we will speak of "functional" vs. "matrix" OOMs, respectively.

We have treated only the discrete-time, discrete-value, stationary case here. However, OOMs can be defined in a similar way also for nonstationary, continuous-time, arbitrary-valued processes (Jaeger, 1999). It turns out that in those cases the resulting observable operators are linear too. In the sense of updating prediction functions, the change of knowledge about a process due to incoming observations is a linear phenomenon.

14.3   From HMMs to OOMs: Matrix Representations of OOMs

If one wishes to carry out concrete computations, one has to work with finite-dimensional matrix representations of OOMs. Instead of deriving them from the abstract definition 14.1, we will introduce matrix representations of OOMs in a very different way, by showing how they can be obtained as a generalization of HMMs.

A basic HMM specifies the distribution of a discrete-time, discrete-value stochastic process $(Y_n)_{n \in \mathbb{N}}$, where the random variables Y_n have outcomes in an alphabet $O = \{a^1, \dots, a^\alpha\}$. To specify $(Y_n)_{n \in \mathbb{N}}$, first a Markov chain $(X_n)_{n \in \mathbb{N}}$ is considered that produces sequences of *hidden* states from a state set $\{s_1, \dots, s_m\}$. Second, when the Markov chain is in state s_j at time n, it "emits" an observable outcome a_i with a time-invariant probability $P(Y_n = a_i \mid X_n = s_j)$.

We now represent a HMM in a matrix formalism that is a bit different from the one customarily found in the literature. The Markov chain state transition probabilities are collected in an $m \times m$ stochastic matrix M which at position (i, j) contains the transition probability from state s_i to s_j. For every $a \in O$, we collect the emission probabilities $P(Y = a \mid X = s_j)$ in the diagonal of an $m \times m$ matrix O_a that is otherwise zero.

In order to fully characterize a HMM, one must supply an initial distribution $w_0 = (P(X_0 = s_1), \dots, P(X_0 = s_m))^\top$ (superscript $^\top$ denotes transpose of vectors and matrices). The process described by the HMM is stationary if w_0 is an invariant distribution of the Markov chain (Doob, 1953), namely, if it satisfies

$$M^\top w_0 = w_0. \qquad (14.5)$$

We consider only stationary processes here. The matrices M, O_a and and the vector w_0 can be used to compute the probability of finite observation sequences. Let $\mathbf{1} = (1, \dots, 1)$ denote the m-dimensional row vector of units, and let $T_a := M^\top O_a$. Then the probability to observe the sequence $a_0 \dots a_r$ among all possible sequences of length $r + 1$ is obtained by

$$P(a_0 \dots a_r) = \mathbf{1} T_{a_r} \cdots T_{a_0} w_0. \qquad (14.6)$$

Equation 14.6 is a matrix notation of the well-known forward algorithm for determining probabilities of observation sequences in HMMs. Proofs may be found in (Ito et al., 1992) and (Ito, 1992).

Matrix M can be recovered from the operators T_a by observing that

$$M^\top = M^\top \cdot \mathbf{id} = M^\top (O_{a^1} + \cdots + O_{a^\alpha}) = T_{a^1} + \cdots + T_{a^\alpha}, \tag{14.7}$$

where $\mathbf{id}$ denotes the identity matrix. Equation (14.6) shows that the distribution of the process (Y_t) is specified by the operators T_a and the vector w_0. Thus, the matrices T_a and w_0 contain the same information as the original HMM specification in terms of M, O_a, and w_0. Namely, one can rewrite a HMM as a structure $(\mathrm{R}^m, (T_a)_{a \in O}, w_0)$, where R^m is the domain of the operators T_a.

From here one arrives at the definition of a finite-dimensional OOM in matrix representation by (i) relaxing the requirement that $M^\top$ be the transpose of a stochastic matrix, to the weaker requirement that the columns of M^T each sum to 1, and by (ii) requiring from w_0 merely that it have a component sum of 1. That is, negative entries are now allowed in matrices and vectors, which are forbidden in the stochastic matrices and probability vectors of HMMs. Using the symbol τ in OOMs in places where T appears in HMMs, and introducing $\mu = \sum_{a \in O} \tau_a$ in analogy to equation 14.7 we get the following definition.

Definition 14.2

An m-dimensional (matrix) OOM is a triple $\mathcal{A} = (\mathrm{R}^m, (\tau_a)_{a \in O}, w_0)$, where $w_0 \in \mathrm{R}^m$ and $\tau_a : \mathrm{R}^m \to \mathrm{R}^m$ are linear maps represented by matrices, satisfying three conditions:

1. $\mathbf{1} w_0 = 1$;

2. $\mu = \sum_{a \in \mathcal{O}} \tau_a$ has column sums equal to 1;

3. for all sequences $a_0 \ldots a_r$ it holds that $\mathbf{1} \tau_{a_r} \cdots \tau_{a_0} w_0 \geq 0$.

Conditions 1 and 2 reflect the relaxations (i) and (ii) mentioned previously, while condition 3 ensures that one obtains nonnegative values when the OOM is used to calculate probabilities. While the nonnegativity of matrix entries in HMMs guarantees nonnegativity of values obtained from the right-hand side (rhs) of 14.6, nonnegativity must be expressedly assured for OOMs. Unfortunately, for given operators $(\tau_a)_{a \in O}$ there exists no known way to decide whether condition 3 holds. This is our first encounter with the central unresolved issue in OOM theory, and we will soon hear more about (and suffer from) it.

Since concatenations of operators like $\tau_{a_r} \cdots \tau_{a_0}$ will be much used in the sequel, we introduce a shorthand notation: for $\tau_{a_r} \cdots \tau_{a_0}$ we also write $\tau_{a_0 \cdots a_r}$ (be aware of the reversal of indices) or even $\tau_{\bar{a}}$.

A matrix-based OOM specifies a stochastic process as in equation 14.4:

Proposition 14.3

Let $\mathcal{A} = (\mathrm{R}^m, (\tau_a)_{a \in O}, w_0)$ be an OOM according to the previous definition. Let $\Omega = O^\infty$ be the set of all right-infinite sequences over O, and A be the σ-algebra generated by all finite-length initial sequences on Ω. Then, if one computes the probabilities of finite-length sequences by

$$P_0(\bar{a}) = \mathbf{1}\tau_{\bar{a}}w_0, \tag{14.8}$$

where the numerical function P_0 can be uniquely extended to a probability measure P on (Ω, A), giving rise to a stochastic process $(\Omega, A, P, (X_n)_{n \in \mathbb{N}})$, where $X_n(a_1 a_2 \dots) = a_n$. If w_0 is an invariant vector of μ, i.e., if $\mu w_0 = w_0$, the process is stationary.

A proof can be found in Jaeger (2000). Since we introduce matrix OOMs here by generalizing away from HMMs, it is clear that every process that can be characterized by a finite-dimensional HMM can also be described by a matrix OOM of dimension at most the number of HMM hidden states.

Conversely, there exist processes that can be described by a matrix OOM, but that cannot be characterized by a finite-dimensional HMM. One way to construct examples of such processes is to design one of the operators τ_a to be a rotation of $\mathbb{R}^m$ by a nonrational angle ϕ. Such a rotation gives rise to a "probability oscillation," that is, the sequence $P(a \,|\, a^n)_{n \geq 0}$ converges to an oscillation with angular velocity ϕ (radian per unit time step). Intuitively, the reason why such a process cannot be modeled by an HMM is that a matrix describing a rotation needs to contain some negative entries. If a HMM for such a process would exist, reinterpreting it as an OOM according to the construction $T_a = M^\top O_a$ would yield a purely nonnegative matrix for the rotating operator, which is impossible. A concrete example of such a process (dubbed the "probability clock") and a proof that it is not a hidden Markov process was given in Jaeger (2000).

In section 14.2 we introduced abstract OOMs in a top-down fashion, by starting from a stochastic process and transforming it into its OOM. In this section we introduced matrix OOMs in a bottom-up fashion by abstracting away from HMMs. These two are related as follows (for proofs, see Jaeger, 2000, 1997a):

- A matrix OOM of matrix dimension m specifies a stochastic process of process dimension $m' \leq m$.

- A process of finite dimension m has matrix OOMs of matrix dimension m.

- A process of finite dimension m has no matrix OOMs of smaller matrix dimension.

When we refer to OOMs in the remainder of this chapter we mean matrix OOMs.

14.4   OOMs as Generators and Predictors

In this section we describe how an OOM can be used to generate a random sequence, and to compute the probabilities of possible continuations of a given initial sequence.

Concretely, assume that an OOM $\mathcal{A} = (\mathbb{R}^m, (\tau_a)_{a \in O}, w_0)$ describes a process $(X_n)_{n \geq 0}$, where $O = \{a^1, \dots, a^\alpha\}$. Then, the task is to use $\mathcal{A}$ to produce at times $n = 0, 1, 2, \ldots$ observations $a_0, a_1, a_2, \ldots$, such that (1) at time $n = 0$, the probability of producing a is equal to $P(X_0 = a)$, and (2) at every time step $n > 0$, the probability of producing a (after $a_0, \ldots, a_{n-1}$ have already been produced) is equal to $P(X_n = a \,|\, X_0 = a_0, \ldots, X_{n-1} = a_{n-1})$. We address conditions 1 and 2 in turn.

1. For generating the first symbol we need the probability vector $\mathbf{p}_0 = (P(X_0 = a^1) \cdots P(X_0 = a^\alpha))^\top$. This could be done by calculating $P(X_0 = a) = \mathbf{1}\tau_a w_0$ for all $a \in O$. A faster way is to precalculate the row vectors $\mathbf{1}\tau_a$ for all a, and assemble them in a matrix

$$\Sigma = \begin{bmatrix} \mathbf{1}\tau_{a^1} \\ \vdots \\ \mathbf{1}\tau_{a^\alpha} \end{bmatrix}, \tag{14.9}$$

and directly obtain

$$\mathbf{p}_0 = \Sigma \, w_0. \tag{14.10}$$

This probability vector is then used to randomly generate the symbol a_0 with the correct distribution.

2. In order to obtain $P(X_n = a \,|\, X_0 = a_0, \ldots, X_{n-1} = a_{n-1})$ we make use of 14.8:

$$P(X_n = a \,|\, X_0 = a_0, \ldots, X_{n-1} = a_{n-1})$$
$$= \mathbf{1}\tau_a \tau_{a_{n-1}} \cdots \tau_{a_0} w_0 \,/\, \mathbf{1}\tau_{a_{n-1}} \cdots \tau_{a_0} w_0$$
$$= \mathbf{1}\tau_a \Big(\frac{\tau_{a_{n-1}} \cdots \tau_{a_0} w_0}{\mathbf{1}\tau_{a_{n-1}} \cdots \tau_{a_0} w_0} \Big). \tag{14.11}$$

Introducing the notation

$$w_{a_0 \ldots a_{n-1}} = \frac{\tau_{a_{n-1}} \cdots \tau_{a_0} w_0}{\mathbf{1}\tau_{a_{n-1}} \cdots \tau_{a_0} w_0}, \tag{14.12}$$

equation 14.11 can be more concisely written as $P(X_n = a \,|\, X_0 = a_0, \ldots, X_{n-1} = a_{n-1}) = \mathbf{1}\tau_a w_{a_0 \ldots a_{n-1}}$. A vector $w_{\bar{a}}$ of the kind (eq. 14.12) that arises after a sequence $\bar{a}$ has been observed is called a *state vector* of an OOM. Note that state vectors have unit component sum. Again we can use Σ to obtain all of the probabilities $P(a^i \,|\, \bar{a})$ in a single operation:

$$\mathbf{p}_n = (P(a^1 \,|\, \bar{a}) \cdots P(a^\alpha \,|\, \bar{a}))^\top = \Sigma \, w_{\bar{a}}. \tag{14.13}$$

Observing that the next state vector can be obtained from the previous one by

$$w_{\bar{a}a} = \tau_a w_{\bar{a}} \,/\, \mathbf{1}\tau_a w_{\bar{a}}, \tag{14.14}$$

the entire generation procedure can be neatly executed as follows:

Step 1: State vector initialization: put $w = w_0$.

Step 2: Assume that at time n a state vector w_n has been computed, then determine the probability vector $\mathbf{p}$ of the $(n+1)$-st symbol as $\Sigma\, w_n$, and choose a_n according to that vector.

Step 3: Update the state vector by $w_{n+1} = \tau_{a_n} w_n \,/\, \mathbf{1}\tau_{a_n} w_n$ and resume at step 2.

Now we consider the task of predicting the probability $P(\bar{b}\,|\,\bar{a})$ of a continuation $\bar{b}$ of an initial sequence $\bar{a}$ that has already been observed. It is easy to see that an iterated application of 14.11 yields

$$P(X_{n+1} = b_{n+1}, \ldots, X_{n+r} = b_{n+r} \,|\, X_0 = a_0, \ldots, X_{n-1} = a_{n-1})$$
$$= \mathbf{1}\tau_{b_{n+r}} \cdots \tau_{b_{n+1}}\; w_{a_0 \cdots a_n}, \tag{14.15}$$

which in our shorthand notation becomes $P(\bar{b}\,|\,\bar{a}) = \mathbf{1}\tau_{\bar{b}}\, w_{\bar{a}}$. If one is interested in repeated predictions of the probability of a particular continuation $\bar{b}$ (for instance, an English word), then it pays to precalculate the row vector $\sigma_{\bar{b}} = \mathbf{1}\tau_{\bar{b}}$ and obtain $P(\bar{b}\,|\,\bar{a}) = \sigma_{\bar{b}}\, w_{\bar{a}}$ by a single inner product computation.

14.5   Understanding Matrix OOMs by Mapping Them to Functional OOMs

OOM states are conceptually quite different from HMM states. This conceptual issue is complicated by the circumstance that the term *state* is used in two different ways for HMMs. First, it may denote the finite set of *physical states* that the target system is assumed to take. Second, it is used for the *current* probability distribution over these physical states that can be inferred from a previous observation sequence. In both cases, the notion is connected to the assumed physical states of the target system. By contrast, OOM states represent the expectation about the system's *future* and *outwardly observable* development given an observed past. In no way do OOM states refer to any assumed physical state structure of the target system—they are purely epistemic, one might say. Incidentally, this agrees with the perspective of modern physics and abstract systems theory: "... a state of a system at any given time is the information needed to determine the behaviour of the system from that time on" (Zadeh, 1969). This perspective was constitutional for the construction of functional OOMs in section 14.2. We will now add further substance to this view by showing how matrix OOMs map to functional OOMs, and thereby how the finite state vectors of matrix OOMs represent the process's future. As by-products our investigation will yield a construction for minimizing the dimension of a matrix OOM, and an algebraic characterization of matrix OOM equivalence.

Definition 14.3
Let $\mathcal{A} = (\mathrm{R}^l, (\tau_a)_{a \in O}, w_0)$ be a matrix OOM of the process $(X_n)_{n \geq 0}$. Let $\mathcal{F} = (F, (t_a)_{a \in O}, f_\varepsilon)$ be the functional OOM of the same process. Let W be the linear subspace of R^l spanned by the state vectors $\{w_{\bar{a}}\,|\,\bar{a} \in O^*\}$. Let $\{w_{\bar{a}_1}, \ldots, w_{\bar{a}_d}\}$ be a basis of W. Define a linear mapping $\pi : w \to F$ through $\pi(w_{\bar{a}_i}) = f_{\bar{a}_i}$ $(i =$

$1, \ldots, d$). This mapping is called the *canonical projection* of $\mathcal{A}$.

This definition is independent of the choice of basis, and the canonical projection has the following properties:

Proposition 14.4

1. $\forall \, \bar{a} \in O^* \;\; \pi(w_{\bar{a}}) = f_{\bar{a}}$.

2. π is surjective.

3. $\forall \, w \in W \;\; \sigma \pi(w) = \mathbf{1}w$.

4. $\forall \, \bar{a} \in O^*, \; w \in W \;\; \pi(\tau_{\bar{a}} w) = t_{\bar{a}} \, \pi(w)$.

The proof of properties 1–3 is given in Jaeger (1997a), the proof of property 4 is in the appendix A. Note that property 2 implies that the matrix dimension l of $\mathcal{A}$ is at least as great as the process dimension m.

Our goal is now to distill from the l-dimensional state vectors of the matrix OOM those parts which are relevant for representing the process's future. Intuitively, if the process dimension is m, only projections of the matrix OOM states on some m-dimensional subspace of R^l contain relevant information.

First observe that a basis $\{w_{\bar{a}_1}, \ldots, w_{\bar{a}_d}\}$ of the linear subspace W can be effectively constructed from $\mathcal{A}$, as follows. Construct a sequence of sets $(S_j)_{j=0,1,\ldots,r}$ of states as follows:

Step 1: Let $S_0 = \{w_0\}$.

Step 2: Obtain S_{j+1} from S_j by first adding to S_j all states from the set $\{\tau_a w \, / \, \mathbf{1}\tau_a w \mid a \in O, w \in S_j\}$, and then deleting from the obtained set as many states as necessary to get a maximal set of linearly independent states.

Step 3: When the size of S_{j+1} is equal to the size of S_j, stop and put $r = j$; else resume at step 2.

It is clear that the size of the sets $(S_j)_{j=0,1,\ldots,r}$ properly grows throughout the sequence, and that the vectors contained in S_r yield the desired basis for W.

To determine the "prediction relevant" portions in the states w, we investigate the kernel, denoted as $ker\,\pi$, of the canonical projection.

Proposition 14.5

$$\forall x \in W : \left(x \in ker\,\pi \;\; \leftrightarrow \;\; \forall \bar{a} \in O^* \;\; \mathbf{1}\tau_{\bar{a}} x = 0 \right). \tag{14.16}$$

The proof is in the appendix B. As a special case we get $\mathbf{1}x = 0$ for all $x \in ker\,\pi$. Using this insight, a basis for $ker\,\pi$ can be constructed from $\mathcal{A}$ as follows. Again build a sequence $(S_j)_{j=0,1,\ldots,s}$ of sets of (row) vectors:

Step 1: Let $S_0 = \{\mathbf{1}\}$.

Step 2: Obtain S_{j+1} from S_j by first adding to S_j all vectors from the set $\{u\tau_a \mid a \in O, u \in S_j\}$, and then delete from the obtained set as many vectors as necessary to get a maximal set of linearly independent vectors.

Step 3: When the size of S_{j+1} is equal to the size of S_j, stop and put $s = j$; else resume at Step 2.

It follows from proposition 14.5 that

$$ker\ \pi = \{x \in W \mid \forall u \in S_s\ \ x \perp u^\top\}, \tag{14.17}$$

from which some orthonormal basis for $ker\ \pi$ is readily constructed. Since π is surjective we have $dim\ ker\ \pi = d - m$. Let $\{x_1, \dots, x_{d-m}\}$ be such a basis. Consider the orthogonal complement of the kernel:

$$V = \{v \in W \mid v \perp ker\ \pi\}, \tag{14.18}$$

where V is a linear subspace of W and has a dimensionality of m. It is an easy exercise to obtain a concrete representation of V through creating an orthonormal basis for V.

For $w \in W$, let $\tilde{w}$ denote the orthogonal projection of w on V. From linearity of orthogonal projections and proposition 14.5 we obtain that

$$\mathbf{1}\tilde{w} = \mathbf{1}w \tag{14.19}$$

for all $w \in W$. Let π_0 be the restriction of π on V. In light of equation 14.19 and proposition 14.4, property 3, π_0 preserves our probability measuring functionals $\mathbf{1}$ (in $\mathcal{A}$) and σ (in $\mathcal{F}$) in the sense that $\mathbf{1}v = \sigma\pi_0(v)$ for all $v \in V$.

Furthermore, define restrictions $\tilde{\tau}_a$ of the observable operators τ_a by

$$\tilde{\tau}_a v = \widetilde{\tau_a v} \tag{14.20}$$

for all $v \in V$. It is easy to see that $\tilde{\tau}_a$ is linear, and a matrix representation for $\tilde{\tau}_a$ is readily obtained from the bases of V and $ker\ \pi$. The projection π_0 maps $\tilde{\tau}_a$ on t_a by virtue of

$$\forall\ v \in V\ \ \pi_0(\tilde{\tau}_a v) = \pi_0(\widetilde{\tau_a v}) = \pi(\tau_a v) = t_a \pi(v), \tag{14.21}$$

where the last equality follows from proposition 14.4, property 4. Assembling our findings we see that

$$\pi_0 : (V, (\tilde{\tau}_a)_{a \in O}, \tilde{w}_0) \cong (F, (t_a)_{a \in O}, f_\varepsilon) \tag{14.22}$$

induces an isomorphism of vector spaces and operators which maps $\mathbf{1}$ on σ. This is just another way of saying that $(V, (\tilde{\tau}_a)_{a \in O}, \tilde{w}_0)$ is an OOM for our process. Note that V is represented here as a linear subspace of R^l and the matrices $\tilde{\tau}_a$ have a size of $l \times l$. A little more elementary linear algebra would finally transform $(V, (\tilde{\tau}_a)_{a \in O}, \tilde{w}_0)$ into an m-dimensional (and thus minimal-dimensional) matrix OOM.

We are now prepared to provide a simple answer to the question of when two matrix OOMs $\mathcal{A}$ and $\mathcal{A}'$ are equivalent in the sense of yielding identical probabilities for finite sequences. To decide the equivalence between $\mathcal{A}$ and $\mathcal{A}'$, we first transform them into minimal-dimensional OOMs of dimension m (if their minimal dimensions turn out not to be identical, they are not equivalent). We then apply the following proposition.

Proposition 14.6

Two minimal-dimensional OOMs $\mathcal{A} = (\mathrm{R}^m, (\tau_a)_{a \in O}, w_0)$ and $\mathcal{A}' = (\mathrm{R}^m, (\tau_a')_{a \in O}, w_0')$ are equivalent if and only if there exists a bijective linear map $\varrho : \mathrm{R}^m \to \mathrm{R}^m$ satisfying the following conditions:

1. $\varrho(w_0) = w_0'$,

2. $\tau_a' = \varrho \tau_a \varrho^{-1}$ for all $a \in \mathcal{O}$,

3. $\mathbf{1}w = \mathbf{1}\varrho w$ for all $w \in \mathrm{R}^m$.

Sketch of Proof (for detailed proof see Jaeger, 1997a)). The "if" direction is a mechanical verification. The interesting direction is to show that if $\mathcal{A}$ and $\mathcal{A}'$ are equivalent then a map ϱ exists. First observe that for minimal-dimensional OOMs, the canonical projection π coincides with π_0 and is an isomorphism of the matrix OOM with the functional OOM. Let π, π' be the canonical projections $\mathcal{A}$ and $\mathcal{A}'$, respectively, then $\varrho = \pi'^{-1} \pi$ satisfies the conditions of the proposition.

A matrix ϱ satisfies condition 3 of proposition 14.6 from the proposition if and only if each column of ϱ sums to unity. Thus, if we have one minimal-dimensional OOM $\mathcal{A}$, we get all the other equivalent ones by applying any transformation matrix ϱ with unit column sum.

14.6   Characterizing OOMs via Convex Cones

The problematic nonnegativity condition3 from definition 14.2 can be equivalently stated in terms of convex cones. This sheds much light on the relationship between OOMs and HMMs, and also allows one to appreciate the difficulty of the issue. I first introduce some cone-theoretic concepts, following the notation of a standard textbook (Berman and Plemmons, 1979).

With a set $S \subseteq \mathrm{R}^n$ we associate the set S^G, the *set generated by S*, which consists of all finite nonnegative linear combinations of elements of S. A set $K \subseteq \mathrm{R}^n$ is defined to be a *convex cone* if $K = K^G$. A convex cone K^G is called *n-polyhedral* if K has n elements. A cone K is *pointed* if for every nonzero $w \in K$, the vector $-w$ is not in K.

Using these concepts, the following proposition gives a condition which is equivalent to condition 3 from definition 14.2, and clarifies the relationship between OOMs and HMMs.

Proposition 14.7

1. Let $\mathcal{A} = (\mathrm{R}^m, (\tau_a)_{a \in O}, w_0)$ be a structure satisfying the first two conditions from definition 14.2, i.e., $\mathbf{1}w_0 = 1$ and $\mu = \sum_{a \in O} \tau_a$ has unit column sums. Then $\mathcal{A}$ is

an OOM if and only if there exists a pointed convex cone $K \subset \mathrm{R}^m$ satisfying the following conditions:

- $\mathbf{1}w \geq 0$ for all $w \in K$,

- $w_0 \in K$,

- $\forall a \in O : \tau_a K \subseteq K$.

2. Assume that $\mathcal{A}$ is an OOM, then there exists a HMM equivalent to $\mathcal{A}$ if and only if a pointed convex cone K according to part 1 exists which is n-polyhedral for some n, where n can be selected such that it is not greater than the minimal state number for HMMs equivalent to $\mathcal{A}$.

Part 1 can be proven by reformulating a similar claim (Jaeger, 2000) that goes back to Heller (1965) and has been renewed in Ito (1992)[2]. Part 2 was shown in Ito (1992). These authors considered a class of stochastic processes called *linearly dependent processes* that is identical to what we introduced as processes with finite dimension m; they did not use observable operators to characterize the processes.

Part 2 has the following interesting implications:

- Every two-dimensional OOM is equivalent to some HMM, because all cones in R^2 are 2-polyhedral. A nice exercise left to the reader is to construct a two-dimensional OOM whose smallest equivalent HMM has four states (hint: derive a two-dimensional OOM from a HMM defined not by emitting observations from states but from state transitions).

- If an OOM contains an operator τ_a that rotates R^m by a nonrational multiple of π, then this OOM has no equivalent HMM because τ_a leaves no polyhedral cone invariant.

- Three-dimensional OOMs can be constructed whose equivalent minimal-size HMMs have at least p states (for any prime $p \geq 3$), by equipping the OOM with an operator that rotates R^3 by $2\pi/p$. This is so because any polyhedral cone left invariant by such an operator is at least p-polyhedral.

Proposition 14.7 is useful to *design* interesting OOMs, starting with a cone K and constructing observable operators satisfying $\tau_a K \subseteq K$. Unfortunately it provides no means to *decide*, for a given structure $\mathcal{A}$, whether $\mathcal{A}$ is a valid OOM, since the proposition is nonconstructive with respect to K.

If one would have effective algebraic methods to decide, for a set of k linear operators on R^m, whether they leave a common cone invariant, then one could decide whether a candidate structure $(\mathrm{R}^m, (\tau_a)_{a \in O}, w_0)$ is a valid OOM. However, this is a difficult and unsolved problem of linear algebra. For a long time, only the case of a single operator ($k = 1$) was understood (Berman and Plemmons, 1979). Recently however there was substantial progress in this matter. In Edwards et al. (2005) interesting subcases of $k = 2$ were solved, namely, the subcases of $m = 2$ and of polyhedral cones.

14.7 Interpretable OOMs

OOM states represent future distributions, but the previous section might have left the impression that this representation is somewhat abstract. We will now see that within the equivalence class of a given minimal-dimensional OOM, there are some members whose states can be interpreted immediately as future distributions—*interpretable* OOMs. Interpretable OOMs are pivotal for OOM learning algorithms.

Because this concept is so important for OOM theory we will first illustrate it with an informal example. Assume we have a 26-dimensional OOM $\mathcal{A}$ over the English alphabet $O = \{a, \ldots, z\}$—the OOM dimension and the alphabet size accidentally coincide. Assume that $\mathcal{A}$ models the distribution of letter sequences in English texts. Utilizing the generation procedure from section 14.4, $\mathcal{A}$ can be run to generate strings of pseudo-English. Remember that at time n, the state w_n is used to compute a 26-dimensional probability vector $\mathbf{p}_{n+1}$ of the nth occurring letter via $\mathbf{p}_n = \Sigma\, w_n$, where Σ's rows are made from the column sums of the 26 observable operators (14.13).

Wouldn't it be convenient if we had $\mathbf{p}_{n+1} = w_n$ and $\Sigma = \mathbf{id}$ (where $\mathbf{id}$ denotes the identity matrix)? Then we could immediately take the next letter probabilities from the current state vector, spare ourselves the computation of $\Sigma\, w_n$, and directly "see" the development of very interesting probabilities in the state evolution.

We will now see that such an *interpretable* OOM can be constructed from $\mathcal{A}$. The definition of interpretable OOMs is more general than this example suggests in that it admits a more comprehensive notion of the future events whose probabilities become the state vector's entries. In our example, these events—which we will call *characteristic events*—were just the singletons $a, \ldots, z$. Here is the general definition of such events:

Definition 14.4

Let $(X_n)_{n \geq 0}$ be an m-dimensional stationary process with observables from O. Let, for some sufficiently large l, $O^l = B_1 \cup \cdots \cup B_m$ be a partition of the set of strings of length l into m disjoint, nonempty sets B_i. Then this partition is called a set of *characteristic events* B_i $(i = 1, \ldots, m)$, if some sequences $\bar{a}_1, \ldots, \bar{a}_m$ exist such that the matrix $(P(B_i | \bar{a}_j))_{1 \leq i,j \leq m}$ is nonsingular.

Here by $P(B_i | \bar{a}_j)$ we mean $\sum_{\bar{b} \in B_i} P(\bar{b} | \bar{a}_j)$. We introduce some further notational commodities. For a state vector w of an OOM $\mathcal{A}$ of $(X_n)_{n \geq 0}$ and a sequence $\bar{b}$ let $P(\bar{b} | w) = \mathbf{1}\tau_{\bar{b}} w$ denote the probability that the OOM will produce $\bar{b}$ when started in state w. Furthermore, let $P(B_i | w) = \sum_{\bar{b} \in B_i} P(\bar{b} | w)$. Now we are equipped to define interpretable OOMs.

Definition 14.5

Let $B_1, \ldots, B_m$ be characteristic events for an m-dimensional process with observables O, and let $\mathcal{A} = (\mathrm{R}^m, (\tau_a)_{a \in O}, w_0)$ be an OOM for that process. Then $\mathcal{A}$ is *interpretable* with respect to $B_1, \dots, B_m$ if the states w of $\mathcal{A}$ have the property

$$w = (P(B_1 \,|\, w) \cdots P(B_m \,|\, w))^\top. \tag{14.23}$$

Here is a method to transform a given OOM $\mathcal{A} = (\mathrm{R}^m, (\tau_a)_{a \in O}, w_0)$ for $(X_n)_{n \geq 0}$ into an OOM that is interpretable with respect to characteristic events $B_1, \dots, B_m$. Define $\tau_{B_i} := \sum_{\bar{b} \in B_i} \tau_{\bar{b}}$. Define a mapping $\varrho : \mathrm{R}^m \to \mathrm{R}^m$ by

$$\varrho(x) := (\mathbf{1}\tau_{B_1} x \cdots \mathbf{1}\tau_{B_m} x)^\top. \tag{14.24}$$

The mapping ϱ is obviously linear. It is also bijective, since according to the definition of characteristic events, sequences $\bar{a}_j$ exist such that the matrix $(P(B_i \,|\, \bar{a}_j)) = (\mathbf{1}\tau_{B_i} x_j)$, where $x_j = \tau_{\bar{a}_j} w_0 / \mathbf{1}\tau_{\bar{a}_j} w_0$, is nonsingular. Furthermore, ϱ preserves component sums of vectors, since for $j = 1, \dots, m$ it holds that

$$\mathbf{1} x_j = 1 = \mathbf{1}(P(B_1 \,|\, x_j) \cdots P(B_m \,|\, x_j))^\top = \mathbf{1}(\mathbf{1}\tau_{B_1} x_j \cdots \mathbf{1}\tau_{B_m} x_j)^\top = \mathbf{1}\varrho(x_j),$$

namely, a linear map preserves component sums if it preserves component sums of basis vectors. Hence ϱ satisfies the conditions of proposition 14.6. We therefore obtain an OOM equivalent to $\mathcal{A}$ by

$$\mathcal{A}' = (\mathrm{R}^m, (\varrho \tau_a \varrho^{-1})_{a \in O}, \varrho w_0) = (\mathrm{R}^m, (\tau_a')_{a \in O}, w_0'). \tag{14.25}$$

Equation 14.23 holds in $\mathcal{A}'$. To see this, let w_n' be a state vector obtained in a generation run of $\mathcal{A}'$ at time n, and w_n the state obtained in $\mathcal{A}$ after the same sequence has been generated. Then it concludes that

$$
\begin{aligned}
w_n' &= \varrho \varrho^{-1} w_n' \\
&= (\mathbf{1}\tau_{B_1}(\varrho^{-1} w_n') \cdots \mathbf{1}\tau_{B_m}(\varrho^{-1} w_n'))^\top \\
&= (\mathbf{1}\tau_{B_1} w_n \cdots \mathbf{1}\tau_{B_m} w_n)^\top \\
&= (P(B_1 \,|\, w_n) \cdots P(B_m \,|\, w_n))^\top \\
&= (P(B_1 \,|\, w_n') \cdots P(B_m \,|\, w_n'))^\top,
\end{aligned}
$$

where the last equality follows from the equivalence of $\mathcal{A}$ and $\mathcal{A}'$.

We will sometimes denote $\mathcal{A}'$ by $\varrho \mathcal{A}$. The $m \times m$ matrix corresponding to ϱ can be obtained from the original OOM $\mathcal{A}$ by observing that

$$\varrho = (\mathbf{1}\tau_{B_i} e_j), \tag{14.26}$$

where e_i is the ith unit vector.

The following fact lies at the heart of the learning algorithm presented in the next section.

Proposition 14.8

In an OOM that is interpretable with respect to $B_1, \dots, B_m$ it holds that

1. $w_0 = (P(B_1) \cdots P(B_m))^\top$,
2. $\tau_{\bar{a}} w_0 = (P(\bar{a} B_1) \cdots P(\bar{a} B_m))^\top$,

where $P(\bar{a}B)$ denotes $\sum_{\bar{b} \in B} P(\bar{a}\bar{b})$. The proof is trivial.

Interpretable OOMs are most often used in a context when they are minimal-dimensional, but sometimes it is useful to generalize the notion by dropping the requirement of minimal dimensionality. An n-dimensional OOM of an m-dimensional process is called interpretable with respect to $B_1, \ldots, B_n$ if the analog of 14.23 holds. An n-dimensional OOM with operators τ_a can be made interpretable by putting $\tau'_a = \varrho\, \tau_a \varrho^{\dagger}$, where again $\varrho = (\mathbf{1}\tau_{B_i} e_j)$ and $\varrho^{\dagger}$ is the pseudo-inverse of ϱ. A special case that we will need to consider later on is obtained when the B_i are all singletons. We introduce some special concepts for this case.

Definition 14.6

Let (X_n) be an m-dimensional process over an observation alphabet O. Fix some $k \in \mathrm{N}, k > 0$, put $\kappa = |O|^k$, and let $\bar{b}_1, \ldots, \bar{b}_\kappa$ be the alphabetical enumeration of O^k. Then these sequences $\bar{b}_i$ are the *characteristic sequences* of length k for (X_n) if m "indicative" sequences $\bar{a}_1, \ldots, \bar{a}_m$ exist that make the $\kappa \times m$ matrix $V = (P(\bar{b}_i | \bar{a}_j))$ have rank m. The minimal k for which such sequences $\bar{a}_1, \ldots, \bar{a}_m$ exist is the *characterizing length* of (X_n).

We list three properties of characteristic sequences (the simple proof is left to the reader).

Proposition 14.9

Let $\bar{b}_i$ be characteristic sequences of $(X_n)_{n \geq 0}$ of length k and let $\kappa = |O|^k$.

1. If $\mathcal{A} = (\mathrm{R}^n, (\tau_a)_{a \in O}, w_0)$ is some (not necessarily minimal-dimensional) OOM for $(X_n)_{n \geq 0}$, then the $\kappa \times n$ matrix $\pi_{\mathcal{A}}$ that has as its i-th row $\mathbf{1}\tau_{\bar{b}_i}$ maps states w of $\mathcal{A}$ to $\pi_{\mathcal{A}}w = (P(\bar{b}_1|w) \cdots P(\bar{b}_\kappa|w))^{\top}$.

2. The characterizing length k_0 of (X_n) is the minimal length of characteristic events for (X_n).

3. The characterizing length k_0 is at most $m - 1$.

Here are some observations concerning interpretable OOMs:

- If an m-dimensional OOM $\mathcal{A}$ has been learned from empirical data, and one chooses disjoint events $B_1, \ldots, B_m$ at random, it is generically the case that some sequences $\bar{a}_1, \ldots, \bar{a}_m$ exist such that the matrix $(P(B_i | \bar{a}_j))_{1 \leq i,j \leq m}$ is nonsingular. The reason is that the matrix composed from rows $(\mathbf{1}\tau_{B_i})$ is a random matrix and as such generically nonsingular. Generally speaking, for arbitrary events $B_1, \ldots, B_m$ being characteristic is the rule, not an exceptional circumstance.

- A given OOM can be transformed into many different equivalent, interpretable OOMs depending on the choice of characteristic events.

- Interpretability yields a very useful way to visualize the state dynamics of an OOM. To see how, first consider the case where the OOM dimension is 3. Interpretable states, being probability vectors, are nonnegative and thus lie in the intersection of the positive orthant of R^3 with the hyperplane $H = \{x \in \mathrm{R}^3 \,|\, \mathbf{1}x = 1\}$. This intersection is a triangular surface. Its corners mark the three unit vectors

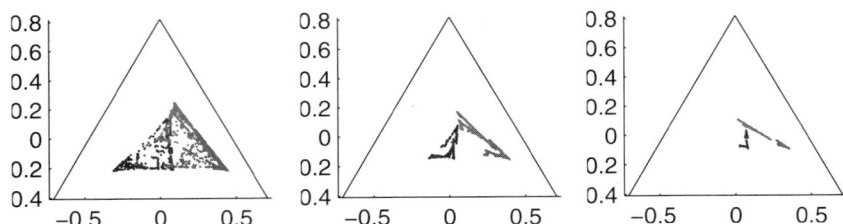

Figure 14.2 State dynamics "fingerprints" of three related interpretable OOMs.

of R^3. This triangle can be conveniently used as a plotting canvas. Figure 14.2 shows three "fingerprint" plots of states obtained from generating runs of three different synthetic three-dimensional OOMs (see appendix C for details) over an observation alphabet of size 3, which were made interpretable with respect to the the same three characteristic events. The states are colored with three colors depending on which of the three operators was used to produce each state. A similar graphical representation of states was first introduced in Smallwood and Sondik (1973) for HMMs. When one wishes to plot states of interpretable OOMs with dimension $m > 3$, one can join some of the characteristic events, until three merged events are left, and create plots as explained above.

■ If one has several nonequivalent OOMs over the same alphabet O, making them interpretable with respect to to a common set of characteristic events is useful for *comparing* them in a meaningful way. This has been done for the three OOMs plotted in fig. 14.2. Their observable operators depended on a control parameter α, which was slightly changed over the three OOMs.

14.8   The Basic Learning Algorithm

We shall address the following learning task. Assume that a realization $S = a_0 a_1 \cdots a_N$ of a stationary, m-dimensional process (X_n) is given; that is, S is generated by some OOM $\mathcal{A}$ of (minimal) dimension m. We assume that m is known but otherwise $\mathcal{A}$ is unknown. We wish to induce from S an estimate $\hat{\mathcal{A}}$ of $\mathcal{A}$ in the sense that the distribution characterized by $\hat{\mathcal{A}}$ comes close to the distribution characterized by $\mathcal{A}$ (the hat $\hat{\cdot}$ will be used throughout this chapter for denoting estimates).

We first collect some observations concerning the unknown generator $\mathcal{A}$. We may assume that $\mathcal{A}$ is interpretable with respect to characteristic events $B_1, \ldots, B_m$. Then the principle of learning OOMs emerges from the following observations:

■ Proposition 14.8, property can be used to procure argument-value pairs for the operator τ_a $(a \in O)$ by exploiting

$$
\begin{aligned}
\tau_a((P(\bar{a}B_1) \cdots P(\bar{a}B_m))^\top) &= \tau_a(\tau_{\bar{a}} w_0) \\
&= \tau_{\bar{a}a} w_0 \\
&= (P(\bar{a}aB_1) \cdots P(\bar{a}aB_m))^\top.
\end{aligned} \tag{14.27}
$$

Such argument-value pairs are vectors that are made from probability values.

▪ A linear operator on R^m is determined by any m argument-value pairs provided the arguments are linearly independent.

▪ Probabilities of the kind $P(\bar{a}B_i)$ that make up the argument-value pairs in equation 14.27 can be estimated from the training string S through the relative frequencies $\hat{P}_S$ of the event $\bar{a}B_i$:

$$
\hat{P}_S(\bar{a}B_i) = \frac{\text{number of occurrences of words } \bar{a}\bar{b} \text{ (where } \bar{b} \in B_i \text{) within } S}{N - |\bar{a}B_i| + 1}, \tag{14.28}
$$

where $|\bar{a}B_i|$ denotes the length of $\bar{a}$ plus the length of the sequences in B_i.

Thus the blueprint for estimating an OOM $\hat{\mathcal{A}}$ from S is clear:

Step 1: Choose characteristic events $B_1, \ldots, B_m$ and *indicative sequences* $\bar{a}_1, \ldots, \bar{a}_m$ such that the matrix $\hat{V} = (\hat{P}_S(\bar{a}_j B_i))_{i,j=1,\ldots,m}$ is non-singular (this matrix contains in its columns m linearly independent argument vectors for the operators τ_a).

Step 2: For each $a \in O$, collect the corresponding value vectors in a matrix $\hat{W}_a = (\hat{P}_S(\bar{a}_j a B_i))_{i,j=1,\ldots,m}$.

Step 3: Obtain an estimate for τ_a by

$$
\hat{\tau}_a = \hat{W}_a \hat{V}^{-1}. \tag{14.29}
$$

If the process (X_n) is ergodic, the estimates $\hat{P}_S(\bar{a}_j B_i), \hat{P}_S(\bar{a}_j a B_i)$ converge with probability 1 to the correct probabilities as the sample size N grows to infinity. This implies that the estimated $\hat{\tau}_a$ will converge to the operators of the true data generator $\mathcal{A}$, assuming that $\mathcal{A}$ is interpretable with respect to the characteristic events $B_1, \ldots, B_m$ used in the learning procedure. In other words, the learning algorithm is asymptotically correct.

The statistical efficiency of the algorithm can be improved if instead of using indicative *sequences* $\bar{a}_j$ one uses indicative *events* A_j that partition O^l into m nonempty, disjoint subsets. Then $\hat{V} = (\hat{P}_S(A_j B_i))_{i,j=1,\ldots,m}$ and $\hat{W}_a = (\hat{P}_S(A_j a B_i))_{i,j=1,\ldots,m}$. If this is done, counting information from *every* subword of S of length $|A_j B_i|$ enters the model estimation, whereas when indicative sequences are used, only those subwords beginning with an indicative sequence are exploited.

A computational simplification of this basic algorithm is obtained if one uses in (14.29) the raw counting matrices

$$
\begin{aligned}
V^{\text{raw}} &= \left(\text{count no. of event } A_j B_i \text{ in } S_{\text{short}} = a_0 \ldots a_{N-1}\right)_{i,j=1,\ldots,m}, \\
W_a^{\text{raw}} &= \left(\text{count no. of event } A_j a B_i \text{ in } S\right)_{i,j=1,\ldots,m}.
\end{aligned} \tag{14.30}
$$

It is easy to see that $W_a^{\text{raw}} (V^{\text{raw}})^{-1} = \hat{W}_a \hat{V}^{-1}$.

The counting matrices can be gleaned in a single sweep of a window of length $| A_j B_i |$ across S, and the computation of equation 14.29 incurs $O(m^3)$ flops. This makes the overall computational cost of the algorithm $O(N + m^3)$.

Note that while the obtained models $\hat{\mathcal{A}}$ converge to an interpretable OOM with increasing sample size, it is not the case that a model obtained from a finite training sample is interpretable with respect to the characteristic events chosen for learning.

The statistical efficiency (model variance) of this basic algorithm depends crucially on the choice of characteristic and indicative events. This can be seen immediately from the basic learning equation 14.29. Depending on the choice of these events, the matrix $\hat{V}$ will have a high or low condition number, that is, its inversion will magnify estimation errors of $\hat{V}$ to a high or low extent, which in turn means a high or low model variance. Several methods of determining characteristic and indicative events that lead to a low condition number of $\hat{V}$ have been devised. The first of these methods is documented in Kretzschmar (2003); another will be presented later in this chapter (it is documented in appendix H).

We assumed here that the *correct* model dimension m is known beforehand. Finding the correct model dimension is, however, an academic question. Real-life processes will hardly ever have a finite dimension. The problem in practical applications is instead to find a model dimension that gives a good compromise in the bias-variance dilemma. The model dimension m should be chosen (1) large enough to enable the model to capture all the properties of the distribution that are statistically revealed in S, and in the meantime (2) small enough to prevent overfitting.

Negotiating this compromise can be effected by the standard techniques of machine learning, for instance cross-validation. But OOM theory suggests a purely algebraic approach to this problem. The key is the matrix $\hat{V}$. Roughly speaking, if it has a low condition number and can thus be stably inverted, model variance will be low and overfitting is avoided. Quantitative bounds on model variance, as well as an algebraic method for finding good characteristic events (of a more general kind than introduced here) that minimize the condition number of $\hat{V}$ for a given model dimension can be found in Kretzschmar (2003).

While the basic learning algorithm is conceptually transparent and computationally cheap, it has two drawbacks that make it ill suited for applications:

1. Even with good characteristic and indicative events for a small condition number of $\hat{V}$, the statistical efficiency of the basic algorithm has turned out to be inferior to that of HMMs estimated via the EM algorithm. The reason is that the EM algorithm implicitly exploits the statistics of arbitrarily long substrings in S, whereas our OOM learning algorithm solely exploits the statistics of substrings of length $|A_j B_i|$.

2. The models returned by this learning method need not be valid OOMs. The nonnegativity condition 3 of definition 14.2 is often violated by the "OOMs" computed via this method.

In sections 14.10 to 14.13 of this chapter, the first of these problems will be completely solved. The second problem will remain unsolved, but practical working solutions will be presented.

14.9 History, Related Work, and Ramifications

Hidden Markov models (HMMs) (Bengio, 1999) of stochastic processes have been investigated in mathematics under the name "functions of Markov chains" long before they became a popular tool in speech processing and engineering. A basic mathematical question was to decide when two HMMs are equivalent, i.e., describe the same distribution (Gilbert, 1959). This problem was tackled by framing HMMs within a more general class of stochastic processes, nowadays termed *linearly dependent processes* (LDPs). Deciding the equivalence of HMMs amounts to characterizing HMM-describable processes as LDPs. This strand of research (Blackwell and Koopmans, 1957; Dharmadhikari, 1963a,b, 1965; Fox and Rubin, 1968, 1969, 1970; Heller, 1965) came to a successful conclusion in the work of Ito et al. (1992), where equivalence of HMMs was characterized algebraically, and a decision algorithm was provided. That article also gives an overview of the work done in this area up to the time of its writing.

The results from Ito et al. (1992) were further elaborated in Balasubramanian (1993), where for the first time matrix representations with negative entries appeared, called "generalized hidden Markov models." The algebraic characterization of HMM equivalence could be expressed more concisely than in the original paper (Ito et al., 1992).

All of this work on HMMs and LDPs was mathematically oriented and did not bear on the practical question of learning models from data.

In 1997, the concept of OOMs was introduced by Jaeger (1997a), including the basic learning algorithm (Jaeger, 1997b). Independently a theory almost identical to the OOM theory presented here was developed by Upper (1997). The only difference is that in that work characteristic *sequences* were utilized for learning instead of characteristic events, which renders the algorithm a bit more complicated.

Unconnected to all of these developments, the idea of describing the observables of a stochastic process as update operators was carried out by Iosifescu and Theodorescu (1969) within a very general mathematical framework. However, it was not perceived that these operators can be assumed to be linear.

Recently we have witnessed a growing interest in observable operator models in the field of optimal decision making/action selection for autonomous agents. Under the name of *predictive state representations* (PSRs) and with explicit connections made to OOMs, a generalization of partially observable Markov decision processes (POMDPs), (see, e.g., Kaelbling et al., 1998) is being explored (see, e.g., James and Singh, 2004; Littman et al., 2001, and try Google on "predictive state representation" to find more). PSRs can be seen as a version of OOMs that models systems with input. Such input-output OOMs (including a variant of the basic learning algorithm) were first described by Jaeger (1998b).

Since their discovery, OOMs have been investigated by the group led by this chapter's first author. The most notable results are (1) matrix OOMs for continuous-valued processes, including a version of the basic learning algorithm (Jaeger, 1998a); (2) a general OOM theory for stochastic processes (nonstationary, continuous time, with arbitrary observation sets) including an algebraic characterization of general processes which reveals fascinating structural similarities between the formalism of quantum mechanics and OOMs; (3) a first solution to the problem of finding characteristic events that optimize statistical efficiency, including bounds on model variance (Kretzschmar, 2003); and (4) the introduction of suffix tree representations for the training string as a tool to improve statistical efficiency (Oberstein, 2002, more about this later). Much effort was spent and wasted on the nonnegativity problem; for the time being we put this to rest. Hopefully, new developments in linear algebra will ultimately help to resolve this issue (Edwards et al., 2005).

Ongoing work in our group focuses on online learning algorithms, heuristics for ascertaining nonnegativity of model-predicted probabilities (more in later sections), and the investigation of *quadratic* OOMs, which arise from replacing the basic equation (14.8) by $P(\bar{a}) = (\sigma\tau_{\bar{a}}w_0)^2$. Non-negativity is clearly a non-issue in quadratic OOMs—this is the prime motivation for considering them—and the basic learning algorithm is easily carried over; however, it is currently not clear which processes can be characterized by quadratic OOMs.

14.10   Overview of the Efficiency Sharpening Algorithm

We have seen that the basic OOM learning algorithm has limited statistical efficiency

1. because only the statistics of substrings of some (small) fixed length are entered in the estimation algorithm, thus much information contained in the training data is ignored, and

2. because it is unclear how to choose the characteristic/indicative events optimally, thus the information that enters the algorithm becomes further degraded when it is agglomerated into (possibly badly adapted) collective events.

Both obstacles can be overcome:

1. Using a suffix tree representation of the training sequence, one can exploit characteristic/indicative sequences of all possible lengths simultaneously. Instead of exploiting a mere m argument-value pairs, the number of used argument-value pairs is in the order of the training data size.

2. We can get rid of characteristic and indicative events altogether. They will only be used for the estimation of an initial model $\hat{\mathcal{A}}^{(0)}$, from which a sequence $\hat{\mathcal{A}}^{(1)}, \hat{\mathcal{A}}^{(2)}, \ldots$ of better models is iteratively obtained without using such events at all. The model improvement is driven by a novel learning principle whose main idea

is to use the model $\hat{\mathcal{A}}^{(n)}$ for improving the statistical efficiency of the estimation procedure that yields $\hat{\mathcal{A}}^{(n+1)}$. We call this the principle of *efficiency sharpening* (ES).

14.11   The Efficiency Sharpening Principle: Main Idea and a Poor Man's ES Learning Algorithm

This is the main section of this chapter. We derive the underlying ideas behind the ES principle, present an elementary instance of an ES-based learning algorithm, and finish with a little simulation study.

The core of the ES principle is to use in each iteration a new set of characteristic events that yields an estimator with a better statistical efficiency. However, a very much generalized version of such events is used:

Definition 14.7

Let $\mathcal{A} = (\mathrm{R}^n, (\tau_a)_{a \in O}, w_0)$ be a (not necessarily minimal-dimensional) OOM of an m-dimensional process (X_n). Let $k \in \mathrm{N}$. A function $c : O^k \to \{r \in \mathrm{R}^n \,|\, \mathbf{1}r = 1\}$ is a *characterizer* of $\mathcal{A}$ (of length k) if

$$\forall \bar{a} \in O^* : \quad w_{\bar{a}} = \sum_{\bar{b} \in O^k} P(\bar{b}\,|\,\bar{a})\, c(\bar{b}), \qquad (14.31)$$

If convenient, we will identify c with the matrix $C = [c(\bar{b}_1) \cdots c(\bar{b}_\kappa)]$, where $\bar{b}_1, \ldots, \bar{b}_\kappa$ is the alphabetical enumeration of O^k.

It is clear that C is a characterizer for $\mathcal{A}$ if and only if every state $w_{\bar{a}}$ of $\mathcal{A}$ can be written as

$$w_{\bar{a}} = C \left(P(\bar{b}_1|\bar{a}) \cdots P(\bar{b}_\kappa|\bar{a}) \right)^\top. \qquad (14.32)$$

The characteristic events introduced in section 14.7 can be regarded as a special characterizer: if $\mathcal{A}$ is interpretable with respect to characteristic events $B_1, \ldots, B_n$ of length k, and if $\bar{b} \in B_i$, then define $c(\bar{b})$ as the binary vector of dimension n that is zero everywhere except at position i. The two conditions from the above definition are easily checked. In matrix form, this gives the *characteristic event characterizer* (apologies for the loopy terminology)

$$C_{B_1, \ldots, B_m} = (c_{ij})_{\substack{i=1, \ldots, m \\ j=1, \ldots, \kappa}}$$
$$\text{where} \qquad\qquad\qquad\qquad\qquad\qquad (14.33)$$
$$c_{ij} = \begin{cases} 1, & \text{if } \bar{b}_j \in B_i \\ 0, & \text{else.} \end{cases}$$

We proceed by investigating other characterizers.

Proposition 14.10

Let κ and $\bar{b}_1, \ldots, \bar{b}_\kappa$ be as in definition 14.7. Given an m-dimensional process (X_n), then an $n \times \kappa$ matrix C whose columns sum to 1 is a characterizer of some n-dimensional OOM for (X_n) if and only if there exist m sequences $\bar{a}_j$ such that the $n \times m$ product matrix $W = CV$ of C and the $\kappa \times m$ matrix $V = (P(\bar{b}_i \mid \bar{a}_j))$ has rank m.

The proof is in appendix D. Now consider two equivalent, minimal-dimensional OOMs $\mathcal{A}$, $\mathcal{A}'$ which are related by $\tau_a' = \varrho \tau_a \varrho^{-1}$ (see 14.25). Then the following holds.

Proposition 14.11

If C is a characterizer of $\mathcal{A}$, then $\varrho \circ C$ is a characterizer of $\mathcal{A}'$.

The reason is that the states $w_{\bar{a}}'$ of $\mathcal{A}'$ are equal to the transforms $\varrho w_{\bar{a}}$ of the respective states of $\mathcal{A}$. A given minimal-dimensional OOM (of dimension m) has many characterizers of length k if $\kappa > m$; if $\kappa = m$ then the characterizer is unique. This is detailed in the following corollary to proposition 14.10 (proof in appendix E).

Proposition 14.12

Let C_0 be a characterizer of length k of a minimal-dimensional OOM $\mathcal{A}$. Let κ and V be as in proposition 14.10. Then C is another characterizer of length k of $\mathcal{A}$ if and only if it can be written as $C = C_0 + G$, where

$$G = [g_1, \cdots, g_{m-1}, -\Sigma_{i=1,\ldots,m-1} g_i]^\top, \tag{14.34}$$

where the g_i are any vectors from $ker\, V^\top$.

An important type of characterizers is obtained from the states of reverse OOMs, that is, OOMs for the time-reversed process. We now describe in more detail the time reversal of OOMs. Given an OOM $\mathcal{A} = (\mathrm{R}^m, (\tau_a)_{a \in O}, w_0)$ with an induced probability distribution $P_\mathcal{A}$, its *reverse* OOM $\mathcal{A}^r$ is characterized by a probability distribution $P_{\mathcal{A}^r}$ satisfying

$$\forall\, a_0 \cdots a_n \in O^* : \quad P_\mathcal{A}(a_0 \cdots a_n) = P_{\mathcal{A}^r}(a_n \cdots a_0). \tag{14.35}$$

The reverse OOM can be computed from the forward OOM observing the following fact, whose proof is in appendix F.

Proposition 14.13

If $\mathcal{A} = (\mathrm{R}^m, (\tau_a)_{a \in O}, w_0)$ is an OOM for a stationary process, and w_0 has no zero entry, then $\mathcal{A}^r = (\mathrm{R}^m, (D\tau_a^\top D^{-1})_{a \in O}, w_0)$ is a reverse OOM to $\mathcal{A}$, where $D = \mathrm{diag}(w_0)$ is a diagonal matrix with w_0 on its diagonal.

Because from an m-dimensional matrix OOM for the "forward" process an m-dimensional matrix OOM for the reverse process can be constructed and vice versa, it follows that the process dimension of the forward process equals the process dimension of the reverse process.

When discussing "forward" and reverse OOMs of a process at the same time, using shorthand notations of the kind $P(\bar{b}_i \mid \bar{a}_j)$ easily leads to confusion. We fix the following conventions:

1. The character b and string shorthands $\bar{b}$ always denote symbols/substrings that follow after symbols/substrings denoted by character a and string shorthands $\bar{a}$—"after" with respect to the forward time direction.

2. We use P to denote probabilities for the forward process and P^r for the reverse process.

3. When using indices i, j for alphabetical enumerations for words $\bar{b}_i, \bar{a}_j$, the enumeration is carried out in the forward direction, even if we denote reverse probabilities. For example, if $O = \{0, 1, 2\}$, and if $\bar{a}_j, \bar{b}_j$ are each the alphabetical enumerations of O^2, and if τ_a, τ_a^r are the observable operators for a forward and a reverse OOM of a process, then $\bar{a}_6 = 12$, $\bar{b}_2 = 01$, and $\mathbf{1}\tau_1\tau_0\tau_2\tau_1 w_0 / \mathbf{1}\tau_2\tau_1 w_0 = P(\bar{b}_2 \mid \bar{a}_6) = P(X_2 = 0, X_3 = 1 \mid X_0 = 1, X_1 = 2) = P^r(X_1^r = 0, X_0^r = 1 \mid X_3^r = 1, X_2^r = 2) = P^r(\bar{b}_2 \mid \bar{a}_6) = \mathbf{1}\tau_1^r\tau_2^r\tau_0^r\tau_1^r w_0^r / \mathbf{1}\tau_1^r\tau_2^r w_0^r$.

4. Likewise, when using $\bar{a}$ as an index to denote a concatenation of operators, the forward direction is always implied for interpreting $\bar{a}$. For example, $\tau_{01} = \tau_1\tau_0$ and $\tau_{01}^r = \tau_0^r\tau_1^r$.

The states of a reverse OOM obtained after sufficiently long reverse words make a characterizer of a forward OOM, as follows.

Proposition 14.14

Let the dimension of (X_n) be m and let $\mathcal{A}^r = (\mathrm{R}^m, (\tau_a^r)_{a \in O}, w_0)$ be a reverse OOM for (X_n) that was derived from a forward OOM $\mathcal{A} = (\mathrm{R}^m, (\tau_a)_{a \in O}, w_0)$ as in proposition 14.13. Let k_0 be the characterizing length of (X_n), let $k \geq k_0$, and let $\kappa = |O^k|$. Then the following two statements hold:

1. $C = [w_{\bar{b}_1}^r \cdots w_{\bar{b}_\kappa}^r]$ is a characterizer of an OOM $\mathcal{A}'$ for (X_n).

2. The states $w_{\bar{a}}'$ of $\mathcal{A}'$ are related to the states $w_{\bar{a}}$ of $\mathcal{A}$ by the transformation $w_{\bar{a}}' = \varrho w_{\bar{a}}$, where $\varrho = C\pi_{\mathcal{A}}$. If in addition $w_0 = (1/m \cdots 1/m)^\top$, then furthermore $\varrho = R^\top R$. The matrices $\pi_{\mathcal{A}}$ and R are

$$\pi_{\mathcal{A}} = \begin{pmatrix} \mathbf{1}\tau_{\bar{b}_1} \\ \vdots \\ \mathbf{1}\tau_{\bar{b}_\kappa} \end{pmatrix}, \quad R = \pi_{\mathcal{A}} \operatorname{diag}\left((m\, P(\bar{b}_1))^{-1/2} \cdots (m\, P(\bar{b}_\kappa))^{-1/2} \right). \quad (14.36)$$

The proof can be found in appendix G. The proposition implies that $\varrho^{-1}C = (C\pi_{\mathcal{A}})^{-1}C =: C_{\mathcal{A}}^r$ is a characterizer for the original forward OOM $\mathcal{A}$. $C_{\mathcal{A}}^r = [\varrho^{-1}w_{\bar{b}_1}^r \cdots \varrho^{-1}w_{\bar{b}_\kappa}^r]$ is the characterizer obtained from the reverse OOM $\varrho^{-1}\mathcal{A}^r = (\mathrm{R}^m, (\varrho^{-1}\tau_a^r\varrho)_{a \in O}, w_0)$, so we may note for later use that every OOM $\mathcal{A}$ has a *reverse characterizer* $C_{\mathcal{A}}^r$ that is made from the states of a suitable reverse OOM.

Among all characterizers of OOMs $\mathcal{A}$ for (X_n), the reverse characterizers minimize a certain measure of variance, an observation which is the key to the ES learning principle. We begin the presentation of this core finding by describing some variants of the basic learning algorithm from section 14.8.

In the basic learning algorithm from 14.29, an estimate $\hat{\tau}_a$ of an m-dimensional OOM was determined from m estimated argument-value pairs for τ_a, which were sorted in the columns of an $m \times m$ matrix $\hat{V} = (\hat{P}(\bar{a}_j B_i))$ (containing the argument vectors) and another $m \times m$ matrix $\hat{W}_a = (\hat{P}(\bar{a}_j a B_i))$ (containing the values), by $\hat{\tau}_a = \hat{W}_a \hat{V}^{-1}$. It is clear that this is equivalent to

$$\hat{\tau}_a = \left(\hat{P}(aB_i|\bar{a}_j) \right)_{i,j=1,\dots,m} \left(\hat{P}(B_i|\bar{a}_j) \right)^{-1}_{i,j=1,\dots,m}. \tag{14.37}$$

The choice of m indicative sequences $\bar{a}_j$ is arbitrary and has the additional drawback that in estimating the argument-value matrices from a training string, only a fraction of the data enters the model estimation—namely, only the counting statistics of substrings beginning with one of the m chosen indicative sequences. The information contained in the data is better exploited if we use *all* indicative sequences $\bar{a}_1, \dots, \bar{a}_\kappa \in O^k$, which yields two $m \times \kappa$ matrices containing the argument and the value vectors, requires the use of the pseudoinverse † instead of the matrix inverse, and turns equation 14.37 into

$$\hat{\tau}_a = \left(\hat{P}(aB_i|\bar{a}_j) \right)_{\substack{i=1,\dots,m \\ j=1,\dots,\kappa}} \left(\hat{P}(aB_i|\bar{a}_j) \right)^{\dagger}_{\substack{i=1,\dots,m \\ j=1,\dots,\kappa}}. \tag{14.38}$$

Let $\underline{V} = \left(P(\bar{b}_i|\bar{a}_j) \right)_{i,j=1,\dots,\kappa}$ be the matrix of all conditional probabilities of length k sequences $\bar{b}$ given length k sequences $\bar{a}$, where i, j index the alphabetical enumeration of O^k (we will always use underlined symbols like $\underline{V}$ to denote "big" matrices of size $\kappa \times \kappa$), and let $\underline{\hat{V}}$ be the estimate of $\underline{V}$ obtained from the training string through the obvious counting procedure. Likewise, let $\underline{W}_a = \left(P(a\bar{b}_i|\bar{a}_j) \right)_{i,j=1,\dots,\kappa}$ and $\underline{\hat{W}}_a$ its estimate. Then equation 14.38 is easily seen to be equivalent to

$$\hat{\tau}_a = C_{B_1,\dots,B_m} \underline{\hat{W}}_a \left(C_{B_1,\dots,B_m} \underline{\hat{V}} \right)^{\dagger}. \tag{14.39}$$

Instead of the characteristic event characterizer $C_{B_1,\dots,B_m}$ one may use any characterizer C, which gives us the following learning equation:

$$\hat{\tau}_a = C \underline{\hat{W}}_a \left(C \underline{\hat{V}} \right)^{\dagger} \quad \text{for any characterizer } C. \tag{14.40}$$

It follows from proposition 14.12 that all characterizers $C + G$, where G is any $m \times \kappa$ matrix with zero column sums and $G\underline{V} = \mathbf{0}$ yield the same state vectors as C, which entails

$$\tau_a = C\underline{W}_a \, C\underline{V}^\dagger = (C + G)\underline{W}_a \, ((C + G)\underline{V})^\dagger \tag{14.41}$$

for any such G. Finally, we observe that if ϱ is an OOM transformation as in proposition 14.6, and if C is a characterizer for some OOM $\mathcal{A}$ and ϱC a characterizer for $\mathcal{A}'$ (cf. proposition 14.11), then it is irrelevant whether we use C or ϱC in the learning equation 14.40, because the estimated OOMs will be equivalent via ϱ:

If $\hat{\tau}_a = C\hat{\underline{W}}_a \, (C\hat{\underline{V}})^\dagger$ and $\hat{\tau}'_a = \varrho C\hat{\underline{W}}_a \, (\varrho C\hat{\underline{V}})^\dagger$

then$\varrho\hat{\tau}_a\varrho^{-1} = \hat{\tau}'_a$. $\tag{14.42}$

After proposition 14.14 we remarked that every OOM $\mathcal{A}$ has a reverse characterizer $C^r_\mathcal{A}$, and proposition 14.11 informs us how transforming OOMs via transformations ϱ is reflected in transforming their characterizers with ϱ. Together with proposition 14.12 and 14.41 we can draw the following overall picture:

■ Call two characterizers *equivalent* if they characterize the same OOM. Then the equivalence class of all characterizers of an OOM $\mathcal{A}$ can be written as $C^r_\mathcal{A} + G$, where G is any matrix as described above.

■ We know empirically that different choices of characteristic events (and hence, different characterizers) yield models of different quality when used in the learning equation 14.40. In order to study such sensitivity of learning with respect to choice of characterizers, equation 14.42 informs us that we may restrict the search for "good" characterizers to a single equivalence class.

■ Concretely, we should analyze the quality of model estimates when G is varied in

$$\hat{\tau}_a = (C^r + G)\hat{\underline{W}}_a \, ((C^r + G)\hat{\underline{V}})^\dagger \tag{14.43}$$

for some reverse characterizer C^r whose choice (and hence, choice of equivalence class) is irrelevant.

In order to explain the ES principle, we concentrate of the role of $(C^r + G)\hat{\underline{V}}$ in this learning equation. We can make the following two observations:

■ The variance of models estimated via equation 14.43 is determined by the variance of $(C^r + G)\hat{\underline{V}}$ across different training sequences. We may ignore the role of variance in $(C^r + G)\hat{\underline{W}}_a$ because either the condition of $(C^r + G)\hat{\underline{V}}$ is significantly larger than one, in which case variance in this matrix becomes magnified through the pseudoinverse operation in 14.43 and the overall variance of 14.43 becomes dominated by the variance of $(C^r + G)\hat{\underline{V}}$. Or, the condition of this matrix is close to one, in which case the variance of both $((C^r + G)\hat{\underline{V}})^\dagger$ and $(C^r + G)\hat{\underline{W}}_a$ will be approximately the same due to the similar makeup of $\hat{\underline{V}}$ and $\hat{\underline{W}}_a$, and again we may focus on $(C^r + G)\hat{\underline{V}}$ alone. (For a detailed analysis of these issues see Kretzschmar, 2003).

- The jth column in the matrix $(C^r + G)\underline{V}$ is the state $w_{\bar{a}_j}$ of an OOM characterized by $(C^r + G)$. This is also the expectation of the j-th column $\hat{\mathbf{v}}_j$ in estimates $(C^r + G)\underline{\hat{V}}$. This column $\hat{\mathbf{v}}_j$ can be computed from the training string S as follows:

  Step 1: Initialize $\hat{\mathbf{v}}_j = \mathbf{0}$.

  Step 2: Sweep an observation window of length $2k$ across S. Whenever the windowed substring begins with $\bar{a}_j$, showing $\bar{a}_j \bar{b}_i$, add the ith column $(C^r + G)(:,i)$ of $(C^r + G)$ to $\hat{\mathbf{v}}_j$.

  Step 3: When the sweep is finished, normalize $\hat{\mathbf{v}}_j$ to unit component sum.

We can interpret each additive update of $\hat{\mathbf{v}}_j$ in step 2 as adding a stochastic approximation $(C^r + G)(:,i)$ of $w_{\bar{a}_j}$ to $\hat{\mathbf{v}}_j$. The variance of $\hat{\mathbf{v}}_j$ will thus grow monotonically with the mean stochastic approximation error. Considering the entire matrix $(C^r + G)\hat{V}$ with all its columns, we see that its variance is monotonically tied to the expected stochastic approximation error

$$\xi_G = \sum_{i,j=1}^{\kappa} P(\bar{a}_i \bar{b}_j) \, \|w_{\bar{a}_i} - (C^r + G)(:,j)\|^2. \tag{14.44}$$

Looking for statistically efficent model estimations via equation 14.43 we thus must ask which choice of G makes ξ_D minimal. Here is the main result of this chapter.

Proposition 14.15

$$\arg \min_G \xi_G = \mathbf{0}. \tag{14.45}$$

That is, the reverse characterizer C^r itself minimizes, within its equivalence class, the variance of the argument matrix $(C^r + G)\hat{V}$. The proof (by M. Zhao) is in appendix H. We would like to point out again that it is irrelevant *which* reverse characterizer (and hence, which equivalence class of characterizers) is used; all reverse characterizers yield equivalent models.

The normalizing step 3. is in fact redundant. Just as in the original learning method (cf. eq 14.30) we may just as well use the "raw" counting matrices $\underline{V}^{\text{raw}} = (\#\bar{a}_j \bar{b}_i)$ and $\underline{W}_a^{\text{raw}} = (\#\bar{a}_j \, a \, \bar{b}_i)$ in place of the normalized matrices $\underline{\hat{V}}$ and $\underline{\hat{W}}_a$ in equation 14.43, saving one normalization operation. Here we use $\#$ to denote the count operator, "number of occurrences in training data."

This finding suggests an iterative learning procedure, with the goal of developing a sequence of characterizers that approaches a reverse characterizer, as follows:

- *Learning task*: Given: a training sequence S of length N over an observation alphabet O of size α, and a desired OOM model dimension m.

  Setup: Choose a characterizing length k (we found that the smallest k satisfying $\kappa = \alpha^k \geq m$ often works best). Construct the $\kappa \times \kappa$ counting matrices $\underline{V}^{\text{raw}} = (\#\bar{a}_j \bar{b}_i)$ and $\underline{W}_a^{\text{raw}} = (\#\bar{a}_j \, a \, \bar{b}_i)$.

- *Initial model estimation*: To get started, use the basic learning algorithm from section 14.8 once. Choose characteristic events $B_1, \ldots, B_m$ and code them in the characteristic event characterizer $C_{B_1,\ldots,B_m}$ (eq. 14.33). The characteristic events should be chosen such that $C_{B_1,\ldots,B_m}\underline{V}^{\text{raw}}$ has a good condition number. A greedy heuristic algorithm for this purpose, which works very well, is detailed in appendix I. Compute an initial model $\hat{\mathcal{A}}^{(0)}$ through $\hat{\tau}_a^{(0)} = C_{B_1,\ldots,B_m}\underline{W}_a^{\text{raw}}(C_{B_1,\ldots,B_m}\underline{V}^{\text{raw}})^\dagger$. The starting state $\hat{w}_0^{(0)}$ can either be computed as the eigenvector to the eigenvalue 1 of the matrix $\hat{\mu}^{(0)} = \sum_{a\in O}\hat{\tau}_a^{(0)}$, or equivalently as the vector of row sums of $C_{B_1,\ldots,B_m}\underline{V}^{\text{raw}}$, normalized to unit component sum.

ES iteration: Assume that $\hat{\mathcal{A}}^{(n)}$ is given. Compute its reverse $\hat{\mathcal{A}}^{r\,(n)}$ and the reverse characterizer $\hat{C}^{(n+1)} = \left(\hat{w}_{\bar{b}_1}^{r\,(n)}\cdots\hat{w}_{\bar{b}_\kappa}^{r\,(n)}\right)$. Compute a new model $\hat{\mathcal{A}}^{(n+1)}$ through $\hat{\tau}_a^{(n+1)} = \hat{C}^{(n+1)}\underline{W}_a^{\text{raw}}(\hat{C}^{(n+1)}\underline{V}^{\text{raw}})^\dagger$. The starting state $\hat{w}_0^{r\,(n+1)}$ can again be computed as the normalized row sum vector of $\hat{C}^{(n+1)}\underline{V}^{\text{raw}}$ or from $\hat{\mu}^{(n+1)}$.

Termination: A standard termination criterium would be to calculate the log-likelihood of each model $\hat{\mathcal{A}}^{(n)}$ on S and stop when this appears to settle on a plateau, which is typically the case after two to five iterations.

The rationale behind the iteration step is that if some model $\hat{\mathcal{A}}^{(n+1)}$ comes closer to the true model than the previous one, then the resulting estimated reverse characterizer $\hat{C}^{(n+1)}$ will come closer to a version of the true reverse characterizer, thereby yielding an estimator with lower variance, which in turn *on average* will yield an even better model, etc. This idea motivated calling the entire approach "efficiency sharpening" (ES). We like to call this particular algorithmic instantiation of the ES principle the "poor man's" ES algorithm because it is simple, cheap, and suboptimal—the latter because it exploits only the statistics of substrings of length $2k$. We will soon see how one can do better in this respect. Here are two optional embellishments of the poor man's algorithm:

■ In each iteration, the model $\hat{\mathcal{A}}^{(n)}$ can be transformed into an equivalent one that is interpretable with respect to the characteristic events used for the initial model estimation, before it is used in the iteration. This has, in principle, no effect on the procedure: a sequence of models each equivalent to the corresponding member in the original sequence of models will be obtained. The benefit of having interpretable models is cosmetic and diagnostic: one can produce state plots for each model which are visually comparable.

■ The computational cost per iteration is dominated by computing the pseudoinverse of $\hat{C}^{(n+1)}\hat{V}^{(0)}$. If this matrix is not too ill conditioned (rule of thumb: with a condition number below 1e10 one is on the safe side when using double precision arithmetics), one may employ the well-known (e.g., Farhang-Boroujeny, 1998), computationally much cheaper Wiener-Hopf equation to compute the desired least-square solution $\hat{\tau}_a^{(n+1)}$ to $(\hat{C}^{(n+1)}\underline{V}^{\text{raw}})^\top X^\top = (\hat{C}^{(n+1)}\underline{W}_a^{\text{raw}})^\top$.

A technical point not directly related to the ES principle: If one uses $\underline{W}_a^{\text{raw}}, \underline{V}^{\text{raw}}$ as suggested here, the pseudoinverse (which minimizes MSE of the obtained argument-value mapping) leads to a solution that disproportionally emphasizes the influence of argument-value pairs that represent a relatively small "mass of evidence" in the sense that the corresponding argument-value pairs in $\underline{V}^{\text{raw}}$ and $\underline{W}_a^{\text{raw}}$ have a small mass. If the jth column of these raw matrices is normalized through dividing by the square root of the total weight of the jth column of $\underline{V}^{\text{raw}}$ (instead of division by the raw total weight), one obtains $\hat{W}_a^{(0)}$, $\hat{V}^{(0)}$, which under the $\hat{\tau}_a = \hat{W}_a \hat{V}^\dagger$ operation behave as if the argument-value pair $(\hat{P}(\bar{b}_1|\bar{a}_j) \cdots \hat{P}(\bar{b}_\kappa|\bar{a}_j))$, $(\hat{P}(a\bar{b}_1|\bar{a}_j) \cdots \hat{P}(a\bar{b}_\kappa|\bar{a}_j))$ would occur in $\underline{\hat{W}}_a^{(0)}$, $\underline{\hat{V}}^{(0)}$ multiple times with a multiplicity proportional to $\hat{P}(\bar{a}_j)$. This more properly reflects the "mass of evidence" represented in each argument-value pair and should be preferred. We omitted this reweighting above for expository reasons.

We conclude this section with a little demonstration of the poor man's algorithm at work. The training sequences were obtained from running a randomly created HMM with four states and three output symbols for 1,000 steps; test sequences were 10,000 steps long. The random creation of Markov transition and emission probabilities was biased toward a few high probabilities and many low ones. The reason for doing so is that if the HMM probabilities were created from a uniform distribution, the resulting processes would typically be close to i.i.d.—only Markov transition and emission matrices with relatively many low and a few high probabilities have enough structure to give "interesting" processes. One hundred train-test sequence pairs from different HMM generators were used to train and test 100 OOMs of dimension 3 with the poor man's algorithm, employing two versions where the raw counting matrices were normalized through division with the column sums (variant A, corresponding to eq. 14.43) and through division with the square root of the column sums (variant B).

For comparison, HMMs with three states were trained with the Baum-Welch algorithm. For HMM training we used a public-domain implementation of Baum-Welch written by K. P. Murphy (http://www.cs.ubc.ca/~murphyk/Software/HMM/hmm.html). The Baum-Welch algorithm was run for at most 100 iterations and stopped earlier when the ratio of two successive training log-likelihoods dropped below 5e-5. Only a single Baum-Welch run was executed per data set with the HMM initialization offered by Murphy's software package. On average Baum-Welch used 40.8 iterations.

Our findings are collected in fig. 14.3. Here is a summary of observations of interest:

- On average, we see a rapid development of training and testing log-likelihoods to a plateau, with the first iteration contributing the bulk of model improvement. A closer inspection of the individual learning runs (not shown here), however, reveals a large variability.

- Interesting things happen to the condition number of the argument matrices $C\underline{\hat{V}}$ (or their square-root-normalized correlates in version B). The first iteration on average leads to a significant decrease of it, steering the learning process into a region where the matrix inversion magnifies estimation error to a lesser degree and thus improves statistical efficiency by an additional mechanism different from the ES mechanism proper. We must concede that all phenomena around this important condition number are not well understood.

- The initial model estimates with the basic learning algorithm are, on average, already quite satisfactory (for variant B, they match the final Baum-Welch outcome). This is due to the heuristic algorithm for finding characteristic events, which not only in this suite of experiments worked to our complete satisfaction.

- Compared to the Baum-Welch trained HMMs, the training log-likelihood of OOMs is higher by about 1%, reflecting the greater expressiveness of OOMs and/or the fact that our learning algorithm cannot be trapped in the local optima. In contrast, the OOM test log-likelihood is significantly lower. This reflects the fact that for this particular kind of data, HMMs possess a built-in bias which prevents these models from overfitting.

- Variant B leads to better training log-likelihoods than variant A. Especially the initial models are superior.

- Even the averaged curves exhibit a nonmonotonic development of likelihoods. Inspection of the individual runs would reveal that sometimes the likelihood development is quite bumpy in the first three steps. This point is worth some extra consideration. The ES principle does not root in a concept of iteratively minimizing training error, as Baum-Welch does (and as most other machine learning algorithms do). In fact, the ES principle comes with no guaranteed mechanism of convergence whatsoever. The ES algorithm only "tries" to find an estimator of better statistical efficiency in each iteration, but there is no guarantee that on a given data set, an estimator of higher efficiency will produce a model with higher likelihood.

- The state fingerprints plotted in fig. 14.3 have been derived from models that were interpretable with respect to the characteristic events used in the initial model computation. The plots exhibit some states which fall outside the triangular region which marks the nonnegative orthant of R^3. Whenever we witness states outside this area in an interpretable OOM, we see an invalid OOM at work, that is, the nonnegativity condition 3 from definition 14.2 is violated. It is unfortunately the rule rather than the exception that trained OOMs are invalid. This is cumbersome in at least two respects. First, if one uses such pseudo-OOMs for computing probabilities of sequences, one may get negative values. Second, and even more critically in our view, invalid OOMs are inherently instable (this idea is not detailed here), that is, if they are used for generating sequences, the states may explode. A fundamental solution to this problem is not in sight (cf. the concluding remarks in section 14.6). We can offer only a heuristic stabilizing mechanism that affords nonexploding states and nonnegative probabilities when an invalid OOM is run, at the expense of slightly "blurred" probabilities computed by such stabilized OOMs. This mechanism is described in appendix J. We used it in this suite of learning experiments for determining the training and testing log-likelihoods of learned OMMs.

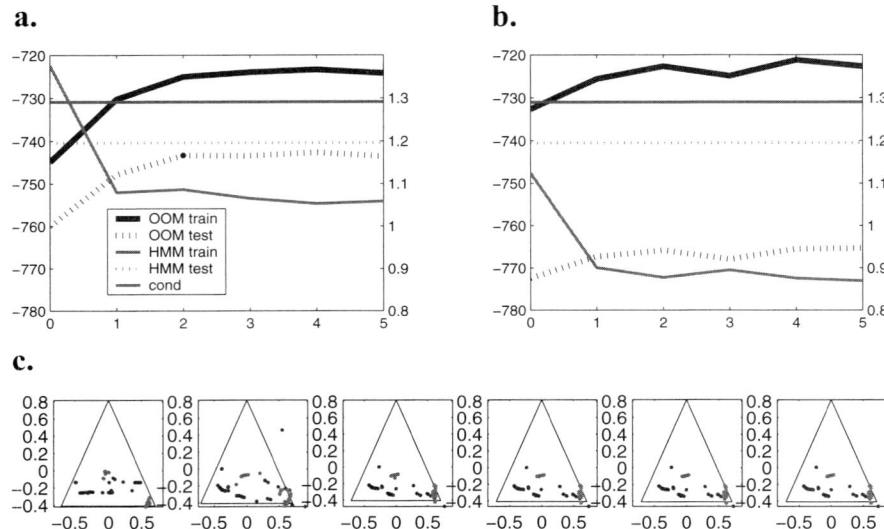

Figure 14.3 (a) Training and testing log-likelihoods (scale on left y-axis) of variant A trained OOMs plotted against the iteration number. The test log-likelihoods were divided by 10 because test strings were longer by this factor, to render them directly comparable with training log-likelihoods. The plot shows the average over 100 training experiments. For comparison, the final train/test log-likelihoods of Baum-Welch trained HMMs is shown by straight horizontal lines. Iteration 0 corresponds to the initial model estimation. In addition, the average of $\log_{10}$ of condition numbers of the matrices $C\hat{\underline{V}}$ are shown (scale on the right y-axis). (b) A similar plot for variant B. (c) "Fingerprints" of the OOM models obtained in the successive iterations in one of the learning runs.

14.12   Essentials of Suffix Trees

We now proceed to describe an improved instantiation of the ES learning principle, using *suffix trees* to represent state sequences. In this section we recapitulate the basic concepts of suffix trees.

The suffix tree T (Weiner, 1973) for a given string S provides a compact representation of all substrings of S while exposing the internal structure of S in an efficient data structure. Moreover, a (compact) suffix tree T can be constructed in linear time $\mathcal{O}(|S|)$. While the suffix tree is a simple enough data structure, linear-time construction algorithms are quite involved (Giegerich and Kurtz, 1997).

Formally, a suffix tree is a *trie* with additional properties. A trie T (Fredkin, 1960)—the name being derived from re*trie*val—is an ordered tree where edges are labeled with strings over an alphabet O such that no two edges from some node k of T to its children have labels beginning with the same symbol $a \in O$. Thus we may speak of the *path* to a node k of T, defined as the concatenation of all edge labels encountered when traveling from the root to k, and we may identify the node k with $\mathsf{path}(k) \in O^*$. The set of words $\mathsf{words}(T) \subset O^*$ represented by a tree is defined by

$$\bar{a} \in \mathsf{words}(T) \quad \Longleftrightarrow \quad \exists k \in T : \ \bar{a} = \mathsf{path}(k)\bar{b}, \qquad (14.46)$$

where $\bar{b}$ is a prefix of an outgoing edge of k. Then, a suffix tree T_S of a string S is a trie that contains all substrings of S:

$$\mathsf{words}(T_S) = \{\bar{a} \in O^* : \bar{a} \text{ is a substring of } S\}. \qquad (14.47)$$

In general there will be more than one trie satisfying the above condition. However if we additionally require that all nodes in T_S are either leaves or have at least two children, the suffix tree of S is uniquely determined (up to ordering). It is called the *compact* suffix tree of S. Compact tries were introduced historically under the name *Patricia trees* (Morrison, 1968). Figure 14.4 illustrates the concept with a compact suffix tree for the string *cocoa*.

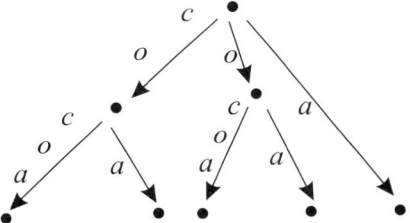

Figure 14.4   The compact suffix tree for *cocoa*. Note that a sentinel is not required because the symbol a appears in the string only at the terminal position and thus acts as a sentinel.

Sometimes it is desirable to have every suffix of S represented by a leaf. This can be enforced by appending a *sentinel* symbol $\$ \notin O$ at the end of S. Then the compact suffix tree for the extended string $S\$$ represents every suffix of $S\$$ and only the suffixes of $S\$$ as leaves. Finally, note that a compact suffix tree of a sequence of length N has at most $2N$ nodes.

14.13   A Suffix-Tree-Based Version of Efficiency Sharpening

An obvious weakness of the poor man's ES algorithm is that the family of indicative sequences $(\bar{a}_i)_{1 \leq \kappa} = O^k$ is not adapted to the training sequence S. Some of the $\bar{a}_i$ might not occur in S at all, entailing a useless computational overhead. Some others may occur only once or twice, yielding very poor estimates of probabilities through relative frequencies. On the other side of the spectrum, if some sequence $\bar{a}_i$

occurs very frequently, learning would clearly benefit from splitting it into α longer sequences by $\bar{a}_i \mapsto \{a\bar{a}_i | a \in O\}$. In this section we show how a compact suffix tree (ST) representation of S can be used to support a choice of indicative sequences which is matched to S. (We will henceforth drop the qualifier "compact", it is tacitly assumed throughout). The idea of representing variable-length "context" for prediction purposes in a suffix tree is known in the literature under various names, e.g., prediction suffix trees or variable-order Markov chains (a concise overview in Dekel et al., 2004).

Implementing suffix trees involves some tedious "indexery witchcraft." We therefore do not attempt a detailed documentation of our suffix-tree-based EM implementation but instead outline the basic ideas, which are quite straightforward.

Let \$ be a sentinel symbol not in O, and let \$$S$ be S prepended by \$. We will still speak of \$$S$ as the training sequence. Suffix trees are brought into the learning game by observing that, first, the words of the ST $T_{(\$S)^r} = T_{S^r\$}$ of the *reverse* training sequence $(\$S)^r = S^r\$$ are the reverse substrings of \$$S$ (this is clear by definition of a ST). But second and more interestingly, it moreover holds that if

- k_1 is a node in $T_{S^r\$}$ and k_2 is a child node of node k_1,
- $\bar{c}_1 = \mathsf{path}(k_1)$ is the path to k_1 and $\bar{c}_1\bar{c}_2$ the path to k_2,
- $\bar{a}_1 = \bar{c}_1^r$, $\bar{a}_2 = \bar{c}_2^r$, $\bar{a}_2\bar{a}_1 = (\bar{c}_1\bar{c}_2)^r$ are the associated forward words, and
- $\bar{a}_2 = \bar{a}_{21}\bar{a}_{22}$ is some split of $\bar{a}_2$,

then wherever $\bar{a}_{22}\bar{a}_1$ occurs in \$$S$, it occurs after $\bar{a}_{21}$.

This can be rephrased as follows. Let, for some word $\bar{a}$, $\mathsf{pos}(\bar{a})$ denote the set of all the position indices n in S such that the sequence from the beginning of S up to position n ends with $\bar{a}$. Furthermore, for some node k of $T_{S^r\$}$, let $\mathsf{pos}(k) = \mathsf{pos}((\mathsf{path}(k))^r)$ be the set of positions in the forward sequence \$$S$ associated with the reverse path of k. Then, if we reuse the notations from the above bullet list, and if $\bar{a}_2 = a_1 \cdots a_l$, then

$$\mathsf{pos}(a_1 a_2 \cdots a_l\,\bar{a}_1) = \mathsf{pos}(a_2 a_3 \cdots a_l\,\bar{a}_1)$$

$$\cdots$$

$$= \mathsf{pos}(a_l\,\bar{a}_1)$$

$$\underset{\neq}{\subseteq} \mathsf{pos}(\bar{a}_1). \tag{14.48}$$

Now think of reverse versions $\bar{a}$ of words $\bar{c}$ from $T_{S^r\$}$ as candidates for indicative sequences. If $\mathsf{pos}(\bar{a}) = \mathsf{pos}(\bar{a}')$, then clearly it makes no sense to collect continuation statistics of the type $\#\bar{a}\bar{b}$ both for $\bar{a}$ and $\bar{a}'$, because they are identical. Therefore, the nodes of $T_{S^r\$}$ correspond to potential indicative sequences that are distinguishable within S with respect to their continuations in S, and we may ignore all words $\bar{a}$ whose reverse does not end in a node of $T_{S^r\$}$. This is the basic idea of suffix-tree-based ES: use as indicative sequences all the words whose reverses correspond to nodes in $T_{S^r\$}$.

We now turn to reverse characterizers. An analysis of the poor man's algorithm reveals that, given a reverse OOM with states $w_{\bar{b}}^r$, we constructed estimates of $w_{\bar{a}}$ through

$$\hat{w}_{\bar{a}} = \frac{\sum_{\bar{b} \in O^k} w_{\bar{b}}^r * (\text{number of occurences of } \bar{a}\bar{b} \text{ in } S)}{\text{number of occurences of } \bar{a} \text{ in } S}. \tag{14.49}$$

If we were to copy this idea identically for use with suffix-tree-managed indicative sequences $\bar{a}$, we would have to collect statistics for continuations by all $\bar{b} \in O^k$, for all our indicative sequences $\bar{a}$. Furthermore, in doing so we would of course have to fix k and thus ignore information provided in S by continuations longer than k. A stronger idea suggests itself here. Let $S = a_1 \ldots a_N$. Instead of precalculating all $w_{\bar{b}}^r$ and collecting the necessary continuation statistics, we simply run the reverse OOM on the reverse training sequence once, obtaining a state sequence $w_{\varepsilon}^r, w_{a_N}^r, w_{a_{N-1} a_N}^r, \ldots, w_S^r$. Reversing in time and renaming yields a state sequence that is more convenient for our purposes, by putting

$$(c_0, c_1, \ldots, c_N) := (w_S^r, \ldots, w_{a_{N-1} a_N}^r, w_{a_N}^r, w_{\varepsilon}^r). \tag{14.50}$$

We interpret c_n as a stochastic approximation to $w_{\bar{a}_n}$, in the following sense. Consider the limit case of $N \to \infty$. Assume that for a right-infinite sequence $\bar{b}_\infty = b_1 b_2 \ldots$ the limit $c_{\bar{b}_\infty} = \lim_{l \to \infty} w_{b_1 \ldots b_l}^r$ exists almost surely (we conjecture that for reverse ergodic processes this always holds). Let $P_{\bar{a}_n}$ be the conditional probability distribution over the set of right-infinite continuations $\bar{b}_\infty$ of $\bar{a}_n$. Then the family $(c_{\bar{b}_\infty})_{\bar{b}_\infty \in O^\infty}$ can be regarded as an (infinite) characterizer by setting

$$w_{\bar{a}} = \int c_{\bar{b}_\infty} dP_{\bar{a}} \tag{14.51}$$

for all $\bar{a} \in O^*$. Because S is finite, interpreting c_n as a stochastic approximation to $w_{\bar{a}_n}$ via equation 14.51 will incur some inaccuracy, which however will be negligible for all n that are not very close to the right end of S. All of this entitles us to change the poor man's strategy (eq. 14.49) to this rich woman's version:

$$\hat{w}_{\bar{a}} = \frac{1}{|\mathsf{pos}(\bar{a})|} \sum_{n \in \mathsf{pos}(\bar{a})} c_n. \tag{14.52}$$

Finally, observe that $T_{S^r\$}$ has $N+1$ leaf nodes, each corresponding uniquely to one position in $\$S$. That is, for a leaf node k, $\mathsf{pos}(k) = \{n\}$, where n is the position within $\$S$ where the reverse path of k ends, started from the beginning of $\$S$. For an internal node k with children $k_1, \ldots, k_x$ it holds that $\mathsf{pos}(k) = \bigcup_{i=1,\ldots,x} \mathsf{pos}(k_i)$.

Now everything is in place for describing the suffix-tree-based ES algorithm.

■ **Task.** Given: a training sequence S of length N and a model dimension m. Wanted: an m-dimensional OOM.

■ **Initialization.**

Learn initial model: Learn an m-dimensional OOM $\hat{\mathcal{A}}^{(0)}$ using the basic learning algorithm, as in the poor man's algorithm.

Construct $T_{S^r\$}$

Procure argument-value pair mapping: Let k_{all} be the leaf that corresponds to the entire sequence. Allocate a map $f : T_{S^r\$} - \{k_{\text{all}}\} \times O \to T_{S^r\$} \cup \{0\}$ and initialize it by all zero values. For each node k except k_{all}, where the reverse path of k is $\bar{a}$, and for each $a \in O$, determine the highest node k' such that $(\bar{a}a)^r$ is a prefix of the path of k' (then $\text{pos}(k') = \text{pos}(\bar{a}a)$). Set $f(k, a) = k'$.

- **ES Iteration.** Input: $\hat{\mathcal{A}}^{(n)}$, $T_{S^r\$}$. Output: $\hat{\mathcal{A}}^{(n+1)}$.

  Procure $\hat{w}_{\bar{a}}$'s: (i) Compute the reverse OOM $\hat{\mathcal{A}}^{r\,(n)}$. (ii) Run it on the reverse training sequence to obtain $(c_0, c_1, \dots, c_N)$ (use the "blurred" stabilizing method for running potentially invalid OOMs that is detailed out in appendix J). (iii) Sort these $N + 1$ states into the leaf nodes of $T_{S^r\$}$, where c_n goes to the leaf node with $\text{pos}(k) = \{n\}$. Formally, for leaves k set $C(k) = c_{\text{pos}(k)}$. (iv) From the leaves upward, for some internal node k for whose children k' $C(k')$ has already been determined, set $C(k) = \sum_{k'\text{is a child of}k} C(k')$. Do this until all nodes have been covered. Then for all nodes k it holds that

  $$C(k) = |\text{pos}(\bar{a})|\, \hat{w}_{\bar{a}} = \sum_{n \in \text{pos}(\bar{a})} c_n. \tag{14.53}$$

where $\bar{a}$ is the reverse path of the node.

Procure argument-value matrices $\hat{V}$ and $\hat{W}_a$: To obtain matrices $\hat{V}$ and $\hat{W}_a$ (each of size $m \times |T_{S^r\$}| - 1$) that play a similar role as we are accustomed to, go through all nodes k of the tree (except k_{all}), write $C(k)$ into the kth column of $\hat{V}$ and $C(f(k, a))$ into the kth column of $\hat{W}_a$.

Reweigh: In analogy to the reweighing scheme described for the poor man's algorithm, divide each column k in $\hat{V}$ and all $\hat{W}_a$ by the square root of the k-th column sum of $\hat{V}$.

Compute new model: Set $\hat{\tau}_a^{(n+1)} = W_a \hat{V}^\dagger$ and $\hat{w}_0^{(n+1)}$ as the eigenvector to the eigenvalue 1 of $\hat{\mu} = \sum_{a \in O} \hat{\tau}_a^{(n+1)}$ (normalized of course to unit component sum), obtaining $\hat{\mathcal{A}}^{(n+1)}$.

- **Termination.** Stop after a predetermined number of iterations or when training log-likelihood seems to saturate.

- **Optional tuning.** One may augment this algorithm in several ways:

  Make models interpretable: Transform each $\hat{\mathcal{A}}^{(n+1)}$ to an interpretable OOM before further use, using the characteristic events employed in the initial model estimation. This gives comparable "fingerprint" plots that are helpful in monitoring the algorithm's progress. More importantly, we found that such renormalization sometimes renders the algorithm more robust when otherwise the condition of V deteriorated over iterations.

  Use subsets of tree: When constructing $\hat{V}$ and $\hat{W}_a$, we may restrict ourselves to using only ST nodes that represent some minimal number of positions, guided by the intuition that nodes representing only very few string positions contain too unreliable stochastic approximations of forward states.

A nasty trick to improve stability: The algorithm depends on a matrix inversion (more correctly, a pseudoinverse computation). We sometimes experienced that the matrix $\hat{V}$ becomes increasingly ill conditioned as iterations progress, completely disrupting the learning process when the ill-conditioning explodes. A powerful but brutal and poorly understood remedy is to transform the reverse OOM $\hat{\mathcal{A}}^{r\,(n)}$ (before using it) into a surely valid OOM by making its operator matrices all-nonnegative. Concretely, set all negative entries in the operator matrices of $\hat{\mathcal{A}}^{r\,(n)}$ to zero and renormalize columns by dividing them through the corresponding column sum of their sum matrix. To our surprise, often the model quality suffered only little from this dramatic operation. Purely intuitively speaking, we might say that the learning process is forced to find a solution that is close to a HMM (where forward and reverse matrices are nonnegative).

All ST-related computations involved in this procedure can be effected in time linear of the ST size, which is at most $2N$. The main cost factors in a suffix-tree ES iteration are the computation of the reverse state sequence, which is $O(Nm^2)$, and the computation of the pseudoinverse. To speed up the computation in cases where $\hat{V}$ is not too ill conditioned, one may use the Wiener-Hopf solution instead, as described for the poor man's version of ES. Then the cost of calculating the operators from $\hat{V}$ and the $\hat{W}_a$'s is $O(\alpha m^2 N)$, which dominates the cost for obtaining the reverse state sequence. In practice we found runtimes per iteration that are somewhat shorter than a Baum-Welch iteration in the same task. However, for the total time of the algorithm we must add the time for the initial model estimation and the computation of the suffix tree. The latter in our implementation takes about one to two times the time of an ES iteration.

14.14   A Case Study: Modeling One Million Pounds

The most demanding task that we have tried so far was to train models on Mark Twain's short story "The £1,000,000 Bank-Note" (e.g., www.eastoftheweb.com/short-stories/UBooks/MilPou.shtml). We preprocessed the text string by deleting all special characters except the blank, changing capital letters to small caps, and coding letters by integers. This gave us 27 symbols: 1 (codes *a*), ..., 26 (*z*), 27 (_). The resulting string was sorted sentence-wise into two substrings of roughly equal length (21,042 and 20,569 symbols, respectively) that were used as training and test sequences.

The suffix-tree based learning algorithm was used with the "brutal" stabilizing method mentioned above, which was necessary for larger model dimensions. Five ES iterations besides the 0th step of estimating an initial model were run for model dimensions $5, 10, 15, \ldots, 50, 60, \ldots, 100, 120, 150$. More details can be found in appendix K.

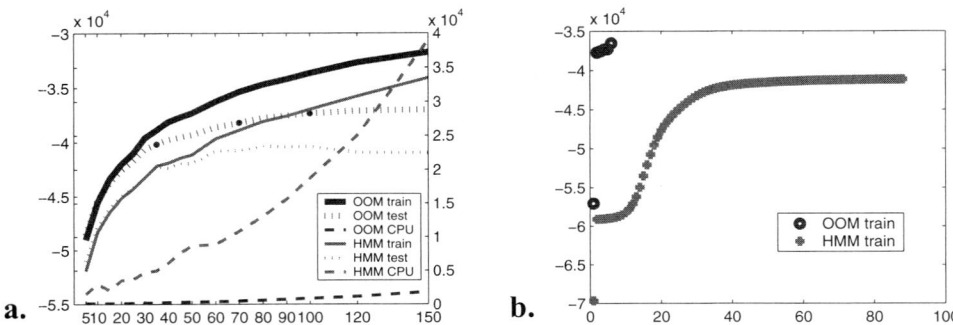

Figure 14.5 (a) Training and test log-likelihoods (scale on left y-axis) of ES-trained OOMs and EM-trained HMMs plotted against model dimension. In addition, the CPU time (in seconds, scale on right y-axis) for both is shown. (b) A close-up on the development of training log-likelihood for the learning of a 50-dimensional OOM and HMM. The EM algorithm here met its termination criterion in the 90th iteration.

For comparison, HMMs of the same dimensions were again trained with K. Murphy's Matlab implementation of Baum-Welch. HMMs were initialized using the default initialization offered by this implementation. Iterations were stopped after 100 iterations or when the ratio of two subsequent training log-likelihoods was smaller than 1e-5, whichever occurred first (almost always 100 iterations were reached). Only a single run per model dimension was carried out; no overhead search methods for finding good local optima were invoked. Thus it remains unclear whether with more search effort, the HMM performance could have been significantly improved. In the light of the fact that across the different model dimension trials the HMM performance develops quite smoothly, this appears unlikely: if significantly better HMMs would exist and could be found by random search, then we would expect that due to the random HMM initialization the performance curve would look much more rugged.

Figure 14.5 shows the obtain results. Computations were done on a notebook PC with a Pentium 330 MHz processor in Matlab.

14.15   Discussion

The findings about the ES algorithm reported in this chapter are not older than three months at the time of writing and far from mature. We could not yet extensively survey the performance of the ES algorithm except for ad hoc tests with some synthetic data (discretized chaotic maps, outputs of FIR filters driven by white-noise input, outputs of random recurrent neural networks, HMM-generated sequences) and a few standard benchmark data sets (sunspot data, Melbourne meteorological data). In all cases, the behavior of ES was similar to the HMM and 1 Million Pound data sets reported here: a rapid development of models toward plateau training log-likelihoods (in 1 to 10 iterations, typically 3), followed by a jittery "hovering" at that plateau. Both training and testing likelihoods at the plateau level were superior to HMM performance (however, these were not optimized) except for HMM-generated data, where HMM models can play out their natural bias for these data. Thus it is fair to say that learning algorithms based on suffix-tree ES clearly have the potential to yield more accurate models of stationary symbol processes, and in a much shorter runtime, than HMMs.

Although it might be tempting, it is misleading to see the ES algorithm as a variant of the EM algorithm. Both algorithms execute an iterative interplay between model reestimation and state generation, but there are two fundamental differences between them:

- The estimator instantiated by each ES step, including the initial model estimator, is asymptotically correct. That is, if the process is in fact m-dimensional and m-dimensional models are trained, the modeling error would go to zero with increasing length of training data almost surely, right from the 0th iteration. This is not the case with the EM algorithm.

- The training log-likelihood does not necessarily grow monotonically under ES, but this behavior is constitutional for EM. The ES principle is not designed to monotonically improve any target quantity.

Contemplating the task of improving an asymptotically correct estimator, the natural target for improvement—actually the only one that we can easily think of— is the statistical efficiency of the estimator. This was the guiding idea that led us to the discovery of the learning methods described in this chapter, and it is borne out by the mathematical analysis presented in section 14.11. However, we are not sure whether this is already the complete explanation of why and how our algorithms work. First, the role of the condition of the $\hat{V}$ matrix has to be clarified—if it is not well conditioned, inverting $\hat{V}$ will magnify estimation errors contained in it and potentially invalidate the reasoning of section 14.11. Interestingly, even when the condition of $\hat{V}$ deteriorates through ES iterations, we mostly find that the model quality increases nonetheless, up to a point where $\hat{V}$ becomes so ill conditioned that numerical errors explode. We confess that this is hard to understand. Second, we have only provided an argument that the reverse characterizer obtained from the correct model yields a maximally efficient estimator. But it remains to be investigated in what sense the sequence of reverse characterizers constructed in ES moves toward this correct model; in other words, how the sequence of reverse characterizers is related to the gradient of ξ at points away from the minimum.

The main problem of current ES implementations is that learning runs sometimes become instable (condition of $\hat{V}$ explodes). This is related to the model dimension: while small dimensions never present difficulties in this respect, learning larger models becomes increasingly prone to instability problems. We currently perceive two sources for instability:

Data-inherent ill-conditioning: If the generating process has a lower dimension than the models one wants to learn, the matrix $\hat{V}$ *must* become ill conditioned with increasing training data size. Now, real-life data sets will hardly come from any low-dimensional generator. But if we would investigate their limiting singular value spectrum (that is, the singular value spectrum of the matrices $V_k = (P(\bar{b}_i|\bar{a}_j))_{\bar{a},\bar{b}\in O^k}$ in the limit of $k \to \infty$) and find a rapidly thinning tail, then for all practical learning purposes such generators behave like low-dimensional generators, intrinsically leading to matrices V of low numerical rank.

Invalid OOMs: Running invalid reverse OOMs to create reverse characterizers is prone to sprinkle the obtained state sequence with outliers, which are hard to detect. Using such contaminated reverse characterizers to feed the input to the linear regression task will even deteriorate the situation because the minimization of MSE will further magnify the impact of outliers—a familiar problem. This clearly happened in some of our more problematic (high model dimension) test runs.

HMM/EM learning does not rely on any matrix (pseudo)inversion and does not suffer from the ensuing stability problems. Although we are fascinated by the promises of ES, for the time being we would prefer HMM/EM over OOM/ES in any safety-critical application. But we hope that the rich repertoire of numerical linear algebra will equip us with the right tools for resolving the stability issue. The rewards should be worth the effort.

Appendix A: Proof of Proposition 14.4, Property 4

We first show $\forall\, a \in O,\; w \in W\;\; \pi(\tau_a w) = t_a\, \pi(w)$. Let $w = \sum_{i=1,\dots,d} \alpha_i w_{\bar{a}_i}$. Then

$$
\begin{aligned}
\pi(\tau_a w) &= \pi\Big(\sum_{i=1,\dots,d} \alpha_i \tau_a w_{\bar{a}_i} \Big) \\
&= \pi\Big(\sum_{i=1,\dots,d} \alpha_i\, \mathbf{1}\tau_a w_{\bar{a}_i} \frac{\tau_a w_{\bar{a}_i}}{\mathbf{1}\tau_a w_{\bar{a}_i}} \Big) \\
&= \sum_{i=1,\dots,d} \alpha_i\, P(a|\bar{a}_i) f_{\bar{a}_i a} \\
&= \sum_{i=1,\dots,d} \alpha_i\, t_a\, f_{\bar{a}_i} = t_a \sum_{i=1,\dots,d} \alpha_i f_{\bar{a}_i} \\
&= t_a\, \pi(w).
\end{aligned}
$$

An iterated application of this finding yields the statement of the proposition.

Appendix B: Proof of Proposition 14.5

"$\Rightarrow$": Let $x \in ker\pi$, $\bar{a} \in O^*$. Then $\mathbf{1}\tau_{\bar{a}}x = \sigma\pi(\tau_{\bar{a}}x) = \sigma t_{\bar{a}}\pi(x) = 0$ by proposition 14.4, properties 3 and 4.

"$\Leftarrow$":

$$\forall \bar{a} \in O^* \quad \mathbf{1}\tau_{\bar{a}}x = 0$$
$$\rightarrow \forall \bar{a} \in O^* \quad \sigma t_{\bar{a}}\pi(x) = 0$$
$$\rightarrow \forall \bar{a} \in O^* \quad (\pi(x))(\bar{a}) = 0$$
$$\rightarrow \pi(x) = \mathbf{0}.$$

Appendix C: The Three OOMs from Figure 14.2

The plotted OOMs were obtained from HMMs over the alphabet $O = \{a, b, c\}$ as described in section 14.3. The Markov transition matrix was

$$M = \begin{pmatrix} 1 - 2\alpha & \alpha & \alpha \\ \alpha & 0 & 1 - \alpha \\ 0 & 1 - \alpha & \alpha \end{pmatrix},$$

where α was set to 0.1, 0.2, and 0.3 respectively for the three plotted OOMs. The symbol emission matrices O_a were

$$O_a = \mathrm{diag}(0.8\ 0.1\ 0.1),\ O_b = \mathrm{diag}(0.1\ 0.8\ 0.3),\ O_c = \mathrm{diag}(0.1\ 0.1\ 0.6)$$

for all HMMs. From these HMMs, OOMs were created that were interpretable with respect to the singleton characteristic events $A_1 = \{a\}, A_2 = \{b\}, A_3 = \{c\}$.

Appendix D: Proof of Proposition 14.10

To see the "if" direction, consider an n-dimensional OOM $\mathcal{A}$ for (X_n) whose states $w_{\bar{a}_j}$ $(j = 1, \ldots, m)$ are the columns of W and whose other states $w_{\bar{a}}$ are linear combinations of the $w_{\bar{a}_j}$ (whereby $\mathcal{A}$ is uniquely determined according to the insights from section 14.5—the column vectors of W span the "prediction-relevant" subspace V from 14.18). It is a mechanical exercise to show that condition 14.31 holds. For the "only if" direction choose any m sequences $\bar{a}_j$ such that V has rank m. Then by the definition of a characterizer, W has the states $w_{\bar{a}_j}$ of the OOM characterized by c as its columns, which must be linearly independent because V has rank m.

Appendix E: Proof of Proposition 14.12

According to proposition 14.10, every characterizer C of $\mathcal{A}$ must satisfy $V^\top C^\top = W^\top$. It is straightforward to derive that conversely, any C with unit column sums satisfying $V^\top C^\top = W^\top$ is a characterizer. Now any $m \times \kappa$ matrix C satisfies $V^\top C^\top = W^\top$ if and only if $C = C_0 + D$, where the rows of D are in $ker\, V^\top$. The additional requirement that the column sums of C must sum to unity is warranted by making the last row of D equal to the negative sum of the other rows.

Appendix F: Proof of Proposition 14.13

Recalling that D is a diagonal matrix with w_0 on its diagonal, the following transformations yield the claim:

$$
\begin{aligned}
P_{\mathcal{A}}(a_0 \cdots a_n) &= \mathbf{1}\tau_{a_n} \cdots \tau_{a_0} w_0 \\
&= w_0^\top \tau_{a_0}^\top \cdots \tau_{a_n}^\top \mathbf{1}^\top \\
&= w_0^\top D^{-1} D \tau_{a_0}^\top D^{-1} \cdots D \tau_{a_n}^\top D^{-1} D \mathbf{1}^\top \\
&= \mathbf{1} D \tau_{a_0}^\top D^{-1} \cdots D \tau_{a_n}^\top D^{-1} w_0 \\
&= P_{\mathcal{A}^r}(a_n \cdots a_0).
\end{aligned}
$$

Appendix G: Proof of Proposition 14.14

We first assume that C is a characterizer and show statement 2. According to 14.32 we have $w_{\bar{a}}' = C \, (P(\bar{b}_1|\bar{a}) \cdots P(\bar{b}_\kappa|\bar{a}))^\top$, which is equal to $C\pi_{\mathcal{A}} w_{\bar{a}}$ by proposition 14.9, property 1. To show that $\varrho = R^\top R$ (assuming now $w_0 = (1/m \cdots 1/m)^\top$), we consider for some $\bar{b} = b_1 \cdots b_k$ a column $c = \tau_{\bar{b}}^r w_0 \,/\, \mathbf{1}\tau_{\bar{b}}^r w_0 = \tau_{\bar{b}}^r w_0 / P(\bar{b})$ of C. Using the terminology from proposition 14.13 and noting that $D = (1/m \cdots 1/m)^\top$, it can be rewritten as follows:

$$
\begin{aligned}
P(\bar{b})c &= \tau_{b_1}^r \cdots \tau_{b_k}^r w_0 \\
&= D \tau_{b_1}^\top \cdots \tau_{b_k}^\top D^{-1} w_0 \\
&= 1/m \;\; \tau_{b_1}^\top \cdots \tau_{b_k}^\top \mathbf{1}^\top \\
&= 1/m \;\; \left(\mathbf{1}\tau_{b_k} \cdots \tau_{b_1} \right)^\top \;\; = \;\; 1/m \;\; (\mathbf{1}\tau_{\bar{b}})^\top.
\end{aligned}
$$

Thus the ith column of C equals the transpose of the ith row of $\pi_{\mathcal{A}}$ up to a factor of $(mP(\bar{b}_i))^{-1}$. Splitting this factor into $(mP(\bar{b}_i))^{-1/2} \, (mP(\bar{b}_i))^{-1/2}$ and redistributing the splits over C and $\pi_{\mathcal{A}}$ yields the statement $\varrho = R^\top R$.

To show the first claim, first notice that C is a characterizer for $\mathcal{A}'$ if and only if $\tilde{\varrho}C$ is a characterizer for $\varrho\,\mathcal{A}'$ for some equivalence transformation $\tilde{\varrho}$ according to propositions 14.6 and 14.11. Because a transformation $\tilde{\varrho}$ can always be found that maps w_0 on $(1/m \cdots 1/m)^\top$ (exercise), we may assume, without loss of generality, that $w_0 = (1/m \cdots 1/m)^\top$. Let

$$R = \left((mP(\bar{b}_1))^{-1/2} (\mathbf{1}_{\tau_{\bar{b}_1}})^\top \;\cdots\; (mP(\bar{b}_\kappa))^{-1/2} (\mathbf{1}_{\tau_{\bar{b}_\kappa}})^\top \right)^\top$$

as above. Define vectors $v_{\bar{a}} = C\pi_{\mathcal{A}} w_{\bar{a}} = R^\top R\, w_{\bar{a}}$. We want to show that the transformation $C\pi_{\mathcal{A}} = R^\top R$ is an OOM equivalence transformation according to proposition 14.6. It is easily checked that $C\pi_{\mathcal{A}}$ has the property 3 listed in proposition 14.6. The critical issue is to show that $C\pi_{\mathcal{A}} = R^\top R$ is bijective, i.e., has rank m. We use that for any matrix A it holds that $\mathrm{rank}(A) = \mathrm{rank}(A^\top A)$. We see that $\mathrm{rank}(R^\top R) = \mathrm{rank}(R) = \mathrm{rank}((\mathbf{1}_{\tau_{\bar{b}_1}})^\top \;\cdots\; (\mathbf{1}_{\tau_{\bar{b}_\kappa}})^\top) = m$, where the last equality is due to the fact that the $\bar{b}_i$ are characterizing sequences. Thus, $C\pi_{\mathcal{A}} = R^\top R$ is an OOM equivalence transformation, and the vectors $v_{\bar{a}} = C\pi_{\mathcal{A}} w_{\bar{a}}$ are the states of an OOM equivalent to $\mathcal{A}$. But considering 14.32, this is just another way of saying that C is a characterizer.

Appendix H: Proof of Proposition 14.15

We first derive some conditions that the matrix G should satisfy. It follows from proposition 14.12 that $\mathbf{1}_m G = \mathbf{0}$ and $G\underline{V} = \mathbf{0}$. As before, here we use $\mathbf{1}$ to denote the row vector of units, but with a subscript specifying its dimension. By the definition of $\underline{V}$ and $\pi_{\mathcal{A}}$ (see eq. 14.36), we have $\underline{V} = \pi_{\mathcal{A}} W$, where $W = [w_{\bar{a}_1} \cdots w_{\bar{a}_\kappa}]$. Because $\mathrm{rank} W = m$, it is clear that $G\underline{V} = \mathbf{0}$ if and only if $G\pi_{\mathcal{A}} = \mathbf{0}$. Thus, it suffices to show that $G = \mathbf{0}$ is a minimizer of the following optimization problem:

$$\min_G J(G) = \frac{1}{2} \sum_{i,j=1}^{\kappa} P(\bar{a}_i \bar{b}_j) \| w_{\bar{a}_i} - (C^r + G)(:,j) \|^2,$$
$$\text{s.t.} \quad \mathbf{1}_m G = \mathbf{0}, \quad G\pi_{\mathcal{A}} = \mathbf{0}. \tag{14.54}$$

The target function $J(G)$ can be rewritten as

$$J(G) = \frac{1}{2} \sum_{i,j=1}^{\kappa} P(\bar{a}_i \bar{b}_j) \left(\| w_{\bar{a}_i} \|^2 + \| (C^r + G)(:,j) \|^2 \right)$$
$$- \sum_{i,j=1}^{\kappa} P(\bar{a}_i \bar{b}_j) \left\langle w_{\bar{a}_i}, (C^r + G)(:,j) \right\rangle, \tag{14.55}$$

where the pair $\langle x, y \rangle$ denotes the inner product of x and y. The second item on the right-hand side of the above equality can be further simplified, as follows:

$$\sum_{i,j=1}^{\kappa} P(\bar{a}_i\bar{b}_j)\Big\langle w_{\bar{a}_i}, (C^r + G)(:,j)\Big\rangle$$

$$= \sum_{i=1}^{\kappa} \Big\langle w_{\bar{a}_i}, \sum_{j=1}^{\kappa} P(\bar{a}_i\bar{b}_j)(C^r + G)(:,j)\Big\rangle$$

$$= \sum_{i=1}^{\kappa} \Big\langle w_{\bar{a}_i}, P(\bar{a}_i) \sum_{j=1}^{\kappa} P(\bar{b}_j|\bar{a}_i)(C^r + G)(:,j)\Big\rangle$$

$$= \sum_{i=1}^{\kappa} \Big\langle w_{\bar{a}_i}, P(\bar{a}_i)(C^r + G)\underline{V}(:,i)\Big\rangle$$

$$= \sum_{i=1}^{\kappa} \Big\langle w_{\bar{a}_i}, P(\bar{a}_i)w_{\bar{a}_i}\Big\rangle$$

$$= \sum_{i=1}^{\kappa} P(\bar{a}_i)\|w_{\bar{a}_i}\|^2,$$

Assuming that the process is stationary, and substituting the above equality into equation 14.55, we get

$$J(G) = \frac{1}{2} \sum_{j=1}^{\kappa} P(\bar{b}_j)\|(C^r + G)(:,j)\|^2 - \frac{1}{2} \sum_{i=1}^{\kappa} P(\bar{a}_i)\|w_{\bar{a}_i}\|^2. \tag{14.56}$$

Since the second item of equation 14.56 is irrelevant to G, minimizing $J(G)$ under the constraints of equation 14.54 is a convex quadratic programming problem. So $G = \mathbf{0}$ is a minimizer of $J(G)$ if and only if it satisfies the following KKT system:

$$\frac{\partial J}{\partial G} = (C^r + G)D_p = \mathbf{1}_m^\top \mu^\top + \lambda \pi_{\mathcal{A}}^\top, \tag{14.57}$$

$$\mathbf{1}_m G = \mathbf{0}, \tag{14.58}$$

$$G\pi_{\mathcal{A}} = \mathbf{0}, \tag{14.59}$$

where $D_p = \text{diag}(P(\bar{b}_1), P(\bar{b}_2), \cdots, P(\bar{b}_\kappa))$; $\mu \in \mathrm{R}^\kappa$ and $\lambda \in \mathrm{R}^{m \times m}$ are Lagrange multipliers. By the definition of C^r (see the paragraph after proposition 14.14), we have

$$C^r D_p = \varrho^{-1}[P(\bar{b}_1)w_{\bar{b}_1}^r, \cdots, P(\bar{b}_\kappa)w_{\bar{b}_\kappa}^r]$$

$$= \varrho^{-1}[\tau_{\bar{b}_1}^r w_0, \cdots, \tau_{\bar{b}_\kappa}^r w_0]$$

$$= \varrho^{-1}[D\tau_{\bar{b}_1}^\top D^{-1}w_0, \cdots, D\tau_{\bar{b}_\kappa}^\top D^{-1}w_0]$$

(see proposition 14.13 and item 4 of the list thereafter)

$$= \varrho^{-1}D[\tau_{\bar{b}_1}^\top \mathbf{1}_m^\top, \cdots, \tau_{\bar{b}_\kappa}^\top \mathbf{1}_m^\top]$$

$$= \varrho^{-1}D\pi_{\mathcal{A}}^\top, \tag{14.60}$$

where $D = \text{diag}(w_0)$ (see proposition 14.13) and $\varrho = C\pi_{\mathcal{A}}$ (see proposition 14.14). And it follows that $(G, \mu, \lambda) = (\mathbf{0}, \mathbf{0}, \varrho^{-1}D)$ is a solver of the KKT system 14.57–14.59.

By the above discussion, we conclude that $G = \mathbf{0}$ is a (global) minimizer of the target function $J(G)$; and is the unique minimizer if $P(\bar{b}_j) > 0$ $(j = 1, \cdots, \kappa)$.

Appendix I: Finding Good Characteristic Events

Given a training sequence $S = a_0 \ldots a_N$ over an alphabet O of size α, and given a desired length k of characteristic events and model dimension m, we use the following heuristic brute-force strategy to construct characteristic events $B_1, \ldots, B_m$ that in all our learning experiments rendered the matrix $\hat{V}$ or V^{raw} (see eq. 14.30) reasonably well behaved with respect to inversion, which is the prime requirement for success with the basic learning algorithm.

Let $\#\bar{a}$ denote the number of occurrences of some word $\bar{a}$ in S, let $\kappa = \alpha^k$ and let $(\bar{a}_j)_{1 \le j \le \kappa}$ and $(\bar{b}_i)_{1 \le i \le \kappa}$ both be the alphabetical enumeration of O^k. Start by constructing a $\kappa \times \kappa$ (often sparse) matrix $V_0{}^{\mathrm{raw}} = (\#\bar{a}_j \bar{b}_i)$. Then it is clear that the matrix V^{raw} is obtained from V_0^{raw} by additively joining rows (to agglomerate characteristic sequences $\bar{b}$ into characteristic events B) and columns (to assemble indicative sequences $\bar{a}$ into indicative events A). We treat only the row-joining operations here; the column joining can be done simultaneously or separately in a similar fashion. So we consider a matrix sequence $V_0^{\mathrm{raw}}, V_1^{\mathrm{raw}}, \ldots, V_{\kappa-m-1}^{\mathrm{raw}}$, where each matrix in the sequence is obtained from the previous by joining two rows. The last matrix $V_{\kappa-m-1}^{\mathrm{raw}}$ then has size $m \times \kappa$; the characteristic sequences of the original rows from V_0^{raw} that are then collected in the ith row of $V_{\kappa-m-1}^{\mathrm{raw}}$ yield the desired characteristic events.

The intuitive strategy is to choose from V_n^{raw} for joining that pair of rows r_x, r_y that have the highest pairwise correlation $r_x/\|r_x\| (r_y/\|r_y\|)^\top$ among all pairs of rows in V_n^{raw}. This greedy strategy will (hopefully) result in characteristic events B that each comprise characteristic sequences $\bar{b}, \bar{b}'$ which are "prediction similar" in the sense that $P(\bar{b}|\bar{a}) \approx P(\bar{b}'|\bar{a})$ for all or most $\bar{a}$—that is, joining $\bar{b}, \bar{b}'$ in B incurs a small loss of to-be-predicted distribution information. In addition we take care that the final characteristic events B_i are reasonably weight-balanced in the sense that $P(B_i) \approx P(B_{i'}) \approx 1/m$, in order to ensure that the estimation accuracy $\hat{P}_S(A_j B_i)$ is roughly similar for all entries of $\hat{V}$. Spelled out in more detail, we get the following joining algorithm:

Step 1. Initialization: Construct V_0^{raw} and a normalized version V_0^{norm} thereof whose rows are either all zero (if the corresponding row in V_0^{raw} is zero) or have unit norm. For all rows of V_0^{raw} whose weight (sum of row entries) already exceeds N/m, put the corresponding row in V_0^{norm} to zero. These *finished* rows will thereby automatically become excluded from further joining. Set f to the number of finished rows. Furthermore, set the remaining mass Q of rows still open for joining to N minus the total entry sum of finished rows.

Step 2. Iteration: $V_n^{\mathrm{raw}}, V_n^{\mathrm{norm}}$, and f are given. If $f = m - 1$, jump to termination by joining all remaining unfinished rows. Else, compute the row correlation matrix

$R = V_n^{\text{norm}}(V_n^{\text{norm}})^\top$ and choose the index (x, y) (where $y > x$) of the maximal off-diagonal entry in R for joining rows r_x, r_y by adding in V_n^{raw} the yth row to the xth and deleting the yth row, obtaining V_{n+1}^{raw}. Normalize the summed row, replace the xth row in V_n^{norm} by it, and zero the yth row in V_n^{norm}. If the entry sum of row x in V_{n+1}^{raw} exceeds $Q/(m - f)$, or if $m - f = \kappa - m - 1 - n$, increment f by one, zero the row x in V_n^{norm}, and decrement Q by the component sum of the x-th row in V_{n+1}^{raw}. The result of these operations on V_n^{norm} yields V_{n+1}^{norm}.

The computationally most expensive operation is $R = V_n^{\text{norm}}(V_n^{\text{norm}})^\top$. It can be effected by reusing the R from the previous step with $O(\kappa^2)$ floating-point operations (the recursion will be easily found). All in all the theoretical cost of this algorithm is $O(\kappa^3) = O(\alpha^{3k})$, but for $\kappa/N > 1$ the concerned matrices quickly become sparse, which could be exploited to greatly reduce the computational load. However, k should be chosen, if possible, such that $\kappa/N \leq 1$. The condition number of matrices $\hat{V}$ that we obtained in numerous experiments with natural and artificial data typically ranges between 2 and 50, which makes the algorithm very useful in practice for an initial model estimation with the basic OOM learning algorithm. Unfortunately it is theoretically not clarified what would be the best possible condition number among all choices of characteristic and indicative events; it may well be the case that the observed condition numbers are close to the optimum (or quite far away from it).

Appendix J: Running Invalid OOMs as Sequence Generators

Here is a modification of the sequence-generation procedure described in section 14.4, which allows us to use invalid OOMs and yet avoid negative probabilities. Notation from that section is reused here without reintroduction. The basic idea is to check, at each generation step, whether the probability vector **p** contains negative entries, and if so, reset them to a predetermined, small positive **margin** (standard range: $0.001 \sim 0.01$), which is one of the three tuning parameters of this method. Furthermore, if the sum of negative entries in **p** falls below a significant **setbackMargin** (standard range: $-0.1 \sim -0.5$), indicating that the generation run is about to become instable, the generation is restarted **setbackLength** (typical setting: 2 or 3) steps earlier with the starting state w_0. Some care has to be taken that the resetting to **margin** leads to probability computations where the summed probability for all sequences of some length k is equal to 1. The method comes in two variants, one for generating random sequences, and the other for computing the probability of a given sequence S. The former has an additional step 3a in the description below. Detailed out, the nth step using this method works as follows.

Input: Fixed parameters: **margin**, **setbackMargin**, **setbackLength**, size α of alphabet, observable operators τ_a, starting state w_0. Variables: the state w_{n-1}, and (if $n \geq$ **setbackLength**) the word $s = a_{n-}$ **setbackMargin** $\cdots a_{n-1}$ of previously processed symbols, and index i_{a_n} of current symbol a_n [*only if used in probability computation mode*].

Output: State w_n, log-probability $L = \log(P(a_n|w_{n-1}))$, and new symbol a_n [*only if used in generation mode*] .

> *Step 1:* Compute $\mathbf{p} = \Sigma\, w_{n-1}$.
>
> *Step 2:* Compute $\delta = \sum_{i\in\{1,\dots,\alpha\},\mathbf{p}(i)\leq 0}(\mathsf{margin} - \mathbf{p}(i))$;
> $p^+ = \sum_{i\in\{1,\dots,\alpha\},\mathbf{p}(i)>0}\mathbf{p}(i)$; $p^- = p^+ - \delta$ and $\nu = p^-/p^+$.
>
> **Step 3:** [*Check for potentially instable state, and act if necessary.*]
> If $\delta < \mathsf{setbackMargin}$ and $n \leq \mathsf{setbackLength} + 1$ [*we encounter a problematic state early in the process*], put $w_{n-1} = w_0$, recompute $\mathbf{p} = \Sigma\, w_{n-1}$ and δ, p^+, p^-, ν as in step 2.
> Else, if $\delta < \mathsf{setbackMargin}$ [*we encounter a problematic state later in the process and restart the generation a few steps earlier*],
> set $w = w_0$;
> for $i = 1$ to $\mathsf{setbackLength}$: $w = \tau_{s(i)}w$; $w = w/\mathbf{1}w$ [*we recompute the last few states from w_0*];
> set $w_{n-1} = w$;
> recompute $\mathbf{p} = \Sigma\, w_{n-1}$, and recompute δ, p^+, p^-, ν as in step 2.
>
> *Step 3a [only executed in the generation variant]:* Randomly choose a_n according to the probability vector $\mathbf{p}$; set i_{a_n} to its index in the alphabet.
>
> *Step 4 [update state and compute "blurred" probability of current symbol]:*
> If $\mathbf{p}(i_{a_n}) \leq 0$ [*current symbol would be assigned a negative probability*], set $L = \log(\mathsf{margin})$ and $w = \tau_{a_n}w_{n-1}$; $w_n = w/\mathbf{1}w$.
> Else [*current symbol is OK but its probability has to be reduced to account for the added probability mass that might have been assigned to other symbols in this step*] set $L = \log(\nu\mathbf{p}(i_{a_n}))$ and $w = \tau_{a_n}w_{n-1}$; $w_n = w/\mathbf{1}w$.

Appendix K: Details of the One Million Pound Learning Experiment

All OOMs computed during the ES iterations were invalid, so we employed the stabilizing method described in appendix I for computing the requisite reverse state sequences. The same method was used to determine the log-likelihoods on the training and testing sequences. The settings (see appendix I) that we used were margin $= 0.001$, setbackMargin $= 0.3$, setbackLength $= 2$. These settings were optimized by hand in preliminary tests.

Only such indicative sequences $\bar{a}$ were gleaned from the suffix tree that occurred at least 10 times in the training sequence.

This little study was carried out before the algorithm for finding good characteristic events described in appendix H was available. Thus we used an inferior method for initial model estimation that we need not detail here. Using the better method for initial model estimation would very likely have resulted in an improved overall performance (the high initial jump in model quality from the initial model to the first ES-estimated model that appears in fig. 14.5b would disappear).

Acknowledgments

The results regarding the OOMs reported in sections 14.1 through 14.9 of this chapter were obtained when the first author worked under a postdoctoral grant from the German National Research Center for Information Technology (GMD), now Fraunhofer Institute for Autonomous Intelligent Systems. The first author feels deeply committed to its director, Thomas Christaller, for his unfaltering support.

Notes

[1] The first author originally devised OOMs while thinking about action selection algorithms for mobile robots.

[2] Heller and Ito used a different definition for HMMs, which yields a different version of the minimality statement in part 2

References

Technical Specification Group 3GPP. Multiplexing and channel coding (FDD). Technical report, June 1999. Technical report TS 25.212 v2.0.0.

D. H. Ackley, G. E. Hinton, and T. J. Sejnowski. A learning algorithm for Boltzmann machines. *Cognitive Science*, 9:147–169, 1985.

R. Ahlswede. The rate distortion region for multiple descriptions without excess rate. *IEEE Transactions on Information Theory*, 31(6):721–726, November 1985.

S. M. Aji and R. J. McEliece. The generalized distributive law. *IEEE Transactions on Information Theory*, 46:325–343, March 2000.

S. M. Alamouti. A simple transmit diversity technique for wireless communications. *IEEE Journal on Selected Areas in Communications*, 16(8):1451–1458, October 1998.

T. W. Allen, T. Bastug, S. Kuyucak, and S. H. Chung. Gramicidin A channel as a test ground for molecular dynamics force fields. *Biophysical Journal*, 84: 2159–2168, 2003.

D. G. Amaral, N. Ishizuka, and B. Claiborne. Neurons, numbers, and the hippocampal network. *Progress in Brain Research*, 83:1–11, 1990.

S. Amari. Characteristics of random nets of analog neuron-like elements. *IEEE Tranactions on Systems, Man, and Cybernetics*, 2:643–657, 1972.

S. Amari, A. Cichocki, and H. H. Yang. A new learning algorithm for blind signal separation. In D. S. Touretzky, M. C. Mozer, and M. E. Hasselmo, editors, *Advances in Neural Information Processing Systems, 8*. MIT Press, Cambridge, MA, 1996.

K. Anand, G. Mathew, and V. U. Reddy. Blind separation of multiple co-channel BPSK signals arriving at an antenna array. *IEEE Signal Processing Letters*, 2 (9):176–178, 1995.

J. Anderson and S. M. Hladick. Tailbiting MAP decoders. *IEEE Journal on Selected Areas in Communications*, 16(2):297–302, February 1998.

S. Andradottir. A method for discrete stochastic optimization. *Management Science*, 41(12):1946–1961, 1995.

S. Andradottir. Accelerating the convergence of random search methods for discrete stochastic optimization. *ACM Transactions on Modeling and Computer Simulation*, 9(4):349–380, 1999.

J. G. Apostolopoulos and M. D. Trott. Path diversity for enhanced media streaming. *IEEE Communications Magazine*, 42(8):80–87, August 2004.

M. Ardakani and F. R. Kschischang. A more accurate one-dimensional analysis and design of irregular LDPC codes. *IEEE Transactions on Communications*, 52:2106–2114, 2004.

M. Arzel, C. Lahuec, M. Jézéquel, and F. Seguin. Analogue decoding of duo-binary codes. In *Proc. of ISITA 2004*, Parma, Italy, October 2004.

J. Atick and N. Redlich. Towards a theory of early visual processing. *Neural Computation*, 2:308–320, 1990.

J. August. The curve indicator random field. PhD thesis, Department of Computer Science, Yale University, 2001.

J. August. Volterra filtering of noisy images. In *Proc. of the European Conference on Computer Vision*, 2002.

L. R. Bahl, J. Cocke, F. Jelinek, and J. Raviv. Optimal decoding of linear codes for minimizing symbol error rate. *IEEE Transactions on Information Theory*, 20:248–287, March 1974.

R. Bajcsy. Active perception. *Proceedings of the IEEE*, 76:996–1005, 1988.

P. Bak, C. Tang, and K. Wiesenfeld. Self-organized criticality. *Physical Review A*, 38(1):364–378, 1988.

V. Balasubramanian. Equivalence and reduction of hidden Markov models. Technical Report 1370, MIT AI Lab, 1993.

D. H. Ballard. Animate vision. *Artificial Intelligence*, 48:57–86, 1988.

S. Bao, V. T. Chan, and M. M. Merzenich. Cortical remodelling induced by activity of ventral tegmental dopamine neurons. *Nature*, 412(6842):79–83, 2001.

H. B. Barlow. Unsupervised learning. *Neural Computation*, 1:295–311, 1989.

H. B. Barlow and P. Földiák. Adaptation and decorrelation in the cortex. In R. Durbin, C. Miall, and G. Mitchison, editors, *The Computing Neuron*, chapter 4, pages 54–72. Addison-Wesley, Reading, MA, 1989.

C. A. Barnes, B. L. McNaughton, S. J. Y. Mizumori, and B. W. Leonard. Comparison of spatial and temporal characteristics of neuronal activity in sequential stages of hippocampal processing. *Progress in Brain Research*, 83:287–300, 1990.

J. R. Barry, A. Kavcic, S. W. McLaughlin, A. Nayak, and W. Zeng. Iterative timing recovery. *IEEE Signal Processing Magazine*, pages 89–102, January 2004.

P. L. Bartlett and W. Maass. Vapnik-Chervonenkis dimension of neural nets. In M. A. Arbib, editor, *The Handbook of Brain Theory and Neural Networks*, pages 1188–1192. MIT Press, Cambridge, MA, 2nd edition, 2003.

G. Battail. Pondération des symboles décodés par l'algorithme de Viterbi. *Annales des Télécommunications*, 42:31–38, January 1987.

J. Baxter, A. Tridgell, and L. Weaver. KnightCap: A chess program that learns by combining TD(λ) with minimax search. Technical report, Department of Systems Engineering, Australian National University, 1997.

R. J. Baxter, editor. *Exactly Solved Models in Statistical Mechanics*. Academic Press, New York, 1982.

L. Becchetti, S. Diggavi, S. Leonardi, A. Marchetti-Spaccamela, S. Muthukrishnan, T. Nandagopal, and A. Vitaletti. Parallel scheduling problems in next generation wireless networks. In *Proc. of ACM Symposium on Parallel Algorithms and Architectures*, pages 238–247, 2002.

S. Becker. A computational principle for hippocampal learning and neurogenesis. *Hippocampus*, 15(6):722–738, 2005.

S. Becker. Mutual information maximization: Models of cortical self-organization. *Network: Computation in Neural Systems*, 7:7–31, 1996.

S. Becker. Implicit learning in 3D object recognition: the importance of temporal context. *Neural Computation*, 10:347–374, 1999.

S. Becker and G. E. Hinton. A self-organizing neural network that discovers surfaces in random-dot stereograms. *Nature*, 355:161–163, 1992.

S. Becker and J. Lim. A computational model of prefrontal control in free recall: strategic memory use in the California verbal learning task. *Journal of Cognitive Neuroscience*, 15(6):1–12, 2003.

A. Bell and T. J. Sejnowski. An information-maximization approach to blind separation and blind deconvolution. *Neural Computation*, 7:1129–1159, 1995.

A. J. Bell and T. J. Sejnowski. The independent components of natural scenes are edge filters. *Vision Research*, 37:3327–3338, 1997.

R. Bellman and S. E. Dreyfus. *Applied Dynamic Programming*. Princeton University Press, Princeton, NJ, 1962.

A. Belouchrani and J.-F. Cardoso. Maximum likelihood source separation for discrete sources. In *Proc. EUSIPCO*, pages 768–771, Edinburgh, Scotland, 1994.

M. Bender, S. Chakrabarti, and S. Muthukrishnan. Flow and stretch metrics for scheduling continuous job streams. In *Proc. of the Annual Symposium on Discrete Algorithms*, pages 270–279, 1998.

P. Bender, P. Black, M. Grob, R. Padovani, N. Sindhushayana, and A. Viterbi. CDMA/HDR: A bandwith-efficient high-speed wireless data service for nomadic users. *IEEE Communications Magazine*, 38(7):70–77, July 2000.

J. Benesty. Adaptive eigenvalue decomposition algorithm for passive acoustic source localization. *Journal of the Acoustical Society of America*, 107(1):384–391, 2000.

Y. Bengio. Markovian models for sequential data. *Neural Computing Surveys*, 2: 129–162, 1999.

T. Berger. Multiterminal source coding. In G. Longo, editor, *The Information Theory Approach to Communications*, pages 172–231. Springer, Berlin, 1977.

A. Berman and R. J. Plemmons. *Nonnegative Matrices in the Mathematical Sciences*. Academic Press, New York, 1979.

C. Berrou and M. Jézéquel. Non binary convolutional codes for turbo coding. *Electronics Letters*, 35(1):39–40, January 1999.

C. Berrou, A. Glavieux, and P. Thitimajshima. Near shannon limit error-correcting coding and decoding: Turbo-codes. In *Proc. of International Conference on Communications (ICC'93)*, pages 1064–1070, Geneva, May 1993.

C. Berrou, C. Douillard, and M. Jézéquel. Multiple parallel concatenation of circular recursive convolutional (CRSC) codes. *Annales des Télécommunications*, 54(3–4):166–172, 1999.

C. Berrou, E.A. Maury, and H. Gonzalez. Which minimum Hamming distance do we really need? In *Proc. of International Symposium on Turbo Codes and Related Topics*, pages 141–148, Brest, France, 2003.

C. Berrou, Y. Saouter, S. Kerouédan, and M. Jézéquel. Designing good permutations for turbo codes: Towards a single model. In *Proc. of International Conference on Communications (ICC'94)*, volume 35, pages 39–40, Paris, France, 2004.

N. Bertschinger and T. Natschläger. Real-time computation at the edge of chaos in recurrent neural networks. *Neural Computation*, 16(7):1413–1436, 2004.

D. P. Bertsekas. *Dynamic Programming and Optimal Control*, volume I,II. Athena Scientific, Belmont, MA, 1995a.

D. P. Bertsekas. *Nonlinear programming*. Athena Scientific, Belmont, MA, 1995b.

D. P. Bertsekas and J. N. Tsitsiklis. *Neuro-Dynamic Programming*. Athena Scientific, Belmont, MA, 1996.

D. P. Bertsimas and J. N. Tsitsiklis. *Introduction to Linear Optimization*. Athena Scientific, Belmont, MA, 1997.

J. Besag and P. J. Green. Spatial statistics and Bayesian computation. *Journal of the Royal Statistical Society, B*, 55(1):25–37, 1993.

P. J. Bickel, Y. Ritov, and T. Rydén. Asymptotic normality of the maximum-likelihood estimator for general hidden Markov models. *Annals of Statistics*, 26:1614–1635, 1998.

E. Bienenstock, S. Geman, and D. Potter. Compositionality, MDL priors and object recognition. In M. Mozer, M. I. Jordan, and T. Petsche, editors, *Advances in Neural Information Processing Systems, 9*. MIT Press, Cambridge, MA, 1998.

E. Biglieri, J. Proakis, and S. Shamai. Fading channels: Information-theoretic and communications aspects. *IEEE Transactions on Information Theory*, 44(6):2619–2692, October 1998.

D. Billings, N. Burch, A. Davidson, R. Holte, J. Schaeffer, T. Schauenberg, and D. Szafron. Approximating game-theoretic optimal strategies for full-scale poker.

In *Proceedings of the International Joint Conference on Artificial Intelligence,* 2003.

C. M. Bishop. *Neural Networks for Pattern Recognition.* Oxford University Press, New York, 1995.

D. Blackwell and L. Koopmans. On the identifiability problem for functions of finite Markov chains. *Annals of Mathematical Statistics,* 38:1011–1015, 1957.

A. Blake and A. Yuille, editors. *Active Vision.* MIT Press, Cambridge, MA, 1992.

A. Blake and A. Zisserman. *Visual Reconstruction.* MIT Press, Cambridge, MA, 1987.

J. Blauert. *Spatial Hearing: The Psychophysics of Human Sound Localization.* MIT Press, Cambridge, MA, 1983.

D. M. Blei and M. I. Jordan. Variational methods for the Dirichlet process. In *Proc. International Conference on Machine Learning.* ACM Press, New York, NY, 2004.

P. Bofill. Sound examples, at http://people.ac.upc.es/pau/shpica/instant.html.

P. Bofill and M. Zibulevsky. Underdetermined blind source separation using sparse representations. *Signal Processing,* 81:2353–2362, 2001.

S. Bornholdt and T. Röhl. Self-organized critical neural networks. *Physical Review E,* 67:066118, 2003.

S. Bornholdt and T. Röhl. Topological evolution of dynamical networks: Global criticality from local dynamics. *Physical Review Letters,* 84(26):6114–6117, 2000.

E. Boutillon and D. Gnaedig. Maximum spread of *d*-dimensional multiple turbo codes. 2005.

M. Bowling. Multiagent learning in the presence of agents with limitations. PhD thesis, Computer Science Department, Carnegie Mellon University, Pittsburgh, PA, 2003.

M. Bowling. Convergence and no-regret in multiagent learning. In *Advances in Neural Information Processing Systems.* MIT Press, Cambridge, MA.

M. Bowling and M. Veloso. Simultaneous adversarial multi-robot learning. In *Proceedings of the Eighteenth International Joint Conference on Artificial Intelligence,* 2003.

S. Boyd and L. O. Chua. Fading memory and the problem of approximating nonlinear oparators with Volterra series. *IEEE Transactions on Circuits and Systems,* 32:1150–1161, 1985.

A. S. Bregman. *Auditory Scene Analysis.* MIT Press, Cambridge, MA, 1990.

A. S. Bregman. Psychological data and computational ASA. In D. F. Rosenthal and H. G. Okuno, editors, *Computational Auditory Scene Analysis.* Lawrence Erlbaum Associates, Mahwah, NJ, 1996.

D. Brennan. Linear diversity combining techniques. *Proceedings of the IEEE,* 47: 1075–1102, June 1959.

N. Brenner, W. Bialek, and R. de Ruyter van Steveninck. Adaptive rescaling maximizes information transmission. *Neuron*, 26:695–702, 2000.

A. Bronkhorst. The cocktail party phenomenon: A review of research on speech intelligibility in multiple-talker conditions. *Acoustica*, 86:117–128, 2000.

G. D. Brown, S. Yamada, and T. J. Sejnowski. Independent component analysis at the neural cocktail party. *Trends in Neuroscience*, 24:54–63, 2001.

G. J. Brown and M. P. Cooke. Computational auditory scene analysis. *Computer Speech and Language*, 8:297–336, 1994.

G. J. Brown and D. L Wang. Modeling the perceptual segregation of concurrent vowels with a network of neural oscillation. *Neural Networks*, 10(9):1547–1558, 1997.

G. W. Brown. Iterative solution of games by fictitious play. In *Activity Analysis of Production and Allocation*. Wiley, New York, 1951.

L. D. Brown. *Fundamentals of Statistical Exponential Families*. Institute of Mathematical Statistics, Hayward, CA, 1986.

M. A. Brown and R. C. Semelka. *MRI: Basic Principles and Applications*. Wiley, New York, 3rd edition, 2003.

E. Brunswik. *Perception and the Representative Design of Psychological Experiments*. University of California Press, Berkeley, CA, 1956.

D. V. Buonomano and M. M. Merzenich. Temporal information transformed into a spatial code by a neural network with realistic properties. *Science*, 267:1028–1030, 1995.

L. Z. Buzi. Design of structured vector quantizers for diversity communication systems. Master's thesis, Department of Electrical Engineering, Texas A&M University, May 1994.

G. Caire and S. Shamai. On the capacity of some channels with channel state information. *IEEE Transactions on Information Theory*, 45(6):2007–2019, September 1999.

B. M. Calhoun and C. E. Schreiner. Spectral envelope coding in cat primary auditory cortex: Linear and non-linear effects of stimulus characteristics. *European Journal of Neuroscience*, 10(3):926–940, 1998.

E. Candes and D. Donoho. Continuous curvelet transform I and II. *Applied and Computational Harmonic Analysis*, 19(2):162–197,198–222, 2005.

O. Cappé, E. Moulines, and T. Rydén. *Inference in Hidden Markov Models*. Springer, Berlin, 2005.

A. Carasso. Singular integrals, image smoothness and the recovery of texture in image deblurring. *SIAM Journal of Applied Mathematics*, 64:1749–1774, 2004.

J.-F. Cardoso. The three easy routes to independent component analysis: contrasts and geometry. In *Proc. ICA2001*, San Diego, CA, 2001.

J.-F. Cardoso. Blind signal separation: statistical principles. *Proceedings of the IEEE*, 86(10):2029–2025, 1998.

J.-F. Cardoso and A. Souloumiac. Blind beamforming for non-Gaussian signals. *IEE Proceedings–F*, 140(6):362–370, 1993.

D. Chandler. *Introduction to Modern Statistical Mechanics*. Oxford University Press, Oxford, 1987.

E. F. Chaponniere, P. J. Black, J. M. Holtzman, and D. N. C. Tse. Transmitter directed, multiple receiver system using path diversity to equitably maximize throughput. United States patent, no. 6,449,490, 2002.

G. Charpiat, O. Faugeras, and R. Keriven. Approximations of shape metrics.

Z. Chen. An odyssey of the cocktail party problem. Technical report, Adaptive Systems Lab, McMaster University, Hamilton, Ontario, Canada, 2003. Available at http://soma.crl.mcmaster.ca/~zhechen/download/cpp.ps.

V. Cherkassky and F. Mulier. *Learning from Data*. Wiley, New York, 1998.

E. C. Cherry. Some experiments on the recognition of speech, with one and two ears. *Journal of the Acoustical of Society of America*, 25:975–979, 1953.

E. C. Cherry. *On Human Communication: A Review, Survey, and a Criticism*. MIT Press, Cambridge, MA, 1957.

E. C. Cherry. Two ears—but one world. In W. A. Rosenblith, editor, *Sensory Communication*, pages 90–117. Wiley, New York, 1961.

E. C. Cherry and B. Sayers. Human "cross-correlation"—A technique for measuring certain parameters of speech perception. *Journal of the Acoustical of Society of America*, 28:889–895, 1956.

E. C. Cherry and B. Sayers. On the mechanism of binaural fusion. *Journal of the Acoustical of Society of America*, 31:535, 1959.

E. C. Cherry and W. K. Taylor. Some further experiments upon the recognition of speech, with one and, with two ears. *Journal of the Acoustical of Society of America*, 26:554–559, 1954.

A. Chindapol, J. A. Ritcey, and X. Li. BIMC-ID for QAM constellations on rayleigh fading channels. In *Proc. of Asilomar Conference on Signals, Systems and Computers*, pages 57–62, 1999.

S. Choi, A. Cichocki, and S. Amari. Flexible independent component analysis. *Journal of VLSI Signal Processing*, 26(1–2):25–38, 2000.

K. Christensen, R. Donangelo, B. Koiller, and K. Sneppen. Evolution of random networks. *Physical Review Letters*, 81:2380, 1998.

S. H. Chung, J. B. Moore, L. Xia, and L. S. Premkumar. Characterization of single channel currents using digital signal processing techniques based on hidden Markov models. *Philosophical Transactions of the Royal Society, B*, 329:256–285, 1990.

S. H. Chung, V. Krishnamurthy, and J. B. Moore. Adaptive processing techniques based on hidden Markov models for characterising very small channel currents buried in noise and deterministic interferences. *Philosophical Transactions of the Royal Society, B*, 334:357–384, 1991.

S.-Y. Chung, T. J. Richardson, and R. L. Urbanke. Analysis of sum-product decoding of low-density parity-check codes using a Gaussian approximation. *IEEE Transactions on Information Theory*, 47(2):657–670, 2001.

A. Cichocki and S. Amari. *Adaptive Blind Signal and Image Processing*. Wiley, New York, 2002.

A. Cichocki, J. Cao, S. Amari, N. Murata, T. Takeda, and H. Endo. Enhancement and blind identification of magnetoencephalographic signals using independent component analysis. In *Proc. of 11th International Conference on Biomagnetism (BIOMAG-98)*, pages 169–172, Sendai, Japan, 1999.

J. M. Cioffi, G. P. Dudevoir, M. V. Eyuboglu, and G. D. Forney. MMSE decision-feedback equalizers and coding: Parts I and II. *IEEE Transactions on Communications*, 43(10):2582–2604, 1995.

A. Clark and C. Eliasmith. Philosophical issues in brain theory and connectionism. In M. Arbib, editor, *The Handbook of Brain Theory and Neural Networks*, pages 886–888. MIT Press, Cambridge, MA, 2nd edition, 2003.

E. F. Codd. *Cellular Automata*. Academic Press, New York, 1968.

P. Comon. Independent component analysis, a new concept? *Signal Processing*, 36:287–314, 1994.

V. Conitzer and T. Sandholm. Complexity results about Nash equilibria. In *Proc. International Journal Conference on Artificial Intelligence*, 2003.

J. H. Conway and N. J. A. Sloane. *Sphere Packings, Lattices, and Groups*. Springer, New York, 3rd edition, 1999.

M Cooke. *Modelling Auditory Processing and Organization*. Cambridge University Press, Cambridge, 1993.

M. P. Cooke and D. Ellis. The auditory organization of speech and other sources in listeners and computational models. *Speech Communication*, 35:141–177, 2001.

T. Cootes, C. Taylor, A. Lanitis, D. Cooper, and J. Graham. Building and using flexible models. In *Proceedings of the International Conference on Computer Vision*, 1993.

B. A. Cornell, V. L. Braach-Maksvytis, L. G. King, P. D. Osman, B. Raguse, L. Wieczorek, and R. J. Pace. A biosensor that uses ion-channel switches. *Nature*, 387:580–583, June 1997.

B. Corry, T. W. Allen, S. Kuyucak, and S. H. Chung. Mechanisms of permeation and selectivity in Calcium channels. *Biophysical Journal*, 80:195–214, 2001.

M. S. Cousins, A. Atherton, L. Turner, and J. D. Salamone. Nucleus accumbens dopamine depletions alter relative response allocation in a T-maze cost/benefit task. *Behavioral Brain Research*, 74(1–2):189–197, 1996.

T. M. Cover. Some advances in broadcast channels. In A. Viterbi, editor, *Advances in Communication Theory*. Academic Press, San Francisco, 1975. Volume 4 of Theory and Applications.

T. M. Cover and A. El Gamal. Capacity theorems for the relay channel. *IEEE Transactions on Information Theory*, 25(5):572–584, September 1979.

T. M. Cover and J. A. Thomas. *Elements of Information Theory*. Wiley, New York, 1991.

J. D. Cowan. Statistical mechanics of nervous nets. In E. R. Caianiello, editor, *Neural Networks*, pages 181–188. Springer, Berlin, 1968.

M. S. Crouse, R. D. Nowak, and R. G. Baraniuk. Wavelet-based statistical signal processing using hidden Markov models. *IEEE Transactions on Signal Processing*, 46:886–902, April 1998.

S. Crozier and P. Guinand. Distance upper bounds and true minimum distance results for turbo-codes with DRP interleavers. In *Proc. International Symposium on Turbo Codes and Related Topics*, pages 169–172, Brest, France, 2003.

I. Csiszar and G. Tusn'ady. Information geometry and alternating minimization procedures. *Statistics and Decisions*, 1(1):205–237, Supplement Issue, 1984.

H. Dai and H.V. Poor. Iterative space-time processing for multiuser detection in multipath CDMA channels. *IEEE Transactions on Signal Processing*, 50(9):2116–2127, 2002.

M. O. Damen, A. Chkeif, and J.-C. Belfiore. Lattice codes decoder for space-time codes. *IEEE Communication Letters*, 4:161–163, May 2000.

A. P. Dawid. Applications of a general propagation algorithm for probabilistic expert systems. *Statistics and Computing*, 2:25–36, 1992.

P. Dayan, G. E. Hinton, R. Neal, and R. S. Zemel. The Helmholtz machine. *Neural Computation*, 7:1022–1037, 1995.

V. R. de Sa. Learning classification with unlabeled data. In J. D. Cowan, G. Tesauro, and J. Alspector, editors, *Advances in Neural Information Processing Systems, 6*, pages 112–119. Morgan Kaufmann, San Mateo, CA, 1994.

G. C. DeAngelis, I. Ohzawa, and R. D. Freeman. Spatiotemporal organization of simple-cell receptive fields in the cat's striate cortex. I. general characteristics and postnatal development. *Journal of Neurophysiology*, 69(4):1118–1135, 1993.

O. Dekel, S. Shalev-Shwartz, and Y. Singer. The power of selective memory: Self-bounded learning of prediction suffix trees. In *Advances in Neural Information Processing Systems, 17*. MIT Press, Cambridge, MA, 2004.

J. W. Demmel. *Applied Numerical Linear Algebra*. SIAM, Philadelphia, 1997.

A. P. Dempster, N. M. Laird, and D. B. Rubin. Maximum likelihood from incomplete data via the EM algorithm. *Journal of the Royal Statistical Society Series B*, 39:1–38, 1977.

B. Derrida and Y. Pomeau. Random networks of automata: A simple annealed approximation. *Europhysics Letters*, 1(2):45–49, 1986.

A. Desolneux, L. Moisan, and J.-M. Morel. Maximal meaningful events and applications to image analysis. *Annals of Statistics*, 31:1822–1851, 2003.

R. A. DeVore and B. J. Lucier. Classifying the smoothness of images. In *Proc. of the IEEE International Conference on Image Processing*, pages 6–10, 1994.

S. W. Dharmadhikari. Functions of finite Markov chains. *Annals of Mathematical Statistics*, 34:1022–1032, 1963a.

S. W. Dharmadhikari. Sufficient conditions for a stationary process to be a function of a finite Markov chain. *Annals of Mathematical Statistics*, 34:1033–1041, 1963b.

S. W. Dharmadhikari. A characterization of a class of functions of finite Markov chains. *Annals of Mathematical Statistics*, 36:524–528, 1965.

K. I. Diamantaras. Blind separation of multiple binary sources using a single linear mixture. In *Proc. of IEEE International Conference on Acoustics, Speech and Signal Processing (ICASSP)*, volume 5, pages 2889–2892, Istanbul, Turkey, June 2000.

K. I. Diamantaras. Blind channel identification based on the geometry of the received signal constellation. *IEEE Transactions on Signal Processing*, 50(5): 1133–1143, May 2002.

K. I. Diamantaras and E. Chassioti. Blind separation of N binary sources from one observation: A deterministic approach. In *Proc. of Independent Component Analysis and Blind Source Separation*, pages 93–98, Helsinki, Finland, June 2000.

K. I. Diamantaras and Th. Papadimitriou. Blind deconvolution of SISO systems with binary source based on recursive channel shortening. In C. G. Puntonet and A. Prieto, editors, *Proc. Independent Component Analysis and Blind Signal Separation (ICA2004)*, Lecture Notes in Computer Science, 3195, pages 548–553. Springer, Granada, Spain, 2004a.

K. I. Diamantaras and Th. Papadimitriou. MIMO blind deconvolution using subspace-based filter deflation. In *Proc. of IEEE International Conference on Acoustic, Speech and Signal Processing (ICASSP2004)*, volume 4, pages 433–436, Montreal, 2004b.

K. I. Diamantaras and Th. Papadimitriou. Blind deconvolution of multi-input single-output systems with binary sources. 2005.

K. I. Diamantaras, A. P. Petropulu, and B. Chen. Blind two-input-two-output FIR channel identification based on second-order statistics. *IEEE Transactions on Signal Processing*, 48(2):534–542, 2000.

S. N. Diggavi. On achievable performance of spatial diversity fading channels. *IEEE Transactions on Information Theory*, 47(1):308–325, 2001.

S. N. Diggavi and D. N. C. Tse. On successive refinement of diversity. In *Proceedings of the Allerton Conference on Communication, Control, and Computing*, 2004.

S. N. Diggavi and V. A. Vaishampayan. On multiple description source coding with decoder side information. In *Proc. of the IEEE Information Theory Workshop*, San Antonio, TX, October 2004.

S. N. Diggavi, N. Al-Dhahir, A. Stamoulis, and A. R. Calderbank. Differential space-time coding for frequency-selective channels. *IEEE Communications Letters*, 6 (6):253–255, June 2002a.

S. N. Diggavi, M. Grossglauser, and D. N. C. Tse. Even one dimensional mobility increases capacity of ad hoc wireless networks. In *Proc. of the IEEE Symposium on Information Theory*, page 388, 2002b.

S. N. Diggavi, N. J. A. Sloane, and V. A. Vaishampayan. Asymmetric multiple description lattice vector quantizers. *IEEE Transactions on Information Theory*, 48(1):174–191, January 2002c.

S. N. Diggavi, N. Al-Dhahir, and A. R. Calderbank. Diversity embedded space-time codes. In *Proc. of the IEEE Global Communications Conference (GLOBECOM)*, pages 1909–1914, 2003.

S. N. Diggavi, N. Al-Dhahir, and A. R. Calderbank. Diversity embedding in multiple antenna communications. In P. Gupta, G. Kramer, and A. J. van Wijngaarden, editors, *Network Information Theory*, pages 285–302. American Mathematical Society, 2004a.

S. N. Diggavi, N. Al-Dhahir, A. Stamoulis, and A. R. Calderbank. Great Expectations: The value of spatial diversity to wireless networks. *Proceedings of the IEEE*, 92(2):217–270, February 2004b.

P. Divenyi, editor. *Speech Separation by Humans and Machines*. Springer, Berlin, 2004.

D. Divsalar, S. Dolinar, and F. Pollara. Low complexity turbo-like codes. In *Proc. of the Second International Symposium on Turbo Codes and Related Topics*, pages 73–80, Brest, France, 2000.

S. Doclo and M. Moonen. Robust adaptive time-delay estimation for speaker localization in noisy and reverberant acoustic environments. *EURASIP Journal of Applied Signal Processing*, 5:2230–2244, 2002.

D. Dong and J. Atick. Temporal decorrelation: A theory of lagged and non-lagged responses in the lateral geniculate nucleus. *Network: Computation in Neural Systems*, 6:159–178, 1995.

R. Dong. Personal communication, 2005.

J. L. Doob. *Stochastic Processes*. Wiley, New York, 1953.

A. Doucet, N. de Freitas, and N. Gordon, editors. *Sequential Monte Carlo Methods in Practice*. Springer, New York, 2001.

S. C. Douglas and X. Sun. Convolutive blind separation of speech mixtures using the natural gradient. *Speech Communication*, 39:65–78, 2003.

C. Douillard, M. Jézéquel, C. Berrou, A. Picart, P. Didier, and A. Glavieux.

Iterative correction of intersymbol interference: Turbo-equalization. *IEEE Signal Processing Magazine*, 6(5):507–511, 1995.

D. A. Doyle, J. M. Cabral, R. A. Pfuetzner, A. Kuo, J. M. Gulbis, S. L. Cohen, B. T. Chait, and R. MacKinnon. The structure of the potassium channel: Molecular basis of K^+ conduction and selectivity. *Science*, 280:69–77, 1998.

L. Duan and B. Rimoldi. The iterative turbo decoding algorithm has fixed points. *IEEE Transactions on Information Theory*, 47(7):2993–2995, November 2001.

R. O. Duda, P. E. Hart, and D. G. Stork. *Pattern Classification*. Wiley, New York, 2nd edition, 2001.

R. Durbin, S. Eddy, A. Krogh, and G. Mitchinson. *Biological Sequence Analysis: Probabilistic Models of Proteins and Nucleic Acids*. Cambridge University Press, Cambridge, 1998.

Digital Video Broadcasting (DVB). Interaction channel for satellite distribution systems. Technical report, ETSI EN 301 790, December 2000. V1.2.2, pp. 21–24.

Digital Video Broadcasting (DVB). Interaction channel for digital terrestrial television. Technical report, ETSI EN 301 958, August 2001. V1.1.1, pp. 28-30.

A. Edelman. Eigenvalues and condition numbers of random matrices. PhD thesis, Massachusetts Institute of Technology, Cambridge, MA, 1989.

R. Edwards, J.J. McDonald, and M.J. Tsatsomeros. On matrices with common invariant cones with applications in neural and gene networks. *Linear Algebra and Its Applications*, 398:37–67, 2005.

B. Efron. The geometry of exponential families. *Annals of Statistics*, 6:362–376, 1978.

A. El Gamal and T. M. Cover. Achievable rates for multiple descriptions. *IEEE Transactions on Information Theory*, 28:851–857, November 1982.

A. El Gamal, J. Mammen, B. Prabhakar, and D. Shah. Throughput-delay trade-off in wireless networks. In *Proceedings of the IEEE INFOCOM*, pages 464–475, 2004.

H. El-Gamal and M. O. Damen. Universal space-time coding. *IEEE Transactions on Information Theory*, 49(5):1097–1119, May 2003.

H. El Gamal and A. R. Hammons. Analyzing the turbo decoder using the Gaussian approximation. *IEEE Transactions on Information Theory*, 47(2):671–686, 2001.

J. Elder and R. Goldberg. Ecological statistics of gestalt laws for the perceptual organization of contours. *Journal of Vision*, 2:324–353, 2002.

R. J. Elliott, L. Aggoun, and J. B. Moore. *Hidden Markov Models: Estimation and Control*. Springer, New York, 1995.

D. Ellis. Prediction-driven computational auditory scene analysis. PhD thesis, Massachusetts Institute of Technology, Cambridge, MA, 1996.

J. L. Elman. Finding structure in time. *Cognitive Science*, 14:179–211, 1990.

Y. Ephraim and N. Merhav. Hidden Markov processes. *IEEE Transactions on Information Theory*, 48:1518–1569, June 2002.

M. Eroa, F.-W. Sun, and L.-N. Lee. DVB-S2 low density parity check codes with near Shannon limit performance. *International Journal of Satellite Communications and Networking*, 22:269–279, 2004.

R. Evans, V. Krishnamurthy, and G. Nair. Sensor adaptive target tracking over variable bandwidth networks. In G. C. Goodwin, editor, *Model Identification and Adaptive Control—A Volume in Honour of Brian Anderson*, pages 115–124. Springer, 2001.

B. Farhang-Boroujeny. *Adaptive Filters: Theory and Applications*. Wiley, New York, 1998.

J. Felsenstein. Evolutionary trees from DNA sequences: A maximum likelihood approach. *Journal of Molecular Evolution*, 17:368–376, 1981.

A. S. Feng and R. Ratnam. Neural basis of hearing in real-world situations. *Annual Review of Psychology*, 51:699–725, 2000.

N. Fertig, R. H. Blick, and J. C. Behrends. The cell patch clamp recording performed on a planar glass chip. *Biophysical Journal*, 82:3056–3062, June 2002.

D. J. Field. Relations between the statistics of natural images and the response properties of cortical cells. *Journal of the Optical Society of America, A*, 4: 2379–2394, 1987.

T. R. Field. Observability of the scattering cross-section through phase decoherence. *Journal of Mathematical Physics*, 46(6):3305, June 2005.

T. R. Field and R. J. A. Tough. Diffusion processes in electromagnetic scattering generating k-distributed noise. *Proceedings of the Royal Society of London*, 45 (9):2169–2193, 2003a.

T. R. Field and R. J. A. Tough. Stochastic dynamics of the scattering amplitude generating k-distributed noise. *Journal of Mathematical Physics*, 44(11):5212–5223, 2003b.

T. R. Field and R. J. A. Tough. Dynamical models of weak scattering. *Journal of Mathematical Physics*, 46(1):13302–13320, 2005.

A. Finkelstein. *Water Movement through Lipid Bilayers, Pores, and Plasma Membranes*. Wiley, New York, 1987.

M. Fleming, Q. Zhao, and M. Effros. Network vector quantization. *IEEE Transactions on Information Theory*, 50(8):1584–1604, August 2004.

F. Forges. Correlated equilibrium in two-person zero-sum games. *Econometrica*, 58 (2), 1990.

F. Forges and B. von Stengel. Computationally efficient coordination in game trees. Technical Report LSE-CDAM-2002-02, London School of Economics and Political Science, 2002.

G. D. Forney, Jr. The Viterbi algorithm. *Proceedings of the IEEE*, 61:268–277, March 1973.

G. D. Forney, Jr. Codes on graphs: Normal realizations. *IEEE Transactions on Information Theory*, 47:520–548, 2001.

D. Forsyth and J. Ponce. *Computer Vision*. Prentice Hall, Upper Saddle River, NJ, 2002.

G. J. Foschini. Layered space-time architecture for wireless communication in a fading environment when using multi-element antennas. *Bell Labs Technical Journal*, 1(2):41–59, September 1996.

C. Foulkes, D. Martin, and J. Malik. Learning affinity functions for image segmentation. In *Proc. of the IEEE Conference on Computer Vision and Pattern Recognition*, 2003.

M. Fox and H. Rubin. Functions of processes with Markovian states. *Annals of Mathematical Statistics*, 39(3):938–946, 1968.

M. Fox and H. Rubin. Functions of processes with Markovian states II. *Annals of Mathematical Statistics*, 40(3):865–869, 1969.

M. Fox and H. Rubin. Functions of processes with Markovian states III. *Annals of Mathematical Statistics*, 41(2):472–479, 1970.

M. Franceschetti, O. Dousse, D. Tse, and P. Thiran. Closing the gap in the capacity of random wireless networks. In *Proc. of the IEEE International Symposium on Information Theory*, 2004.

E. Fredkin. Trie memory. *Communications of ACM*, 3(9):490–499, September 1960.

W. J. Freeman. Mesoscopic neurodynamics: From neuron to brain. *Journal of Physiology*, 94:303–320, 2000.

W. J. Freeman. *Mass Action in the Nervous System*. Academic Press (New York), 1975.

W. T. Freeman, E. C. Pasztor, and O. T. Carmichael. Learning low-level vision. *International Journal of Computer Vision*, 40(1):25–47, 2000.

G. Frenkel, E. Katzav, M. Schwartz, and N. Sochen. Distribution of anomalous exponents of natural images. *Physical Review Letters*. Submitted.

Y. Freund and R. E. Schapire. Game theory, on-line prediction and boosting. In *Proc. of the Ninth Annual Conference on Computational Learning Theory*, 1996.

B. Frey, R. Koetter, and N. Petrovic. Very loopy belief propagation for unwrapping phase images. In *Advances in Neural Information Processing Systems, 14*. MIT Press, Cambridge, MA, 2001.

A. Fridman. Mixed markov models. *Proceedings of the National Academy of Science, USA*, 100:8092–8096, 2003.

A. Friedman. *Stochastic Differential Equations and Applications*. Academic Press, San Diego, CA, 1975.

F-W. Fu and R. Yeung. On the rate-distortion region for multiple descriptions. *IEEE Transactions on Information Theory*, 48(7):2012–2021, July 2002.

D. Fudenberg and D. K. Levine. *The Theory of Learning in Games*. MIT Press, 1998.

R. G. Gallager. *Low-Density Parity-Check Codes*. MIT Press, Cambridge, MA, 1963.

M. Galun, E. Sharon, R. Basri, and A. Brandt. Texture segmentation by multiscale aggregation of filter responses and shape elements. In *Proc. of the International Conference on Computer Vision*, pages 716–723, Nice, France, 2003.

V. C. Gaudet and P. G. Gulak. A 13.3-mb/s 0.35μm CMOS analog turbo decoder IC with a configurable interleaver. *IEEE Journal of Solid-State Circuits*, 38(11): 2010–2015, November 2003.

D. Geiger, H-K. Pao, and N. Rubin. Salient and multiple illusory surfaces. In *Proc. of IEEE Conference on Computer Vision and Pattern Recognition*, 1998.

A. Geisler, J. Perry, B. Super, and D. Gallogly. Edge co-occurence in natural images predicts contour grouping performance. *Vision Research*, 41:711–724, 2001.

D. Geman and A. Koloydenko. Invariant statistics and coding of natural microimages. In *Proceedings of the IEEE Workshop on Statistical and Computational Theories of Vision*, 1999.

S. Geman and D. Geman. Stochastic relaxation, Gibbs distributions and Bayesian restoration of images. *IEEE Transactions on Pattern Analysis and Machine Intelligence*, 6:721–741, 1984.

S. Geman and C. Graffigne. Markov random field image models and their applications to computer vision. In *Proceedings of the International Congress of Mathematicans*, pages 1496–1517, 1986.

S. Geman, D. Potter, and Z. Chi. Composition systems. *Quarterly of Applied Mathematics*, 60:707–736, 2002.

A. Gersho and R. M. Gray. *Vector quantization and signal compression*. Kluwer, Boston, MA, 1992.

W. Gerstner and L. F. Abbott. Learning navigational maps through potentiation and modulation of hippocampal place cells. *Journal of Computational Neuroscience*, 4:79–94, 1997.

Z. Ghahramani and M. I. Jordan. Factorial hidden Markov models. *Machine Learning*, 29:245–273, 1997.

B. Gidas and D. Mumford. Stochastic models for generic images. *Quarterly of Applied Mathematics*, 59:85–111, 2001.

R. Giegerich and S. Kurtz. From Ukkonen to McCreight and Weiner: A unifying view of linear-time suffix tree construction. *Algorithmica*, 19(3):331–353, 1997.

I. I. Gihman and A. V. Skorohod. *Stochastic Differential Equations*. Springer, Berlin, 1972.

B. Gilbert. A precise four-quadrant multiplier with subnanosecond response. *IEEE Journal of Solid-State Circuits*, 34(16):365–373, December 1968.

E. J. Gilbert. On the identifiability problem for functions of finite Markov chains. *Annals of Mathematical Statistics*, 30:688–697, 1959.

W. Gilks, S. Richardson, and D. Spiegelhalter, editors. *Markov Chain Monte Carlo in Practice*. Chapman and Hall, New York, 1996.

B. Girod and N. Farber. Wireless video. In A. Reibman and M.-T. Sun, editors, *Compressed Video over Networks*. Marcel Dekker, New York, 2000.

J. C. Gittins. *Multi-armed Bandit Allocation Indices*. Wiley, 1989.

A. Glavieux, C. Laot, and J. Labat. Turbo equalization over a frequency selective channel. In *Proc. of International Symposium on Turbo Codes and Related Topics*, pages 96–102, Brest, France, 1997.

D. N. Godard. Self-recovering equalization and carrier tacking in two-dimensional data communication systems. *IEEE Transactions on Communications*, 28(11): 1867–1875, 1980.

A. Goldsmith and P. Varaiya. Capacity of fading channels with channel side information. *IEEE Transactions on Information Theory*, 43(6):1986–1992, November 1997.

G. J. Gordon. No-regret algorithms for structured prediction problems. Draft manuscript, available at www.cs.cmu.edu/~ggordon, 2005.

Y. Gousseau. Morphological statistics of natural images. In *Proc. SPIE: Wavelet Applications in Signal and Image Processing*, pages 208–214, 2000.

V. K. Goyal and J. Kovacevic. Generalized multiple description coding with correlating transforms. *IEEE Transactions on Information Theory*, 47(6):2199–2224, 2001.

I. S. Gradshteyn and I. M. Ryzhik. *Table of Integrals, Series, and Products*. Academic Press, San Diego, CA, 1994.

R. M. Gray and D. L. Neuhoff. Quantization. *IEEE Transactions on Information Theory*, 44:2325–2383, October 1998.

A. Greenwald and A. Jafari. A class of no-regret algorithms and game-theoretic equilibria. In *Proc. Conference on Computational Learning Theory*, pages 1–11, 2003.

O. Grellier and P. Comon. Blind separation of discrete sources. *IEEE Signal Processing Letters*, 5(8):212–214, August 1998.

U. Grenander and A. Srivastava. Probability models for clutter in natural images. *IEEE Transactions on Pattern Analysis and Machine Intelligence*, 23:424–429, 2001.

S. Grossberg. Nonlinear difference-differential equations in prediction and learning theory. *Proceedings of the National Academy of Sciences, USA*, 58:1329–1334, 1967.

M. Grossglauser and D. N. C. Tse. Mobility increases the capacity of ad-hoc wireless networks. *IEEE/ACM Transactions on Networking*, 10(4):477–486, August 2002.

E. Guàrdia, R. Rey, and J. A. Padró. Potential of mean force by constrained molecular dynamics: A sodium chloride ion-pair in water. *Chemical Physics*, 155:187–195, 1991a.

E. Guàrdia, R. Rey, and J. A. Padró. Na^+-Na^+ and Cl^--Cl^- ion pairs in water: mean force potentials by constrained molecular dynamics. *Chemical Physics*, 95: 2823–2831, 1991b.

J-C. Guey, M P. Fitz, M R. Bell, and W-Y Kuo. Signal design for transmitter diversity wireless communication systems over Rayleigh fading channels. *IEEE Transactions on Communications*, 47(4):527–537, 1999.

P. Gupta and P. R. Kumar. The capacity of wireless networks. *IEEE Transactions on Information Theory*, 46(2):388–404, March 2000.

P. Gupta and P. R. Kumar. Towards an information theory of large networks: An ahievable rate region. *IEEE Transactions on Information Theory*, 49(8): 1877–1894, August 2003.

J. Hagenauer. Der analoge decoder, 1997a.

J. Hagenauer. The turbo principle: Tutorial introduction and state of the art. In *Proc. of International Symposium on Turbo Codes and Related Topics*, pages 1–11, Brest, France, 1997b.

J. Hagenauer and P. Hoeher. A Viterbi algorithm with soft-decision outputs and its applications. In *Proc. of GLOBECOM'89*, pages 47.11–47.17, Dallas, TX, 1989.

P. Hallinan, G. Gordon, A. Yuille, P. Giblin, and D. Mumford. *Two- and three-dimensional patterns of the face*. A K Peters, Wellesley, MA, 1999.

O. P. Hamill, A. Marty, E. Neher, B. Sakmann, and F. J. Sigworth. Improved patch-clamp techniques for high-resoluton current recodring from cells and cell-free membrane patches. *European Journal of Physiology*, 391:85–100, 1981.

J. Harsanyi. Games of incomplete information played by Bayesian players. *Management Science*, 14:159, 1967.

S. Hart and A. Mas-Colell. A general class of adaptive strategies. *Journal of Economic Theory*, 98(1):26–54, 2001.

M. E. Hasselmo and E. Schnell. Laminar selectivity of the cholinergic suppression of synaptic transmission in rat hippocampal region CA1: computational modeling and brain slice physiology. *Journal of Neuroscience*, 14(6):3898–3914, 1994.

M. E. Hasselmo, B. P. Wyble, and G. V. Wallenstein. Encoding the retrieval of episodic memories: Role of cholinergic and GABAergic modulation in the hippocampus. *Hippocampus*, 6:693–708, 1996.

B. Hassibi and B. Hochwald. High-rate codes that are linear in space and time. *IEEE Transactions on Information Theory*, pages 1804–1824, 2002.

B. Hassibi and T. L. Marzetta. Multiple-antennas and isotropically random unitary inputs: The received signal density in closed form. *IEEE Transactions on Information Theory*, 48(6):1473–1484, June 2002.

M. L. Hawley, R. Y. Litovsky, and H. S. Colburn. Speech intelligibility and localization in a multisource environment. *Journal of the Acoustical Society of America*, 105(8):3436–3448, 1999.

S. Haykin. *Unsupervised Adaptive Filtering Volume 1: Blind Source Separation*. Wiley, New York, 2001a.

S. Haykin. *Unsupervised Adaptive Filtering Volume 2: Blind Deconvolution*. Wiley, New York, 2001b.

S. Haykin. *Neural Networks: A Comprehensive Foundation*. Macmillan, New York, 1994.

S. Haykin and Z. Chen. The cocktail party problem. *Neural Computation*, 17(9), 2005. 1875–1902.

S. Haykin, M. Sellathurai, Y. de Jong, and T. Willink. Turbo-MIMO for wireless communications. *IEEE Communications Magazine*, pages 48–53, October 2004.

Ibn Al Haytham. *Kitab al-Manazir*. c. 1000. In Arabic; translation by A. Sabra, *The Optics of Ibn Al-Haytham*, London: The Warburg Institute, 1989.

R. Hecht-Nielsen. A theory of cerebral cortex. In *Proceedings of the 1998 International Conference on Neural Information Processing*, pages 1459–1464. IOS Press, Burke, VA, 1998.

A. Heller. On stochastic processes derived from Markov chains. *Annals of Mathematical Statistics*, 36:1286–1291, 1965.

T. Heskes, K. Albers, and B. Kappen. Approximate inference and constrained optimization. In *Proc. of Uncertainty in Artificial Intelligence*.

G. E. Hinton. Training products of experts by minimizing contrastive divergence. Technical Report GCNU TR 2000-004, 2000.

G. E. Hinton and A. Brown. Spiking Boltzmann machines. In *Advances in Neural Information Processing Systems, 12*. MIT Press, Cambridge, MA, 2000.

J. Hiriart-Urruty and C. Lemaréchal. *Convex Analysis and Minimization Algorithms*. Springer, New York, 1993.

A. Hiroike, F. Adachi, and N. Nakajima. Combined effects of phase sweeping transmitter diversity and channel coding. *IEEE Transactions on Vehicular Technology*, 41(5):170–176, May 1992.

B. Hochwald and T. L. Marzetta. Unitary space-time modulation for multiple-antenna communications in Rayleigh flat fading . *IEEE Transactions on Information Theory*, 46(2):543–564, 2000.

B. Hochwald and T. L. Marzetta. Capacity of a mobile multiple-antenna communication link in Rayleigh flat fading. *IEEE Transactions on Information Theory*, 45(1):139–157, January 1999.

B. Hochwald and W. Sweldens. Differential unitary space-time modulation. *IEEE Transactions on Communications*, pages 2041–2052, December 2000.

T. Hofmann, J. Puzicha, and J. M. Buhmann. Unsupervised texture segmentation in a deterministic annealing framework. *IEEE Transactions on Pattern Analysis and Machine Intelligence*, 20(8):803–818, 1998.

J. J. Hopfield. Neural networks and physical systems with emergent collective computational abilities. *Proceedings of the National Academy of Sciences, USA*, 79:2554–2558, 1982.

J. J. Hopfield. Neurons with graded response have collective computational properties like those of two-state neurons. *Proceedings of the National Academy of Sciences, USA*, 81:3088–3092, 1984.

J. J. Hopfield and D. W. Tank. "Neural" computation of decisions in optimization problems. *Biological Cybernetics*, 52:141–152, 1985.

J. J. Hopfield and D. W. Tank. Computing with neural circuits: A model. *Science*, 233:625–633, 1986.

R. A. Horn and C. R. Johnson. *Matrix Analysis*. Cambridge University Press, Cambridge, 1985.

K. Hornik, M. Stinchcombe, and H. White. Multi-layer feedforward networks are universal approximators. In H. White, editor, *Artificial Neural Networks: Approximation and Learning Theory*. 1992.

T.-C Hou and V. O. K. Li. Transmission range control in multihop packet radio networks. *IEEE Transactions on Communications*, 34(1):38–44, January 1986.

J. Huang. Statistics of natural images and models. PhD thesis, Division of Applied Mathematics, Brown University, 2000. Available at www.dam.brown.edu/people/mumford/Papers/Huangthesis.pdf.

J. Huang and D. Mumford. Statistics of natural images and models. In *Proc. of IEEE Conference on Computer Vision and Pattern Recognition*, 1999.

D. H. Hubel and T. N. Wiesel. Receptive fields and functional architecture of monkey striate cortex. *Journal of Physiology*, 195(1):215–243, 1968.

B. L. Hughes. Differential space-time modulation. *IEEE Transactions on Information Theory*, 46(7):2567–2578, 2000.

A. Hyvärinen and E. Oja. A fast fixed-point algorithm for independent component analysis. *Neural Computation*, 9:1483–1492, 1997.

A. Hyvärinen, J. Karhunen, and E. Oja. *Independent Component Analysis*. Wiley, New York, 2001.

F. Iida, R. Pfeifer, L. Steels, and Y. Kuniyoshi, editors. *Embodied Artificial Intelligence*. Springer, Berlin, 2004.

A. Ingle and V. A. Vaishampayan. DPCM system design for diversity systems with applications to packetized speech. *IEEE Transactions on Speech and Audio Processing*, 1:48–58, January 1995.

M. Iosifescu and R. Theodorescu. *Random Processes and Learning*, volume 150. Springer, 1969.

H. Ito. An algebraic study of discrete stochastic systems. PhD thesis, Department of Mathematical Engineering and Information Physics, University of Toyko, 1992.

H. Ito, S.-I. Amari, and K. Kobayashi. Identifiability of hidden Markov information sources and their minimum degrees of freedom. *IEEE Transactions on Information Theory*, 38(2):324–333, 1992.

T. S. Jaakkola. Tutorial on variational approximation methods. In M. Opper and D. Saad, editors, *Advanced Mean Field Methods: Theory and Practice*, pages 129–160. MIT Press, Cambridge, MA, 2001.

R. A. Jacobs, M. I. Jordan, S. J. Nowlan, and G. E. Hinton. Adaptive mixtures of local experts. *Neural Computation*, 3(1):79–87, 1991.

H. Jaeger. Observable operator models for discrete stochastic time series. *Neural Computation*, 12(6):1371–1398, 2000.

H. Jaeger. Modeling and learning continuous-valued stochastic processes with OOMs. GMD Report 42, GMD, Sankt Augustin, 1998a. Available at http://www.faculty.iu-bremen.de/hjaeger/pubs/jaeger.00.tr.contoom.pdf.

H. Jaeger. Observable operator models and conditioned continuation representations. Arbeitspapiere der GMD 1043, GMD, Sankt Augustin, 1997a.

H. Jaeger. Observable operator models II: Interpretable models and model induction. Arbeitspapiere der GMD 1083, GMD, Sankt Augustin, Available at http://www.faculty.iu-bremen.de/hjaeger/pubs/jaeger.97.oom2.pdf, 1997b.

H. Jaeger. Discrete-time, discrete-valued observable operator models: A tutorial. GMD Report 42, GMD, Sankt Augustin, 1998b. Available at http://www.faculty.iu-bremen.de/hjaeger/pubs/oom_tutorial.pdf.

H. Jaeger. Characterizing distributions of stochastic processes by linear operators. GMD Report 62, German National Research Center for Information Technology, 1999.

H. Jaeger. Short term memory in echo state networks. GMD Report 152, German National Research Center for Information Technology, 2002.

H. Jaeger and H. Haas. Harnessing nonlinearity: Predicting chaotic systems and saving energy in wireless communication. *Science*, 304:78–80, 2004.

H. Jafarkhani and V. Tarokh. Multiple description trellis coded quantizers. *IEEE Transactions on Communications*, 47:799–803, June 1999.

E. Jakeman. On the statistics of k-distributed noise. *Journal of Physics, A*, 13: 31–48, 1980.

W. C. Jakes. *Microwave Mobile Communications*. IEEE Press, New York, 1974.

A. Jalali, R. Padovani, and R. Pankaj. Data throughput of CDMA-HDR a high efficiency high data rate personal communication wireless system. In *Proceedings of Vehicular Technology Conference (VTC)*, pages 1854–1858, 2000.

M. James and S. Singh. Learning and discovery of predictive state representations in dynamical systems with reset. In *Proc. the Twenty-first International Conference on Machine Learning*, pages 417–424, 2004.

M. R. James, V. Krishnamurthy, and F. LeGland. Time discretization of continuous-time filters and smoothers for HMM parameter estimation. *IEEE Transactions on Information Theory*, 42(2):593–605, March 1996.

N. D. Jayant and P. Noll. *Digital Coding of Waveforms*. Prentice Hall, Englewood Cliffs, NJ, 1984.

N. S. Jayant. Sub-sampling of a DPCM speech channel to provide two "self-contained" half rate channels. *Bell Systems Technical Journal*, 60(4):501–509, April 1981.

L. A. Jeffress. A place theory of sound localization. *Journal of Computational Physiology and Psychiatry*, 41:35–39, 1948.

F. Jelinek. *Statistical Methods for Speech Recognition*. MIT Press, Cambridge, MA, 1998.

M. Jones and W. Yee. Attending to auditory events: The role of temporal organization. In S. McAdams and E. Bigand, editors, *Thinking in Sound*, pages 69–106. Clarendon Press, Oxford, 1996.

M. I. Jordan, Z. Ghahramani, T. S. Jaakkola, and L. Saul. An introduction to variational methods for graphical models. In M. I. Jordan, editor, *Learning in Graphical Models*, pages 105–161. MIT Press, Cambridge, MA, 1999.

P. Joshi and W. Maass. Movement generation with circuits of spiking neurons. *Neural Computation*, 17(8):1715–1738, 2005.

B. Julesz. Textons, the elements of texture perception. *Nature*, 290:91–97, 1981.

B. Julesz and I. J. Hirsh. Visual and auditory perception: An essay in comparison. In E. David and P. B. Denes, editors, *Human Communication: A Unified View*. McGraw-Hill, New York, 1972.

A. Jung, F. Theis, C. G. Puntonet, and E. W. Lang. FASTGEO—A histogram based approach to linear geometric ICA. In *Proc. of Independent Component Analysis and Blind Source Separation*, pages 349–354, San Diego, CA, 2001.

M. W. Jung and B. L. McNaughton. Spatial selectivity of unit activity in the hippocampal granular layer. *Hippocampus*, 3(2):165–182, 1993.

T.-P. Jung, C. Humphries, T.-W. Lee, S. Makeig, M. J. McKeown, V. Iragui, and T. J. Sejnowski. Extended ICA removes artifacts from electroencephalographic recordings. In *Advances in Neural Information Processing Systems, 10*, Cambridge, MA, 1998. MIT Press.

C. Jutten and J. Herault. Blind separation of sources, part I-III. *Signal Processing*, 24:1–29, 1991.

L. P. Kaelbling, editor. *Learning in Embedded Systems*. MIT Press, Cambridge, MA, 1993.

L. P. Kaelbling, M. L. Littman, and A. R. Cassandra. Planning and acting in partially observable stochastic domains. *Artificial Intelligence*, 101:99–134, 1998.

T. Kailath. Channel characterization: Time-variant dispersive channels. In E. J. Baghdady, editor, *Lectures on Communication System Theory*, pages 95–123. McGraw-Hill, New York, 1961.

T. Kailath, A. H. Sayed, and B. Hassibi. *Linear Estimation*. Prentice Hall, Upper Saddle River, NJ, 2000.

A. Kalai and S. Vempala. Geometric algorithms for online optimization. Technical Report MIT-LCS-TR-861, Massachusetts Institute of Technology, 2002.

S. Kali and P. Dayan. The involvement of recurrent connections in area CA3 in establishing the properties of place fields: A model. *Journal of Neuroscience*, 20: 7463–7477, 2000.

R. E. Kalman. A new approach to linear filtering and prediction problems. *Transactions of the ASME, Journal of Basic Engineering*, 82:35–45, March 1960.

E. R. Kandel, J. H. Schwartz, and T. M. Jessell, editors. *Principles of Neural Science*. McGraw-Hill, New York, 4th edition, 2000.

G. Kanizsa. *Grammatica del vedere*. Società Editrice il Mulino, 1980. French translation, *La Grammaire du Voir*, Diderot, 1997.

A. Kannan and V. U. Reddy. Maximum likelihood estimation of constellation vectors for blind separation of co-channel BPSK signals and its performance analysis. *IEEE Transactions on Signal Processing*, 45(7):1736–1741, July 1997.

H. Kantz and T. Schreiber. *Nonlinear Time Series Analysis*. Cambridge University Press, Cambridge, 1997.

D. Kaplan and L. Glass. *Understanding Nonlinear Dynamics*. Springer, Berlin, 1995.

H. Kappen and P. Rodriguez. Efficient learning in Boltzmann machines using linear response theory. *Neural Computation*, 10:1137–1156, 1998.

S. Karlin and W. Studden. *Tchebycheff Systems, with Applications in Analysis and Statistics*. Interscience Publishers, New York, NY, 1966.

K. Kashino, K. Nakadai, T. Kinoshita, and H. Tanaka. Application of the Bayesian probability network to music scene analysis. In D. F. Rosenthal and H. G. Okuno, editors, *Computational Auditory Scene Analysis*, pages 115–137. Lawrence Erlbaum, Mahwah, NJ, 1998.

S. A. Kauffman. Metabolic stability and epigenesis in randomly connected nets. *Journal of Theoretical Biology*, 22:437, 1969.

S. A. Kauffman. *The Origins of Order: Self-Organization and Selection in Evolution*. Oxford University Press, New York, 1993.

J. Kay. Feature discovery under contextual supervision using mutual information. In *Proceedings of the International Joint Conference on Neural Networks*, volume 4, pages 79–84, 1992.

S. Keshav. *An Engineering Approach to Computer Networking.* Addison Wesley, Reading, MA, 1997.

P. E. Kloeden and E. Platen. *Numerical Solution of Stochastic Differential Equations.* Springer, Berlin, 1992.

C. H. Knappand and G. C. Carter. The generalized correlation method for estimation of time delay. *IEEE Transactions on Acoustics, Speech, and Signal Processing,* 24(4):320–327, 1976.

R. Knopp and P. Humblet. Information capacity and power control in single-cell multiuser communications. In *Proc. of the IEEE International Conference on Communications,* pages 331–335, 1995.

T. Kohonen. *Self-Organization and Associative Memory.* Springer, 3rd edition, 1989.

M. Kojima, N. Megiddo, T. Noma, and A. Yoshise. *A Unified Approach to Interior Point Algorithms for Linear Complementarity Problems.* Springer, Berlin, 1991.

J. F. Kolen. Dynamical systems and iterated function systems. In J. F. Kolen and S. Kremer, editors, *A Field Guide to Dynamical Recurrent Networks,* pages 57–81, 2001.

J. F. Kolen and S. Kremer, editors. *A Field Guide to Dynamical Recurrent Networks.* IEEE Press, New York, 2001.

D. Koller, N. Meggido, and B. von Stengel. Efficient computation of equilibria for extensive two-person games. *Games and Economic Behaviour,* 14(2), 1996.

P. König and A. K. Engel. Correlated firing in sensory-motor systems. *Current Opinions in Neurobiology,* 5:511–519, 1995.

P. König, A. K. Engel, and W. Singer. Integrator or coincidence detector? the role of the cortical neuron revisited. *Trends in Neuroscience,* 19(4):130–137, 1996.

N. Kowalski, D. A. Depireux, and S. Shamma. Analysis of dynamic spectra in ferret primary auditory cortex. I. Characteristics of single-unit responses to moving ripple spectra. *Journal of Neurophysiology,* 76(5):3503–3523, 1996.

K. Kretzschmar. Learning symbol sequences with Observable Operator Models. GMD Report 161, Fraunhofer Institute AIS, 2003. Available at http://omk.sourceforge.net/files/OomLearn.pdf.

V. Krishnamurthy. Algorithms for optimal scheduling and management of hidden markov model sensors. *IEEE Transactions on Signal Processing,* 50(6):1382–1397, June 2002.

V. Krishnamurthy. Dynamic ion channel activation in patch clamp on a chip. *IEEE Transactions on NanoBioScience,* 3(3):217–224, September 2004.

V. Krishnamurthy. Decentralized emission management for low probability of intercept sensor platforms in network centric warfare—a multi-armed bandit approach. *IEEE Transactions on Aerospace and Electronic Systems,* 41(1):133–152, 2005.

V. Krishnamurthy and S. H. Chung. Adaptive learning algorithms for nernst potential and current-voltage curves in nerve cell membrane ion channels. *IEEE Transactions on NanoBioScience*, 2(4):266–278, December 2003.

V. Krishnamurthy and S. H. Chung. Brownian dynamics simulation for modeling ion permeation across bio-nanotubes. *IEEE Transactions on NanoBioScience*, 4, 2005.

V. Krishnamurthy and S. H. Chung. Exact formulation of adaptive controlled brownian dynamics for studying the permeation dynamics in ion channels. *Biophysical Journal*, a.

V. Krishnamurthy and S. H. Chung. Adaptive controlled Brownian dynamics simulation approach for estimating the potential mean profile in Gramicidin A ion channels. *Biophysical Journal*, b.

V. Krishnamurthy and G. Yin. Recursive algorithms for estimation of hidden Markov models and autoregressive models with Markov regime. *IEEE Transactions on Information Theory*, 48(2):458–476, February 2002.

V. Krishnamurthy, J. B. Moore, and S. H. Chung. Hidden Markov model signal processing in the presence of unknown deterministic interferences. *IEEE Transactions on Automatic Control*, 38(1):146–152, Janaury 1993.

V. Krishnamurthy, X. Wang, and G. Yin. Spreading code optimization and adaptation in CDMA via discrete stochastic approximation. *IEEE Transactions on Information Theory*, 50(9):1927–1949, 2004.

V. Krishnamurthy, S. H. Chung, and G. Dumont, editors. *Ion Channels: Bio Nanotubes—Special Issue of IEEE Transactions on NanoBioScience*. March 2005.

F. Kschischang and B. Frey. Iterative decoding of compound codes by probability propagation in graphical models. *IEEE Journal of Selected Areas in Communications*, 16(2):219–230, February 1998.

F. R. Kschischang, B. J. Frey, and H.-A. Loeliger. Factor graphs and the sum-product algorithm. *IEEE Transactions on Information Theory*, 47(2):498–519, 2001.

H. J. Kushner and G. Yin. *Stochastic Approximation Algorithms and Applications.* Springer, New York, 1997.

C. Langlais and M. Hélard. Phase carrier for turbo codes over a satellite link with the help of tentative decisions. In *Proc. of International Symposium on Turbo Codes and Related Topics*, pages 439–442, September 2000.

C. G. Langton. Computation at the edge of chaos. *Physica D*, 42:12–37, 1990.

A. Lapidoth and S. Shamai. Fading channels: How perfect need "perfect side information" be? *IEEE Transactions on Information Theory*, 48(5):1118–1134, May 2002.

J. B. Lasserre. Global optimization with polynomials and the problem of moments. *SIAM Journal on Optimization*, 11(3):796–817, 2001.

S. L. Lauritzen. *Graphical Models*. Oxford University Press, New York, 1996.

S. L. Lauritzen and D. J. Spiegelhalter. Local computations with probabilities on graphical structures and their application to expert systems (with discussion). *Journal of the Royal Statistical Society B*, 50:155–224, 1988.

Y. LeCun, J. S. Denker, and S. A. Solla. Optimal brain damage. In D. S. Touretzky, editor, *Advances in Neural Information Procesing Systems, 2*, pages 598–605. Morgan Kaufmann, San Mateo, CA, 1990.

A. Lee, D. Mumford, and J. Huang. Occlusion models for natural images. *International Journal of Computer Vision*, 41:35–59, 2001.

A. Lee, K. Pedersen, and D. Mumford. The nonlinear statistics of high-contrast patches in natural images. *International Journal of Computer Vision*, 54:83–103, 2003a.

A. Lee, E. Simoncelli, A. Srivastava, and S.-C. Zhu. On advances in statistical modeling of natural images. *Journal of Mathematical Imaging and Vision*, 18: 17–33, 2003b.

T. S. Lee and D. Mumford. Hierarchical Bayesian inference in the visual cortex. *Journal of the Optical Society of America, A*, 20(7):1434–1448, July 2003.

T. S. Lee, D. Mumford, and A. Yuille. Texture segmentation by minimizing vector-valued energy functionals. In *Proceedings of the European Conference Computer Vision*, pages 165–173, 1992.

R. A. Legenstein, H. Markram, and W. Maass. Input prediction and autonomous movement analysis in recurrent circuits of spiking neurons. *Reviews in the Neurosciences*, 14(1–2):5–19, 2003.

F. Lehmann and G. M. Maggio. An approximate analytical model of the message passing decoder of LDPC codes. In *Proc. IEEE International Symposium on Information Theory*, page 31, Lausanne, Switzerland, 2002.

M. A. R. Leisink and H. J. Kappen. A tighter bound for graphical models. In S. A. Solla, T. K. Leen, and K.-R. Müller, editors, *Advances in Neural Information Processing Systems, 13*, pages 266–272. MIT Press, Cambridge, MA, 2001.

C. E. Lemke. Bimatrix equilibrium points and mathematical programming. *Management Science*, 11:681–689, 1965.

O. Leveque and E. Telatar. Information theoretic upper bounds on the capacity of large extended ad hoc wireless networks. *IEEE Transactions on Information Theory*, March 2005.

W. B. Levy. A sequence predicting CA3 is a flexible associator that learns and uses context to solve hippocampal-like tasks. *Hippocampus*, 6:579–590, 1996.

W. B. Levy, C. M. Colbert, and N. L. Desmond. Elemental adaptive processes of neurons and synapses: A statistical/computational perspective. In M. Gluck and D. Rumelhart, editors, *Neuroscience and Connectionist Models*, pages 187–235. Lawrence Erlbaum Associates, Hillsdale, NJ, 1990.

L. Li and A. Goldsmith. Optimal resource allocation for fading broadcast channels—Part I: Ergodic capacity. *IEEE Transactions on Information Theory*, 47(3):1083–1102, March 2001.

Y. Li, A. Cichocki, and L. Zhang. Blind separation and extraction of binary sources. *IEICE Transactions on Fundamentals*, E86-A(3):580–589, March 2003.

Z. Li and J. Atick. Efficient stereo coding in the multiscale representation. *Network: Computation in Neural Systems*, 5:157–174, 1994.

S. Lin and D. J. Costello, Jr. *Error Control Coding.* Pearson Prentice Hall, 2nd edition, 2004.

R. Linsker. From basic network principles to neural architecture: Emergence of spatial opponent cells. *Proceedings of the National Academy of Sciences USA*, 83:7508–7512, 1986a.

R. Linsker. From basic network principles to neural architecture: Emergence of orientation-selective cells. *Proceedings of the National Academy of Sciences USA*, 83:8390–8394, 1986b.

R. Linsker. From basic network principles to neural architecture: Emergence of orientation columns. *Proceedings of the National Academy of Sciences USA*, 83: 8779–8783, 1986c.

R. Linsker. Self-organization in a perceptual network. *IEEE Computer*, 21:105–117, March 1988.

R. Linsker. How to generate ordered maps by maximizing the mutual information between input and output signals. *Neural Computation*, 1(3), 1989.

W. A. Little. The existence of persistent states in the brain. *Mathematical Biosciences*, 19:101–120, 1974.

N. Littlestone and M. Warmuth. The weighted majority algorithm. Technical Report UCSC-CRL-91-28, University of California, Santa Cruz, 1992.

M. L. Littman. Friend or foe Q-learning in general-sum Markov games. In *Proceedings of the Eighteenth International Conference on Machine Learning*, pages 322–328, 2001.

M. L. Littman, R. S. Sutton, and S. Singh. Predictive representation of state. In *Advances in Neural Information Processing Systems, 14*, pages 1555–1561. MIT Press, Cambridge, MA, 2001.

S. P. Lloyd. Least squares quantization in PCM. *IEEE Transactions on Information Theory*, 28:129–137, March 1982.

H.-A. Loeliger. An introduction to factor graphs. *IEEE Signal Processing Magazine*, 21:28–41, 2004.

V. Lottici and M. Luise. Carrier phase recovery for turbo-coded linear modulations. In *Proc. of International Conference on Communications*, pages 1543–1545, New York, USA, 2002.

W. S. Lovejoy. A survey of algorithmic methods for partially observable markov decision processes. *Annals of Operations Research*, 28:47–66, 1991.

H. F. Lu and P. V. Kumar. Rate-diversity trade-off if space-time codes with fixed alphabet and optimal constructions for PSK modulation. *IEEE Transactions on Information Theory*, 49(10):2747–2752, October 2003.

M. Luettgen, W. Karl, and A. Willsky. Efficient multiscale regularization with application to optical flow. *IEEE Transactions Image Processing*, 3(1):41–64, 1994.

S. Luke. Evolving soccerbots: A retrospective. In *Proc. 12th Annual Conference on Japanese Society for Artificial Intelligence*, 1998.

F. Lustenberger, M. Helfenstein, H. A. Loeliger, T. Tarkoy, and G. S. Moschytz. An analog VLSI decoding technique for digital codes. In *Proc. of IEEE International Symposium on Circuits and Systems*, volume 2, pages 424–427, 1999.

W. Maass and H. Markram. Theory of the computational function of microcircuit dynamics. In *Proceedings of the 2004 Dahlem Workshop on Microcircuits*, Cambridge, MA, 2005. MIT Press. Available at http://www.igi.tugraz.at/maass/psfiles/157_v2_web.pdf.

W. Maass, T. Natschläger, and H. Markram. Real-time computing without stable states: A new framework for neural computation based on perturbations. *Neural Computation*, 14(11):2531–2560, 2002.

W. Maass, T. Natschläger, and H. Markram. Computational models for generic cortical microcircuits. In J. Feng, editor, *Computational Neuroscience: A Comprehensive Approach*, chapter 18, pages 575–605. Chapman and Hall/CRC, 2004.

W. Maass, R. A. Legenstein, and N. Bertschinger. Methods for estimating the computational power and generalization capability of neural microcircuits. In L. K. Saual, Y. Weiss, and L. Bottou, editors, *Advances in Neural Information Processing Systems, 17*. MIT Press, Cambridge, MA, 2005.

D. J. C. MacKay. *Information Theory, Inference, and Learning Algorithms*. Cambridge University Press, Cambridge, 2003.

W. R. MacLean. On the acoustics of cocktail parties. *Journal of the Acoustical Society of America*, 31(1):79–80, 1959.

J. MacQueen. Some methods for classification and analysis of multivariate observations. In L. M. LeCam and J. Neyman, editors, *Proc. 5th Berkeley Symposium on Mathematical Statistics and Probability*, pages 281–297. University of California Press, Berkely, CA, 1967.

S. Makeig, A. J. Bell, T.-P. Jung, and T. J. Sejnowski. Independent component analysis of electroencephalographic data. In D. Touretzky and M. Mozer, editors, *Advances in Neural Information Processing Systems, 8*. MIT Press, Cambridge, MA, 1995.

S. Makeig, A. J. Bell, T.-P. Jung, D. Ghahremani, and T. J. Sejnowski. Blind separation of auditory evoked potentials into independent components. *Proceedings of the National Academy of Science, USA*, 94:10797–10984, 1997.

J. Malik, S. Belongie, J. Shi, and T. Leung. Textons, contours and regions. In *Proceedings of International Conference on Computer Vision*, pages 918–925, 1999.

S. Mallat. *A Wavelet Tour of Signal Processing*. Academic Press, San Diego, CA, 1999.

D. P. Mandic and J. A. Chambers. *Recurrent Neural Networks for Prediction: Learning Algorithms, Architectures, and Stability*. Wiley, New York, 2001.

A. Mansour, C. G. Puntonet, and N. Ohnishi. A simple ica algorithm based on geometrical approach. In *Proc. International Symposium on Signal Processing and Its Applications*, pages 9–12, Kuala Lumpur, Malaysia, August 2001.

H. Markram, Y. Wang, and M. Tsodyks. Differential signaling via the same axon of neocortical pyramidal neurons. *Proceedings of the National Academy of Sciences, USA*, 95:5323–5328, 1998.

D. Marr. Simple memory: A theory for archicortex. *Philosophical Transactions of the Royal Society of London, Series B*, 262:23–81, 1971.

D. Marr. *Vision: A Computational Investigation into the Human Representation and Processing of Visual Information*. Freeman, San Francisco, CA, 1982.

T. M. Martinetz, S. G. Berkovich, and K. J. Schulten. "Neural-gas" network for vector quantization and its application to time-series prediction. *IEEE Transactions on Neural Networks*, 4(4):558–569, July 1993.

A. Matache, S. Dolinar, and F. Pollara. Stopping rules for turbo decoders. Technical Report 42–142, TMO progress report, Jet Propulsion Lab, NASA, August 2000.

N. Maxemchuk. Dispersity routing in store and forward networks. PhD thesis, University of Pennsylvania, Philadelphia, 1975.

J. Mayor and W. Gerstner. Signal buffering in random networks of spiking neurons: Microscopic vs. macroscopic phenomena, 2005.

J. L. McClelland, B. L. McNaughton, and R. C. O'Reilly. Why there are complementary learning systems in the hippocampus and neocortex: Insights from the successes and failures of connectionist models of learning and memory. *Psychological Review*, 102(3):419–457, 1995.

R. J. McEliece and M. Yildirim. Belief propagation on partially ordered sets. In D. Gilliam and J. Rosenthal, editors, *Mathematical Theory of Systems and Networks*. Institute for Mathematics and its Applications, 2002.

R. J. McEliece, D. J. C. MacKay, , and J-F. Cheng. Turbo decoding as an instance of Pearl's belief propagation algorithm. *IEEE Journal on Selected Areas in Communications*, 16(2):140–152, February 1998.

R.J. McEliece, D.J.C. McKay, and J.F. Cheng. Turbo decoding as an instance of Pearl's belief propagation algorithm. *IEEE Journal of Selected Areas in Commmunications*, 16(2):140–152, February 1998.

H. McGurk and J. MacDonald. Hearing lips and seeing voices. *Nature*, 264:746–748, 1976.

R. D. McKelvey and A. McLennan. Computation of equilibria in finite games. In *The Handbook of Computational Economics*.

M. J. McKeown, S. Makeig, G. G. Brown, T.-P. Jung, S. S. Kinderman, A. J. Bell, and T. J. Sejnowski. Analysis of fMRI data by blind spearation into independent spatial components. *Human Brain Mapping*, 6:160–188, 1998.

B. L. McNaughton and R. G. M. Morris. Hippocampal synaptic enhancement and information storage within a distributed memory systems. *Trends in Neurosciences*, 10:408–415, 1987.

L. R. Medsker and L. C. Jain. *Recurrent Neural Networks: Design and Application.* CRC Press, Roca Raton, FL, 1999.

P. Michor and D. Mumford. Riemannian geometries on spaces of plane curves. *European Journal of Mathematics*, 8:1–48, 2006.

P. Michor and D. Mumford. Vanishing geodesic distance on spaces of submanifolds and diffeomorphisms. *Documenta Mathematica*, 10:217–245, 2005.

M. Miller. On the metrics of Euler-Lagrange equations of computational anatomy. In *Annual Reviews of Biomedical Engineering*, pages 375–405. Annual Reviews Press, 2002.

M. Miller and L. Younes. Group actions, homeomorphisms, and matching. *International Journal of Computer Vision*, 41:61–84, 2001.

T. P. Minka. *A family of algorithms for approximate Bayesian inference.* PhD thesis, Massachusetts Institute of Technology, 2001.

M. Mishkin, W. A. Suzuki, D. G. Gadian, and F. Vargha-Khadem. Hierarchical organization of cognitive memory. *Philosophical Transactions of the Royal Society of London B: Biological Sciences*, 352(1360):1461–1467, 1997.

M. Mitchell, P. T. Hraber, and J. P. Crutchfield. Revisiting the edge of chaos: Evolving cellular automata to perform computations. *Complex Systems*, 7:89–130, 1993.

M. Moerz, T. Gabara, R. Yan, and J. Hagenauer. An analog 0.25μm BiCMOS tailbiting map decoder. In *Proc. of ISCC*, pages 356–357, 2000.

P. R. Montague, P. Dayan, and T. J. Sejnowski. A framework for mesencephalic dopamine systems based on predictive Hebbian learning. *Neuroscience*, 16:1936–1947, 1996.

J. Moody and C. J. Darken. Fast learning in networks of locally tuned processing units. *Neural Computation*, 1(2):281–284, 1989.

T. K. Moon. The expectation-maximization algorithm. *IEEE Signal Processing Magazine*, 13(6):47–60, November 1996.

B. C. J. Moore. *An Introduction to the Psychology of Hearing.* Academic Press, London, 4th edition, 1997.

D. R. Morrison. PATRICIA: Practical algorithm to retrieve information coded alphanumeric. *Journal of the ACM,* 15(4):514–534, October 1968.

R. J. Muirhead. *Aspects of Multivariate Statistical Theory.* Wiley, New York, 1982.

D. Mumford. Pattern theory: A unifying perspective. In D. Knill and W. Richards, editors, *Perception as Bayesian Inference,* pages 25–62. Cambridge University Press, Cambridge, 1996.

D. Mumford. Pattern theory: The mathematics of perception. In *Proceedings of International Congress of Mathematics.* Beijing, China, 2002.

D. Mumford. Elastica and computer vision. In C. Bajaj, editor, *Algebraic Geometry and Its Applications,* pages 491–506. Springer, Berlin, 1992.

D. Mumford and A. Desolneux. *Pattern Theory through Examples.* In preparation. Drafts available at www.dam.brown.edu/people/mumford/Papers/IHP.

D. Mumford and J. Shah. Optimal approximations of piecewise smooth functions and associated variational problems. *Communications in Pure and Applied Mathematics,* 42:577–685, 1989.

D. Mumford and E. Sharon. 2D shape analysis using conformal mappings. In *Proc. of the IEEE Conference on Computer Vision and Pattern Recognition,* 2004.

J. J. Murphy. *Intermarket Technical Analysis.* New York Institute of Finance, New York, 1999.

A. Naguib, V. Tarokh, N. Seshadri, and A. R. Calderbank. A space-time coding modem for high-data-rate wireless communications. *IEEE Journal on Selected Areas in Communications,* 16(8):1459–1477, October 1998.

T. Natschläger, N. Bertschinger, and R. Legenstein. At the edge of chaos: Real-time computations and self-organized criticality in recurrent neural networks. In L. K. Saual, Y. Weiss, and L. Bottou, editors, *Advances in Neural Information Processing Systems, 17.* MIT Press, Cambridge, MA, 2005.

R. Nau, S. G. Canovas, and P. Hansen. On the geometry of Nash equilibria and correlated equilibria. *International Journal of Game Theory,* 2004.

R. Neal and G. E. Hinton. A view of the EM algorithm that justifies incremental, sparse, and other variants. In M. I. Jordan, editor, *Learning in Graphical Models.* MIT Press, Cambridge, MA, 1999.

F. D. Neeser and J. L. Massey. Proper complex random processes with applications to information theory. *IEEE Transactions on Information Theory,* 39:1293–1302, July 1993.

E. Neher and B. Sakmann. Single-channel currents recorded from membrane of denervated frog muscle fibres. *Nature,* 260:799–802, 1976.

R. Neuneier and H. G. Zimmermann. How to train neural networks. In *Neural Networks: Tricks of the Trade,* pages 373–423. Springer, Berlin, 1998.

M. Nitzberg, D. Mumford, and T. Shiota. *Filtering, Segmentation and Depth.* Springer, Berlin, 1992.

J. Nix, M. Kleinschmidt, and V. Hohmann. Computational scene analysis of cocktail-party situations based on sequential Monte Carlo methods. In *Proc. 37th Asilomar Conference on Signals, Systems and Computers*, pages 735–739, 2003.

S. J. Nowlan. Maximum likelihood competitive learning. In D. S. Touretzky, editor, *Advances in Neural Information Processing Systems, 2*, pages 574–582. Morgan Kaufmann, San Mateo, CA, 1990.

T. Oberstein. Efficient training of observable operator models. Master thesis, Köln University, Germany, 2002. Available at http://omk.sourceforge.net/files/eloom.pdf.

K. Okajima. Binocular disparity encoding cells generated through an Infomax based learning algorithm. *Neural Networks*, 17(7):953–962, 2004.

B. Oksendal. *Stochastic Differential Equations: An Introduction with Applications.* Springer, New York, 5th edition, 1998.

B. Olshausen and D. Field. Emergence of simple cell receptive field properties by learning a sparse code for natural images. *Nature*, 381:607–609, 1996.

M. O'Mara, P. H. Barry, and S. H. Chung. A model of the glycine receptor deduced from Brownian dynamics studies. *Proceedings of the National Academy of Sciences USA*, 100(4310–4315), 2003.

M. Opper and D. Saad. Adaptive TAP equations. In M. Opper and D. Saad, editors, *Advanced Mean Field Methods: Theory and Practice*, pages 85–98. MIT Press, 2001.

R. C. O'Reilly and J. W. Rudy. Conjunctive representations in learning and memory: Principles of cortical and hippocampal function. *Psychological Review*, 108:311–345, 2001.

L. Ozarow. On a source coding problem with two channels and three receivers. *Bell Systems of Technical Journal*, 59:1909–1921, December 1980.

L. H. Ozarow, S. Shamai, and A. D. Wyner. Information theoretic considerations for cellular mobile radio. *IEEE Transactions on Vehicular Technology*, 43(2): 359–378, May 1994.

N. Packard. Adaption towards the edge of chaos. In J. A. S. Kelso, A. J. Mandell, and M. F. Shlesinger, editors, *Dynamic Patterns in Complex Systems*, pages 293–301. World Scientific, Singapore, 1988.

P. Pajunen. Blind separation of binary sources with less sensors than sources. In *Proc. of International Conference on Neural Networks*, pages 1994–1997, Houston, TX, June 1997.

P. Pakzad and V. Anantharam. Iterative algorithms and free energy minimization. In *Proc. Conference on Information Sciences and Systems*, March 2002.

C. B. Papadias and A. Paulraj. A constant modulus algorithm for multi-user signal separation in presence of delay spread using antenna arrays. *IEEE Signal Processing Letters*, 4(6):178–181, 1997.

P. Parent and S. Zucker. Trace inference, curvature consistency and curve detection. *IEEE Transactions on Pattern Analysis and Machine Intelligence*, 11:823–839, 1989.

L. Parra and C. Spence. Convolutive blind source separation of non-stationary sources. *IEEE Transactions on Speech and Audio Processing*, 8(3):320–327, May 2000.

L. C. Parra and C. V. Alvino. Geometric source separation: Merging convolutive source separation with geometric beamforming. *IEEE Transactions on Speech and Audio Processing*, 10(6):352–362, September 2002.

P. Parrilo. Semidefinite programming relaxations for semialgebraic problems. *Mathematical Programming, Series B*, 96:293–320, 2003.

P. Patel and J. Holtzman. Analysis of a simple successive interference cancellation scheme in a DS/CDMA system. *IEEE Journal of Selected Areas in Communications*, 12(5):796–807, June 1994.

A. Paulraj and C. Papadias. Space-time processing for wireless communications. *IEEE Signal Processing Magazine*, 14(6):49–83, 1997.

J. Pearl. *Probabilistic Reasoning in Intelligent Systems*. Morgan Kaufmann, San Mateo, CA, 1988.

B. Pearlmutter. Gradient calculations for dynamic recurrent neural networks. In J. F. Kolen and S. Kremer, editors, *A Field Guide to Dynamical Recurrent Networks*, pages 179–206. IEEE Press, New York, 2001.

B. Pearlmutter. Gradient calculations for dynamic recurrent neural networks: A survey. *IEEE Transactions on Neural Networks*, 6(5):1212–1228, 1995.

K. Pedersen and A. Lee. Toward a full probability model of edges in natural images. In *Lecture Notes in Computer Science*, volume 2350, page 328. Springer, 2002.

R. Pfeifer and C. Scheier. *Understanding Intelligence*. MIT Press, Cambridge, MA, 1999.

G. Pflug. *Optimization of Stochastic Models: The Interface between Simulation and Optimization*. Kluwer, Dordrecht, 1996.

D. T. Pham and J. F. Cardoso. Blind separation of instantaneous mixtures of non-stationary sources. *IEEE Transactions on Signal Processing*, 49(9):1837–1848, 2001.

W. A. Phillips, D. Floreano, and J. Kay. Contextually guided unsupervised learning using local multivariate binary processors. *Neural Networks*, 11(1):117–140, 1998.

T. Plefka. Convergence condition of the TAP equation for the infinite-ranged Ising model. *Journal of Physics A*, 15(6):1971–1978, 1982.

R. Podemski, W. Holubowicz, C. Berrou, and G. Battail. Hamming distance spectra of turbo-codes. *Annales des Télécommunications*, 50:790–797, 1995.

H. V. Poor. Iterative multiuser detection. *IEEE Signal Processing Magazine*, pages 81–88, 2004.

G. J. Pottie and W. J. Kaiser. Wireless integrated network sensors. *Communications of the ACM*, 43(2):51–58, May 2000.

S. S. Pradhan, J. Kusuma, and K. Ramchandran. Distributed compression in a dense microsensor network. *IEEE Signal Processing Magazine*, 2:51–60, March 2002.

S. S. Pradhan, R. Puri, and K. Ramchandran. n-channel symmetric multiple descriptions—part i: (n, k) source-channel erasure codes. *IEEE Transactions on Information Theory*, 50(1):47–61, January 2004.

F. P. Preparata and M. I. Shamos. *Computational Geometry: An Introduction.* Springer, New York, 1985.

J. G. Proakis. *Digital Communications.* McGraw-Hill, New York, 4th edition, 2000.

J. G. Proakis. *Digital Communications.* McGraw-Hill, New York, 3rd edition, 1995.

C. Puntonet, A. Mansour, and C. Jutten. A geometrical algorithm for blind separation of sources. In *Actes du XVeme Colloque GRETSI*, pages 273–276, Juan-Les-Pins, France, 1995.

R. Puri and K. Ramchandran. PRISM: An uplink-friendly multimedia coding paradigm. In *Proc. of the IEEE International Conference on Acoustics, Speech, and Signal Processing*, pages 856–859, 2003.

L. R. Rabiner. A tutorial on hidden Markov models and selected applications in speech recognition. In A. Waibel and K.-F. Lee, editors, *Readings in Speech Recognition*, pages 267–296. Morgan Kaufmann, San Mateo, CA, 1990.

L. R. Rabiner and B. H. Juang. *Fundamentals of Speech Recognition.* Prentice Hall, Englewood Cliffs, NJ, 1993.

D. V. Rabinkin, R. J. Renomeron, A. J. Dahl, J. C. French, J. L. Flanagan, and M. H. Bianchi. A DSP implementation of source location using microphone arrays. In *Proc. SPIE*, volume 2846, pages 88–99, Denver, CO, 1996.

G. Raleigh, S. N. Diggavi, A. F. Naguib, and A. Paulraj. Characterization of fast fading vector channels for multi-antenna communication systems. In *Proc. 28th Asilomar Conference on Signals, Systems, and Computers*, pages 853–857, 1994.

R. P. N. Rao and D. H. Ballard. An active vision architecture based on iconic representations. *AI Journal*, 78(1):461–505, October 1995.

R. P. N. Rao and D. H. Ballard. Predictive coding in the visual cortex: a functional interpretation of some extra-classical receptive-field effects. *Nature Neuroscience*, 2(1):79–87, 1999.

T. Rappaport. *Wireless Communications.* IEEE Press, New York, 1996.

A. Reibman and M.-T. Sun. *Compressed Video over Networks*. Marcel Dekker, New York, 2000.

X. Ren and J. Malik. A probabilistic multi-scale model for contour completion based on image statistics. In *Proc. of European Conference on Computer Vision*, 2002.

X. Ren, C. Fowlkes, and J. Malik. Mid-level cues improve boundary detection. In *Proc. of IEEE Conference on Computer Vision and Pattern Recognition*, 2005.

H. Resnikoff. *The Illusion of Reality*. Springer, 1989.

T. J. Richardson and R. L. Urbanke. The capacity of low-density parity-check codes under message-passing decoding. *IEEE Transactions on Information Theory*, 47 (2):599–618, 2001.

T. J. Richardson and R. L. Urbanke. The renaissance of Gallager's low-density parity-check codes. *IEEE Communications Magazine*, 41:126–131, 2003.

J. Rissanen. Modeling by shortest data description. *Automatica*, 14:465–471, 1978.

P. Robertson and T. Wörz. A novel bandwidth efficient coding scheme employing turbo codes. In *Proc. of ICC'96*, 1996.

P. Robertson, P. Höher, and E. Villebrun. Optimal and suboptimal maximum a posteriori algorithms suitable for turbo decoding. *European Transactions on Telecommunications*, 8:119–125, 1997.

G. Rockafellar. *Convex Analysis*. Princeton University Press, Princeton, 1970.

E. Rolls. Functions of neural networks in the hippocampus and neocortex in memory. In J. H. Byrne and W. O. Berry, editors, *Neural Models of Plasticity: Theoretical and Empirical Approaches*, pages 240–265. Academic Press, San Diego, CA, 1989.

D. F. Rosenthal and H. G. Okuno, editors. *Computational Auditory Scene Analysis*. Lawrence Erlbaum, Mahwah, NJ, 1998.

D. Ruderman and W. Bialek. Statistics of natural images: Scaling in the woods. *Physical Review Letters*, 73.

D. E. Rumelhart, G. E. Hinton, and R. J. Williams. Learning representations by back-propagating errors. *Nature*, 323(9):533–536, October 1986.

S. Sagi, S. C. Nemat-Nasser, R. Kerr, R. Hayek, C. Downing, and R. Hecht-Nielsen. A biologically motivated solution to the cocktail party problem. *Neural Computation*, 13:1575–1602, 2001.

L. K. Saul and M. I. Jordan. Exploiting tractable substructures in intractable networks. In *Advances in Neural Information Processing Systems, 8*, pages 486–492. MIT Press, Cambridge, MA, 1996.

B. Sayers and E. C. Cherry. Mechanism of binaural fusion in the hearing of speech. *Journal of the Acoustical Society of America*, 31:535, 1957.

F. Schuermann, K. Meier, and J. Schemmel. Edge of chaos computation in mixed-mode VLSI: A hard liquid. In L. K. Saual, Y. Weiss, and L. Bottou, editors,

Advances in Neural Information Processing Systems, 17. MIT Press, Cambridge, MA, 2005.

S. R. Schultz, H. D. R. Golledge, and S. Panzeri. Synchronization, binding and the role of correlated firing in fast information transmission. In *Emergent Neural Computational Architectures Based on Neuroscience.* Springer, Berlin, 2000.

F. Seguin, C. Lahuec, J. Lebert, M. Arzel, and M. Jezequel. Analogue 16-qam demodulator. *Electronics Letters*, 40(18):1138–1139, September 2004.

J. Serra. *Image Analysis and Mathematical Morphology*, volume I and II. Academic Press, San Diego, CA, 1983 and 1988.

B. A. Sethuraman, B. S. Rajan, and V. Shashidhar. Full-diversity, high-rate space-time block codes from division algebras. *IEEE Transactions on Information Theory*, 49(10):2596–2616, 2003.

O. Shalvi and E. Weinstein. New criteria for blind deconvolution of nonminimum phase systems (channels). *IEEE Transactions on Information Theory*, 36:312–321, 1990.

S. Shamma. On the role of space and time in auditory processing. *Trends in Cognitive Sciences*, 5(8):340–348, 2001.

S. Shamsunder and G. Giannakis. Multichannel blind signal separation and reconstruction. *IEEE Transactions on Speech and Audio Processing*, 5:515–528, 1997.

C. E. Shannon. A mathematical theory of communications. *Bell Systems Technical Journal*, 27:379–423, 623–656, 1948.

C. E. Shannon. Channels with side-information at the transmitter. *IBM Journal of Research and Development*, 2:289–293, October 1958a.

C. E. Shannon. Coding theorems for a discrete source with a fidelity criterion. In *IRE National Convention Record, Part 4*, pages 142–163. 1958b.

R. M. Shapley and J. D. Victor. The contrast gain conrol of the cat retina. *Vision Research*, 19:431–434, 1979.

E. Sharon, S. Litsyn, and J. Goldberger. An efficient message-passing schedule for LDPC decoding. In *Proceedings of the Twenty-third IEEE Convention on Electrical and Electronics Engineering in Israel*, pages 223–226, 2004.

A. Shashua and S. Ullman. Structural saliency. In *Proc. of International Conference on Computer Vision*, pages 321–327, 1988.

J. Shi and M. L. Littman. Abstraction methods for game theoretic poker. In *Computers and Games*, pages 333–345. Springer, Berlin, 2001.

F. J. Sigworth and K. G. Klemic. Patch clamp on a chip. *Biophysical Journal*, 82: 2831–2832, June 2002.

E. Simoncelli. Modeling the joint statistics of images in the wavelet domain. In *Proceedings of the Forty-fourth Annual Meeting of SPIE*, pages 188–195, 1999.

W. Singer. Synchronization of cortical activity and its putative role in information processing and learning. *Annual Review of Physiology*, 55:349–374, 1993.

W. Singer. Synchronization of neural responses as putative binding mechanism. In M. Arbib, editor, *The Handbook of Brain Theory and Neural Networks*, pages 960–964. MIT Press, Cambridge, MA, 1995.

S. P. Singh, M. J. Kearns, and Y. Mansour. Nash convergence of gradient dynamics in general-sum games. In *Proceedings of Sixteenth Conference on Uncertainty in Artificial Intelligence*, pages 541–548, 2000.

D. Slepian and J. K. Wolf. Noiseless coding of correlated information sources. *IEEE Transactions on Information Theory*, 19(4):471–480, July 1973.

R. D. Smallwood and E. J. Sondik. The optimal control of partially observable Markov processes over a finite horizon. *Operations Research*, 21:1071–1088, 1973.

A. Smith, S. Becker, and S. Kapur. A computational model of the selective role of the striatal D2-receptor in the expression of previously acquired behaviours. *Neural Computation*, 17(2):361–395, 2005.

A. R. Smith. Simple computation-universal cellular spaces. *Journal of the ACM*, 18(3):339–353, 1971.

A. Soofi and L. Cao. *Modeling and Forecasting Financial Data: Techniques of Nonlinear Dynamics*. Kluwer, Boston, 2002.

J. Spall. *Introduction to Stochastic Search and Optimization*. Wiley, Hoboken, NJ, 2003.

O. Sporns. Embodied cognition. In M. Arbib, editor, *The Handbook of Brain Theory and Neural Networks*, pages 395–398. MIT Press, Cambridge, MA, 2nd edition, 2003.

F. H. Stillinger and A. Rahman. Improved simulation of liquid water by molecular dynamics. *Journal of Chemical Physics*, 60:1545–1557, 1974.

J. Stone. Learning perceptually salient visual parameters using spatiotemporal smoothness constraints. *Neural Computation*, 8:1463–1492, 1996.

P. Stone and M. Veloso. Beating a defender in robotic soccer: Memory-based learning of a continuous function. In *Advances in Neural Information Processing Systems, 8*, pages 896–902. MIT Press, 1996.

P. Stone, R. S. Sutton, and G. Kuhlmann. Reinforcement learning for RoboCup-soccer keepaway. *Adaptive Behavior*, 13(3):165–188, 2005.

S. M. Stringer, E. T. Rolls, T. P. Trappenberg, and I. T. de Araujo. Self-organizing continuous attractor networks and path integration: Two-dimensional models of place cells. *Network*, 13(4):429–446, 2002.

S. H. Strogatz. *Nonlinear Dynamics and Chaos: With Applications in Physics, Biology, Chemistry, and Engineering*. Addison-Wesley, Reading, MA, 1994.

R. S. Sutton. Learning to predict by the methods of temporal differences. *Machine Learning*, 3:9–44, 1988.

R. S. Sutton and A. G. Barto. Toward a modern theory of adaptive networks: Expectation and prediction. *Psychology Review*, 88:135–170, 1981.

R. S. Sutton and A. G. Barto. *Reinforcement Learning: An Introduction*. MIT Press, Cambridge, MA, 1998.

R. S. Sutton, D. McAllester, S. Singh, and Y. Mansour. Policy gradient methods for reinforcement learning with function approximation. In *Advances in Neural Information Processing Systems, 12*, pages 1057–1063. MIT Press, Cambridge, MA, 2000.

Y. V. Svirid. Weight distributions and bounds for turbo-codes. *European Transactions on Telecommunications*, 6(5):543–555, 1995.

J. R. Swisher, S. H. Jacobson, P. D. Hyden, and L. W. Schruben. A survey of simulation optimization techniques and procedures. In *Proc. of the 2000 Winter Simulation Conference*, Orlando, FL, 2000.

R. Szeliski. Bayesian modeling of uncertainty in low-level vision. *International Journal of Computer Vision*, 5(3):271–301, 1990.

S. Talwar, M. Viberg, and A. Paulraj. Blind estimation of multiple co-channel digital signals using an antenna array. *IEEE Signal Processing Letters*, 1(2): 29–31, 1994.

S. Talwar, M. Viberg, and A. Paulraj. Blind separation of synchronous co-channel digital signals using an antenna array—Part I: algorithms. *IEEE Transactions on Signal Processing*, 44(5):1184–1197, 1996.

V. Tarokh and H. Jafarkhani. A differential detection scheme for transmit diversity. *IEEE Journal on Selected Areas in Communications*, 18(7):1169–1174, July 2000.

V. Tarokh, N. Seshadri, and A. R. Calderbank. Space-time codes for high data rate wireless communications: Performance criterion and code construction. *IEEE Transactions on Information Theory*, 44(2):744–765, March 1998.

V. Tarokh, H. Jafarkhani, and A. R. Calderbank. Space-time block codes from orthogonal designs. *IEEE Transactions on Information Theory*, 45:1456–1467, July 1999.

I. E. Telatar. Capacity of multiple antenna Gaussian channels. *European Transactions on Telecommunications*, 10(6):585–595, 1999.

S. ten Brink. Iterative decoding trajectories of parallel concatenated codes. In *Proc. of the Third IEE/ITG Conference on Source and Channel Coding*, pages 75–80, Munich, Germany, 2000.

S. ten Brink. Convergence behavior of iteratively decoded parallel concatenated codes. *IEEE Transactions on Communications*, 49:1727–1737, 2001.

S. ten Brink and Gerhard Kramer. Design of repeat-accumulate codes for iterative detection and decoding. *IEEE Transactions on Signal Processing*, 51:2764–2772, 2003.

S. ten Brink, J. Speidel, and R. Yan. Iterative demapping and decoding for multilevel modulations. In *Proc. of GLOBECOM*, New York, 1998.

S. ten Brink, G. Kramer, and A. Ashikhmin. Design of low-density parity-check codes for multi-antenna modulation and detection. *IEEE Transactions on Communications*, 52:670–678, 2004.

J. M. Tenenbaum. Accommodation in computer vision. PhD thesis, Department of Computer Science, Stanford University, Stanford, CA, 1970.

G. Tesauro. TD-Gammon, a self-teaching backgammon program, achieves master-level play. *Neural Computation*, 6:215–219, 1994.

F. J. Theis, A. Jung, C. G. Puntonet, and E. W. Lang. Linear geometric ica: Fundamentals and algorithms. *Neural Computation*, 15(2):419–439, 2003a.

F. J. Theis, C. G. Puntonet, and E. W. Lang. A histogram-based overcomplete ica algorithm. In *Proc. of Independent Component Analysis and Blind Source Separation*, pages 1071–1076, Nara, Japan, 2003b.

S. Theodoridis and K. Koutroubas. *Pattern Recognition*. Academic Press, London, 1998.

TIA/EIA/IS. Physical layer standard for CDMA2000 spread spectrum systems. Technical Report 2000-2, 1999.

D. M. Titterington, A. F. M. Smith, and U. E. Makov, editors. *Statistical Analysis of Finite Mixture Distributions*. Wiley, New York, 1986.

L. Tong, G. Xu, and T. Kailath. Blind identification and equalization based on second-order statistics: A time domain approach. *IEEE Transactions on Information Theory*, 40(2):340–349, 1994.

M. Torlak and G. Xu. Blind multiuser channel estimation in asynchronous CDMA systems. *IEEE Transactions on Signal Processing*, 45(1):137–147, 1997.

J. R. Treichler and M. G. Agee. A new approach to multipath correction of constant modulus signals. *IEEE Transactions on Acoustics, Speech, and Signal Processing*, 31(2):459–472, 1983.

A. Treves and E. T. Rolls. Computational constraints suggest the need for two distinct input systems to the hippocampal CA3 network. *Hippocampus*, 2:189–200, 1992.

M. K. Tsatsanis and G. B. Giannakis. Transmitter induced cyclostationarity for blind channel equalization. *IEEE Transactions on Signal Processing*, 45(7):1785–1794, July 1997.

D. N. C. Tse. Optimal power allocation over parallel Gaussian broadcast channels. In *Proc. of IEEE International Symposium on Information Theory*, page 27, Ulm, Germany, 1997.

J. N. Tsitsiklis. Asynchronous stochastic approximation and q-learning. *Machine Learning*, 16:185–202, 1994.

Z. Tu, X. Chen, A. Yuille, and S-C. Zhu. Image parsing: Segmentation, detection, and recognition. In *Proc. of International Conference on Computer Vision*, Nice, France, 2003.

M. Tuechler and J. Hagenauer. EXIT charts and irregular codes. In *Proceedings of the Conference on Information Sciences and Systems*, pages 748–753, Princeton, NJ, 2002.

S. Ullman, M. Vidal-Naquet, and E. Sali. Visual features of intermediate complexity and their use in classification. *Nature Neuroscience*, 5:1–6, 2002.

G. Ungerboeck. Channel coding with multilevel/phase signals. *IEEE Transactions on Information Theory*, 28(1):55–67, January 1982.

D. R. Upper. Theory and algorithms for hidden Markov models and generalized hidden Markov models. PhD thesis, Department of Mathematics, University of California at Berkeley, 1997. Available at http://www.santafe.edu/projects/CompMech/papers/TAHMMGHMM.html.

V. A. Vaishampayan. Design of multiple description scalar quantizers. *IEEE Transactions on Information Theory*, 39:821–834, May 1993.

V. A. Vaishampayan, N. J. A. Sloane, and S. Servetto. Multiple description vector quantization with lattice codebooks: Design and analysis. *IEEE Transactions on Information Theory*, 47(5):1718–1734, 2001.

E. C. van der Meulen. A survey of multiway channels in information theory. *IEEE Transactions on Information Theory*, 23(1):1–37, January 1977.

A.-J. van der Veen. Analytical method for blind binary signal separation. *IEEE Transactions on Signal Processing*, 45(4):1078–1082, April 1997.

A.-J. van der Veen and A. Paulraj. An analytical constant modulus algorithm. *IEEE Transactions on Signal Processing*, 44(5):1136–1155, May 1996.

A.-J. van der Veen, S. Talwar, and A. Paulraj. Blind estimation of multiple digital signals transmitted over FIR channels. *IEEE Signal Processing Letters*, 2(5): 99–102, 1995.

W. F. van Gunsteren, H. J. Berendsen, and J. A. C. Rullmann. Stochastic dynamics for molecules with constraints: Brownian dynamics of n-alkalines. *Molecular Physics*, 44(1):69–95, 1981.

H. van Hateren. The image database at: hlab.phys.rug.nl/imlib, 1998.

B. D. Van Veen and K. M. Buckley. Beamforming: A versatile approach to spatial filtering. *IEEE ASSP Magazine*, 5(2):4–24, 1988.

B. D. Van Veen and K. M. Buckley. Beamforming techniques for spatial filtering. In V. K. Madisetti and D. B. Williams, editors, *Digital Signal Processing Handbook*. CRC Press, Boca Raton, FL, 1997.

L. Vandenberghe, S. Boyd, and S. Wu. Determinant maximization with linear matrix inequality constraints. *SIAM Journal on Matrix Analysis and Applications*, 19:499–533, 1998.

V. N. Vapnik. *Statistical Learning Theory*. Wiley, New York, 1998.

M. Varanasi and T. Guess. Optimum decision feedback multiuser equalization with successive decoding achieves the total capacity of the Gaussian multiple-access channel. In *Proc. of the Asilomar Conference on Signals, Systems and Computers*, pages 1405–1409, 1997.

F. Varela, E. Thompson, and E. Rosch. *The Embodied Mind: Cognitive Science and Human Experience*. MIT Press, Cambridge, MA, 1991.

L. Venkataramanan, R. Kuc, and F. J. Sigworth. Identification of hidden Markov models for ion channel currents—Part II: Bandlimited, sampled data. *IEEE Transactions on Signal Processing*, 48(2):376–385, 2000.

R. Venkataramani, G. Kramer, and V. K. Goyal. Multiple description coding with many channels. *IEEE Transactions on Information Theory*, 49(9):2106–2114, September 2003.

S. Verdu. *Multiuser Detection*. Cambridge University Press, Cambridge, 1998.

R. Vigário, J. Särelä, V. Jousmäki, M. Hämäläinen, and E. Oja. Independent component approach to the analysis of EEG and MEG recordings. *IEEE Transactions on Biomedical Engineering*, 47(5):589–593, 2000.

V. Virsu, B. B. Lee, and O. D. Creutzfeldt. Dark adaptation and receptive field organisation of cells in the cat lateral geniculate nucleus. *Experimental Brain Research*, 27(1):35–50, 1977.

P. Viswanath, R. Laroia, and D. N. C. Tse. Methods and apparatus for transmitting information between a basestation and multiple mobile stations. United States patent, no. 6,694,147, 2004.

P. Viswanath, D. N. C. Tse, and R. Laroia. Opportunistic beamforming using dumb antennas. *IEEE Transactions on Information Theory*, 48(6):1277–1294, 2002.

H. Viswanathan. Capacity of Markov channels with receiver CSI and delayed feedback. *IEEE Transactions on Information Theory*, 45(2):761–770, March 1999.

C. von der Malsburg. The correlation theory of brain function. Internal Report 81-2, Department of Neurobiology, Max Plank Institute for Biophysical Chemistry, G 1981.

C. von der Malsburg. The what and why of binding: The modeler's perspective. *Neuron*, 24:95–104, 1999.

C. von der Malsburg and W. Schneider. A neural cocktail-party processor. *Biological Cybernetics*, 54:29–40, 1986.

J. von Neumann. *Theory of Self-Reproducing Automata*. University of Illinois Press, Urbana, IL, 1966.

M. Wainwright and E. Simoncelli. Scale mixtures of Gaussians and the statistics of natural images. In *Advances in Neural Information Processing Systems, 12*. MIT Press, Cambridge, MA, 2000.

M. J. Wainwright and M. I. Jordan. Semidefinite relaxations for approximate inference on graphs with cycles. Technical report, UCB/CSD-3-1226, University of California, Berkeley, January 2003a.

M. J. Wainwright and M. I. Jordan. Graphical models, exponential families, and variational inference. Technical report, no. 649, Department of Statistics, University of California, Berkeley, September 2003b.

M. J. Wainwright, T. S. Jaakkola, and A. S. Willsky. MAP estimation via agreement on (hyper)trees: Message-passing and linear programming approaches. In *Proceedings of the Allerton Conference on Communication, Control, and Computing*, October 2002.

M. J. Wainwright, T. S. Jaakkola, and A. S. Willsky. Exact MAP estimates via agreement on (hyper)trees: Linear programming and message-passing approaches. Technical report, UCB/CSD-3-1269, University of California, Berkeley, August 2003a.

M. J. Wainwright, T. S. Jaakkola, and A. S. Willsky. Tree-based reparameterization framework for analysis of sum-product and related algorithms. *IEEE Transactions Information Theory*, 49(5):1120–1146, 2003b.

M. J. Wainwright, T. S. Jaakkola, and A. S. Willsky. Tree consistency and bounds on the max-product algorithm and its generalizations. *Statistics and Computing*, 14:143–166, April 2004.

G. V. Wallenstein and M. E. Hasselmo. GABAergic modulation of hippocampal place cell activity: Sequence learning, place field development, and the phase precession effect. *Journal of Neurophysiology*, 78:393–408, 1997.

D. L. Wang and G. J. Brown. Separation of speech from interfering sounds based on oscillatory correlation. *IEEE Transactions on Neural Networks*, 10(3):684–697, 1999.

D. L. Wang, J. Buhmann, and C. von der Malsburg. Pattern segmentation in associative memory. *Neural Computation*, 2:94–106, 1990.

C. J. C. H. Watkins. Learning from delayed rewards. PhD thesis, King's College, Cambridge University, 1989.

C. J. C. H. Watkins and P. Dayan. Q-learning. *Machine Learning*, 8:279–292, 1992.

V. Weerackody. Characteristics of a simulated fast fading indoor radio channel. In *Proc. of the IEEE Conference on Vehicular Technology*, pages 231–235, 1993.

P. Weiner. Linear pattern matching algorithms. In *Proceedings of the Fourteenth Symposium on Switching and Automata Theory*, pages 1–11, 1973.

C. Weiss, C. Bettstetter, and S. Riedel. Code construction and decoding of parallel concatenated tail-biting codes. *IEEE Transactions on Information Theory*, 47(1):366–386, January 2001.

Y. Weiss and W. T. Freeman. On the optimality of solutions of the max-product belief-propagation algorithm in arbitrary graphs. *IEEE Transactions on Information Theory*, 47(2):736–744, February 2001.

Y. Weiss and W. T. Freeman. Correctness of belief propagation in Gaussian graphical models of arbitrary topology. In *Advances in Neural Information Processing Systems, 12*, pages 673–679. MIT Press, Cambridge, MA, 2000.

M. Welling and Y. Teh. Belief optimization: A stable alternative to loopy belief propagation. In *Uncertainty in Artificial Intelligence*, July 2001.

P. J. Werbos. Beyond regression: New tools for prediction and analysis in the behavioral sciences. PhD thesis, Harvard University, Cambridge, MA, 1974.

P. Whittle. Multi-armed bandits and the Gittins index. *Journal of the Royal Society of Statistics, B*, 42(2):143–149, 1980.

W. Wiegerinck. Variational approximations between mean field theory and the junction tree algorithm. In *Uncertainty in Artificial Intelligence (UAI 2000)*. Morgan Kaufmann, San Francisco, CA, 2000.

L. Williams and D. Jacobs. Stochastic completion fields. *Neural Computation*, 9: 837–858, 1997.

A. S. Willsky. Multiresolution Markov models for signal and image processing. *Proceedings of the IEEE*, 90(8):1396–1458, 2002.

H. Witsenhausen and A. D. Wyner. Interframe coder for video signals. United States patent, no. 4,191,970, 1980.

S. Wolfram. Universality and complexity in cellular automata. *Physica D*, 10:1–35, 1984.

E. Wong and B. Hajek. *Stochastic Processes in Engineering Systems*. Springer, Berlin, 2nd edition, 1985.

W. Woods, M. Hansen, T. Wittkop, and B. Kollmeier. A simple architecture for using multiple cues in sound separation. In *Proc. of International Conference on Spoken Language Processing*, volume 2, pages 909–912, Philadelphia, 1996.

G. Wornell and M. Trott. Efficient signal processing techniques for exploiting transmit antenna diversity on fading channels. *IEEE Transactions on Signal Processing*, 45(1):191–205, Jan 1997.

Y. Wu, S.-C. Zhu, and C. Guo. From information scaling of natural images to regimes of statistical models. *Journal of the American Statistical Association*, 2006, in revision.

A. D. Wyner. Recent results in the Shannon theory. *IEEE Transactions on Information Theory*, 20(1):2–10, 1974.

A. D. Wyner and J. Ziv. The rate distortion function for source coding with side information at the decoder. *IEEE Transactions on Information Theory*, 22:1–10, January 1976.

H. Xiao and A. H. Banihashemi. Graph-based message-passing schedules for decoding LDPC codes. *IEEE Transactions on Comunications*, 52:2098–2105, 2004.

L.-L. Xie and P. R. Kumar. A network information theory for wireless communication: Scaling laws and optimal operation. *IEEE Transactions on Information Theory*, 50(5):748–767, 2004.

J. Xu, X. Wang, B. Ensign, and M. Li. Ion-channel assay technologies. *Drug Discovery Today*, 6:1278–1287, 2001.

J. S. Yedidia, W. T. Freeman, and Y. Weiss. Generalized belief propagation. In *Advances in Neural Information Processing Systems, 13*, pages 689–695. MIT Press, Cambridge, MA, 2001.

J. S. Yedidia, W. T. Freeman, and Y. Weiss. Understanding belief propagation and its generalizations. Technical Report TR2001-22, Mitsubishi Electric Research Labs, January 2002.

D. Yellin and B. Porat. Blind identification of FIR systems excited by discrete alphabet inputs. *IEEE Transactions on Signal Processing*, 41(3):1331–1339, March 1993.

D. Yellin and E. Weinstein. Multi-channel signal separation: Methods and analysis. *IEEE Transactions on Signal Processing*, 44(1):106–118, 1996.

G. Yin, V. Krishnamurthy, and C. Ion. Regime switching stochastic approximation algorithms with application to adaptive discrete stochastic optimization. *SIAM Journal on Optimization*, 14(4):117–1215, 2004.

W. A. Yost. *Fundamentals of Hearing: An Introduction*. Academic Press, San Diego, CA, 2000.

W. A. Yost and G. Gourevitch, editors. *Directional Hearing*. Springer, New York, 1987.

A. Yuille. CCCP algorithms to minimize the Bethe and Kikuchi free energies: Convergent alternatives to belief propagation. *Neural Computation*, 14:1691–1722, 2002.

L. A. Zadeh. The concept of system, aggregate, and state in system theory. In L. A. Zadeh and E. Polak, editors, *System Theory*, volume 8, pages 3–42. McGraw-Hill, New York, 1969.

R. S. Zemel and G. E. Hinton. Discovering viewpoint-invariant relationships that characterize objects. In R. P. Lippmann, J. E. Moody, and D. S. Touretzky, editors, *Advances in Neural Information Processing Systems, 3*, pages 299–305. Morgan Kaufmann, San Mateo, CA, 1991.

R. S. Zemel and G. E. Hinton. Developing population codes by minimizing description length. *Neural Computation*, 7:549–564, 1995.

J. Zhang. The application of the Gibbs-Bogoliubov-Feynman inequality in mean-field calculations for Markov random-fields. *IEEE Transactions on Image Processing*, 5(7):1208–1214, July 1996.

Z. Zhang and T. Berger. New results in binary multiple descriptions. *IEEE Transactions on Information Theory*, 33:502–521, July 1987.

L. Zheng and D. N. C. Tse. Communication on the Grassmann manifold: A geometric approach to the noncoherent multiple-antenna channel. *IEEE Transactions on Information Theory*, 48(2):359–383, February 2002.

L. Zheng and D. N. C. Tse. Diversity and multiplexing: A fundamental tradeoff in multiple antenna channels. *IEEE Transactions on Information Theory*, 49(5): 1073–1096, May 2003.

S. C. Zhu. Embedding Gestalt laws in Markov random fields. *IEEE Transactions on Pattern Analysis and Machine Intelligence*, 21:1170–1187, 1999.

S. C. Zhu, Y. Wu, and D. Mumford. Minimax entropy principle and its application to texture modeling. *Neural Computation*, 9:1627–1660, 1997.

H. G. Zimmermann and R. Neuneier. Neural network architectures for the modeling of dynamical systems. In J. F. Kolen and S. Kremer, editors, *A Field Guide to Dynamical Recurrent Networks*, pages 311–350. IEEE Press, New York, 2001.

H. G. Zimmermann and R. Neuneier. The observer-observation dilemma in neuro-forecasting. In *Advances in Neural Information Processing Systems, 10*, pages 179–206. MIT Press, Cambridge, MA, 1998.

H. G. Zimmermann, R. Grothmann, and Ch. Tietz. Yield curve forecasting by error correction neural networks and partial learning. In M. Verleysen, editor, *Proceedings of the 2002 European Symposium on Artificial Neural Networks*, pages 407–412, 2002a.

H. G. Zimmermann, R. Neuneier, and R. Grothmann. Modeling of dynamical systems by error correction neural networks. In A. Soofi and L. Cao, editors, *Modeling and Forecasting Financial Data: Techniques of Nonlinear Dynamics*, pages 237–263. Kluwer, Boston, 2002b.

M. Zinkevich. Online convex programming and generalized infinitesimal gradient ascent. In *Proc. 20th International Conference on Machine Learning*, 2003.

Contributors

Masoud Ardakani
Department of Electrical and Computer Engineering
University of Toronto
Toronto, ON, Canada

Suzanna Becker
Department of Psychology
McMaster University
Hamilton, ON, Canada

Claude Berrou
ENST Bretagne
Brest, France

Zhe Chen
RIKEN Brain Science Institute
Wako-shi, Saitama, Japan

Konstantinos Diamantaras
Department of Informatics
TEI of Thessaloniki
Sindos, GR, Greece

Suhas N. Diggavi
School of Computer and Communication Sciences
Laboratory of Information and Communication Systems (LICOS)
École Polytechnique Fédérale de Lausanne (EPFL)
Lausanne, Switzerland

Timothy R. Field
Department of Electrical and Computer Engineering
and Department of Mathematics and Brain Body Institute
McMaster University
Hamilton, ON, Canada

Geoffrey J. Gordon
Center for Automated Learning and Discovery
School of Computer Science
Carnegie Mellon University
Pittsburgh, PA

Ralph Grothmann
Corporate Technology
Information & Communication Division
Neural Computation
Siemens AG, Munich, Germany

Simon Haykin
Adaptive Systems Laboratory
Department of Electrical and Computer Engineering
McMaster University
Hamilton, ON, Canada

Herbert Jaeger
International University Bremen
Bremen, Germany

Michael I. Jordan
Department of Electrical Engineering and Computer Science
Department of Statistics
University of California, Berkeley
Berkeley, CA

Andreas Kolling
International University Bremen
Bremen, Germany

Klaus Kretzschmar
SAP, Walldorf, Germany

Vikram Krishnamurthy
Department of Electrical and Computer Engineering
University of British Columbia
Vancouver, BC, Canada

Frank R. Kschischang
Department of Electrical and Computer Engineering
University of Toronto
Toronto, ON, Canada

Charlotte Langlais
ENST Bretagne
Brest, France

Robert Legenstein
Institute for Theoretical Computer Science
Technische Universität Graz
Graz, Austria

Wolfgang Maass
Institute for Theoretical Computer Science
Technische Universität Graz
Graz, Austria

David Mumford
Division of Applied Mathematics
Brown University
Providence, RI

Anton Maximilian Schäfer
Corporate Technology
Information & Communication Division
Neural Computation
Siemens AG, Munich, Germany

Tobias Oberstein
Tavendo Gmlh, Herzogenaurach, Germany

Dan Popovici
Université de Montréal
Montreal, QC, Canada

Fabrice Seguin
ENST Bretagne
Brest, France

Christoph Tietz
Corporate Technology
Information & Communication Division
Neural Computation
Siemens AG, Munich, Germany

Martin J. Wainwright
Department of Electrical Engineering and Computer Science
Department of Statistics
University of California, Berkeley
Berkeley, CA

Hans-Georg Zimmermann
Corporate Technology
Information & Communication Division
Neural Computation
Siemens AG, Munich, Germany

Mingjie Zhao
International University Bremen
Bremen, Germany

Index